Lecture Notes in Control and Information Sciences

Edited by M. Thoma and A. Wyner

Vol. 62: Analysis and Optimization
of Systems
Proceedings of the Sixth International
Conference on Analysis and Optimization
of Systems
Nice, June 19–22, 1984
Edited by A. Bensoussan, J. L. Lions
XIX, 591 pages. 1984.

Vol. 63: Analysis and Optimization
of Systems
Proceedings of the Sixth International
Conference on Analysis and Optimization
of Systems
Nice, June 19–22, 1984
Edited by A. Bensoussan, J. L. Lions
XIX, 700 pages. 1984.

Vol. 64: Arunabha Bagchi
Stackelberg Differential Games
in Economic Models
VIII, 203 pages, 1984

Vol. 65: Yaakov Yavin
Numerical Studies
in Nonlinear Filtering
VIII, 273 pages, 1985.

Vol. 66: Systems and Optimization
Proceedings of the Twente Workshop
Enschede, The Netherlands, April 16–18, 1984
Edited by A. Bagchi, H. Th. Jongen
X, 206 pages, 1985.

Vol. 67: Real Time Control of Large Scale Systems
Proceedings of the First European Workshop
University of Patras, Greece, Juli 9–12, 1984
Edited by G. Schmidt, M. Singh, A. Titli,
S. Tzafestas
XI, 650 pages, 1985.

Vol. 68: T. Kaczorek
Two-Dimensional Linear Systems
IX, 397 pages, 1985.

Vol. 69: Stochastic Differential Systems –
Filtering and Control
Proceedings of the IFIP-WG 7/1 Working Conference
Marseille-Luminy, France, March 12-17, 1984
Edited by M. Metivier, E. Pardoux
X, 310 pages, 1985.

Vol. 70: Uncertainty and Control
Proceedings of a DFVLR International Colloquium
Bonn, Germany, March, 1985
Edited by J. Ackermann
IV, 236 pages, 1985.

Vol. 71: N. Baba
New Topics in Learning Automata
Theory and Applications
VII, 231 pages, 1985.

Vol. 72: A. Isidori
Nonlinear Control Systems:
An Introduction
VI, 297 pages, 1985.

Vol. 73: J. Zarzycki
Nonlinear Prediction
Ladder-Filters for Higher-Order
Stochastic Sequences
V, 132 pages, 1985.

Vol. 74: K. Ichikawa
Control System Design based on
Exact Model Matching Techniques
VII, 129 pages, 1985.

Vol. 75: Distributed Parameter
Systems
Proceedings of the 2nd International
Conference, Vorau, Austria 1984
Edited by F. Kappel, K. Kunisch,
W. Schappacher
VIII, 460 pages, 1985.

Vol. 76: Stochastic Programming
Edited by F. Archetti, G. Di Pillo,
M. Lucertini
V, 285 pages, 1986.

Vol. 77: Detection of
Abrupt Changes in Signals
and Dynamical Systems
Edited by M. Basseville,
A. Benveniste
X, 373 pages, 1986.

Vol. 78: Stochastic
Differential Systems
Proceedings of the 3rd Bad Honnef
Conference, June 3–7, 1985
Edited by N. Christopeit, K. Helmes,
M. Kohlmann
V, 372 pages, 1986.

Vol. 79: Signal
Processing for Control
Edited by K. Godfrey, P. Jones
XVIII, 413 pages, 1986.

Vol. 80: Artificial Intelligence
and Man-Machine Systems
Edited by H. Winter
IV, 211 pages, 1986.

For information about Vols. 1–61 please contact your bookseller or Springer-Verlag.

Lecture Notes in Control and Information Sciences

Edited by M. Thoma and A. Wyner

121

A. Blaquière (Editor)

Modeling and Control of Systems

in Engineering, Quantum Mechanics, Economics and Biosciences

Proceedings of the Bellman Continuum Workshop 1988, June 13–14, Sophia Antipolis, France

Springer-Verlag Berlin Heidelberg GmbH

Editor
Austin Blaquière
Laboratoire d'Automatique Théorique
Tour 14-24
Université de Paris 7
2, Place Jussieu
75251 Paris Cedex 05
France

Library of Congress Cataloging in Publication Data

Bellman Continuum Workshop (1988 : Sophia-Antipolis, France)
Modeling and control of systems in engineering,
quantum mechanics, economics and biosciences :
proceedings of the Bellman Continuum Workshop 1988, June 13-14, Sophia Antipolis, France /
A. Blaquière, editor.
(Lecture notes in control and information sciences ; 121)
ISBN 978-3-540-50790-1 ISBN 978-3-540-46087-9 (eBook)
DOI 10.1007/978-3-540-46087-9
1. Automatic control – Congresses. 2. Control theory – Congresses. 3. System analysis – Congresses.
I. Blaquière, Austin. II. Title. III. Series.
TJ212.2.B45 1988
629.8-dc19 89-4153

Originally published by Springer-Verlag Berlin Heidelberg New York in 1989.

2161/3020-543210

FOREWORD

Richard Bellman, a most prolific and renowned mathematician of the United States, has made major contributions in pure mathematics and in numerous areas of applications : engineering, economics, medicine, energy, water resources, mathematical physics, operations research, management sciences, psychology and sociology. This breadth of interests and this ability to contribute to so many fields at such a high level is rare indeed. Throughout his years in science, he had a large number of scientific friends, students and followers. Among them, after Professor Bellman has passed away, a group of scientists of the United States attempted to preserve his School. As a mechanism for achieving this goal, they suggested an annual or biennial workshop : *the Bellman Continuum*. This workshop was envisioned as being interdisciplinary in nature, as the achievement of Richard Bellman was.

The first meeting was held at the University of Michigan, Ann Arbor, Michigan,in 1985 and the second was hosted by the Georgia Institute of Technology, Atlanta, Georgia, in 1986. They correspond to a formative stage : they attracted a few scientific friends of Richard Bellman and some scientists whose work has connections with his School.

The third Bellman Continuum, sponsored by IFAC and AFCET (french NMO of IFAC), has been organized by the Institut National de Recherche en Informatique et en Automatique (I.N.R.I.A.) and the University Paris 7, on June 13-14, 1988, at the INRIA Research Center of Sophia-Antipolis, on the french Riviera, 6 miles North-West Antibes. It immediately followed the INRIA Eight International Conference Analysis and Optimization of Systems, held at the Palais des Congrès of Antibes on June 8-10, 1988, and it immediately preceeded the third International Symposium on Differential Games and Applications, held at INRIA-Sophia on June 16-17, 1988.

The program included invited and contributed lectures in the following areas where research is very active and promising :

- Deterministic Approach to the Control of Uncertain Dynamical Systems.
- Control and Non-Linear Filtering of Quantum Mechanical Processes.
- Models and Control Policies in Economics.
- Models and Control Policies for Biological Systems and Ecosystems.

Key-note speakers were Prof. G. Leitmann, University of California, Berkeley, U.S.A., and Prof. S.K. Mitter, Massachusetts Institute of Technology, U.S.A.

The areas above correspond to the four main sections of this book. An additional section has been devoted to computational bearings. They have in common the fact that they all deal with *uncertain systems*.

The third Bellman Continuum was attended by 73 participants and observers from 17 different countries (Austria, Brazil, Canada, Finland, France, FRG, Great-Britain, Hungary, Israël, Italy, Japan, the Netherlands, Poland, Switzerland, U.R.S.S., U.S.A., Yugoslavia) and has been considered very successful in highlighting the current trend and perspectives of the new questions set forth in its program. The interdisciplinary exchange of ideas was much in the honor and spirit of Richard Bellman.

The papers collected here speak for themselves; there is no point in attempting to summarize their content. However it is, perhaps, appropriate to briefly outline the main scientific directions defined by the choice of the above topics, whose unifying scheme is the *modelling and control of uncertain systems.*

Many systems in the "real" world are subject to human intervention and control. The first step in devising a control policy or strategy for the accomplishment of a desired end is the abstraction of the perceived salient features of the actual (physical, chemical,engineering, biological, economic, etc ...) system. Such an abstraction is usually embodied in a *mathematical model,* e.g., ordinary differential equations, finite difference equations, partial differential equations, and so on. Mathematical models are *uncertain,* partly because they are approximations involving unknown or partially known elements, and partly because they include elements which model uncertain effects in the real world.

Two avenues are open to the system analyst dealing with such uncertain mathematical models, a *statistical approach* and a *deterministic one*.

Part I of this book is devoted to the latter. It is centered on the deterministic approach to uncertain dynamical systems initiated by G. Leitmann : on the basis of known nominal model and bounds on uncertainties, feedback schemes are determined which force the system output to track a given signal. The operative controllers are obtained via a constructive use of Lyapunov functions. In this book, the subject matter has been more specifically oriented towards the most recent results concerning *robustness*, i.e., the ability of a system to retain certain performance measures in the presence of perturbations, systems with *two time scale* structure and related treatment utilizing *singular perturbation* analysis.

Physics literature is a rich source of interesting mathematical questions. In connection with system theory we have seen in the past few years a growing realization of the interconnections between estimation theory and quantum physics, between stochastic models and quantum mechanical ones, between Hamilton-Jacobi theory, stochastic control and the evolution in time of quantum systems, etc ... "This reunification suggests that what we have seen so far may be just the initial part of a long term trend".[1]

Part II is mainly concerned with systems in which uncertainty comes out through quantum mechanical rules. It owes its origin to a work of S.K. Mitter highlighting the analogy between quantum physics and mathematical problems of nonlinear filtering.

It collects papers in stochastic control, nonlinear filtering, and in the new area of quantum filtering and control approached from different points of view and with different mathematical techniques.

---

[1] Borrowed from R.W. Brockett, Ricerche di Automatica, Vol X, Dec. 1979, n° 2.

A quantum mechanical control system is a quantum mechanical system with a time varying part considered as a perturbation. Different kinds of problems can be studied : one can be interested in the time varying part as a signal to be extracted from the measurements on the system. This is the *quantum filtering problematic,* usually associated with the concept of *non demolition measurements.* Also the time varying part can be considered as a purposeful control on the system, a *control problem* stricto sensu.

Quantum mechanical control theory is an essential step on the way from quantum physics to quantum technology.

One of the fields of applications of the mathematical theory of systems is in the overlapping areas of mathematical economics, econometrics, social sciences, and management science. The papers of Part III report recent results in mathematical economics, in the framework of dynamic optimization, continuous or sequential, with finite or infinite horizon, and in uncertain stochastic environment. The wide area of differential-game theoretical approaches has been more specifically and extensively explored in the third International Symposium on Differential Games and Applications which immediately followed the third Bellman Continuum. Its Proceedings are being published by SPRINGER-VERLAG in the same Series.

Beginning in the 1960's, Bellman recognized that many biological systems display a number of characteristics similar to those of the decision processes to which he had devoted much attention. He then turned his talents towards developing models and control policies for these systems. He published many excellent papers in this field and achieved recognition as one of the pioneers in bringing the strength of mathematics and computer science into the medical area. His original motivation was the cancer problem. Part IV is along the line of this part of Bellman's work.

Part V collects papers which do not pertain to one of the categories above but which are attached to several of them through Bellman's general approach to system science and related computational bearings.

The third international workshop of the Bellman Continuum could not have taken place without the technical and financial assistance of INRIA to whom we express our gratitude. In particular I would like to take this opportunity to thank his President Professor Alain Bensoussan and the Director of the INRIA-Sophia Antipolis Research Center Professor Pierre Bernhard.

We are indebted to the staff of the Public Relations Department of INRIA for the job they have carried out in the organization of the workshop. I personnally address special thanks to Thérèse Bricheteau who, at the head of this Department, took care of all the myriad details of organization so efficiently and ably. We are most grateful to the expert assistance of Catherine Juncker and her staff who took care of the organization at Sophia and of the local arrangements.

The workshop was financially supported by the organizing Institutions : INRIA and the University Paris 7, and by the following national and intergovernmental Organizations : AFCET (France), CNRS (France), ERO United States Army (U.S.A.), the french Ministaries of Affaires Etrangères, Education Nationale, and Recherche et Enseignement Supérieur, UNESCO. Additional fellowships to the participants were provided by various organizations listed separately, to whom, as well as to the above mentioned organizations, we express our gratitude.

We also would like to extend our gratitude to :

- the participants who have shown their interest in this workshop,
- the many referees who have accepted the difficult task of selecting papers,
- the chairpersons for the different sessions,
- our colleagues of the Organizing Committee,
- Gilbert Mallet and his staff who, at INRIA, produced preprints of the conference,
- Professor M. Thoma who has accepted to publish the Proceedings of the workshop in the Series : Lecture Notes in Control and Information Sciences,
- Mr. Albrecht von Hagen, Engineering Editor,
- the Publisher SPRINGER-VERLAG.

Austin Blaquière

Workshop Chairman

PREFACE

Richard Bellman, un des mathématiciens les plus féconds et les plus renommés des Etats-Unis, a apporté des contributions majeures aux mathématiques pures et à de nombreux domaines d'applications : sciences de l'ingénieur, économie, médecine, énergie, gestion des ressources en eau, physique mathématique, recherche opérationnelle, sciences de la gestion, psychologie et sociologie. Une telle variété des domaines abordés et des moyens mis en oeuvre pour approfondir ces domaines avec une telle pénétration se rencontre rarement en science. Tout au long du développement de son oeuvre, il eut un grand nombre d'amis, d'élèves et de correspondants portés vers les mêmes centres d'intérêt. Parmi eux, après la disparation du Professeur Bellman, un groupe de scientifiques des Etats-Unis s'est efforcé de perpétuer son Ecole. Dans ce but ils ont proposé d'organiser un Colloque annuel ou bi-annuel : *le Bellman Continuum*. Ce Colloque devait être de nature interdisciplinaire, comme l'était l'oeuvre de Richard Bellman.

Le premier congrès s'est tenu à l'Université du Michigan, Ann Arbor, Michigan, en 1985 et le second a été accueilli par l'Institut de Technologie de Georgie, Atlanta, Georgie, en 1986. Ces efforts préliminaires ont réuni quelques scientifiques, amis de Richard Bellman ou dont le travail présente des liens avec son Ecole.

Le troisième Bellman Continuum, patronné par l'IFAC et l'AFCET (OMN française de l'IFAC) a été organisé par l'Institut National de Recherche en Informatique et en Automatique (I.N.R.I.A.) et l'Université Paris 7, les 13 et 14 Juin 1988, au Centre de Recherche de l'INRIA à Sophia-Antipolis, sur la Côte d'Azur, à une dizaine de kilomètres au Nord-Ouest d'Antibes. Il succédait à la huitième Conférence Internationale Analyse et Optimisation des Systèmes de l'INRIA, tenue au Palais des Congrès d'Antibes les 8-10 Juin 1988, et précédait le troisième Symposium International sur les Jeux Différentiels et leurs Applications, tenu à INRIA-Sophia les 16 et 17 Juin 1988.

Le programme comportait des conférences sur invitation et des rapports destinés à la présentation de travaux récents, dans les domaines suivants où la recherche est très active et en expansion :

- Approche déterministe de la commande des systèmes dynamiques incertains.
- Commande et filtrage non-linéaire des processus en Mécanique quantique.
- Modélisation et commande en Economie.
- Modélisation et commande des systèmes biologiques et des écosystèmes.

Les conférenciers d'ouverture de sessions étaient le Prof. G. Leitmann, University of California, Berkeley, U.S.A., et le Prof. S.K. Mitter, Massachusetts Institute of Technology, U.S.A.

Les domaines mentionnés ci-dessus correspondent aux quatre parties principales de ce livre. Une cinquième partie a été réservée à l'aspect calcul. Elles ont en commun le fait qu'elles traitent toutes de *systèmes incertains*.

Le troisième Bellman Continuum a réuni 73 participants et auditeurs de 17 pays différents (Autriche, Brésil, Canada, Finlande, France, Grande-Bretagne, Hongrie, Israël, Italie, Japon, Pays-Bas, Pologne, R.F.A., Suisse, U.R.S.S., U.S.A., Yougoslavie) et a atteint son objectif avec succès : celui de mettre en lumière les orientations et les perspectives des questions nouvelles mises en avant par son programme, et de susciter des échanges d'idées de nature interdisciplinaire dans l'esprit et à l'honneur de Richard Bellman.

Les papiers réunis ici sont suffisamment explicites pour qu'il n'y ait pas lieu d'analyser leur contenu. Peut-être, cependant, sera-t-il utile d'indiquer dans leurs grandes lignes les orientations scientifiques définies par le choix des sujets traités, dont le thème unificateur est *la modélisation et la commande des systèmes incertains*.

De nombreux systèmes du monde réel sont sujets à l'intervention humaine, et sont commandés. La première étape dans l'élaboration d'un système de commande est l'établissement d'un *modèle mathématique*, abstraction qui résume l'information jugée intéressante pour l'étude proposée, utile au mathématicien : systèmes d'équations différentielles ordinaires, d'équations aux différences finies, d'équations aux dérivées partielles, etc ... Les modèles mathématiques sont *incertains*, en partie parce qu'ils sont tributaires d'approximations dûes à une connaissance imparfaite des données, en partie parce qu'ils contiennent des éléments représentant des facteurs aléatoires du monde réel.

Deux types de méthodes s'offrent à l'analyste : une *approche statistique* et une *approche déterministe*.

La *première partie* du livre est consacrée à cette dernière. L'accent y est mis sur l'approche déterministe des systèmes dynamiques incertains introduite par G. Leitmann : la connaissance du modèle nominal et des bornes sur les incertitudes permettent de déterminer une rétroaction qui force la sortie du système à suivre un signal donné, rétroaction obtenue par la construction d'une fonction de Lyapunov. Dans ce livre, le sujet a été plus particulièrement orienté vers les résultats les plus récents relatifs à la *robustesse* : aptitude d'un système à conserver certaines performances en présence de perturbations, aux systèmes à *deux échelles de temps* et aux *perturbations singulières*.

La Physique est une riche source de questions mathématiques intéressantes. En liaison avec la théorie des systèmes, nous avons vu au cours des années passées une prise de conscience de plus en plus nette des interconnexions entre la théorie de l'estimation et la Physique quantique, entre les modèles stochastiques et ceux de la Mécanique quantique, entre la théorie d'Hamilton-Jacobi, la commande stochastique et l'évolution au cours du temps des systèmes quantiques, etc ... "Cette réunification suggère que ce que nous avons vu jusque là est peut-être le signe précurseur d'une tendance à long terme".[1]

La *deuxième partie* du livre est plus particulièrement concernée par les systèmes dans lesquels les incertitudes sont d'origine quantique. Elle doit son origine à un travail de S.K. Mitter qui met en lumière l'analogie entre la Physique quantique et

---

[1] R.W. Brockett, Ricerche di Automatica, Vol. X, Dec. 1979, n° 2.

certains problèmes mathématiques de filtrage non-linéaire. Elle rassemble des papiers en commande stochastique, filtrage non-linéaire, et dans le domaine nouveau du filtrage et de la commande quantiquesexaminés de différents points de vue et avec différentes techniques mathématiques.

Un système de commande en Mécanique quantique est un système quantique ayant une partie variable au cours du temps, considérée comme une perturbation. On peut s'intéresser à la partie variable au cours du temps en tant que signal à extraire d'un ensemble de mesures. C'est le problème du *filtrage quantique*. On peut aussi considérer cette partie variable comme une commande appliquée à dessein et se proposer de la déterminer en fonction du but recherché. C'est le problème inverse du précédent, c'est un problème de *commande* stricto sensu.

La théorie de la commande en mécanique quantique est une étape essentielle sur le chemin menant de la physique quantique à la *technologie quantique*.

L'un des champs d'application importants de la théorie mathématique des systèmes est celui où se recouvrent en partie les domaines de l'Economie mathématique, l'Econométrie, les Sciences sociales, la Recherche opérationnelle. Les papiers de la *troisième partie* présentent des travaux récents en Economie mathématique, dans le cadre de l'optimisation dynamique, continue ou séquentielle, en horizon fini ou infini, et dans un environnement incertain. Le vaste domaine des approches théoriques reposant sur les jeux différentiels a été exploré de façon plus étendue et plus spécifique dans le troisième Symposium International sur les Jeux Différentiels et leurs Applications venant juste à la suite du troisième Bellman Continuum. Nous reportons le lecteur à ses actes publiés par SPRINGER-VERLAG dans la même Série.

Dès le début des années 60, Bellman a reconnu que des systèmes biologiques très divers présentent un certain nombre de traits caractéristiques semblables à ceux des processus de décision auxquels il avait consacré une partie importante de son oeuvre. Ceci le conduisit à orienter ses efforts et son talent vers le développement de modèles et de lois de commande pour ces systèmes. Il a publié plusieurs excellents papiers sur ces questions, qui l'ont fait reconnaître comme l'un des pionniers ayant introduit la puissance des mathématiques et de la science des machines à calculer dans le domaine médical. Sa motivation première était le problème du cancer. La *quatrième partie* du livre est dans le droit fil de cette partie de l'oeuvre de Bellman.

La *cinquième partie* rassemble des papiers qui n'appartiennent pas à l'une des catégories ci-dessus mais qui se rattachent à plusieurs d'entr'elles dans la ligne des méthodes générales de Bellman en science des systèmes. Elle est plus particulièrement consacrée à l'aspect calcul.

Le troisième Colloque International du Bellman Continuum n'aurait pu avoir lieu sans le soutien technique et financier de l'INRIA à qui nous exprimons notre gratitude. En particulier, qu'il me soit permis de remercier ici son Président le Professeur Alain Bensoussan et le Directeur du Centre de Recherche de l'INRIA-Sophia Antipolis le Professeur Pierre Bernhard.

Nous tenons à remercier les personnes du Service des Relations Extérieures de l'INRIA qui ont organisé ce Colloque. J'adresse personnellement des remerciements tout particuliers à Thérèse Bricheteau qui, à la tête de ce Service, a pris soin si efficacement de la multitude des problèmes d'organisation, et nous a fait profiter de sa grande expérience. Nous sommes très reconnaissants à Catherine Juncker qui a pris en main de façon experte l'organisation du congrès à Sophia et son implantation sur le site, ainsi qu'aux personnes de son Service.

Ce Colloque international a reçu le soutien financier de l'INRIA et de l'Université Paris 7 qui l'ont co-organisé, et des organismes nationaux et intergouvernementaux suivants : AFCET (France), CNRS (France), ERO United States Army (U.S.A.), Ministère des Affaires Etrangères (France), Ministère de l'Education Nationale (France), Ministère de la Recherche et de l'Enseignement Supérieur (France), UNESCO. D'autres subventions ont été attribuées aux participants par divers organismes mentionnés séparément, auxquels, comme aux organismes cités dans cette préface nous exprimons notre gratitude.

Nos remerciements s'adressent également :

- aux participants qui ont manifesté leur intérêt pour ce Colloque,
- aux nombreux experts qui ont accepté la difficile tâche de sélectionner les communications,

- aux présidents de sessions,
- à nos collègues du Comité d'Organisation,
- à Gilbert Mallet et aux personnes de son Service qui, à l'INRIA, ont publié les actes provisoires de la conférence,
- au Professeur M. Thoma qui a accepté de publier les actes définitifs du Colloque dans la série qu'il dirige : Lecture Notes in Control and Information Sciences,
- à Mr. Albrecht von Hagen, Engineering Editor,
- à l'éditeur SPRINGER-VERLAG.

Austin Blaquière

Président du Colloque

COMITE D'ORGANISATION

ORGANIZING COMMITTEE

A. BLAQUIERE,	Chairman, Université Paris 7, France
N. BELLMAN,	22 Latimer Road, Santa Monica, CA 90402, USA
A. BENSOUSSAN,	Université Paris-Dauphine / INRIA, France
P. BERNHARD,	INRIA-Sophia-Antipolis, France
Th. BRICHETEAU,	INRIA, France
A. ESOGBUE,	GA Institute of Technology, USA
G. FEICHTINGER,	Technical University Vienna, Austria
M. FLIESS,	Laboratoire des Signaux et Systèmes, CNRS-ESE, France
A. FOSSARD,	ENSAE, Toulouse, France
S. LEE,	Kansas State University, USA
G. LEITMANN,	University of California, Berkeley, USA
S. MEERKOV,	University of Michigan, USA
M. THOMA,	Technische Universität, Hannover, FRG
L. ZADEH,	University of California, Berkeley, USA

LIST OF SPONSORS

LISTE DES ORGANISMES APPORTANT LEURS PATRONAGES OU CONCOURS FINANCIERS

AFCET	Association Française pour la Cybernétique Economique et Technique
Collège Mathématiques Appliquées	
CNRS	Centre National de la Recherche Scientifique
ERO USA	European Research Office, United States Army
IFAC	International Federation of Automatic Control
Technical Committee of Mathematics of Control	
Technical Committee on Theory	
INRIA	Institut National de Recherche en Informatique et en Automatique
MAE	Ministère des Affaires Etrangères
MEN	Ministère de l'Education Nationale
MRES	Ministère de la Recherche et de l'Enseignement Supérieur
UNESCO	United Nations Educational, Scientific and Cultural Organization
UP7	Université Paris 7

C O N T R I B U T O R S

G. ADOMIAN	Center for Applied Mathematics The University of Georgia Athens, Georgia 30602 (USA)
J.P. AUBIN	CEREMADE, Université de Paris-Dauphine Place du Maréchal de Lattre de Tassigny 75016 Paris (France)
T. BAŞAR	University of Illinois at Urbana-Champaign Coordinated Science Laboratory 1101 West Springfield Avenue Urbana, Il 61801 (USA)
V.P. BELAVKIN	M.I.E.M. B. Vuzovski Ul. 3/12 109 028 Moscow (URSS)
E. BENOIT	Centre de Mathématiques Appliquées, Ecole des Mines de Paris, Sophia Antipolis 06565 Valbonne Cedex et G.R. Automatique-E.N.S.I.E.G., B.P. 46 38402 Saint-Martin d'Hères (France)
D. BENSOUSSAN	Université du Québec - Ecole de Technologie Supérieure 4750 Rue Henri Julien Case Postale 1000, Succursale E Montréal H2T 1R0 (Canada)

J. BENTSMAN	University of Illinois at Urbana-Champaign Department of Mechanical and Industrial Engineering 144 Mechanical Engineering Building 1206 West Green Street Urbana, Il 61801 (USA)
A. BLAQUIERE	Laboratoire d'Automatique Théorique, Tour 14-24 Université Paris 7 2 Place Jussieu 75251 Paris Cedex 05 (France)
G. BOJADZIEV	Department of Mathematics and Statistics Simon Fraser University Burnaby B.C. (Canada V5A 1S6)
B. CANDELPERGHER	Laboratoire de Mathématiques, U.A. 168 Parc Valrose 06000 Nice (France) et G.R. Automatique - E.N.S.I.E.G., B.P. 46 38402 Saint-Martin d'Hères (France)
J.W. CLARK	McDonnell Center for the Space Sciences and Department of Physics Washington University Saint-Louis, Missouri 63130 (USA)
D. CLAUDE	Laboratoire des Signaux et Systèmes C.N.R.S. - E.S.E. Plateau du Moulon 91190 Gif-sur-Yvette (France)
M. CORLESS	School of Aero/Astron. Purdue University West Lafayette, Indiana 47907 (USA)

D. DUBOIS
Langages et Systèmes Informatiques
Université Paul Sabatier - C.N.R.S.
118 Route de Narbonne
31062 Toulouse Cedex (France)

W. DÜCHTING
Universität-Gesamthochschule-Siegen
Fachbereich 13 - Elektrotechnik I
Regelungstechnik
Hölderlinstrasse 3, D-5900
Siegen (F.R.G.)

H. EHTAMO
Systems Analysis Laboratory
Helsinki University of Technology
Otakaari IM, 02150 Espoo (Finland)

W.H. FLEMING
Brown University
Division of Applied Mathematics
Providence, Rhode Island 02912 (USA)

H. FRANKOWSKA
CEREMADE, Université de Paris-Dauphine
Place du Maréchal de Lattre de Tassigny
75016 Paris (France)

F. GAROFALO
Dipt. di Informatica e Sistemistica
Università di Napoli
Napoli (Italy)

B.K. GHOSH
Department of Systems Science and Mathematics
School of Engineering and Applied Science
Washington University
Saint-Louis, Missouri 63130 (USA)

L. GLIELMO
Dipt. di Informatica e Sistemistica
Università di Napoli
Napoli (Italy)

R. GONZALEZ — Facultad de Ciencias Exactas e Ingenieria
Av. Pellegrini 250
(2000) Rosario (Argentina)

M. HACHED — School of Electrical Engineering
Purdue University
West Lafayette, Indiana 47907 (USA)

R.P. HÄMÄLÄINEN — Systems Analysis Laboratory
Helsinki University of Technology
Otakaari IM, 02150 Espoo (Finland)

B. HANNON — University of Illinois at Urbana-Champaign
Department of Geography
Urbana, Il 61801 (USA)

M. HAZEWINKEL — Mathematical Center
P.O. Box 4079
1009 AB Amsterdam (The Netherlands)

G.M. HUANG — Department of Electrical Engineering
Texas A & M University
College Station, Texas 77843 (USA)

S. IWAMOTO — Department of Economic Engineering
Faculty of Economics
Kyushu University 27
Fukuoka (Japan)

N. KIEFER — Department of Economics
Uris Hall
Cornell University
Ithaca, N.Y. 14853 (USA)

K. KIME — Department of Mathematics and Statistics
Case Western Reserve University
Cleveland, Ohio 44106 (USA)

V.P. KROTOV — Institute of Control Sciences
Profsouznaja Ul. 65
117 342 Moscow (URSS)

E.S. LEE — Department of Industrial Engineering
Durland Hall
Kansas State University
Manhattan, Kansas 66506 (USA)

G. LEITMANN — University of California
Department of Mechanical Engineering
Berkeley, California 94720 (USA)

C. LOBRY — Université de Nice
06000 Nice (France)

S.K. MITTER — Department of Electrical Engineering and Computer Science
and Center for Intelligent Control Systems
Massachusetts Institute of Technology
Cambridge, Mass. 02139 (USA)

Y. NYARKO — Department of Economics
Brown University
Providence, Rhode Island 02912 (USA)

H. PRADE — Langages et Systèmes Informatiques
Université Paul Sabatier - C.N.R.S.
118 Route de Narbonne
31062 Toulouse Cedex (France)

E. ROFMAN I.N.R.I.A.
Domaine de Voluceau, Rocquencourt
78153 Le Chesnay Cedex (France)

J. RUUSUNEN Systems Analysis Laboratory
Helsinki University of Technology
Otakaari IM, 02150 Espoo (Finland)

E.P. RYAN School of Mathematical Sciences
University of Bath
Bath BAZ 7AY (Great-Britain)

G. SONNEVEND Institüt für Mathematik
Universität Würzburg am Hubland
Würzburg D 8700 (R.F.A.)

H. STALFORD Virginia Polytechnic Institute and State University
Aerospace and Ocean Engineering
Blacksburg, Virginia 24061 (USA)

T.J. TARN Washington University
Department of Systems Science and Mathematics
Saint-Louis, Missouri 63130 (USA)

K.M. WANG Department of Industrial Engineering
Tsinghua University
Taiwan (China)

Z.B. YAACOB School of Mathematical Sciences
University of Bath
Bath BAZ 7AY (Great-Britain)

S.H. ŻAK School of Electrical Engineering
Purdue University
West Lafayette, Indiana 47907 (USA)

J.C. ZAMBRINI Math. Institüt
Ruhr-Universität, D-4630
Bochum 1, (F.R.G.)

TABLE OF CONTENTS

PART I

COMMANDE DES SYSTEMES DYNAMIQUES INCERTAINS

CONTROL OF UNCERTAIN DYNAMICAL SYSTEMS

PART II

SYSTEMES STOCHASTIQUES ET QUANTIQUES

STOCHASTIC AND QUANTUM SYSTEMS

PART III

MODELISATION ET COMMANDE EN ECONOMIE

MODELS AND CONTROL POLICIES IN ECONOMICS

PART IV

MODELISATION ET COMMANDE DES SYSTEMES BIOLOGIQUES ET DES ECOSYSTEMES

MODELS AND CONTROL POLICIES FOR BIOLOGICAL SYSTEMS AND ECOSYSTEMS

COMMANDE DES SYSTEMES DYNAMIQUES INCERTAINS

CONTROL OF UNCERTAIN DYNAMICAL SYSTEMS

CONTROLLING SINGULARLY PERTURBED UNCERTAIN DYNAMICAL SYSTEMS[1]

G. Leitmann

College of Engineering, University of California
Berkeley, California 94720, USA

INTRODUCTION

The prototype for the class of systems considered in this chapter is depicted in Figure 1 and consists of a dynamical process P (imperfectly known) controlled by a (judiciously designed) feedback law (operator F) acting on state data generated by sensor S and implemented via actuator A.

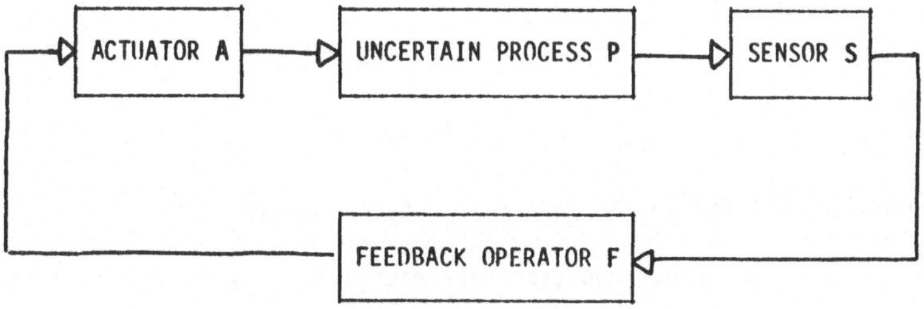

Figure 1. Prototype System

We assume (realistically) that the sensor and actuator are <u>dynamic</u> elements of the feedback loop; furthermore, we adopt the viewpoint that these dynamics are "fast" relative to those of the process P to be controlled. If this is not the case, then, at the modelling stage, the sensor and actuator should be explicitly incorporated as an integral part of the process to be controlled.

We recognize, of course, that in the context of nonlinear systems, the concept of "fastness" is difficult to quantify. Here we use the term loosely to indicate that the overall system exhibits a "two time scale" structure as described in the next section.

THE FULL-ORDER SYSTEM

The above prototype typifies a general class of singularly perturbed uncertain systems which can be decomposed, by means of a scalar parameter μ, into two coupled

---

[1] Based on research supported by the NSF and AFOSR. This paper deals with a special case of the problem considered in [8] and [9].

subsystems which henceforth will be referred to as the "slow" subsystem (with state $x(t)$) and the "fast" subsystem (with state $y(t)$). The parameter μ, henceforth referred to as the singular perturbation parameter, can be interpreted as some measure of the ratio of characteristic times of the fast and slow subsystems.

We model this general class of systems by the following coupled pair of differential equations.

$$\dot{x}(t) = X(t,x(t),y(t),u(t)), \quad x(t) \in R^n, \quad u(t) \in R^m \tag{1a}$$

$$\mu\dot{y}(t) = Y(t,x(t),y(t),u(t),\mu), \quad y(t) \in R^p, \quad \mu \in (0,\infty) \tag{1b}$$

with measured output

$$z(t) = Sx(t) + Ty(t), \quad z(t) \in R^n \tag{1c}$$

where X and Y are uncertain functions with the following structure:

$$X(t,x,y,u) = A_{11}x + A_{12}y + B_1 u + g_1(t,x,y,u) \tag{2a}$$

$$Y(t,x,y,u,\mu) = C(t)[A_{21}x + y + B_2 u] + g_2(t,x,y,u,\mu) . \tag{2b}$$

A_{ij}, B_i, S and T are known constant real matrices; C is an uncertain measurable matrix-valued function; g_1 and g_2 are uncertain Caratheodory functions (i.e. measurable in their first argument, continuous in their other arguments and integrably bounded on compact sets).

Note that we require that the dimension of the output space coincides with the dimension of the slow subsystem state space. We refer to system (1)-(2) as the <u>full-order system</u> (a dynamical system on R^{n+p}).

Now suppose that the dynamics of the fast subsystem are neglected, i.e. suppose that μ is set to zero, in which case (1b) reduces to an algebraic constraint on (1a). This procedure yields the <u>reduced-order system</u> (a dynamical system on R^n). Suppose further that a feedback strategy is designed which guarantees some stability property P for the uncertain reduced-order system. (One such design is proposed in §5 and analysed in §6, using the deterministic framework developed in e.g. [1-7]). Then the essential question to be addressed is that of structural stability of property P with respect to singular perturbation, i.e. does property P persist when the fast dynamics are re-introduced? More usefully, does there exist a calculable threshold value $\mu^* > 0$ such that property P persists for all values of the singular perturbation parameter in the interval $(0,\mu^*)$?

Our objective is to answer such questions affirmatively, under additional hypotheses on the full-order system. The first of these is an assumption which ensures that a well-defined reduced order system results from setting $\mu = 0$ in (1b).

Assumption A1

(i) $C(\cdot) = C_0 + \Delta C(\cdot)$, where $C_0 \in R^{p \times p}$ is known with spectrum $\sigma(C_0) \subset \mathbb{C}^-$ (the open left half complex plane) and $\Delta C: R \to R^{p \times p}$ is an unknown measurable function with known bound κ_c (sufficiently small), viz. for all $t, \|\Delta C(t)\| \leq \kappa_c < 1/2\|P\|^{-1}$, where $P > 0$ (symmetric) solves the Lyapunov equation $PC_0 + C_0^T P + I = 0$;

(ii) $g_2(\cdot,\cdot,\cdot,\cdot,0) = 0$.

THE REDUCED-ORDER SYSTEM

Solving the algebraic equation $Y(t,x,y,u,0) = 0$ for y (uniquely in view of Assumption A1) determines the function

$$(x,u) \mapsto H(x,u) \triangleq -[A_{21}x + B_2u] \; . \tag{3}$$

The reduced-order system associated with (1) is now defined as

$$\dot{x}(t) = X_r(t,x(t),u(t)), \qquad x(t) \in R^n \tag{4a}$$

with output

$$z(t) = Sx(t) + TH(x(t),u(t)), \qquad z(t) \in R^n \tag{4b}$$

where

$$X_r(t,x,u) \triangleq X(t,x,H(x,u),u) = \bar{A}x + \bar{B}u + \bar{g}(t,x,u) \tag{5a}$$

and

$$\bar{A} \triangleq A_{11} - A_{12}A_{21}, \quad \bar{B} \triangleq B_1 - A_{12}B_2, \quad \bar{g}(t,x,u) \triangleq g_1(t,x,H(x,u),u) \; . \tag{5b}$$

At this stage, we loosely define our preliminary goal as that of rendering, by feedback, some acceptably small compact neighborhood of the zero state of (4) globally attractive. Thus, it is not unreasonable to require the following of the nominal linear system pair $(\bar{A},\bar{B})$:

Assumption A2

(i) $(\bar{A},\bar{B})$ is a stabilizable pair,

(ii) $S - TA_{21}$ is non-singular.

Now, let $(Q,\gamma_0) \in R^{n \times n} \times R^+$ $(R^+ \triangleq [0,\infty))$ be a pair of design parameters with the properties (i) Q is symmetric and positive definite (ii) $\gamma_0 > 0$ if $\sigma(\bar{A}) \not\subset \mathbb{C}^-$.

These properties, in conjunction with A2, ensure that the Riccati equation

$$K\bar{A} + \bar{A}^T K + Q - 2\gamma_0 K\bar{B}\bar{B}^T K = 0 \tag{6}$$

admits a unique real positive-definite symmetric solution $K > 0$. Hence, for example, in the absence of uncertainty ($\bar{g} \equiv 0$) and if $S = I$ and $T = 0$, the output feedback law $u = -\gamma_0 \bar{B}^T Kz$ renders the zero state of (4) asympstotically stable.

We now impose some additional structure and bounds on the system uncertainty.

Assumption A3

There exist known non-negative real numbers c_1, c_2, c_3, and unknown Caratheodory function $e: R \times R^n \times R^m \to R^m$ such that:

(i) $\bar{g} = \bar{B}e$;

and, for all $(t,x,u) \in R \times R^n \times R^m$,

(ii) $\|e(t,x,u)\| \leq c_1 + c_2 \|x\| + c_3 \|u\|$.

In the familiar terminology, the uncertainty is assumed to be matched and cone-bounded. The more general case of unmatched and non-conebounded uncertainty is considered in [8] and [9], albeit at the expense of a considerably more complicated controller design.

Define $A: R \to R^{n \times n}$ and $\Gamma_1, \Gamma_2 \subset R$ as follows:

$$A(\gamma) \triangleq A_{21} - \gamma B_2 \bar{B}^T K \tag{7a}$$

$$\Gamma_1 \triangleq \begin{cases} [\underline{\gamma}, \infty); & c_2 = 0 \\ (\underline{\gamma}, \infty); & c_2 > 0 \end{cases} ; \quad \underline{\gamma} \triangleq (1 - c_3)^{-1}[\gamma_0 + c_2\|Q^{-1}\|] \tag{7b}$$

$$\Gamma_2 \triangleq \{\gamma : |S - TA(\gamma)| \neq 0; \quad \kappa(\gamma) < (1 - 2\kappa_c\|P\|)/2\|PC_0\| + 2\kappa_c\|P\|)\} \tag{7c}$$

where

$$\kappa(\gamma) \triangleq \gamma\|B_2\bar{B}^T K[S - TA(\gamma)]^{-1}T\| . \tag{7d}$$

Then the following additional assumption is required.

Assumption A4

$\Gamma^* \triangleq \Gamma_1 \cap \Gamma_2 \neq \emptyset$.

PROBLEM FORMULATION

Suppose a (time-dependent) output feedback control function $(t,z) \mapsto q(t,z)$ is designed which guarantees that the feedback-controlled reduced-order system (viz. $u(t) = - q(t,z(t))$ in (4)) possesses some desired stability property P, then the basic question to be addressed is that of robustness of P with respect to singular perturbation, where the singularly perturbed system is defined by (1) with $u(t) = - q(t,z(t))$; in particular, does there exist a (calculable) constant $\mu^* > 0$ such that the full system (1), under output feedback control $u(t) = - q(t,z(t))$, possesses property P for all values $\mu \in (0,\mu^*)$?

Here, we take the desired property P to be the existence of a compact set $\Sigma \subset R^n$ (respectively $\Sigma \subset R^{n+p}$) containing the origin which is a global uniform attractor for the reduced-order system (respectively, the full-order system) in the following sense.

Definition 1

A compact set $\Sigma \subset R^q$ is a global uniform attractor for the system

$$\dot{w}(t) = \Xi(t,w(t)), \quad w(t) \in R^q \qquad (*)$$

if the following properties hold:

(i) Existence and continuation of solutions: For each pair $(t_0,w^0) \in R \times R^q$ there exists a solution $w: [t_0,t_1) \to R^q$ (absolutely continuous function satisfying (*) almost everywhere) with $w(t_0) = w^0$ and every such solution can be extended into a solution on $[t_0,\infty)$;

(ii) Uniform boundedness of solutions: For each $r > 0$ there exists $R(r) > 0$ such that $\|w(t)\| < R(r)$ for all t on every solution $w: [t_0,\infty) \to R^q$ of (*) with $\|w(t_0)\| < r$, where $t_0 \in R$ is arbitrary;

(iii) Uniform stability of Σ: For each $d > 0$ there exists $D(d) > 0$ such that $w(t) \in \Sigma + d\mathcal{B}$ for all t on every solution $w: [t_0,\infty) \to R^q$ of (*) with $w(t_0) \in \Sigma + D(d)\mathcal{B}$ where t_0 is arbitrary (note, $\mathcal{B}$ denotes the open unit ball in R^q and, for $\delta > 0$, $\Sigma + \delta\mathcal{B}$ denotes the set $\{\sigma + \rho: \sigma \in \Sigma; \|\rho\| < \delta\}$);

(iv) Global uniform attractivity of Σ: For each $d > 0$ and $r > 0$ there exists $\tau(d,r) > 0$ such that $w(t) \in \Sigma + r\mathcal{B}$ for all $t > t_0 + \tau(d,r)$ on every solution $w: [t_0,\infty) \to R^q$ of (*) with $w(t_0) \in \Sigma + d\,\mathcal{B}$, where $t_0 \in R$ is arbitrary.

In the next section, we construct a feedback strategy which ensures property P for the reduced-order system (4).

NONLINEAR OUTPUT FEEDBACK

Choose $\varepsilon_1, \varepsilon_2 > 0$; these are design parameters and can be chosen arbitrarily small. Define $p: R \times R^n \to R^m$ as

$$p(t,x) \triangleq p_0(x) + p_1(x) \, . \qquad (8a)$$

The function p_0 is linear and is given by

$$p_0(x) \triangleq \gamma_1 \bar{B}^T K x \qquad (8b)$$

where $\gamma_1 \in R^+$ satisfies

$$\gamma_1 \in \Gamma^* \, . \qquad (8c)$$

The function p_1 is nonlinear and bounded and is given by

$$p_1(x) \triangleq \begin{cases} \rho_1 \phi_1(\rho_1 \bar{B}^T K x) & \text{if } T = 0 \text{ or } B_2 = 0 \\ 0 & \text{otherwise} \end{cases} \qquad (8d)$$

where $\rho_1 \in R^+$ satisfies

$$\rho_1 > (1 - c_3)c_1 \qquad (8e)$$

and $\phi_1: R^m \to R^m$ is any smooth (C^1) function which satisfies

$$\|\phi_1(\nu)\| < 1 \, , \quad \langle \nu, \phi_1(\nu) \rangle > \|\nu\| - \varepsilon_1 \quad \forall \nu \in R^m \qquad (8f)$$

and which has bounded derivative $D\phi_1$; i.e., there exists $\kappa_\phi \in R^+$ such that $\|D\phi_1(\nu)\| < \kappa_\phi$ for all $\nu \in R^m$. The proposed output feedback control function $q: R \times R^n \to R^m$ is now defined by

$$q(t,z) \triangleq p(t, [S - TA(\gamma_1)]^{-1} z) \, . \qquad (9)$$

Loosely speaking, the linear component (8b) of the control stabilizes (if necessary) the nominal linear system and counteracts part of the uncertainty e while nonlinear component (8d) (when active) counteracts the remaining part of e.

As an example of a function ϕ_1, satisfying the above requirements, consider the function

$$\phi_1 : \nu \mapsto \|\nu\| + \varepsilon_1]^{-1}\nu$$

for which (8f) clearly holds, and moreover, ϕ_1 is C^1 with $\|D\phi_1(\nu)\| \leq \varepsilon_1^{-1}$ for all $\nu \in R^m$.

A COMPACT ATTRACTOR FOR THE OUTPUT FEEDBACK CONTROLLED REDUCED-ORDER SYSTEM

For the reduced-order system (4), it may be verified that $q(t,z(t)) = p(t,x(t))$. Hence, setting $u(t) = - q(t,z(t))$ in (4a) yields the system

$$\dot{x}(t) = F_r(t,x(t)), \qquad x(t) \in R^n \tag{10a}$$

with

$$F_r(t,x) \triangleq \bar{A}x - \bar{B}p(t,x) + \bar{g}(t,x, - p(t,x)). \tag{10b}$$

As shown in [9], system (10) possesses stability property P .

To this end, we define $V: R^n \to R^+$ (a Lyapunov function candidate) by

$$V(x) \triangleq \langle x,Kx \rangle \ . \tag{11}$$

Theorem 1.

There exists a closed ellipsoid

$$\Sigma_{r_0} \triangleq \{x \in R^n : \ V(x) \leq r_0^2\} \ ,$$

where r_0 is defined in [9], which is a global uniform attractor for system (10).

Our next objective is to show that property P is not destroyed by the re-introduction of the fast dynamics.

A COMPACT ATTRACTOR FOR THE OUTPUT FEEDBACK CONTROLLED FULL-ORDER SYSTEM

Define

$$h(x) \triangleq H(x, - p(t,x)) = - A(\gamma_1)x + B_2 p_1(x). \tag{12}$$

Our final assumption is now made.

Assumption A5

(i) For all (t,x),

$$\|g_1(t,x,y_1,-q(t,Sx+Ty_1)) - g_1(t,x,y_2,-q(t,Sx+Ty_2))\| \leq \lambda\|y_1-y_2\| \qquad \forall\, y_1,y_2$$

where $\lambda > 0$ is a known constant;

(ii) for all (t,x,y) and $\mu > 0$,

$$\|g_2(t,x,y,-q(t,Sx+Ty),\mu)\| \leq \mu[\kappa_1\|y-h(x)\| + \kappa_2\|x\| + \kappa_3]$$

where $\kappa_1, \kappa_2, \kappa_3 > 0$ are known constants.

While Assumptions 1 to 5 might appear somewhat esoteric, it is stressed that the class of systems which satisfy these hypotheses is far from trivial; for example, the assumptions hold for a class of uncertain systems with parasitic actuator and sensor dynamics considered in [10].

Let functions $F: R\times R^n\times R^p \to R^n$ and $G: R\times R^n\times R^p\times R^+ \to R^p$ be given by

$$\begin{aligned} F(t,x,y) &\triangleq A_{11}x + A_{12}y - B_1q(t,Sx+Ty) + g_1(t,x,y,-q(t,Sx+Ty)) \qquad (13)\\ &= F_r(t,x) + A_{12}[y-h(x)] + B_1[p(t,x)-q(t,Sx+Ty)]\\ &\quad + g_1(t,x,y,-q(t,Sx+Ty)) - g_1(t,x,h(x),-p(t,x)) \end{aligned}$$

$$\begin{aligned} G(t,x,y,\mu) &\triangleq C(t)[A_{21}x + y - B_2q(t,Sx+Ty)] + g_2(t,x,y,-q(t,Sx+Ty),\mu) \qquad (14)\\ &= C(t)[y-h(x)] + C(t)B_2[p(t,x)-q(t,Sx+Ty)] + g_2(t,x,y,-q(t,Sx+Ty),\mu). \end{aligned}$$

Then the problem under consideration reduces to that of determining a threshold value $\mu^* > 0$ (if such exists) such that the system (two coupled subsystems):

$$\dot{x}(t) = F(t,x(t),y(t)) \qquad (15a)$$

$$\mu\dot{y}(t) = G(t,x(t),y(t),\mu) \qquad (15b)$$

possesses stability property P for all $\mu\in(0,\mu^*)$. We resolve this question via an analysis akin to that of [11].

As stated in [8] and shown in [9], the following theorem establishes property P for the full order system under output feedback control.

Theorem 2.

There exists a $\mu^* > 0$ such that, for all $\mu\in(0,\mu^*)$, a certain ellipsoid is a global uniform attractor for system (15); the value of μ^* and the definition of the attracting ellipsoid are given in [8] and [9]. Moreover, the reduced order dynamical behavior is recovered as $\mu \to 0$.[2]

EXAMPLE: UNCERTAIN SYSTEM WITH ACTUATOR AND SENSOR DYNAMICS

Consider the uncertain system

$$\dot{x}(t) = Ax(t) + [B + \Delta B(t)]y_1(t) + d(t,x(t)), \qquad x(t) \in R^n \tag{16a}$$

with actuator dynamics

$$\mu\dot{y}_1(t) = [C_1 + \Delta C_1(t)](y_1(t) - u(t)), \quad y_1(t), u(t) \in R^m \tag{16b}$$

and sensor dynamics

$$\mu\dot{y}_2(t) = [C_2 + \Delta C_2(t)](y_2(t) - x(t)), \quad y_2(t) \in R^n \tag{16c}$$

where the known nominal system matrices A, B, C_1, C_2 satisfy the following:

H1

(i) (A,B) is a stabilizable pair;

(ii) $\sigma(C_1) \subset \mathbb{C}^-$;

(iii) $\sigma(C_2) \subset \mathbb{C}^-$.

The uncertain functions $\Delta B(\cdot)$ and $d(\cdot,\cdot)$ are assumed to satisfy

H2

(i) $\Delta B(\cdot) = BE(\cdot)$, where $E(\cdot)$(unknown) is measurable with $\|E(t)\| \leq \beta < 1 \ \forall t$;

(ii) $d(\cdot,\cdot) = Bg(\cdot,\cdot)$, where $g(\cdot,\cdot)$ is a Caratheodory function with $\|g(t,x)\| \leq \alpha_1\|x\| + \alpha_2 \quad \forall(t,x)$ and where α_1, α_2, β are known constants.

Let P (symmetric and positive definite) denote the unique solution of

$$P\begin{bmatrix} C_1 & 0 \\ 0 & C_2 \end{bmatrix} + \begin{bmatrix} C_1^T & 0 \\ 0 & C_2^T \end{bmatrix} P + I = 0. \tag{17}$$

---

[2]Loosely speaking, in the sense that the projection of the attracting ellipsoid onto R^n approaches the attracting ellipsoid $\sum_{r_0}$ of the reduced order system.

Then the uncertain functions $\Delta C_1(\cdot)$ and $\Delta C_2(\cdot)$ are assumed to satisfy

H3

$\|\mathrm{diag}\{\Delta C_1(t), \Delta C_2(t)\}\| \leq \kappa_c < 1/2\|P\|^{-1}$ $\forall t$, where κ_c is a known constant.

The above can be interpreted in the context of system (1)-(2) by making the following identifications:

$$y = \begin{bmatrix} y_1 \\ y_2 \end{bmatrix} \in R^p, \quad p \triangleq m + n \tag{18a}$$

$$A_{11} = A, \quad A_{12} = [B \vdots 0], \quad A_{21} = \begin{bmatrix} 0 \\ -I \end{bmatrix} \tag{18b}$$

$$B_1 = 0, \quad B_2 = \begin{bmatrix} -I \\ 0 \end{bmatrix}, \quad S = 0, \; T = [0 \vdots I] \tag{18c}$$

$$C(t) = C_o + \Delta C(t), \quad C_o = \mathrm{diag}\{C_1, C_2\}, \quad \Delta C(t) = \mathrm{diag}\{\Delta C_1(t), \Delta C_2(t)\} \tag{18d}$$

$$g_1(t,x,y,u) = d(t,x) + BE(t)[I \vdots 0]y \tag{18e}$$

$$g_2 \equiv 0 . \tag{18f}$$

In view of H1(ii),(iii) and H3, it is clear that Assumption A1 holds for this system.

Now,

$$\bar{A} = A_{11} - A_{12}A_{21} = A_{11} = A \tag{19a}$$

$$\bar{B} = B_1 - A_{12}B_2 = - A_{12}B_2 = B \tag{19b}$$

and hence, in view of H1(i), it follows that Assumption A2 holds.

Also,

$$H(x,u) = - [A_{21}x + B_2u] = \begin{bmatrix} u \\ x \end{bmatrix} \tag{20}$$

and

$$\bar{g}(t,x,u) = g_1(t,x,H(t,x),u) = Be(t,x,u) \tag{21a}$$

where (21b)

$$e(t,x,u) = g(t,x) + E(t)u .$$

Thus, in view of H2, it is clear that Assumption A3 holds with $c_3 = \beta$. Proceeding,

$$A(\gamma) = A_{21} - \gamma B_2\bar{B}^T K = \begin{bmatrix} \gamma B^T K \\ -I \end{bmatrix} \tag{22a}$$

$$S - TA(\gamma) = I, \quad \kappa(\gamma) = \gamma \| B^T K \| \tag{22b}$$

$$\Gamma_1 = (-\infty, (1-2\kappa_c\|P\|)(2\|P C_o\| + 2\kappa_c\|P\|)^{-1}\|B^T K\|^{-1}) \subset R . \tag{22c}$$

Assumption A4 now reduces to the following:

$$A4^*: \ \underline{\gamma} < (1-2\kappa_c\|P\|)(1+2\kappa_c\|P\|)^{-1}\|B^T K\|^{-1} \quad .$$

Finally, it is readily verified that Assumption A5(ii) holds trivially (since $g_2 \equiv 0$) and A5(i) holds with $\lambda = \beta\|B\|$.

A specific example of this subclass of systems is considered in detail in [9].

OTHER METHODS

An approach, differing from the one proposed here, can be found in [12-15]. In these references, the design procedure requires the sequential construction of controllers which assure existence of global uniform attractors for (i) an approximation of the reduced order ("slow") subsystem, and (ii) the "fast" subsystem under the influence of the slow uncertainties. The controller for the full system is then obtained as the sum of these subsystem controllers.

REFERENCES

[1] S. Gutman and G. Leitmann, "Stabilizing feedback control for dynamical systems with bounded uncertainty," Proc. IEEE Conference on Decision and Control (1976).

[2] G. Leitmann, "Deterministic control of uncertain systems," Astronautica Acta, 7(1980), pp. 1457-1461.

[3] G. Leitmann, "On the efficacy of nonlinear control in uncertain linear systems," J. Dynamic Systems Meas. Control, 103(1981), pp. 95-102.

[4] M. Corless and G. Leitmann, "Continuous state feedback guaranteeing uniform ultimate boundedness for uncertain dynamic systems," IEEE Trans. Autom. Control, AC-26 (1981), pp. 1139-1144.

[5] B.R. Barmish and G. Leitmann, "On ultimate boundedness control of uncertain systems in the absence of matching conditions," IEEE Trans. Autom. Control, AC-27 (1982), pp. 153-158.

[6] B.R. Barmish, M. Corless and G. Leitmann, "A new class of stabilizing controllers for uncertain dynamical systems," SIAM J. Control and Optimization, 21 (1983), pp. 246-255.

[7] E.P. Ryan and M. Corless, "Ultimate boundedness and asymptotic stability of a class of uncertain dynamical systems via continuous and discontinuous feedback control," IMA J. Math. Control and Info., 1 (1984), pp. 223-243.

[8] G. Leitmann and E.P. Ryan, "Output feedback control of a class of singularly perturbed uncertain dynamical systems," Proceed. American Control Conference (1987), pp. 1590-1594.

[9] M. Corless, G. Leitmann and E.P. Ryan, "Control of uncertain systems with neglected dynamics," in "Variable Structure Control Systems", edited by A.I.S. Zinober, IEE Publ., London (in preparation).

[10] G. Leitmann, E.P. Ryan and A. Steinberg, "Feedback control of uncertain systems: robustness with respect to neglected actuator and sensor dynamics," Int. J. Control, 43 (1986), pp. 1243-1256.

[11] A. Saberi and H.K. Khalil, Quadratic-type Lyapunov functions for singularly perturbed systems," IEEE Trans. Autom. Contro, AC-29 (1984), pp. 542-550.

[12] F. Garofalo, "Composite control of a singularly perturbed uncertain system with slow uncertainties," Int. J. Control (to appear).

[13] F. Garofalo and G. Leitmann, "Nonlinear composite control of a nominally linear singularly perturbed uncertain system," Proceed. 12th IMACS World Congress (1988).

[14] F. Garofalo and G. Leitmann, "Nonlinear composite control of a class of nominally linear singularly perturbed uncertain systems," in "Variable Structure Control Systems", edited by A.I.S. Zinober, IEE Publ., London (in preparation).

[15] F. Garofalo and G. Leitmann, "Composite control of nonlinear, singularly perturbed uncertain systems," Proceed. Control 88, Oxford University (1988).

On Robust Control of Uncertain Linear Systems in the Absence of Matching Conditions

Harold Stalford

Aerospace and Ocean Engineering
Interdisciplinary Center for Applied Mathematics
Virginia Polytechnic Institute and State University
Blacksburg, Virginia 24061

ABSTRACT

We establish a general robust control result for linear time-invariant uncertain systems using the Lyapunov approach initiated by Leitmann and Gutman. We show that systems satisfying matching conditions are handled by this result. We give necessary and sufficient conditions for the existence of a robust sliding mode controller. We show that its existence implies the existence of a robust linear controller. A counter example is provided to establish that the converse does not hold. The feedback controllers treated are functions of the complete state without any dynamic compensation.

1. INTRODUCTION

The Lyapunov approach to uncertain systems received an initial thrust by Leitmann and Gutman, [1] - [7], for systems satisfying matching conditions. They are joined by numerous authors (e.g. [8] - [33]) in extending the Lyapunov approach to handle more general systems since it is well suited for addressing structured uncertainty. Our work herein focuses on applying the Lyapunov approach to systems which have constant uncertainties but do not necessarily satisfy the matching conditions. It builds on the work of [9], [14], and [20] - [33]. Our main objective is to establish a robust control result based on the Lyapunov approach which generalizes some of the past work on linear uncertain systems with constant uncertainties. We specifically consider linear and sliding mode controllers and give necessary and sufficient conditions for their existence. We prove that the existence of a robust stabilizing sliding mode controller implies the existence of a robust stabilizing linear controller. The converse does not hold. We provide a counter example showing the existence of a robust linear controller in the absence of such a sliding mode controller. Herein, we use the term stability to mean that the poles are in the left-half plane, i.e., asymptotic stability or, equivalently, that the characteristic polynomial is Hurwitz. We say that a controller is robust if it asymptotically stabilizes the system for all uncertainties. We treat both the scalar input and the multi-input problems.

We investigate the robust control of linear time-invariant uncertain systems that are not required necessarily to satisfied matching conditions:

$$\dot{x} = A(\gamma)x + B(\gamma)u, \quad \gamma \in \Gamma \tag{1}$$

where $A(\gamma)$ is a nxn uncertain matrix, $B(\gamma)$ is an nxm uncertain matrix with full rank ($m \leq n$) and γ belongs to a set of uncertainties Γ where Γ is a simply connected, compact subset of p-dimensional Euclidean space E^p. We assume that $A(\gamma)$ and $B(\gamma)$ are continuous with respect to the uncertainty argument $\gamma \in \Gamma$. In this paper we consider only full state feedback controllers $u(x)$, i.e., those which are functions of the state x only. That is, we do not address dynamic compensation as part of the feedback controller. We require that system (1) satisfy the controllability assumption:

ASSUMPTION I. For each $\gamma \in \Gamma$ the pair $(A(\gamma), B(\gamma))$ is controllable.

The controllability assumption is equivalent to the assumption that closed-loop poles can be arbitrarily placed by a suitable gain matrix. We state this equivalent assumption:

ASSUMPTION I′. For each $\gamma \in \Gamma$ and prescribed eigenvalues $\Lambda(\gamma) = (\lambda_1(\gamma), \ldots, \lambda_n(\gamma))$ in which imaginary eigenvalues occur in complex conjugate pairs there exists a real gain matrix $K(\gamma)$ such that the closed-loop matrix

$$\bar{A}(\gamma) = A(\gamma) - B(\gamma)K(\gamma) \tag{2}$$

has the prescribed eigenvalues $\Lambda(\gamma)$.

For arbitrarily prescribed eigenvalues $\Lambda(\gamma)$, $\gamma \in \Gamma$, we can rewrite (1) as

$$\dot{x} = \bar{A}(\gamma)x + B(\gamma)[K(\gamma)x + u] \tag{3}$$

where $K(\gamma)$ is the corresponding gain matrix and $\bar{A}(\gamma)$ satisfies (2).

The next assumption makes it possible to define a control law with which to stabilize (1) in the presence of uncertainties $\gamma \in \Gamma$.

ASSUMPTION II. For each $\gamma \in \Gamma$ there exist an mxn gain matrix $K(\gamma)$, an invertible mxm matrix $R(\gamma)$ and an nxn symmetric, positive definite matrix $Q(\gamma)$ such that

(*i*) $\bar{A}(\gamma) = A(\gamma) - B(\gamma)K(\gamma)$ is asymptotically stable

(*ii*) $F = R^{-1}(\gamma)B^T(\gamma)P(\gamma)$ is a constant mxn matrix where $P(\gamma)$ is the symmetric, positive definite solution of Lyapunov equation

$$P(\gamma)\bar{A}(\gamma) + \bar{A}^T(\gamma)P(\gamma) + Q(\gamma) = 0 \tag{4}$$

We make the following assumption on the mxm matrix $R(\gamma)$ which is defined in Assumption II.

ASSUMPTION III. For $\gamma \in \Gamma$ the matrix $\Phi(\gamma)$ defined as

$$\Phi(\gamma) = \frac{R^T(\gamma) + R(\gamma)}{2} \tag{5a}$$

is positive definite and has the square root form

$$\Phi(\gamma) = S^T(\gamma)S(\gamma) \tag{5b}$$

where $S(\gamma)$ is invertible. The following upper bound exists and is finite

$$h = \max_{\gamma \in \Gamma}\left[\|S^{-1}(\gamma)\| \; \|S^{-T}(\gamma) \; R^T(\gamma)\|\right] \tag{6}$$

In Sections 2-4 and 6 we show how to use the constant matrix F in establishing a robust controller.

2. MAIN ROBUST CONTROL RESULT

Assumptions I - III permit the development of a robust control law that is discontinuous in nature. This is established in the next theorem.

THEOREM 1: If system (1) satisfies Assumptions I - III then the discontinuous controller

$$u(x) = -\frac{Fx}{\|Fx\|}\rho(x), \quad Fx \neq 0 \tag{7}$$

stabilizes (1) for all $\gamma \varepsilon \Gamma$ where $\rho(x)$ satisfies

$$\rho(x) = h \max_{\gamma \varepsilon \Gamma} \|K(\gamma)\, x\| \tag{8}$$

The scalar h is given by (6) and the gain matrix $K(\gamma)$ is defined in Assumption II.

PROOF: For $\gamma \varepsilon \Gamma$ let $K(\gamma)$, $R(\gamma)$, $Q(\gamma)$, $P(\gamma)$ and F be the matrices described in Assumption II. Define the Lyapunov function

$$V(\gamma) = x^T P(\gamma)x \tag{9}$$

It has the time derivative

$$\dot{V}(\gamma) = -x^T Q(\gamma)x + 2[B^T(\gamma)P(\gamma)x]^T\,[K(\gamma)x + u] \tag{10}$$

Using property (*ii*) of Assumption II this derivative becomes

$$\dot{V}(\gamma) = -x^T Q(\gamma)x + 2[Fx]^T\,R^T(\gamma)\,[K(\gamma)x + u(x)] \tag{11}$$

We show that the control law (7) yields

$$\dot{V}(\gamma) \leq -x^T Q(\gamma)x, \quad \gamma \varepsilon \Gamma \tag{12}$$

Since $Q(\gamma) > 0$ (i.e., positive definite) it suffices to show that $W(\gamma)$ is nonpositive:

$$W(\gamma) = 2[Fx]^T R^T(\gamma)[K(\gamma)x + u(x)] \leq 0 \tag{13}$$

Consider a control law of the form (7) in which the scalar function $\rho(x)$ is defined by (8). substitution of (7) into (13) yields

$$W(\gamma) = 2W_1(\gamma) - 2W_2(\gamma) \leq 0 \tag{14}$$

where

$$W_1(\gamma) = [Fx]^T R^T(\gamma) K(\gamma)x \tag{15a}$$

$$W_2(\gamma) = [Fx]^T R^T(\gamma)\frac{Fx}{\|Fx\|}\rho(x), \quad Fx \neq 0 \tag{15b}$$

Eq. (15b) can be rewritten as

$$W_2(\gamma) = [Fx]^T \Phi(\gamma)\frac{Fx}{\|Fx\|}\rho(x) \tag{16}$$

or, equivalently as,

$$W_2(\gamma) = [S(\gamma)Fx]^T \frac{S(\gamma)\, Fx}{\|Fx\|}\rho(x) \tag{17}$$

where $\Phi(\gamma)$ and $S(\gamma)$ are defined by (5) and (6). Making the vector definition

$$y(\gamma) = S(\gamma)\, Fx \tag{18}$$

Eq. (17) becomes

$$W_2(\gamma) = \frac{y^T(\gamma)\, y(\gamma)}{\|Fx\|}\rho(x) \tag{19}$$

Eq. (15a) can be rewritten as

$$W_1(\gamma) = y^T(\gamma) z(\gamma) \tag{20}$$

where

$$z(\gamma) = S(\gamma)[\Phi(\gamma)]^{-1} R^T(\gamma) K(\gamma)x \tag{21}$$

Inequality (14) is met provided

$$W_1(\gamma) \leq W_2(\gamma) \tag{22}$$

In terms of (19) and (20) this inequality is given by

$$y^T(\gamma)z(\gamma) \leq \frac{y^T(\gamma)y(\gamma)}{\|Fx\|}\rho(x), \quad Fx \neq 0 \tag{23}$$

This inequality is met provided

$$\|z(\gamma)\| \leq \frac{\|y(\gamma)\|}{\|Fx\|}\rho(x), \quad Fx \neq 0 \tag{24}$$

Taking the norm of (21) yields

$$\|z(\gamma)\| \leq \|S^{-T}(\gamma)\, R^T(\gamma)\|\ \|K(\gamma)x\| \tag{25}$$

Multiplying both sides by the norm $\|S^{-1}(\gamma)\|$ gives

$$\|S^{-1}(\gamma)\|\ \|z(\gamma)\| \leq \rho(x) \tag{26}$$

Observe that

$$\|Fx\| = \|S^{-1}(\gamma)S(\gamma)Fx\| \leq \|S^{-1}(\gamma)\|\ \|y(\gamma)\| \tag{27}$$

from which it follows that

$$1 \leq \frac{\|S^{-1}(\gamma)\| \, \|\nu(\gamma)\|}{\|Fx\|}, \quad Fx \neq 0 \tag{28}$$

Multiplying both sides by $\rho(x)$ yields

$$\rho(x) \leq \frac{\|S^{-1}(\gamma)\| \, \|\nu(\gamma)\|}{\|Fx\|} \rho(x), \quad Fx \neq 0 \tag{29}$$

The inequalities (26) and (29) yield

$$\|z(\gamma)\| \leq \frac{\|\nu(\gamma)\|}{\|Fx\|} \rho(x), \quad Fx \neq 0 \tag{30}$$

This verifies (24) which establishes (12). By the theory of Lyapunov, the control law (7) stabilizes (1) for each uncertainty $\gamma \in \Gamma$.

3. ROBUST CONTROL IN THE PRESENCE OF MATCHING CONDITIONS

Systems which satisfy the matching conditions of linear uncertain systems, [2] - [7], satisfy Assumptions I - III. This result is given by the next theorem.

THEOREM 2: Let system (1) satisfy the following matching conditions: There exist an nxn matrix A and an nxm matrix B and for each $\gamma \in \Gamma$ there exist an mxn gain matrix $D(\gamma)$ and an invertible mxm matrix $\Pi(\gamma)$ such that

(a) $A(\gamma) = A + BD(\gamma)$.

(b) $B(\gamma) = B\Pi(\gamma)$.

(c) (A, B) is a controllable pair

(d) $\Phi(\gamma)$ is an mxm positive definite matrix where

$$\Phi(\gamma) = \frac{\Pi^T(\gamma) + \Pi(\gamma)}{2} \tag{31}$$

Then Assumptions I - III are met. As a consequence of Theorem 1, there exists a robust stabilizing control law of the form

$$u = -Kx + \frac{Fx}{\|Fx\|} \rho(x) \tag{32}$$

such that

$$\bar{A} = A - BK \tag{33}$$

is asymptotically stable and such that

$$F = B^T P \tag{34}$$

where P is the symmetric, positive definite solution of the Lyapunov equation

$$P\bar{A} + \bar{A}^T P + Q = 0 \tag{35}$$

in which $Q > 0$ is arbitrarily chosen.

PROOF: Conditions (a) - (c) imply that $(A(\gamma), B(\gamma))$ is controllable for $\gamma \varepsilon \Gamma$. Controllability is invariant under linear feedback and coordinate transformation on the input, [34]. Thus Assumption I is met. Since (A, B) is controllable there exists a gain matrix K such that $\bar{A}$ of (33) is asymptotically stable. Define the uncertain gain matrix

$$K(\gamma) = \Pi^{-1}(\gamma)\,[D(\gamma) + K] \tag{36}$$

Using conditions (a) and (b) we find $\bar{A}(\gamma)$ of condition (*i*) of Assumption II reduces to

$$\bar{A}(\gamma) = A - BK \tag{37}$$

and is, therefore, asymptotically stable for $\gamma \varepsilon \Gamma$. Select any $Q > 0$. Let P be the solution of (36) and let F be defined by (34). For $\gamma \varepsilon \Gamma$ define

$$R(\gamma) = \Pi^T(\gamma) \tag{38}$$

The matrix F of condition (*ii*) of Assumption II and that of (34) are identical. That is, (34) can be rewritten as

$$F = \Pi^{-T}(\gamma)\,[B\,\Pi(\gamma)]^T P \tag{39}$$

which, in view of condition (b) and (38), is equivalent to

$$F = R^{-1}(\gamma)\,B^T(\gamma)\,P \tag{40}$$

Thus, condition (*ii*) of Assumption II is met with

$$P(\gamma) \equiv P \tag{41}$$

Condition (d) implies Assumption III since $B(\gamma)$ is continuous and Γ is compact. That is, h exists and is finite. Since all conditions of Theorem 1 are met, the existence of the stabilizing control law (32) follows with

$$\rho(x) = h \max_{\gamma \varepsilon \Gamma} \|K(\gamma)x\| \tag{42}$$

where $K(\gamma)$ is defined by (36).

4. ROBUST CONTROL IN THE ABSENCE OF MATCHING CONDITIONS: SCALAR INPUT

We show that the robust control assumptions presented in [29] for scalar control satisfy the assumptions of Theorem 1.

Consider system (1) with scalar control. The input matrix $B(\gamma)$ is a column vector. The work in [29] assumes that the system (1) is controllable, Assumption I. Under this assumption there is a unique coordinate transformation $T(\gamma)$

$$z = T(\gamma)x \tag{43}$$

of (1) to the following controllable companion form, [34],

$$\dot{z} = A_z(a(\gamma))z + B_z u(x) \tag{44}$$

where

$$A_z(a(\gamma)) = \begin{bmatrix} -a_1(\gamma) & -a_2(\gamma) & \dots & -a_{n-1}(\gamma) & -a_n(\gamma) \\ 1 & 0 & \dots & 0 & 0 \\ 0 & 1 & \dots & 0 & 0 \\ 0 & 0 & \dots & 1 & 0 \end{bmatrix} \tag{45a}$$

$$A_z(a(\gamma)) = T(\gamma)\, A(\gamma)\, T^{-1}(\gamma) \tag{45b}$$

and

$$B_z = [1, 0, 0, \dots, 0]^T \tag{46a}$$

$$B_z = T(\gamma)\, B(\gamma) \tag{46b}$$

The vector $a(\gamma) = (a_1(\gamma), \dots, a_n(\gamma))$ is the coefficient vector of the open-loop characteristic polynomial:

$$a_\gamma(s) = \det[sI - A(\gamma)] \tag{47}$$

We need the following definition in order to introduce the next assumption of [29]-[31].

DEFINITION 1: The row vector $P_1 = (P_{11}, P_{12}, \dots, P_{1n})$ is said to be $n-1$ stable provide $P_{11} > 0$ and the polynomial

$$P_{11}\lambda^{n-1} + P_{12}\lambda^{n-2} + \dots + P_{1n} = 0 \tag{48}$$

is Hurwitz (i.e., all eigenvalues are in left-half plane).

ASSUMPTION IV: There exist an uncetain $\gamma_0 \in \Gamma$ and an $n-1$ stable row vector $P_1(\gamma_0)$ such that

$$P_1(\gamma) = P_1(\gamma_0)\, T(\gamma_0)\, T^{-1}(\gamma) \tag{49}$$

is $n-1$ stable for all $\gamma \in \Gamma$

The concept of a vector being $n-1$ stable is fundamental in the asymptotically stable solution of Lyapunov equation. This result is presented in the next lemma. Its proof is given in [30].

LEMMA 1. Let $a = (a_1, \ldots, a_n)$. Define $A(a)$ to be in the controllable companion form (45). Let P be the solution to the Lyapunov equation

$$P\,A(a) + A^T(a)\,P + Q = 0 \tag{50}$$

where $Q > 0$ and $Q = Q^T$. Then $A(a)$ is stable if, and only if, P_1 is $n-1$ stable where P_1 is the first row of P.

PROOF: See [30].

The next lemma is a consequence of Lemma 1.

LEMMA 2. Suppose Assumption IV holds. For each $\gamma \in \Gamma$ define $Q(\gamma) > 0$, $Q(\gamma) = Q^T(\gamma)$. Then for each γ, there is a unique stable coefficient vector $\hat{a}(\gamma)$ satisfying Lyapunov equation

$$P(\gamma)A(\hat{a}(\gamma)) + A^T(\hat{a}(\gamma))P(\gamma) + Q(\gamma) = 0 \tag{51}$$

where $P_1(\gamma)$, the first row of $P(\gamma)$, is prescribed under Assumption IV. That is, $A(\hat{a}(\gamma))$ is stable for $\gamma \in \Gamma$.

The above lemmas are used in the next theorem to establish a stabilizing controller for system (1).

THEOREM 3. If Assumptions I and IV hold then there is a stabilizing controller for system (1) having the form

$$u = -\frac{Fx}{\|Fx\|}\rho(x), \quad Fx \neq 0 \tag{52}$$

where F is a constant row vector and $\rho(x)$ is a nonnegative scalar function of the state x.

PROOF: Since system (1) is controllable for each uncertainty $\gamma \in \Gamma$ it can be transformed to the controllable companion form (44). Assumption IV implies there is a stable coefficient vector $\hat{a}(\gamma)$ for $\gamma \in \Gamma$ such that (51) is satisfied. Define $\sigma(\gamma)$ to be the difference between the stable coefficient vector $\hat{a}(\gamma)$ and the open-loop characteristic polymonial coefficient vector $a(\gamma)$ of System (1)

$$\sigma(\gamma) = \hat{a}(\gamma) - a(\gamma) \tag{53}$$

Note that the negative of $a(\gamma)$ is contained in the first row of (45). Substitution of (53) into (44) yields

$$\dot{z} = A_z(\hat{a}(\gamma))z + B_z[\sigma(\gamma)T(\gamma)x + u(x)] \tag{54}$$

after making use of (43). We use the symmetric, positive definite solution $P(\gamma)$ of (51) to construct the Lyapunov function

$$V(\gamma) = z^T P(\gamma) z \tag{55}$$

Taking its derivative gives

$$\dot{V}(\gamma) = -z^T Q(\gamma)z + 2[Fx]^T[\sigma(\gamma)\, T(\gamma)x + u(x)] \tag{56}$$

where F satisfies

$$F = P_1(\gamma_0)T(\gamma_0) \tag{57a}$$

and as a consequence of Assumption IV we have

$$F = P_1(\gamma)T(\gamma) \tag{57b}$$

or, equivalently,

$$F = B_z^T P(\gamma)T(\gamma) \tag{57c}$$

where $P(\gamma)$ satisfies (51) and $T(\gamma)$ satisfies (43). Any admissible control law $u(x)$ satisfying

$$u(x) \leq -\max_{\gamma \varepsilon \Gamma}[\sigma(\gamma)T(\gamma)x]\ , \quad Fx > 0 \tag{58a}$$

$$u(x) \geq \max_{\gamma \varepsilon \Gamma}[\sigma(\gamma)T(\gamma)x]\ , \quad Fx < 0 \tag{58b}$$

stabilizes (1) since for such a control law

$$\dot{V}(\gamma) \leq -z^T Q(\gamma)z\,, \quad \gamma \ \varepsilon \ \Gamma \tag{59}$$

The maxima of (58) exist since Γ is compact and since the functions $\sigma(\gamma)$ and $T(\gamma)$ are continuous on Γ. An admissible control law satisfying (58) is (52) where

$$\rho(x) = \max_{\gamma \varepsilon \Gamma} \|\sigma(\gamma)\ T(\gamma)\ x\| \tag{60}$$

and F is given by (57). In the next theorem we establish that a system satisfying Assumption IV also satisfies Assumption II.

THEOREM 4: If the system (1) satisfies Assumptions I and IV then Assumptions II and III are met.

PROOF: We make the following identifications

$$\bar{A}(\gamma) = T^{-1}(\gamma)\, A(\hat{a}(\gamma))\, T(\gamma) \tag{61a}$$

$$\bar{P}(\gamma) = T^{T}(\gamma)\, P(\gamma)\, T(\gamma) \tag{61b}$$

$$\bar{Q}(\gamma) = T^{T}(\gamma)\, Q(\gamma)\, T(\gamma) \tag{61c}$$

$$K(\gamma) = \sigma(\gamma)\, T(\gamma) \tag{61d}$$

where $T(\gamma)$ is defined by (43), where $P(\gamma)$, $Q(\gamma)$and$A(\hat{a}(\gamma))$ are defined by (51) and where $\sigma(\gamma)$ is defined by (53). The matrix $A(\hat{a}(\gamma))$ is asymptotically stable. This follows from Lemma 2 and the fact that eigenvalues are invariant under coordinate transformation. From (44), (45), (53) and (54) it follows that

$$\bar{A}(\gamma) = A(\gamma) - B(\gamma)\, K(\gamma) \tag{62}$$

so that condition (i) of Assumption III is met. The vector F of (57) satisfies

$$F = B^{T}(\gamma)\, \bar{P}(\gamma) \tag{63}$$

where $\bar{P}(\gamma)$ is the solution of the Lyapunov equation

$$\bar{P}(\gamma)\, \bar{A}(\gamma) + \bar{A}^{T}(\gamma)\, \bar{P}(\gamma) + \bar{Q}(\gamma) = 0 \tag{64}$$

which shows that condition (ii) of Assumption II is met. Here, the scalar R = 1. Thus Assumption III is also met.

Theorems 3 and 4 establish that Assumption IV implies Assumption II. The converse need not hold. Thus Assumption IV is a stronger assumption. Assumption IV admits a sliding mode controller (52). From the next theorem we see that it also admits a stabilizing linear controller.

THEOREM 5: If Assumption I and IV hold then there exists a stabilizing linear control

$$u = -cFx \tag{65}$$

where F is defined as in Theorem 3 and the scalar c satisfies

$$c > \frac{1}{2} \max_{\gamma \in \Gamma} \|Q^{-1}(\gamma)\| \max_{\gamma \in \Gamma} \|K(\gamma)\|^2 \tag{66}$$

where $Q(\gamma)$, $\gamma \in \Gamma$, is defined as in Lemma 2 and where $K(\gamma)$ is given by (61d).

PROOF: See [31].

The maxima of (66) exist since $K(\gamma)$ is continuous, Γ is compact and the matrices $Q(\gamma)$ are chosen in a continuous manner. Usually $Q(\gamma)$ is set to be the identity I or it is computed from

$$Q(\gamma) = T^{-T}(\gamma)\, Q\, T^{-1}(\gamma) \tag{67}$$

where Q is a prescribed symmetric, positive definite matrix. The next result gives an equivalence between Assumption IV and a minimum phase condition on the system.

THEOREM 6: Assumption IV is met if, and only if, there is a row vector F such that

$$F[sI - A(\gamma)]^{-1}\, B(\gamma), \quad \gamma \in \Gamma \tag{68}$$

is minimum phase with n-1 transmission zeros where I is the nxn identity matrix. That is, the determinant

$$\det\begin{bmatrix} A(\gamma) - sI & B(\gamma) \\ F & 0 \end{bmatrix} = 0, \quad \gamma \in \Gamma \tag{69}$$

is Hurwitz.

PROOF: From (49) of Assumption IV

$$P_1(\gamma) = F\,T^{-1}(\gamma), \quad \gamma \in \Gamma \tag{70}$$

where $P_1(\gamma)$ is n-1 stable with polynomial Eq. (48) can be rewritten as

$$P_1(\gamma)[s^{n-1}\, s^{n-2} \ldots s\ 1]^T = 0 \tag{71}$$

where $s = \sigma + j\omega = \lambda$. Multiplying (70) on both sides by $[s^{n-1}\, s^{n-2} \ldots s\ 1]^T$ gives

$$FT^{-1}(\gamma)[s^{n-1}\, s^{n-2} \ldots s\ 1]^T = 0 \tag{72}$$

The open-loop characteristic polynomial $a_\gamma(s)$, (47), is given by

$$a_\gamma(s) = s^n + a_1(\gamma)s^{n-1} + \ldots + a_{n-1}(\gamma)s + a_n(\gamma) = 0 \tag{73}$$

Since $A_z(a(\gamma))$andB_z are in the controller companion form (45) and (46) we have the following identity from linear system theory, [34]:

$$[sI - A_z(a(\gamma))]^{-1}B_z = \frac{[s^{n-1} s^{n-2} \ldots s\ 1]^T}{a_\gamma(s)} \tag{74}$$

Substitution from (45b) and (46b) into (74) gives

$$T(\gamma)[sI - A(\gamma)]^{-1}B(\gamma) = \frac{[s^{n-1}s^{n-2} \dots s\ 1]^T}{a_\gamma(s)} \tag{75}$$

Multiplying both sides by $FT^{-1}(\gamma)$ yields

$$F[sI - A(\gamma)]^{-1}B(\gamma) = \frac{FT^{-1}(\gamma)[s^{n-1}s^{n-2} \dots s\ 1]^T}{a_\gamma(s)} = 0 \tag{76}$$

after making use of (72). The transmission zeros, [35] , of (76) are the n-1 stable eigenvalues of the (n-1) stable $P_1(\gamma)$ row vector of (70). This proves that (68) is minimum phase. From (76) we have

$$\det[sI - A(\gamma)]\ F[sI - A(\gamma)]^{-1}B(\gamma) = 0 \tag{77}$$

A reciprocal form of (77) is given by, [34],

$$\det\begin{bmatrix} sI - A(\gamma) & B(\gamma) \\ -F & 0 \end{bmatrix} = 0 \tag{78}$$

which yields (69). Since $P_1(\gamma)$ is n-1 stable it follows that (71) is Hurwitz. Thus (69) is Hurwitz.

Conversely, if there exists an F such that (69) is Hurwitz then the vector $P_1(\gamma)$ defined by (70) is n-1 stable and Assumption IV is met. From the above theorem we have the corollary.

COROLLARY 1. A necessary and sufficient conditions for the existence of a stabilizing sliding mode controller

$$u = -\frac{Fx}{\|Fx\|}\rho(x), \quad Fx \neq 0 \tag{79}$$

of (1) is the existence of a row-vector F such that (69) is Hurwitz for all $\gamma \varepsilon \Gamma$.

The existence of a stabilizing linear controller

$$u = -Kx \tag{80}$$

does not imply the existence of a stabilizing sliding mode controller (79). Before this is illustrated by an example we give necessary and sufficient conditions for the existence of a linear controller (80).

THEOREM 7: A necessary and sufficient condition that there exist a stabilizing linear controller (80) of system (1) is that there exists a row vector K such that the following determinant is Hurwitz:

$$\det\begin{bmatrix} sI - A(\gamma) & B(\gamma) \\ -K & 1 \end{bmatrix} = 0, \quad \gamma \in \Gamma \tag{81}$$

PROOF: Suppose there is a row vector K such that (80) asymptotically stabilizes (1). The feedback matrix

$$A_c(\gamma) = A(\gamma) - B(\gamma)K, \quad \gamma \in \Gamma \tag{82}$$

is asymptotically stable and the determinant

$$a_\gamma^c(s) = \det[sI - A_c(\gamma)] = 0 \tag{83}$$

is Hurwitz. Eq. (83) can be written as the following series of identities, [34],

$$a_\gamma^c(s) = \det\{[sI - A(\gamma)]\,[I + [sI - A(\gamma)]^{-1}B(\gamma)K]\} \tag{84a}$$

$$a_\gamma^c(s) = a_\gamma(s)\,\det[I + [sI - A(\gamma)]^{-1}B(\gamma)K] \tag{84b}$$

$$a_\gamma^c(s) = a_\gamma(s)\,[1 + K[sI - A(\gamma)]^{-1}B(\gamma)] \tag{84c}$$

where $a_\gamma(s)$ is the open-loop characteristic polynomial (47). The reciprocal form of (81) is (84c), [34]. That is, (81) and (84c) are identities. Therefore, (81) is Hurwitz if, and only if, (83) is Hurwitz. Eq. (84c) can be used to prove Theorem 5. If (76) is Hurwitz then with

$$K = cF \tag{85}$$

Eq. (84c) becomes

$$a_\gamma^c(s) = a_\gamma(s)\,[1 + cF[sI - A(\gamma)]^{-1}B(\gamma)] \tag{86}$$

which is Hurwitz for sufficiently large c. That is, in view of (71) -(76), Eq. (86) can be rewritten as

$$a_\gamma^c(s) = s^n + c\,P_1(\gamma)[s^{n-1}, s^{n-2} \ldots s\ 1]^T + [a_\gamma(s) - s^n] \tag{87}$$

in which the last term is an n-1 order polynomial that is dominated by the middle term for large c. The first two terms give a Hurwitz polynomial for sufficiently large c. As a consequence, the existence of a robust stabilizing sliding mode controller (52) implies the existence of a robust stabilizing linear controller (65). In general, the converse does not hold as is illustrated by the following example.

5. EXAMPLE OF ROBUST LINEAR CONTROLLER WITHOUT SLIDING MODE CONTROLLER

Consider the uncertain system

$$\dot{x} = \begin{bmatrix} 0 & -1 \\ 1 & 0 \end{bmatrix} x + \begin{bmatrix} \gamma \\ 1 \end{bmatrix} u, \quad \gamma \, \varepsilon \, \Gamma \tag{88}$$

where $\Gamma = [-M, M]$ and where M is a positive scalar greater than 1

$$M \geq 1 \tag{89}$$

The determinant of the controllability matrix $[B(\gamma), AB(\gamma)]$ is given by $\gamma^2 + 1$ which satisfies the equality

$$\gamma^2 + 1 > 0 \quad \forall \, \gamma \, \varepsilon (-\infty, \infty) \tag{90}$$

The system (88) is controllable for all uncertainties γ. Thus Assumption I is satisfied. The requirement for the existence of a stable sliding mode surface

$$Fx = 0 \tag{91}$$

depends on (69) being Hurwitz. For our example system (88) Eq. (69) reduces to the first order polynomial

$$(F_1\gamma + F_2)\lambda + (\gamma F_2 - F_1) = 0 \tag{92}$$

which is Hurwitz for $\gamma \, \varepsilon \, \Gamma$ provided the coefficients are positive

$$F_1\gamma + F_2 > 0, \quad \gamma \, \varepsilon \, \Gamma \tag{93a}$$

$$\gamma F_2 - F_1 > 0, \quad \gamma \, \varepsilon \, \Gamma \tag{93b}$$

Evaluating the first inequality at $\gamma = 1$ and the second at $\gamma = -1$ give the contradicting inequalities

$$F_2 > -F_1 \tag{94a}$$

$$F_2 < -F_1 \tag{94b}$$

That is, there exists no $F = (F_1, F_2)$ satisfying (69) for $\gamma \varepsilon [-1, 1]$ which is a subset of Γ. Consequently, there is no stable sliding mode surface (91) on which a robust sliding mode controller (52) can be designed for $\gamma \varepsilon [-1, 1]$.

The requirement for the existence of a robust stabilizing linear feedback controller (80) is that (81) is Hurwitz. The characteristic polynomial of (81) is given by

$$\lambda^2 + a_1(\gamma)\lambda + a_2(\gamma) = 0 \tag{95}$$

where robustness follows from positiveness of the coefficients

$$a_1(\gamma) = K_2 + \gamma K_1 > 0, \quad \gamma \in \Gamma \tag{96a}$$

$$a_2(\gamma) = \gamma K_2 - K_1 + 1 > 0, \quad \gamma \in \Gamma \tag{96b}$$

The following gain vector $K = (K_1, K_2)$ provides a robust linear controller (80)

$$K_1 = 0 \tag{97a}$$

$$K_2 = \frac{1}{M + \varepsilon} \tag{97b}$$

where $\varepsilon > 0$. Substitution of the gain vector (97) into (96) gives

$$K_2 > 0, \quad \gamma \in \Gamma \tag{98a}$$

$$\gamma > -(M + \varepsilon), \quad \gamma \in \Gamma \tag{98b}$$

The inequalities (96) are met. Thus (81) is Hurwitz which implies that the linear controller defined by (97) robustly stabilizes (88). Consequently, (88) has a robust stabilizing linear controller but no stabilizing sliding mode controller.

6. ROBUST CONTROL IN THE ABSENCE OF MATCHING CONDITIONS: MULTI-INPUT

The multi-input case parallels that of the scalar case, Section 4. We consider a condition similar to (69) and show that it leads to necessary and sufficient conditions for the existence of a sliding mode controller (7). In this section $B(\gamma), \gamma \in \Gamma$, is an nxm uncertain matrix with full rank ($m \leq n$) We consider system (1) for which Assumption I holds. Our main result for a robust sliding mode controller is given in the next theorem.

THEOREM 8. A robust stabilizing sliding mode controller (7) exists for system (1) in which Assumption I holds if, and only if, the following determinant is Hurwitz:

$$\det\begin{bmatrix} A(\gamma) - sI & B(\gamma) \\ F & 0 \end{bmatrix} = 0, \quad \gamma \in \Gamma \tag{99}$$

PROOF: The reciprocal form of (99) is

$$a_\gamma(s)\ \det\left[F[sI - A(\gamma)]^{-1}B(\gamma)\right] = 0, \quad \gamma\ \varepsilon\ \Gamma \tag{100}$$

where $a_\gamma(s)$, defined by (73), is the determinant of $[sI - A(\gamma)]$ which is the open-loop characteristic polynomial of $A(\gamma)$. Since Assumption I holds there is a coordinate transformation $T(\gamma)$

$$z = T(\gamma)x \tag{101}$$

which takes (1) into a controllable companion form, [34],

$$\dot{z} = A_z(a(\gamma))z + B_z u(x) \tag{102}$$

where

$$A_z(a(\gamma)) = T(\gamma)\,A(\gamma)\,T^{-1}(\gamma) \tag{103a}$$

$$B_z(\gamma) = T(\gamma)\,B(\gamma) \tag{103b}$$

The mxm matrix $B_z(\gamma)$ is formed from m columns of the nxn identity matrix. The dependence of $B_z(\gamma)$ on the uncertainty γ follows from the fact that the selection of the m columns may depend on $\gamma\varepsilon\Gamma$. The nxn matrix $A_z(a(\gamma))$ is in block controllable companion form. Such companion forms are described in [32]-[34]. In view of the Transformation (101) we can rewrite (100) as

$$\det\left[\left[FT^{-1}(\gamma)\right]\ a_\gamma(s)\ T(\gamma)[sI - A(\gamma)]^{-1}B(\gamma)\right] = 0 \tag{104}$$

Consider the last two factors

$$a_\gamma(s)\ \left[T(\gamma)[sI - A(\gamma)]^{-1}B(\gamma)\right] \tag{105}$$

which in z-coordinates is given by

$$a_\gamma(s)\ \left[[sI - A_z(a(\gamma))]^{-1}B_z(\gamma)\right] \tag{106}$$

which is equivalent to

$$Adj[sI - A_z(a(\gamma))]B_z(\gamma) \tag{107}$$

where Adj is the matrix adjoint operation. Consider the definition of an nxn symmetric, positive definite matrix $P(\gamma)$ and the definition of an mxm symmetric, positive definite matrix $R(\gamma)$ such that

$$R^{-1}(\gamma)B_z^T(\gamma)P(\gamma)T(\gamma) = F, \quad \gamma\ \varepsilon\ \Gamma \tag{108}$$

That is, $P(\gamma)$ must be such that

$$R^{-1}(\gamma)B_z^T(\gamma)P(\gamma) = FT^{-1}(\gamma), \quad \gamma\ \varepsilon\ \Gamma \tag{109}$$

Furthermore, consider the Lyapunov equation

$$P(\gamma)A_z(\hat{a}(\gamma)) + A_z^T(\hat{a}(\gamma))P(\gamma) + Q(\gamma) = 0, \quad \gamma \in \Gamma \tag{110}$$

where $Q(\gamma) > 0$ and $Q(\gamma) = Q^T(\gamma), \gamma \in \Gamma$. A necessary and sufficient condition that $A_z(\hat{a}(\gamma))$ be asymptotically stable and $P(\gamma)$ be symmetric, positive definite and satisfy the constraint (109) is that the determinant of the following mxm matrix (111a) be Hurwitz and that the following mxm matrix (111b) be positive definite , [32]:

$$B_z^T(\gamma)P(\gamma)Adj[sI - A_z(\gamma)]B_z(\gamma), \quad \gamma \in \Gamma \tag{111a}$$

$$B_z^T(\gamma)P(\gamma)B_z(\gamma) > 0, \quad \gamma \in \Gamma \tag{111b}$$

From (104), (107), (109) and (111) it follows that (99) is necessary and sufficient in order that for each $\gamma \epsilon \Gamma$ there exist a symmetric, positive definite $P(\gamma)$ satisfying (109) and a stable $A_z(\hat{a}(\gamma))$ such that the Lyapunov equation (110) is satisfied. The theorem now follows from Theorem 1. Define $\sigma(\gamma), \gamma \in \Gamma$

$$\sigma(\gamma) = B_z^T(\gamma)\left[A_z(a(\gamma)) - A_z(\hat{a}(\gamma))\right] \tag{112}$$

By the canonical form of A_z and B_z it follows that

$$A_z(\hat{a}(\gamma)) = A_z(a(\gamma)) - B_z(\gamma)\sigma(\gamma) \tag{113}$$

Define $K(\gamma), \gamma \in \Gamma$, as

$$K(\gamma) = \sigma(\gamma)T(\gamma) \tag{114}$$

Transforming (113) from z-coordinates to x-coordinates using (101) yields the following asymptotically stable matrix.

$$\bar{A}(\gamma) = A(\gamma) - B(\gamma)K(\gamma) \tag{115}$$

Thus condition (i) of Assumption II is met. Transforming (108) from z-coordinates to x-coordinates using (101) gives

$$F = R^{-1}(\gamma)B^T(\gamma)\bar{P}(\gamma) \tag{116}$$

where $\bar{P}(\gamma)$ satisfies the Lyapunov equation which is transformed from (110)

$$\bar{P}(\gamma)\bar{A}(\gamma) + \bar{A}^T(\gamma)\bar{P}(\gamma) + \bar{Q}(\gamma) = 0, \quad \gamma \in \Gamma \tag{117}$$

where

$$\bar{P}(\gamma) = T^T(\gamma)P(\gamma)T(\gamma) \tag{118a}$$

$$\bar{Q}(\gamma) = T^T(\gamma)Q(\gamma)T(\gamma) \tag{118b}$$

Thus condition (ii) of Assumption II is met. Consequently all conditions of Theorem 1 are satisfied. The existence of a robust sliding mode controller (7) now follows.

The existence of a robust stabilizing sliding mode controller implies the existence of a robust linear controller. This result is given in the next theorem which parallels the scalar result, Theorem 7:

THEOREM 9: The existence of a stabilizing sliding mode controller (7) for system (1) implies the existence of a robust stabilizing linear controller

$$u = -Kx \tag{119}$$

PROOF: A necessary and sufficient condition for the existence of a robust stabilizing linear controller is that the determinant

$$\det\begin{bmatrix} sI - A(\gamma) & B(\gamma) \\ -K & I_m \end{bmatrix} = 0, \quad \gamma \varepsilon \Gamma \tag{120}$$

is Hurwitz where I_m is the mxm identity matrix. Paralleling the developement (81) - (84) the determinant (120) is Hurwitz if, and only if, the mxm matrix

$$a_\gamma^c(s) = a_\gamma(s) \det[I_m + K\,[sI - A(\gamma)]^{-1} B(\gamma)], \quad \gamma \varepsilon \Gamma \tag{121}$$

is Hurwitz. If a robust stabilizing sliding mode controller (7) exists then there exists an mxn matrix F such that (99) is Hurwitz. Consequently, (100) is Hurwitz. For an arbitrary mxm matrix C define the gain matrix

$$K = CF \tag{122}$$

substitution of (122) into (121) gives

$$a_\gamma^c(s) = a_\gamma(s) \det[I_m + CF\,[sI - A(\gamma)]^{-1} B(\gamma)], \quad \gamma \varepsilon \Gamma \tag{123}$$

In view of (73) we can rewrite (123) as

$$a_\gamma^c(s) = \det[\lambda^n I_m + CF\,Adj[sI - A(\gamma)]\,B(\gamma) + (a_\gamma(s) - \lambda^n)\,I_m], \quad \gamma \in \Gamma \tag{124}$$

Since the mxm matrix (100) is Hurwitz, it follows that there exists an mxm matrix C with sufficiently "large elements" such that (124) is Hurwitz, [33]. The last term is dominated by the second term. The control law (119) robustly stabilizes (1) for a " sufficiently large" C matrix in (122).

7. SUMMARY

A linear time-invariant uncertain system is investigated for robust stabilization. The uncertainties belong to a compact subset of multi-dimensional Euclidean space. The dynamics and input matrices are continuous functions of uncertainty. The system is controllable for each uncertainty, Assumption I. In Assumption II two general conditions are stated which involve an uncertain Lyapunov equation. The first condition deals with the existence of an uncertain gain matrix for stabilizing the system. The second deals with the existence of a constant F matrix which has the appearance of a Riccati gain matrix. F is the product of three uncertain quantities one of which is the uncertain solution $P(\gamma)$ of the Lyapunov equation. Another is the $R(\gamma)$ matrix which is assumed in Assumption III to form a positive definite matrix when added to its transpose.

A general robustness result is established in Theorem 1. It states that a robust stabilizing sliding mode controller exists under the general Assumptions I - III. In Theorem 2 we prove that the matching conditions of uncertain systems satisfy the Assumptions I - III.

Robust control in the absence of matching conditions is examined in Theorems 3, 4 and 5 for scalar control input. For such systems necessary and sufficient conditions are given for the existence of robust stabilizing sliding mode controllers. In Theorem 4 we show that systems satisfying such conditions also meet Assumptions I - III. Theorem 5 goes one step further and shows the existence of a robust linear control for such systems. The existence of a robust sliding mode controller is shown to depend on a minimum phase condition, Theorem 6. In Section 5 we give an example of a simple system which admits a robust linear controller but no robust sliding mode controller that stabilizes the system.

In Section 6 we investigate robust control in the absence of matching conditions for multi-input systems. In Theorem 8 we show that a certain determinant being Hurwitz is necessary and sufficient for the existence of a sliding mode controller. A similar condition is stated in Theorem 9 for the existence of a robust linear controller.

REFERENCES

1. GUTMAN, S. and LEITMANN, G., "On a class of Linear Differential Games", Journal of Optimization Theory and Applications, Vol. 17, Nos 5-6, pp. 511-522, 1975.

2. LEITMANN, G., "Stabilization of Dynamical Systems under Bounded Input Disturbance and Parameter Uncertainty," Proceed. of 2nd Kingston Conference on Differential Games and Control Theory II, M. Dekker, New York, 1976.

3. GUTMAN, S. and LEITMANN, G., "Stabilizing Feedback Control for Dynamical Systems with Bounded Uncertainty," Proceedings of IEEE Conference on Decision and Control, 1976.

4. LEITMANN, G., "Guaranteed Ultimate Boundedness for a Case of Uncertain Linear Dynamical Systems," IEEE Transactions on Automatic Control, Vol. AC-23, No. 6, 1978.

5. GUTMAN, S., "Uncertain Dynamical Systems - A Lyapunov Min-Max Approach," IEEE Transactions on Automatic Control, Vol. AC-24, No. 3, pp. 437-443, June 1979.

6. LEITMANN, G., "Guaranteed Asymptotic Stability for Some Linear Systems with Bounded Uncertainties," J. Dynamic Systems, Measurement and Control, Vol. 101, No. 3, pp. 212-216, 1979.

7. LEITMANN, G., "On the Efficacy of Nonlinear Control in Uncertain Linear Systems," Dynam. Syst., Meas. Cont., Vol. 102, No. 2, pp. 95-102 1981.

8. CORLESS, M.J. and LEITMANN, G., "Continuous State Feedback Guaranteeing Uniform Ultimate Boundedness for Uncertain Dynamic Systems," IEEE Transactions on Automatic Control, Vol. AC-26, No. 5, pp. 1139-1144, October 1981.

9. BARMISH, B.R. and LEITMANN, G., "On Ultimate Boundedness Control of Uncertain Systems in the Absence of Matching Assumptions," IEEE Transactions on Automatic Control, Vol. AC-27, No. 1, pp.153-158, February 1982.

10. GUTMAN, S. and PALMOR, Z., "Properties of Min-Max Controllers in Uncertain Dynamical Systems," SIAM J. Control and Optimization, Vol. 20, No. 6, pp. 850-861, November 1982.

11. BARMISH, B.R., CORLESS, M. and LEITMANN, G., "A New Class of Stabilizing Controllers for Uncertain Dynamical Systems," SIAM J. Control and Optimization, Vol. 21, No. 2, pp. 246-255, March 1983.

12. LEITMANN, G., "Deterministic Control of Uncertain Systems," The Fourth International Conference: Mathematical Modeling in Science and Technology, Zurich, August 1983.

13. BARMISH, B.R., PETERSEN, I.R. and FEUER, A., "Linear Ultimate Boundedness Control of Uncertain Dynamical Systems," Automatica, Vol. 19, No. 5, pp. 523-532, September 1983.

14. BARMISH, B.R., "Necessary and Sufficient Conditions for Quadratic Stabilizability of an Uncertain System", Journal of Optimization Theory and Applications, Vol. 46, No. 4. pp. 399-408, August 1985.

15. PETERSEN, I.R., "Structural Stabilization of Uncertain Systems: Necessity of the Matching Condition," Proceed. of 20th Allerton Conference on Communication, Control, and Computing, 1982, also in SIAM J. Control and Opimization, Vol. 23, No. 2, pp. 286-296, March 1985.

16. GALIMIDI, Alberato R. and BARMISH, B.Ross, "The Constrained Lyaponov Problem and its Application to Robust Output Feedback Stabilization", IEEE Transactions on A.C., Vol. AC-31, No. 5, pp. 410-418, May 1986.

17. PETERSEN, I.R., "Quadratic Stabilizability of Uncertain Linear Systems: Existence of a Nonlinear Stabilizing Control Does Not Imply Existence of a Linear Stabilizing Control," IEEE Transactions on Automatic Control, Vol. AC-30, No.3, pp. 291-293, March 1985.

18. PETERSEN, I.R., "Nonlinear Versus Linear Control in the Direct Output Feedback Stabilization of Linear Systems," IEEE Transactions on Automatic Control, Vol. AC-30, No. 8, pp. 799-802, August 1985.

19. STALFORD, H.L., "Necessary and Sufficient Conditions for Matching Conditions in Uncertain Systems: Scalar Input," Proceed. of the 1987 American Control Conference, pp. 879-903, June 10-12, 1987.

20. SINGH, S.N. and COELHO, A.A.R. "Ultimate Boundedness Control of Set Points of Mismatched Uncertain Linear Systems," Int. J. Systems SCI., 1983, Vol. 14, No. 7, pp.693-710.

21. SINGH, S.N. and COELHO, A.A.R., "Nonlinear Control of Mismatched Uncertain Linear Systems and Application to Control of Aircraft," Journal of Dynamic Systems, Measurement, and Control, September 1984, Vol. 106, pp. 203-210.

22. LEITMANN, G., RYAN, E.P. and STEINBERG, A., "Feedback Control of Uncertain Systems: Robustness with respect to neglected actuator and sensor dynamics," Int. J. Control, Vol. 43, No. 4, pp. 1243-1256, 1986.

23. SCHMITENDORF, W.E. and BARMISH, B.R., "Robust Asymptotic Tracking for Linear Systems with Unknown Parameters," Automatica, Vol. 22, No. 3, pp. 355-360, 1986.

24. CHEN, Y.H. and LEITMANN, G., "Robustness of Uncertain Systems in the Absence of Matching Conditions," Int. J. Control, Vol. 45, No. 5, pp. 1527-1542, 1987.

25. SCHMITENDORF, W.E. and BARMISH, B.R., "Guaranteed Asymptotic Output Stability for Systems with Constant Disturbances", Transactions of the ASME, Vol. 109, pp. 186-189, June 1987.

26. STALFORD, H. and GARRETT, F. Jr., "Robust Nonlinear Control for High Angle-of-Attack Flight," Presented at the AIAA 25th Aerospace Sciences Meeting, Reno, Nevada, paper AIAA-87-0346, January 12-15, 1987.

27. STALFORD, H., "On Robust Control of Wing Rock Using Nonlinear Control," Proceed. 1987 American Control Conference, Minneapolis Minnesota, June 10-12, 1987.

28. STALFORD, H., "Tracking at High α Using Certain Robust Nonlinear Controllers," Proceed. AIAA Guidance, Navigation and Control Conference, Monterey, California, August 17-19, 1987.

29. STALFORD, H.L., "Robust Control of Uncertain Systems in the Absence of Matching Conditions: Scalar Input," 1987 Conference on Decision and Control, Los Angeles, California, December 8-10, 1987.

30. STALFORD, H.L. and CHAO, C.-H., "Necessary and Sufficient Condition in Lyapunov Robust Control," submitted for publication, December, 1987.

31. STALFORD, H.L. and CHAO, C.-H., "On the Robustness of Linear Stabilizing Feedback for Linear Uncertain Systems," submitted for publication, December, 1987.

32. STALFORD, H. L. and CHAO, Chien-Hsiang, "A Necessary and Sufficient Condition in Lyapunov Robust Control: Multi-Input," Submitted for possible presentation at the 27th IEEE Conference on Decision and Control, Austin, Texas, December 7-9, 1988.

33. CHAO, Chien-Hsiang and STALFORD, H. L., "On the Robustness of Linear Stabilizing Feedback Control for Linear Uncertain Systems: Multi-Input," submitted for possible presentation at the 27th IEEE Conference on Decision and Control, Austin, Texas, December 7-9, 1988.

34. KAILATH, T., "Linear Systems", Prenctice-Hall, Inc., New Jersey, 1980.

35. DAVISON, E.J. and WANG, S.H., "Properties and Calculations of Transmission Zeros of Linear Multivariable System," Control System Design by Pole-Zero Assignment , F. Fallside, Editor, Academic Press, New York, pp. 16-42, 1977.

SINGULARLY PERTURBED UNCERTAIN SYSTEMS AND DYNAMIC OUTPUT FEEDBACK CONTROL

E P Ryan and Z B Yaacob

School of Mathematical Sciences
University of Bath
Bath BA2 7AY
United Kingdom

ABSTRACT

A dynamic output feedback strategy is proposed for a class of uncertain systems. Using a singular perturbation approach, a threshold measure of "fastness" of the feedback dynamics, to ensure overall system stability, is derived. This threshold is calculable in terms of known bounds on the system uncertainties but may be conservative in practice. To circumvent this drawback and to allow for bounded uncertainties with unknown bounds, an adaptive version of the strategy is then developed.

1. Introduction

We address the problem of design of dynamic output feedback controls for a class of uncertain nonlinearly perturbed linear multivariable systems. The approach is similar in concept to that of [1], and fundamentally stems from the deterministic theory developed in, for example, [2-8] (see also bibliographies therein).

Initially considering a hypothetical output $y^{\#}$ for the system, a (generally unrealizable) stabilizing static output feedback control is established. This static control is then approximated by a realizable compensator (with parameter $\mu \geq 0$) which filters the true system output y. Physically, the parameter μ is a measure of "fastness" for the filter dynamics; analytically, μ plays the role of a singular perturbation parameter. Using a singular perturbation analysis akin to that of [9,10], a threshold measure μ^* of "fastness" of the compensator dynamics, to ensure overall system stability, is then derived. The threshold is explicitly calculable from known system data but corresponds to a "worst-case" value and consequently may be conservative. To counteract this inherent conservatism (and to allow for bounded uncertainties with unknown bounds) an adaptive version of the compensator is also developed by an approach which is essentially that of [11] (see also [12-16] and related work in [17-23]).

2. The system

We consider uncertain nonlinearly perturbed linear systems of the form

$$\dot{x}(t) = Ax(t) + B[u(t) + g(t,x(t),u(t))], \quad x(t) \in I\!R^n, \quad u(t) \in I\!R^m \tag{1}$$

for which the only available state information is provided by the output

$$y(t) = Cx(t), \quad y(t) \in I\!R^p, \quad m \leq p \leq n . \tag{2}$$

The triple (C,A,B), which defines the nominal linear system, is assumed to satisfy the following.

Assumption 1: (A,B) is a controllable pair and rank $B = m$.

Assumption 2:

There exist known integer $r \geq 1$ and known matrices $F_1, F_2, \cdots, F_r \in \mathbb{R}^{m \times p}$, such that

(i) for $i = 1,2,\cdots, r-1$, im $CA^{i-1}B \subset \bigcap_{j=i+1}^{r} \ker F_j$;

moreover, the matrix $C_r := F_1C + F_2CA + \cdots + F_rCA^{r-1}$ is such that

(ii) $|C_rB| \neq 0$, and

(iii) the transmission zeros of the m-input m-output linear system (C_r, A, B) lie in $\mathbb{C}^-$ (the open left half complex plane).

Example 1: If $A = \begin{bmatrix} 0 & 1 & 0 \\ 0 & 0 & 1 \\ 0 & 0 & 0 \end{bmatrix}$, $B = \begin{bmatrix} 0 \\ 0 \\ 1 \end{bmatrix}$, $C = \begin{bmatrix} 1 & 0 & 0 \\ 0 & 0 & 1 \end{bmatrix}$, then the above assumptions hold with $r = 2$, $F_1 = [1 \;\; 1]$ and $F_2 = [1 \;\; 0]$.

Finally, we impose some structure on the uncertain function g.

Assumption 3:

g: $\mathbb{R} \times \mathbb{R}^n \times \mathbb{R}^m \to \mathbb{R}^m$ is (i) Carathéodory, with (ii) $\|g(t,x,u)\| \leq \alpha\|x\| + \beta\|u\|$ for all (t,x,u), where α and β are known constants with $\beta < 1$, and (iii) if $r \geq 2$, then g is uniformly Lipschitz in its final argument (with known Lipschitz constant λ), i.e. (if $r \geq 2$) there exists known λ, independent of (t,x), such that, for all u and v, $\|g(t,x,u)-g(t,x,v)\| \leq \lambda\|u-v\|$.

The outline of the paper is as follows:

Firstly, the problem of designing a (dynamic) output feedback compensator for system (1,2) is addressed. This is accomplished by initially considering system (1) with hypothetical output

$$y^{\#}(t) = C_r x(t) \tag{3}$$

where C_r is as in Assumption 2. Note that, if $r = 1$ then $y^{\#}(t) = F_1 y(t)$ and hence is realizable; however, if $r \geq 2$ then $y^{\#}(t)$ is unavailable to the controller, hence the qualifier "hypothetical". For the system (1,3) so defined, (ii) and (iii) of Assumption 2 in essence play the role of "relative degree one" and "minimum phase" conditions on the hypothetical nominal linear system triple (C_r, A, B). Under such conditions, it is known (see, for example, [11-13]) that the zero state of system (1,3) can be rendered globally uniformly asymptotically stable by static output feedback; this is reiterated in Theorem 1. However, with the exception of the case $r = 1$, such static output feedback is unrealizable in the context of the true system (1,2). Therefore, in §3, a realizable dynamic compensator is constructed for the cases $r \geq 2$, which filters the actual output y. This filter can be interpreted as providing a realizable approximation to the static hypothetical output feedback; moreover, it is shown in Theorem 2 that global uniform asymptotic stability of the zero state of (1,2) is guaranteed provided that the filter dynamics are sufficiently fast (a calculable threshold measure of fastness is provided).

Secondly, in §4, an adaptive version of the dynamic compensator is developed, which counteracts conservatism (induced by crude estimates in the analysis) inherent in the non-adaptive filter and which also dispenses with the requirement that the uncertainty parameters α, β and λ in Assumption 3 be known (however, the assumption that $\beta < 1$ remains in force and, moreover, if $r \geq 2$ then g is assumed to depend linearly on x).

3. Stabilizing static output feedback control for hypothetical system

Let $T_1 \in \mathbb{R}^{(n-m)\times n}$ be such that $\ker T_1 = \mathrm{im}\, B$, then

$$T = \begin{bmatrix} T_1 \\ (C_r B)^{-1} C_r \end{bmatrix} \quad \text{with inverse} \quad T^{-1} = [S_1 \,\vdots\, B]$$

is a similarity transformation which takes system (1,3) into the form

$$\dot{\tilde{x}}(t) = A_{11}\tilde{x}(t) + A_{12}\tilde{y}(t)\,, \quad \tilde{x}(t) \in \mathbb{R}^{n-m} \tag{4a}$$

$$\dot{\tilde{y}}(t) = A_{21}\tilde{x}(t) + A_{22}\tilde{y}(t) + u(t) + \tilde{g}(t,\tilde{x}(t),\tilde{y}(t),u(t))\,, \quad \tilde{y}(t) \in \mathbb{R}^{m} \tag{4b}$$

$$\tilde{g}(t,\tilde{x},\tilde{y},u) := g(t,S_1\tilde{x}+B\tilde{y},u) \tag{4c}$$

with output

$$y^{*}(t) = (C_r B)\tilde{y}(t)\,. \tag{5}$$

Note that the eigenvalues of A_{11} coincide with the transmission zeros of (C_r, A, B); thus, by virtue of Assumption 2(iii), $\sigma(A_{11}) \subset \mathbb{C}^-$.

Let $P_1 > 0$ be the unique positive definite solution of the Lyapunov equation

$$P_1 A_{11} + A_{11}^T P_1 + I = 0 \tag{6}$$

then we state our first result.

Theorem 1:

Define $\kappa^* := \|A_{22}\| + \alpha\|B\| + \frac{1}{2}[\|P_1A_{12}+A_{21}^T\|+\alpha\|S_1\|]^2$, then, for each fixed $\hat{k} > \kappa^*(1-\beta)^{-1}$, the static output feedback

$$u(t) = -\hat{k}(C_r B)^{-1} y^{*}(t) = -\hat{k}\,\tilde{y}(t) \tag{7}$$

renders the zero state of the hypothetical system (1,3) globally uniformly asymptotically stable.

Proof: Let $V: (\tilde{x},\tilde{y}) \mapsto \frac{1}{2}\langle \tilde{x}, P_1\tilde{x}\rangle + \frac{1}{2}\|\tilde{y}\|^2$, then a straightforward calculation reveals that, along solutions $(\tilde{x}(\cdot),\tilde{y}(\cdot))$ of (4,5,7) (equivalent to (1,3,7)), the following holds almost everywhere

$$\frac{d}{dt} V(\tilde{x}(t),\tilde{y}(t)) \leq -U(\tilde{x}(t),\tilde{y}(t))$$

where

$$U(\tilde{x},\tilde{y}) := \frac{1}{2}\left\langle \begin{bmatrix} \|\tilde{x}\| \\ \|\tilde{y}\| \end{bmatrix}, M \begin{bmatrix} \|\tilde{x}\| \\ \|\tilde{y}\| \end{bmatrix} \right\rangle, \quad M := \begin{bmatrix} 1 & -[\|P_1A_{12}+A_{21}^T\|+\alpha\|S_1\|] \\ -[\|P_1A_{12}+A_{21}^T\|+\alpha\|S_1\|] & 2[\hat{k}(1-\beta)-\|A_{22}\|-\alpha\|B\|] \end{bmatrix}.$$

Noting the M is positive definite, the result follows. □

In the context of the true system (1,2), if $r = 1$, then the static feedback (7) is realizable as

$$u(t) = -\hat{k}(C_rB)^{-1}F_1y(t) \tag{8}$$

whence:-

Corollary 1:

Let $\hat{k}$ be as in Theorem 1. If $r = 1$ then the static output feedback (8) renders the zero state of the true system (1,2) globally uniformly asymptotically stable.

However, in all other cases ($r \geq 2$), the feedback (7) is unrealizable for the true system (1,2); in its place, we will develop a realizable dynamic compensator in the next section.

4. Cases r ≥ 2: Stabilizing dynamic output feedback for the true system (1,2)

In view of Assumption 2(i), we note that

$$y^*(t) = C_rx(t) = F_1y(t) + F_2\dot{y}(t) + \cdots + F_ry^{(r-1)}(t)$$

which can be interpreted in the frequency domain as

$$\bar{y}^*(s) = [F_1 + N(s)]\bar{y}(s) \ ,$$

where

$$N(s) = sF_2 + \cdots + s^{r-1}F_r$$

is physically unrealizable. Our approach is to replace $N(s)$ by a physically realizable transfer matrix (filter) of the form $H_\mu(s)N(s)$ with appropriately chosen $H_\mu(s)$. To this end, let $d_i \leq r-1$ denote the degree of the highest-degree polynomial in the ith row of $N(s)$. Let constants $a_j^i > 0$, $j=2,\cdots,d_i$, be such that

$$\pi_i(s) = s^{d_i} + a_{d_i}^i s^{d_i-1} + \cdots + a_2^i s + 1, \quad i = 1,2,\cdots,m$$

is Hurwitz (i.e. with all its roots lying in the open left half complex plane $\mathbb{C}^-$). For $i = 1,2,\cdots,m$, define $h_i^\mu(s)$, parameterized by $\mu > 0$, as

$$h_i^\mu(s) = \frac{1}{\pi_i(\mu s)}$$

which, interpreted as a transfer function, has minimal realization $(c_i^T, \mu^{-1}A_i, \mu^{-1}b_i)$, where

$$A_i = \begin{bmatrix} 0 & 1 & 0 & \cdots & 0 \\ 0 & 0 & 1 & \cdots & 0 \\ \vdots & \vdots & \vdots & \ddots & \vdots \\ 0 & 0 & 0 & \cdots & 1 \\ -1 & -a_2^i & -a_3^i & \cdots & -a_{d_i}^i \end{bmatrix} \in \mathbb{R}^{d_i \times d_i}, \quad b_i = \begin{bmatrix} 0 \\ 0 \\ \vdots \\ 0 \\ 1 \end{bmatrix} \in \mathbb{R}^{d_i}, \quad c_i = \begin{bmatrix} 1 \\ 0 \\ \vdots \\ 0 \\ 0 \end{bmatrix} \in \mathbb{R}^{d_i}.$$

We now introduce the transfer matrix

$$H_\mu(s) := \mathrm{diag}\{h_i^\mu(s)\}$$

which clearly has minimal realization $(C^*, \mu^{-1}A^*, \mu^{-1}B^*)$, where

$$A^* = \text{diag}\,\{A_i\} \in \mathbb{R}^{q\times q}\ ,\quad B^* = \text{diag}\,\{b_i\} \in \mathbb{R}^{q\times m}\ ,\quad C^* = \text{diag}\,\{c_i^T\} \in \mathbb{R}^{m\times q}\ ,\quad \text{with } q := \sum_{i=1}^{m} d_i\ .$$

We note, in passing, that $\sigma(A^*) \subset \mathbb{C}^-$ and that $C^*(A^*)^{-1}B^* = -I$.

Let κ^* be as in Theorem 1, then, for fixed $\hat{k} > \kappa^*(1-\beta)^{-1}$, the proposed physically realizable compensator (which filters the actual output y) for system (1,2) is parameterized by μ, and has frequency domain characterization:

$$G_\mu(s) = -\hat{k}(C_rB)^{-1}[F_1+H_\mu(s)N(s)]\ . \tag{9}$$

For notational convenience we introduce functions φ, f_1, f_2, Δf_2 and f_3, defined as follows.

$$\varphi\colon (\tilde{x},\tilde{y},\tilde{z}) \mapsto -\hat{k}(C_rB)^{-1}\,[F_1C[S_1\tilde{x}+B\tilde{y}] + C^*\tilde{z}]$$

$$f_1\colon (\tilde{x},\tilde{y}) \mapsto A_{11}\tilde{x} + A_{12}\tilde{y}$$

$$f_2\colon (t,\tilde{x},\tilde{y}) \mapsto A_{21}\tilde{x} + A_{22}\tilde{y} - \hat{k}\tilde{y} + \tilde{g}(t,\tilde{x},\tilde{y},-\hat{k}\tilde{y})$$

$$\Delta f_2\colon (t,\tilde{x},\tilde{y},\tilde{z}) \mapsto \hat{k}\tilde{y} + \varphi(\tilde{x},\tilde{y},\tilde{z}) + \tilde{g}(t,\tilde{x},\tilde{y},\varphi(\tilde{x},\tilde{y},\tilde{z})) - \tilde{g}(t,\tilde{x},\tilde{y},-\hat{k}\tilde{y})$$

$$f_3\colon (\tilde{x},\tilde{y},\tilde{z}) \mapsto A^*\tilde{z} + B^*[\,C_rB\tilde{y}-F_1C[S_1\tilde{x}+B\tilde{y}]]\ .$$

Then it is readily verified that, in the time domain and under state transformation T, the differential equations governing the dynamic output feedback controlled system may now be expressed in the form:

$$\dot{\tilde{x}}(t) = f_1(\tilde{x}(t),\tilde{y}(t))\ ,\quad \tilde{x}(t) \in \mathbb{R}^{n-m} \tag{10a}$$

$$\dot{\tilde{y}}(t) = f_2(t,\tilde{x}(t),\tilde{y}(t)) + \Delta f_2(t,\tilde{x}(t),\tilde{y}(t),\tilde{z}(t))\ ,\quad \tilde{y}(t) \in \mathbb{R}^{m} \tag{10b}$$

$$\mu\dot{\tilde{z}}(t) = f_3(\tilde{x}(t),\tilde{y}(t),\tilde{z}(t))\ ,\quad \tilde{z}(t) \in \mathbb{R}^{q}\ . \tag{10c}$$

In analysing the stability of this system, we regard μ as a singular perturbation parameter. Recalling that $C^*(A^*)^{-1}B^* = -I$, we note that system (4) with control (7) is recovered on setting $\mu = 0$ in (10); thus, in the usual terminology [9,10,24], system (4,7) may be interpreted as the reduced-order system associated with the singularly perturbed system (10). The ensuing approach is akin to that of [9,10], our objective being to determine a threshold value $\mu^* > 0$ such that, for all $\mu \in (0,\mu^*)$, the zero state of system (10) is globally uniformly asymptotically stable.

Recalling that $\sigma(A^*) \subset \mathbb{C}^-$, let $P^* > 0$ be the unique symmetric positive definite solution of the Lyapunov equation

$$P^*A^* + (A^*)^TP^* + I = 0\ . \tag{11}$$

Define $W\colon \mathbb{R}^{n-m}\times\mathbb{R}^{m}\times\mathbb{R}^{q} \to [0,\infty)$ by

$$W(\tilde{x},\tilde{y},\tilde{z}) := \tfrac{1}{2}\langle w(\tilde{x},\tilde{y},\tilde{z}),\, P^*w(\tilde{x},\tilde{y},\tilde{z})\rangle \tag{12a}$$

where

$$\begin{aligned} w(\tilde{x},\tilde{y},\tilde{z}) &:= \tilde{z} + (A^*)^{-1}B^*[\,C_rB\tilde{y}-F_1C[S_1\tilde{x}+B\tilde{y}]] \\ &= (A^*)^{-1}f_3(\tilde{x},\tilde{y},\tilde{z})\ . \end{aligned} \tag{12b}$$

We now establish some preliminary lemmas.

The first is implicit in the proof of Theorem 1.

Lemma 1:

$$\langle \nabla_{\tilde{x}} V(\tilde{x},\tilde{y}), f_1(\tilde{x},\tilde{y})\rangle + \langle \nabla_{\tilde{y}} V(\tilde{x},\tilde{y}), f_2(t,\tilde{x},\tilde{y})\rangle \le -\alpha_0 V(\tilde{x},\tilde{y}) \quad \text{where} \quad \alpha_0 := [\|M^{-1}\|[\|P_1\|+1]]^{-1} > 0.$$

Lemma 2: $\langle \nabla_{\tilde{z}} W(\tilde{x},\tilde{y},\tilde{z}), f_3(\tilde{x},\tilde{y},\tilde{z})\rangle \le -\beta_0 W(\tilde{x},\tilde{y},\tilde{z})$ where $\beta_0 := \|P^*\|^{-1} > 0$.

Proof:

$$\begin{aligned}\langle \nabla_{\tilde{z}} W(\tilde{x},\tilde{y},\tilde{z}),\ f_3(\tilde{x},\tilde{y},\tilde{z})\rangle &= \langle P^* w(\tilde{x},\tilde{y},\tilde{z}),\ f_3(\tilde{x},\tilde{y},zy)\rangle \\ &= \langle P^* w(\tilde{x},\tilde{y},\tilde{z}),\ A^* w(\tilde{x},\tilde{y},\tilde{z})\rangle \\ &= -\tfrac{1}{2}\|w(\tilde{x},\tilde{y},\tilde{z})\|^2 \\ &\le -\|P^*\|^{-1} W(\tilde{x},\tilde{y},\tilde{z}) . \quad \square\end{aligned}$$

Clearly, the function $\|f_1\|$ is bounded above by a calculable scalar multiple of the function $V^{\frac{1}{2}}$. In view of Assumption 3(ii), $\|f_2\|$ is also bounded above by a calculable scalar multiple of $V^{\frac{1}{2}}$. By Assumption 3(iii), $\tilde{g}$ is uniformly Lipschitz in its final argument (with known Lipschitz constant λ); hence,

$$\|\Delta f_2(t,\tilde{x},\tilde{y},\tilde{z})\| \le (1+\lambda)\|\hat{k}\,\tilde{y} + \varphi(\tilde{x},\tilde{y},\tilde{z})\| \quad \text{for all } (t,\tilde{x},\tilde{y},\tilde{z})$$

and, since $\hat{k}\,\tilde{y} + \varphi(\tilde{x},\tilde{y},\tilde{z}) = -\hat{k}(C,B)^{-1}C^* w(\tilde{x},\tilde{y},\tilde{z})$, it follows that $\|\Delta f_2\|$ is bounded above by a calculable scalar multiple of $W^{\frac{1}{2}}$. Therefore, we may conclude:

Lemma 3:

There exist calculable constants θ_0, ψ_1, ψ_2 and η_0 such that, for all $(t,\tilde{x},\tilde{y},\tilde{z})$,

(i) $\langle \nabla_{\tilde{x}} W(\tilde{x},\tilde{y},\tilde{z}),\ f_1(\tilde{x},\tilde{y})\rangle \le \theta_0 V^{\frac{1}{2}}(\tilde{x},\tilde{y}) W^{\frac{1}{2}}(\tilde{x},\tilde{y},\tilde{z})$,

(ii) $\langle \nabla_{\tilde{y}} W(\tilde{x},\tilde{y},\tilde{z}),\ f_2(t,\tilde{x},\tilde{y}) + \Delta f_2(t,\tilde{x},\tilde{y},\tilde{z})\rangle \le \psi_1 W(\tilde{x},\tilde{y},\tilde{z}) + \psi_2 V^{\frac{1}{2}}(\tilde{x},\tilde{y}) W^{\frac{1}{2}}(\tilde{x},\tilde{y},\tilde{z})$,

(iii) $\langle \nabla_{\tilde{y}} V(\tilde{x},\tilde{y}),\ \Delta f_2(t,\tilde{x},\tilde{y},\tilde{z})\rangle \le \eta_0 V^{\frac{1}{2}}(\tilde{x},\tilde{y}) W^{\frac{1}{2}}(\tilde{x},\tilde{y},\tilde{z})$.

The next theorem demonstrates that system (10) is asymptotically stable for all $\mu > 0$ sufficiently small.

Theorem 2:

Let κ^* be as in Theorem 1 and define $\mu^* := \alpha_0\beta_0[\alpha_0\psi_1+\eta_0(\theta_0+\psi_2)]^{-1} > 0$. Then, for each fixed $\hat{k} > \kappa^*(1-\beta)^{-1}$ and fixed $\mu \in (0,\mu^*)$, the zero state of system (10) is globally uniformly asymptotically stable.

Proof: Define the positive definite quadratic form (Lyapunov function candidate) $\mathcal{W}$ by

$$\mathcal{W}(\tilde{x},\tilde{y},\tilde{z}) := V(\tilde{x},\tilde{y}) + [\theta_0+\psi_2]^{-1}\eta_0 W(\tilde{x},\tilde{y},\tilde{z})$$

then, invoking Lemmas 1, 2 and 3, the following holds almost everywhere along solutions $(\tilde{x}(\cdot),\tilde{y}(\cdot),\tilde{z}(\cdot))$ of (10):

$$\frac{d}{dt}\,\mathcal{W}(\tilde{x}(t),\tilde{y}(t),\tilde{z}(t)) \le -\left\langle \begin{bmatrix} V^{\frac{1}{2}}(\tilde{x}(t),\tilde{y}(t)) \\ W^{\frac{1}{2}}(\tilde{x}(t),\tilde{y}(t),\tilde{z}(t)) \end{bmatrix},\ \mathcal{M}\begin{bmatrix} V^{\frac{1}{2}}(\tilde{x}(t),\tilde{y}(t)) \\ W^{\frac{1}{2}}(\tilde{x}(t),\tilde{y}(t),\tilde{z}(t)) \end{bmatrix}\right\rangle$$

where

$$\mathcal{M} := \begin{bmatrix} \alpha_0 & -\eta_0 \\ -\eta_0 & (\mu^{-1}\beta_0-\psi_1)(\theta_0+\psi_2)^{-1}\eta_0 \end{bmatrix}.$$

Noting that $\mathcal{M}$ is positive definite, the result follows. □

In practice, the component $H_\mu(s)N(s)$ of the proposed compensator is realized by constructing a total of mp filters of the form $n_{ij}(s)h_i^\mu(s)$, where n_{ij} denotes the ij-th element of N. It follows that $H_\mu(s)N(s)$ has a state space realization in the form of a p-input, m-output linear system $\Sigma_\mu = (\mathcal{F}_1(\mu), \mathcal{F}_2(\mu), \mu^{-1}\mathcal{A}, \mu^{-1}\mathcal{B})$ with state dimension $\bar{q} = pq$ for which $\sigma(\mathcal{A}) \subset \mathbb{C}^-$ and the pair $(\mathcal{F}_1(\mu), \mathcal{F}_2(\mu))$ determines the output map, $\mathcal{F}_1(\mu)$ being a feedforward operator. Therefore, the overall controlled system has the structure shown below.

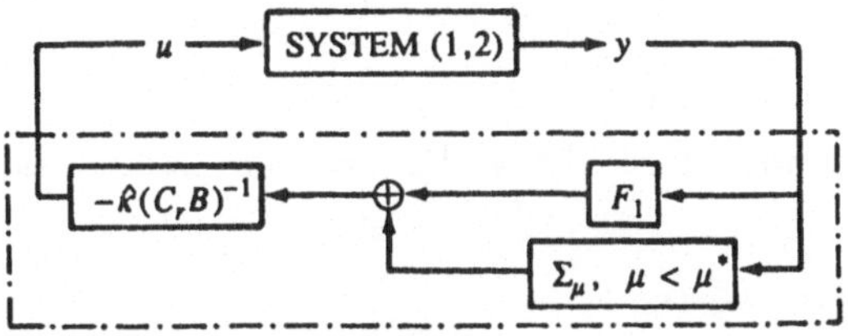

The governing equations (equivalent to (10)) can be expressed as

$$\dot{x}(t) = Ax(t) + B[u(t) + g(t,x(t),u(t))], \quad x(t) \in \mathbb{R}^n \tag{13a}$$

$$\mu\dot{z}(t) = \mathcal{A}z(t) + \mathcal{B}y(t), \quad z(t) \in \mathbb{R}^{\bar{q}}, \quad \mu < \mu^*, \tag{13b}$$

$$y(t) = Cx(t) \in \mathbb{R}^p \tag{13c}$$

$$u(t) = -\hat{k}(C_rB)^{-1}[F_1y(t) + \mathcal{F}_1(\mu)y(t) + \mathcal{F}_2(\mu)z(t)] \in \mathbb{R}^m, \quad \hat{k} > \kappa^*(1-\beta)^{-1}. \tag{13d}$$

Clearly, the threshold values κ^* and μ^* are central to this design. Since these values are determined via a "worst-case" analysis, it is to be expected that, in practice, the compensator will be conservative. In the next section, a stabilizing *adaptive* version of the compensator is developed; however, in the case $r \geq 2$, this is achieved at the expense of imposing further structure on the uncertain function g.

5. Adaptive compensator

5.1 Case I: $r = 1$

If Assumption 2 holds with $r = 1$ then, by Corollary 1, system (1,2) is asymptotically stabilized by the static output feedback (8) with $\hat{k} > \kappa^*(1-\beta)^{-1}$ provided, of course, that F_1 and C_rB are known and that sufficient *a priori* information is avilable to compute the (conservative) gain threshold $\kappa^*(1-\beta)^{-1}$. We now consider the case for which the latter information is unavailable, i.e. we only assume knowledge of F_1 and C_rB and, in particular, the constants α and $\beta < 1$ in Assumption 3 may be unknown. All other assumptions remain in force.

Replace fixed $\hat{\kappa}$ in (8) by variable $\kappa(t)$ to yield

$$u(t) = -\kappa(t)(C_rB)^{-1}F_1y(t) \tag{14a}$$

and let $\kappa(t)$ evolve according to the adaptation law

$$\dot{\kappa}(t) = \|(C_rB)^{-1}F_1y(t)\|^2 \ , \tag{14b}$$

then:-

Theorem 3:

For all initial data $(t_0,x(t_0),\kappa(t_0)) \in \mathbb{R}\times\mathbb{R}^n\times[0,\infty)$, the adaptively controlled system (1,2,14) exhibits the following properties:

(i) $\lim_{t\to\infty} \kappa(t)$ exists and is finite;

(ii) $\lim_{t\to\infty} \|x(t)\| = 0.$

Proof: For fixed (but unknown) $\hat{\kappa} > \kappa^*(1-\beta)^{-1}$ and under the similarity transformation T, system (1,2,14) may be expressed as

$$\dot{\tilde{x}}(t) = A_{11}\tilde{x}(t) + A_{12}\tilde{y}(t) \tag{15a}$$

$$\dot{\tilde{y}}(t) = A_{21}\tilde{x}(t) + A_{22}\tilde{y}(t) - \hat{\kappa}\tilde{y}(t) - [\kappa(t)-\hat{\kappa}]\tilde{y}(t) + \tilde{g}(t,\tilde{x}(t),\tilde{y}(t),-\kappa(t)\tilde{y}(t)) \tag{15b}$$

$$\dot{\kappa}(t) = \|\tilde{y}(t)\|^2 \ . \tag{15c}$$

Let U and V be as in the proof of Theorem 1 and define the positive definite (since $\beta < 1$) function

$$\mathcal{V}:\ (\tilde{x},\tilde{y},\kappa) \mapsto V(\tilde{x},\tilde{y}) + \tfrac{1}{2}(\kappa-\hat{\kappa})^2 - \tfrac{1}{2}\beta(\kappa-\hat{\kappa})|\kappa-\hat{\kappa}| \ .$$

Then, along solutions $(\tilde{x}(\cdot),\tilde{y}(\cdot),\kappa(\cdot))$ of (15), the following holds almost everywhere

$$\begin{aligned} \frac{d}{dt}\mathcal{V}(\tilde{x}(t),\tilde{y}(t),\kappa(t)) &\le -U(\tilde{x}(t),\tilde{y}(t)) - \beta\hat{\kappa}\|\tilde{y}(t)\|^2 - (\kappa(t)-\hat{\kappa})\|\tilde{y}(t)\|^2 + \beta\kappa(t)\|\tilde{y}(t)\|^2 \\ &\quad + [(\kappa(t)-\hat{\kappa})-\beta|\kappa(t)-\hat{\kappa}|]\|\tilde{y}(t)\|^2 \\ &\le -U(\tilde{x}(t),\tilde{y}(t)) \ . \end{aligned} \tag{16}$$

Since U is positive definite, we conclude that $t \mapsto (\tilde{x}(t),\tilde{y}(t),\kappa(t))$ is bounded and since $t \mapsto \kappa(t)$ is also monotonic, assertion (i) of the theorem follows. Furthermore, in view of (16), we have $\int_{t_0}^{\infty} U(\tilde{x}(t),\tilde{y}(t))dt \le \mathcal{V}(\tilde{x}(t_0,\tilde{y}(t_0),\kappa(t_0)) < \infty$ and hence, since U and V are positive definite quadratic forms, $\int_{t_0}^{\infty} V(\tilde{x}(t),\tilde{y}(t))dt < \infty$; moreover, $\dot{V}(\tilde{x}(\cdot),\tilde{y}(\cdot))$ is essentially bounded from above. Therefore, we conclude that $V(\tilde{x}(t),\tilde{y}(t)) \to 0$ as $t \to \infty$ (see Lemma 6.3 of [22]), whence assertion (ii) of the theorem. □

5.2 Case II: $r \ge 2$

Before describing the adaptive strategy in this case, it is remarked that the argument used in establishing Theorem 3 cannot be carried over directly. Instead, we will base our approach on that of Mårtensson [11]. For this reason, further conditions are imposed on the uncertain function g. In particular, Assumption 3 is now replaced by:

Assumption 3′:

There exist a bounded continuous function $\Delta A: \mathbb{R} \to \mathbb{R}^{m\times n}$, a Carathéodory function $g_a: \mathbb{R}\times\mathbb{R}^m \to \mathbb{R}^m$ which is uniformly Lipschitz in its second argument, and a constant $\beta < 1$ such that

(i) $g(t,x,u) = \Delta A(t)x + g_a(t,u)$, for all (t,x,u),

(ii) $\|g_a(t,u)\| \le \beta\|u\|$, for all (t,u),

and

(iii) $(C,A+B\Delta A(\cdot))$ is uniformly completely observable in the sense of [25].

Note that, if Assumption 3′ holds, then Assumption 3 holds *a fortiori* with $\alpha = \sup_t \|\Delta A(t)\|$ provided that α, β and the Lipschitz constant for $g_a(t,\cdot)$ are known. However, knowledge of these constants is *not* required here.

Example 2: With (C,A,B) defined as in Example 1 of §2, Assumption 3′(i) holds for any bounded continuous $\Delta A: t \mapsto (\Delta a_1(t),\Delta a_2(t),\Delta a_3(t))$.

Now replace fixed $\hat{k}$ in (13d) by variable $\kappa(t) > 0$ and replace fixed μ in (13b) by $(\delta\kappa(t))^{-1}$, where $\delta > 0$ is a constant (design parameter) and let $\kappa(t)$ evolve according to the adaptation law (other adaptation laws may be feasible, as discussed in [20])

$$\dot{\kappa}(t) = \|y(t)\|^2 + \|z(t)\|^2 .$$

Writing (as in [11])

$$x^\dagger(t) = \begin{bmatrix} x(t) \\ z(t) \end{bmatrix}, \quad u^\dagger(t) = \begin{bmatrix} u(t) \\ \dot{z}(t) \end{bmatrix}, \quad y^\dagger(t) = \begin{bmatrix} y(t) \\ z(t) \end{bmatrix},$$

then the overall adaptively controlled system may be expressed in the form

$$\dot{x}^\dagger(t) = A^\dagger(t)x^\dagger(t) + B^\dagger[u^\dagger(t) + g^\dagger(t,u^\dagger(t))] , \quad x^\dagger(t) \in \mathbb{R}^{n+\bar{q}} , \tag{17a}$$

$$y^\dagger(t) = C^\dagger x^\dagger(t) \in \mathbb{R}^{p+\bar{q}} , \tag{17b}$$

$$u^\dagger(t) = -\kappa(t)K^\dagger(\kappa(t))y^\dagger(t) \in \mathbb{R}^{m+\bar{q}} , \tag{17c}$$

$$\dot{\kappa}(t) = \|y^\dagger(t)\|^2 , \tag{17d}$$

where

$$A^\dagger(t) := \begin{bmatrix} A+B\Delta A(t) & 0 \\ 0 & 0 \end{bmatrix}, \quad B^\dagger := \begin{bmatrix} B & 0 \\ 0 & I \end{bmatrix}, \quad C^\dagger := \begin{bmatrix} C & 0 \\ 0 & I \end{bmatrix}, \tag{17e}$$

and

$$K^\dagger(\kappa) := \begin{bmatrix} (C_rB)^{-1}(F_1+\mathcal{F}_1((\delta\kappa)^{-1})) & (C_rB)^{-1}\mathcal{F}_2((\delta\kappa)^{-1}) \\ -\delta\mathcal{B} & -\delta\mathcal{A} \end{bmatrix}, \quad g^\dagger(t,u^\dagger) := \begin{bmatrix} g_a(t,u) \\ 0 \end{bmatrix}. \tag{17f}$$

The stability of system (17) will now be investigated. We first require the following lemma (essentially a non-autonomous version of Mårtensson's lemma [11]).

Lemma 4:

Let $x^\dagger : I\!R \to I\!R^{n+\bar{q}}$ satisfy

$$\dot{x}^\dagger(t) = A^\dagger(t)x^\dagger(t) + B^\dagger[v(t) + g^\dagger(t,v(t))]$$

where $v: I\!R \to I\!R^{m+\bar{q}}$ is measurable. Then, there exist constants c, $\tau > 0$ such that, for all t,

$$\|x^\dagger(t)\|^2 \le c \int_{t-\tau}^{t} [\|y^\dagger(s)\|^2 + \|v(t)\|^2]\, ds \,.$$

Proof: Let $\Phi(\cdot,\cdot)$ denote the state transition matrix function generated by $A+B\Delta A(\cdot)$ and define the observability Gramian for the pair $(C, A+B\Delta A(\cdot))$ in the usual manner, that is,

$$\Gamma(t,s) := \int_s^t \Phi^T(\sigma,s)C^TC\Phi(\sigma,s)\, d\sigma \,.$$

Now, for some constants k_1 and ω, we have $\|\exp At\| \le k_1 e^{\omega t}$ and, since $\Delta A(\cdot)$ is bounded (by assumption), there exists constant k_2 such that $\|B\Delta A(t)\| \le k_2$. By standard perturbation theory, we conclude that

$$\|\Phi(t,s)\| \le k_1 e^{(\omega+k_1k_2)(t-s)} \quad \text{for all } t,s \,.$$

Clearly, the state transition matrix function $\Phi^\dagger(\cdot,\cdot)$ generated by $A^\dagger(\cdot)$ is given by

$$\Phi^\dagger(t,s) = \begin{bmatrix} \Phi(t,s) & 0 \\ 0 & I \end{bmatrix},$$

whence

$$\|\Phi^\dagger(t,s)\| \le c_1(t-s) \quad \text{for all } t,s, \tag{18a}$$

where

$$c_1 : \sigma \mapsto 1 + k_1 e^{(\omega+k_1k_2)\sigma} \,. \tag{18b}$$

The observability Gramian for the pair $(C^\dagger, A^\dagger(\cdot))$ is given by

$$\Gamma^\dagger(t,s) := \begin{bmatrix} \Gamma(t,s) & 0 \\ 0 & (t-s)I \end{bmatrix},$$

and, since $(C, A+B\Delta A(\cdot))$ is uniformly completely observable (by assumption), we may conclude (see [25]) that there exist positive constants τ, c_2 and c_3 such that, for all t,

$$c_2\|\zeta\|^2 \le \langle \zeta \,,\, \Gamma^\dagger(t,t-\tau)\zeta\rangle \le c_3\|\zeta\|^2 \quad \forall\, \zeta \in I\!R^{n+\bar{q}} \,. \tag{19}$$

Now define the measurable function $v^\dagger : t \mapsto v(t)+g^\dagger(t,v(t))$ and note that $\|v^\dagger(t)\| \le (1+\beta)\|v(t)\|$. Then,

$$x^\dagger(t) = \Phi^\dagger(t,t-\tau)x^\dagger(t-\tau) + \int_{t-\tau}^{t} \Phi^\dagger(t,s)B^\dagger v^\dagger(s)\, ds$$

whence

$$\begin{aligned} \|x^\dagger(t)\|^2 &\le 2\|\Phi^\dagger(t,t-\tau)x^\dagger(t-\tau)\|^2 + 2\|\int_{t-\tau}^{t} \Phi^\dagger(t,s)B^\dagger v^\dagger(s)\, ds\|^2 \\ &\le 2c_4\|x^\dagger(t-\tau)\|^2 + 2c_5(1+\beta)^2\|B^\dagger\|^2 \int_{t-\tau}^{t} \|v(s)\|^2 ds \,, \end{aligned} \tag{20a}$$

wherein (18) has been used, and

$$c_4 := c_1^2(\tau), \quad c_5 := \int_0^\tau c_1^2(s)\, ds\ . \tag{20b}$$

Also, invoking both (18) and (19),

$$\begin{aligned}\|x^\dagger(t-\tau)\|^2 &\le c_2^{-1} \langle x^\dagger(t-\tau), \Gamma^\dagger(t,t-\tau)x^\dagger(t-\tau)\rangle \\ &= c_2^{-1}\int_{t-\tau}^t \| y^\dagger(s) - C^\dagger \int_{t-\tau}^s \Phi^\dagger(s,\sigma)B^\dagger v^\dagger(\sigma)\, d\sigma\, \|^2\, ds \\ &\le 2c_2^{-1}[\, \int_{t-\tau}^t \|y^\dagger(s)\|^2 ds + c_6\tau(1+\beta)^2\|C^\dagger\|^2\|B^\dagger\|^2\int_{t-\tau}^t \|v(s)\|^2 ds]\ ,\end{aligned} \tag{21a}$$

where

$$c_6 := \int_0^\tau\int_0^s c_1^2(\sigma) d\sigma\, ds\ . \tag{21b}$$

Combining (20) and (21) yields the required result. □

Theorem 4:

For all initial data $(t_0, x^\dagger(t_0), \kappa(t_0)) \in \mathbb{R}\times\mathbb{R}^{n+\bar{q}}\times(0,\infty)$, system (17) exhibits the following properties:

(i) $\lim_{t\to\infty} \kappa(t)$ exists and is finite;

(ii) $\lim_{t\to\infty} \|x^\dagger(t)\| = 0$.

Proof: Seeking a contradiction to (i), suppose that the monotonically increasing function $t \mapsto \kappa(t)$ is unbounded. Then, for some $t_1 \in [0,\infty)$, $\kappa(t_0+t_1) = \hat{k} > \kappa^*(1-\beta)^{-1}$ and $(\delta\kappa(t_0+t_1))^{-1} = \mu < \mu^*$. Now, an argument similar to that used in the proof of Theorem 2 can be adopted to establish that $x(\cdot)$ (and hence $y(\cdot) = Cx(\cdot)$) must ultimately tend exponentially to zero (and hence are square integrable on $[t_0,\infty)$). Since $\sigma(\mathcal{A}) \subset \mathbb{C}^-$, we may conclude from (13b) that z is bounded and so there exists constant c_0 such that $\kappa(t) \le \kappa(t_0)+c_0(t-t_0)$ for all $t \ge t_0$. Let K (with inverse K^{-1}) denote the monotonic function $t \mapsto \int_{t_0}^t \delta\kappa(s)ds$. It is readily verified that the function $y(K^{-1}(\cdot))$ ultimately satisfies $\|y(K^{-1}(s))\| \le c_1\exp(c_2-\sqrt{(c_2^2+c_3 s)})$ for some positive constants c_i, and so is square integrable. Solving (13b), we have

$$z(t) = \exp(\mathcal{A}K(t))z(t_0) + \int_0^{K(t)} \exp(\mathcal{A}(K(t)-s))\mathcal{B}y(K^{-1}(s))\, ds$$

from which we may conclude that $z(\cdot)$ is square integrable on $[t_0,\infty)$. Thus $x^\dagger\colon [t_0,\infty) \to \mathbb{R}^{n+\bar{q}}$ and *a fortiori* $y^\dagger$ are square integrable which, in view of (17d), contradicts our supposition that the function κ is unbounded. This establishes assertion (i) of the theorem.

It remains to show that $x^\dagger(t) \to 0$ as $t \to \infty$. Clearly, (i) ensures that $y^\dagger$ is square integrable on $[t_0,\infty)$ and, in view of (17c), that $u^\dagger$ is a bounded linear transformation of $y^\dagger$. Thus, we may conclude that $u^\dagger$ is also square integrable. Now, by Lemma 4, we have

$$\begin{aligned}\|x^\dagger(t)\|^2 &\le c\int_{t-\tau}^t [\|y^\dagger(s)\|^2 + \|u^\dagger(s)\|^2]\, ds \\ &= c\int_{t_0}^t [\|y^\dagger(s)\|^2+\|u^\dagger(s)\|^2]\, ds \;-\; c\int_{t_0}^{t-\tau}[\|y^\dagger(s)\|^2+\|u^\dagger(s)\|^2]\, ds\ .\end{aligned}$$

Therefore, $\|x^\dagger(t)\| \to 0$ as $t \to \infty$. □

6. Discontinuous feedback

In this final section, some possible generalizations of the proposed compensators are briefly discussed. In [23] and for the case $r = 1$ only, a wider class of uncertain functions g is studied; specifically, Assumption 3 (ii) is replaced by the condition

$$\|g(t,x,u)\| \le \alpha\|x\| + \beta\|u\| + \gamma\xi(Cx) \quad \text{for all } (t,x,u)$$

with α and $\beta < 1$ as before and where γ is a constant (assumed known in the non-adaptive case) and ξ is a known continuous function. Thus, loosely speaking, in [23] a non-cone-bounded component of uncertainty is allowed but this is required to be bounded by a function of the system output y. In the context of this more general class of systems, the assertion of Corollary 1 of the present paper remains true for fixed $\hat{k} > (1-\beta)^{-1}\max\{\kappa^*,\gamma\}$ if (8) is replaced by the generalized feedback

$$u(t) \in -\hat{k}\left[(C_rB)^{-1}F_1y(t) + \xi(y(t))\mathcal{N}(y(t))\right], \tag{22a}$$

where the set-valued map $y \mapsto \mathcal{N}(y) \subset \mathbb{R}^m$ in essence models a discontinuous control component and is given by

$$\mathcal{N}(y) := \begin{cases} \{\|(C_rB)^{-1}F_1y\|^{-1}(C_rB)^{-1}F_1y\}; & F_1y \neq 0 \\ \{v:\ \|v\| \le 1\}; & F_1y = 0\,, \end{cases} \tag{22b}$$

and the overall controlled system is consequently interpreted in the generalized sense of a controlled differential inclusion [26]. Furthermore, the assertions of Theorem 3 of the present paper remain true if (22) is replaced by the adaptive control

$$u(t) \in -\kappa(t)\left[(C_rB)^{-1}F_1y + \xi(y(t))\mathcal{N}(y(t))\right]$$

where $\kappa(t)$ evolves according to (14b).

In the cases $r \ge 2$, preliminary investigations indicate that again a non-cone-bounded component of uncertainty (although considerably less general than that of the preceding paragraph) can be tolerated in g and counteracted by augmenting the compensator (13d) (or its adaptive counterpart implicit in (17c,d)) with an appropriately chosen set-valued map (again essentially modelling a discontinuous control component). However, the requisite structural conditions on the non-cone-bounded uncertainty are, as might be expected, of a rather restrictive and technical nature (akin to those in [10]) and are not detailed here.

7. References

[1] A. Steinberg and E.P. Ryan, *Dynamic output feedback control of a class of uncertain systems*, IEEE Trans. Autom. Control, AC-31 (1986), pp. 1163-1165.

[2] S. Gutman, *Uncertain dynamical systems – A Lyapunov min–max approach*, IEEE Trans. Autom. Control, AC–24 (1979), pp. 437-443.

[3] G. Leitmann, *Deterministic control of uncertain systems*, Astronautica Acta, 7 (1980), pp. 1457-1461.

[4] G. Leitmann, *On the efficacy of nonlinear control in uncertain linear systems*, J. Dynamic Systems Meas. Control, 103 (1981), pp. 95-102.

[5] M. Corless and G. Leitmann, *Continuous state feedback guaranteeing uniform ultimate boundedness for uncertain dynamic systems*, IEEE Trans. Autom. Control, **AC-26** (1981), pp. 1139-1144.

[6] B.R. Barmish and G. Leitmann, *On ultimate boundedness control of uncertain systems in the absence of matching conditions*, IEEE Trans. Autom. Control, **AC−27** (1982), pp. 153-158.

[7] B.R. Barmish, M. Corless and G. Leitmann, *A new class of stabilizing controllers for uncertain dynamical systems*, SIAM J. Control & Optimization, **21** (1983), pp. 246-255.

[8] E.P. Ryan and M. Corless, *Ultimate boundedness and asymptotic stability of a class of uncertain dynamical systems via continuous and discontinuous feedback control*, IMA J. Math. Control & Info., **1** (1984), pp. 223-242.

[9] A. Saberi and H.K. Khalil, *Quadratic−type Lyapunov functions* for *singularly perturbed systems*, IEEE Trans. Autom. Control, **AC−29** (1984), pp. 542-550.

[10] M. Corless, G. Leitmann and E.P. Ryan, *Control of uncertain systems with neglected dynamics*, preprint (1988).

[11] B. Mårtensson, *The order of any stabilizing regulator is sufficient a priori information for adaptive stabilization*, Systems & Control Letters **6** (1985), pp. 87-91.

[12] C.I. Byrnes and A. Isidori, *A frequency domain philosophy for nonlinear systems, with applications to stabilization and to adaptive control*, Proc. 23rd IEEE Conf. on Decision & Control, Las Vegas (1984), pp. 1569-1573.

[13] C.I. Byrnes and J.C. Willems, *Adaptive stabilization of multivariable linear systems*, Proc. 23rd IEEE Conf. on Decision & Control, Las Vegas (1984), pp. 1574-1577.

[14] A.S. Morse, *A three-dimensional universal controller for the adaptive stabilization of any strictly proper minimum-phase system with relative degree not exceeding two*, IEEE Trans. Autom. Control, **AC-30** (1985), pp. 1188-1191.

[15] D.R. Mudgett and A.S. Morse, *Adaptive stabilization of linear systems with unknown high frequency gains*, IEEE Trans. Autom. Control, **AC-30** (1985), pp. 549-554.

[16] R.D. Nussbaum, *Some remarks on a conjecture in parameter adaptive control*, Systems & Control Letters **3** (1983), pp. 243-246.

[17] M. Fu and B. Ross Barmish, *Adaptive stabilization of linear systems via switching control*, IEEE Trans. Autom. Control, **AC-31** (1986), pp. 1097-1103.

[18] P. Ioannou, *Adaptive stabilization of not necessarily minimum phase plants*, Systems & Control Letters **7** (1986), pp. 281-287.

[19] H. Khalil and A. Saberi, *Adaptive stabilization of a class of nonlinear systems using high-gain feedback*, IEEE Trans. Autom. Control, **AC-32** (1987), pp. 1031-1035.

[20] A. Ilchmann, D.H. Owens and D. Prätzel-Wolters, *High-gain robust adaptive controllers for multivariable systems*, Systems & Control Letters, **8** (1987), pp. 397-404.

[21] M. Corless and G. Leitmann, *Adaptive control of systems containing uncertain functions and unknown functions with uncertain bounds*, JOTA **41** (1983), pp. 155-168.

[22] M. Corless and G. Leitmann, *Adaptive control for uncertain dynamical systems*, in Dynamical Systems and Microphysics (eds: A Blaquière & G Leitmann) (Academic Press, New York, 1984).

[23] E.P. Ryan, *Adaptive stabilization of a class of uncertain nonlinear systems: A differential inclusion approach*, Systems & Control Letters, **10** (1988), pp. 95-101.

[24] P.V. Kokotovic, R.E. O'Malley Jr. and P. Sannuti, *Singular perturbations and order reduction in control theory - an overview*, Automatica, **12** (1976), pp. 123-132.

[25] B.D.O. Anderson, *Exponential stability of linear equations arising in adaptive identification*, IEEE Trans. Autom. Control, AC-22 (1977), pp. 83-88.

[26] J.P. Aubin and A. Cellina, *Differential Inclusions*, (Springer-Verlag, New York, 1984).

CONTROL OF UNCERTAIN MECHANICAL SYSTEMS WITH ROBUSTNESS IN THE PRESENCE OF UNMODELLED FLEXIBILITIES†

Martin Corless
School of Aeronautics and Astronautics
Purdue University
West Lafayette, Indiana 47907
USA

ABSTRACT

We consider a class of uncertain mechanical systems containing flexible elements and subject to memoryless output-feedback controllers. The damping and stiffness properties of some of the flexible elements are parameterized linearly in μ^{-1} and μ^{-2}, respectively, where $\mu > 0$ and these components become more rigid as μ approaches zero. We propose a class of "stabilizing" controllers for a system model in which the above components are rigid. Subject to a "linear growth condition," the controllers also stabilize the model in which the components are flexible, provided $\mu > 0$ is sufficiently small. The results are illustrated by an example.

1. INTRODUCTION

The effect of the flexibility of mechanical elements is becoming more significant in engineering applications, e.g., light high-speed robotic manipulators and flexible space structures. We consider here the problem of obtaining memoryless, stabilizing, feedback controllers for a class of uncertain mechanical systems with flexible elements. These elements are not rigid and can deform. The uncertainties are characterized deterministically rather than stochastically. An example of a system with a deterministic uncertainty is one which contains an uncertain disturbance input or an uncertain parameter about which the only information available is an upper bound on its magnitude.

In general, if one models some of the flexible elements as rigid components, a simpler model results and controller design is simplified. However, one should then assure that the stability properties of the feedback-controlled system are robust in the presence of the previously unmodelled flexibilities.

† Based on research supported by the U.S. National Science Foundation under grant MSM-8706927.

In this paper, we present "stabilizing" controllers whose designs are based on a model of the mechanical system in which some of the flexible elements are modelled as rigid components. These controllers also have the following robustness property. Consider a model of the system in which the above components are treated as flexible components whose damping and stiffness properties are parameterized linearly in μ^{-1} and μ^{-2}, respectively, where $\mu > 0$ and these components become more rigid as μ approaches zero. Then the controllers also "stabilize" this model, provided μ is sufficiently small.

Controller design is based on the constructive use of Lyapunov functions; see, e.g., [1-5, 9-15].

The results are illustrated by a simple example in Section 6.

2. PROBLEM STATEMENT

Consider a mechanical system which at each instant of time $t \in \mathbb{R}$ is subject to a control input $u(t) \in \mathbb{R}^m$. Suppose the system contains certain flexible components (hereafter called the *neglected components)* whose flexibilities are neglected in the design of a feedback controller generating u(t), i.e., they are modelled as rigid components for controller design.

Letting $q(t) \in \mathbb{R}^N$ denote a vector of generalized coordinates which describe the configuration of the mechanical system at t, we assume that, when modelled as rigid bodies, the neglected components give rise to a linear constraint

$$Sq = 0 \tag{2.1}$$

where $S \in \mathbb{R}^{L \times N}$ has rank $L < N$; see the example in Sec. 6. Also, we suppose that there are no other possible kinematical constraints on the system.

We model all uncertainty in the system by a *lumped uncertain element* ω. The only information assumed available on ω is the knowledge of a non-empty set Ω to which it belongs.

Letting

$$\dot{q}(t) \triangleq \frac{dq}{dt}(t),$$

we suppose that the kinetic energy of the system is equal to[(1)]

$$\frac{1}{2}\dot{q}^T M(\omega)\dot{q}$$

where the *system mass matrix* $M(\omega) \in \mathbb{R}^{N \times N}$ is symmetric and positive definite.

---

(1) Sometimes we omit arguments.

Modelling the neglected components a la [6], the motion of the system can be described by

$$M(\omega)\ddot{q} = \chi(t, q, \dot{q}, u, \omega) + S^T\lambda \tag{2.2}$$

where $S^T\lambda$ represents the sum of the generalized forces exerted by the neglected components and χ represents the sum of all the other generalized forces. We assume that for each $\omega \in \Omega$, $\chi(\cdot,\omega)$: $\mathbb{R}\times\mathbb{R}^N\times\mathbb{R}^N\times\mathbb{R}^m \to \mathbb{R}^m$ is continuous. We suppose that the measurement vector $z(t) \in \mathbb{R}^l$ available for feedback control is given by

$$z = D(t, q, \dot{q}, \omega) \tag{2.3}$$

where $D(\cdot,\omega)$ is continuous.

Consider first the situation in which the neglected components are modelled as rigid bodies. Then (2.1) holds. Without loss of generality, we suppose that the coordinates have been chosen so that

$$S = [0\ I]\,, \tag{2.4}$$

i.e., (2.1) can be written as

$$\theta = 0$$

where

$$\begin{bmatrix} \phi \\ \theta \end{bmatrix} = q \tag{2.5}$$

with $\theta \in \mathbb{R}^L$, $\phi \in \mathbb{R}^{\bar{N}}$, and $\bar{N} \triangleq N-L$. Utilizing (2.2), (2.3), this model can be described by

$$M_{11}(\omega)\ddot{\phi} = \bar{\chi}_1(t, \phi, \dot{\phi}, u, \omega) \tag{2.6a}$$

$$z = \bar{D}(t, \phi, \dot{\phi}, \omega) \tag{2.6b}$$

with

$$\bar{\chi}_i(t, \phi, \dot{\phi}, u, \omega) \triangleq \chi_i(t, (\phi, 0), (\dot{\phi}, 0), u, \omega)\,, \qquad i = 1, 2 \tag{2.7a}$$

$$\bar{D}(t, \phi, \dot{\phi}, \omega) \triangleq D(t, (\phi, 0), (\dot{\phi}, 0), \omega) \tag{2.7b}$$

where $M_{11}(\omega) \in \mathbb{R}^{\bar{N}\times\bar{N}}$ and

$$\begin{bmatrix} M_{11} & M_{12} \\ M_{21} & M_{22} \end{bmatrix} = M\,, \quad \begin{bmatrix} \chi_1 \\ \chi_2 \end{bmatrix} = \chi\,. \tag{2.8}$$

Also, λ is given by

$$\lambda = M_{21}(\omega)\ddot{\phi} - \bar{\chi}_2(t, \phi, \dot{\phi}, u, \omega)\,. \tag{2.9}$$

Although the model described by (2.6) may contain other flexible elements we shall, for convenience, refer to it as the *"rigid" model.*

Suppose now the neglected components are considered flexible, i.e., they are not rigid and can deform; hence constraint (2.1) no longer holds. Following [6] and assuming the components to be linear, their effect on the system can be represented by letting

$$\lambda = -CS\dot{q} - KSq$$
$$= -C\dot{\theta} - K\theta \tag{2.10}$$

in (2.2). The matrix $K \in \mathbb{R}^{L \times L}$, which is assumed symmetric and positive definite, represents the stiffness properties of the components and $C \in \mathbb{R}^{L \times L}$, which is assumed positive definite, represents the damping properties of the components. For robustness considerations we shall let

$$K = \mu^{-2} K^{\circ}, \quad C = \mu^{-1} C^{\circ}, \tag{2.11}$$

where $\mu > 0$, and consider behavior for sufficiently small μ. Substituting (2.10), (2.11) into (2.2)-(2.3), the system is now described

$$M(\omega)\ddot{q} = \chi(t, q, \dot{q}, u, \omega) - \mu^{-1} S^T C^{\circ} S\dot{q} - \mu^{-2} S^T K^{\circ} Sq, \tag{2.12a}$$
$$z = D(t, q, \dot{q}, \omega). \tag{2.12b}$$

We shall refer to (2.12) as the *flexible model.*

The following assumption is made.

Assumption A1.[(2)] For each $\omega \in \Omega$, there is a real number $k \geq 0$ such that for all $t \in \mathbb{R}$, ϕ, $\dot{\phi} \in \mathbb{R}^{\bar{N}}$, $u \in \mathbb{R}^m$:

(i)

$$\|\bar{\chi}_i\|, \ \|\frac{\partial \bar{\chi}_i}{\partial t}\| \leq k[1 + \|\phi\| + \|\dot{\phi}\| + \|u\|], \quad i = 1, 2,$$
$$\|\bar{D}\|, \ \|\frac{\partial \bar{D}}{\partial t}\| \leq k[1 + \|\phi\| + \|\dot{\phi}\|],$$

$$\|\frac{\partial \bar{\chi}_i}{\partial \phi}\|, \ \|\frac{\partial \bar{\chi}_i}{\partial \dot{\phi}}\|, \ \|\frac{\partial \bar{\chi}_i}{\partial u}\|, \ \|\frac{\partial \bar{D}}{\partial \phi}\|, \ \|\frac{\partial \bar{D}}{\partial \dot{\phi}}\| \leq k, \quad i = 1, 2;$$

(ii) for all $\theta^1, \theta^2, \dot{\theta}^1, \dot{\theta}^2 \in \mathbb{R}^L$,

$$\|\chi_i(t, (\phi, \theta^2), (\dot{\phi}, \dot{\theta}^2), u, \omega) - \chi_i(t, (\phi, \theta^1), (\dot{\phi}, \dot{\theta}^1), u, \omega)\| \leq k[\|\theta^2 - \theta^1\| + \|\dot{\theta}^2 - \dot{\theta}^1\|], \quad i = 1, 2,$$
$$\|D(t, (\phi, \theta^2), (\dot{\phi}, \dot{\theta}^2), \omega) - D(t, (\phi, \theta^1), (\dot{\phi}, \dot{\theta}^1), \omega)\| \leq k[\|\theta^2 - \theta^1\| + \|\dot{\theta}^2 - \dot{\theta}^1\|].$$

---

(2) If a derivative appears in a condition, this implicitly assumes that the derivative exists.

Note that the above assumption is readily satisfied by a linear system whose time-varying coefficients are bounded and have bounded derivatives.

We shall consider the control u(t) to be given by a *memoryless feedback controller* $p: \mathbb{R}\times\mathbb{R}^{l} \to \mathbb{R}^{m}$ operating on z(t), i.e.,

$$u(t) = p(t, z(t)) \, . \tag{2.13}$$

Roughly speaking, the problem we wish to consider is as follows. Utilizing *only the information available on the "rigid" model*, obtain a feedback controller p whose utilization assures that

(i) the feedback-controlled "rigid" model is "stable" about zero, and

(ii) the feedback-controlled flexible model is "stable" about zero, provided $\mu > 0$ is sufficiently small.

Ideally "stable" means asymptotic stability. However, for systems with uncertain disturbance inputs, asymptotic stability may not be achievable, so, we content ourselves with "stable" behavior which is close to asymptotic stability.

To obtain a more precise problem statement, we introduce state vectors

$$x \triangleq \begin{bmatrix} \phi \\ \dot{\phi} \end{bmatrix}, \quad \xi \triangleq \begin{bmatrix} q \\ \dot{q} \end{bmatrix}. \tag{2.14}$$

The "rigid" model is described by

$$\dot{x} = \bar{F}(t, x, u, \omega) \, , \tag{2.15a}$$

$$z = \bar{d}(t, x, \omega) \, , \tag{2.15b}$$

where

$$\bar{F}(t, x, u, \omega) \triangleq \begin{bmatrix} \dot{\phi} \\ M_{11}(\omega)^{-1}\bar{\chi}_1(t, \phi, \dot{\phi}, u, \omega) \end{bmatrix}, \tag{2.16a}$$

$$\bar{d}(t, x, \omega) \triangleq \bar{D}(t, \phi, \dot{\phi}, \omega) \, ; \tag{2.16b}$$

the flexible model is described by

$$\dot{\xi} = F(t, \xi, u, \mu, \omega) \, , \tag{2.17a}$$

$$z = d(t, \xi, \omega) \, , \tag{2.17b}$$

where

$$F(t, \xi, u, \mu, \omega) \triangleq \begin{bmatrix} \dot{q} \\ M(\omega)^{-1}[\chi(t, q, \dot{q}, u, \omega) - \mu^{-1}S^{T}C^{o}S\dot{q} - \mu^{-2}S^{T}K^{o}Sq] \end{bmatrix}, \tag{2.18a}$$

$$d(t, \xi, \omega) \triangleq D(t, q, \dot{q}, \omega) \, . \tag{2.18b}$$

The feedback-controlled "rigid" model is described by

$$\dot{x} = \bar{F}(t, x, p(t, \bar{d}(t, x, \omega)), \omega) \tag{2.19}$$

and the feedback-controlled flexible model is described by

$$\dot{\xi} = F(t, \xi, p(t, d(t, \xi, \omega)), \mu, \omega) . \tag{2.20}$$

The problem is as follows. Using only the information available on the "rigid" model, obtain a function $p: \mathbf{R}\times\mathbf{R}^l \rightarrow \mathbf{R}^m$ which assures that

(i) system (2.19) asymptotically tracks[3] 0 to within a bounded set, and

(ii) system (2.20) asymptotically tracks 0 to within a bounded set, provided μ is sufficiently small.

3. PROPOSED CONTROLLERS

The following assumption yields a "stabilizing" controller for the "rigid" model.

Assumption A2. There exists a continuous function $p: \mathbf{R}\times\mathbf{R}^l \rightarrow \mathbf{R}^m$ such that for some symmetric, positive-definite matrices[4] $P, Q \in \mathbf{R}^{n\times n}$ and non-negative numbers[5] a, b,

$$x^T P \bar{F}(t, x, p(t, \bar{d}(t, x, \omega)), \omega) \leq -\|x\|_Q^2 + a\|x\|_Q + b \tag{3.1}$$

for all $t \in \mathbf{R}$, $x \in \mathbf{R}^n$, and $\omega \in \Omega$.

Roughly speaking, the following theorem states that any function p which assures satisfaction of A2 is a "stabilizing" controller for the "rigid" model.

Theorem 3.1. Consider an uncertain "rigid" model described by (2.6) or (2.15), satisfying Assumption A2, and subject to feedback control given by (2.13) where p assures A2.

Then, the feedback-controlled "rigid" model, (2.19), asymptotically tracks 0 to within the set

$$B \triangleq \{x \in \mathbf{R}^n \mid \|x\|_P \leq \underline{d}\} \tag{3.2}$$

where[6]

$$\underline{d} \triangleq [\lambda_{max}(Q^{-1}P)]^{1/2}[a/2 + (a^2/4 + b)^{1/2}] . \tag{3.3}$$

Proof. The proof proceeds by considering the function $V: \mathbf{R}^n \rightarrow \mathbf{R}$, given by

---

(3) The appendix contains a definition.

(4) $n \triangleq 2(N-L)$

(5) If $Q \in \mathbf{R}^{n\times n}$ is symmetric and positive-definite and $x \in \mathbf{R}^n$, $\|x\|_Q \triangleq (x^T Q x)^{1/2}$.

(6) If all the eigenvalues of $M \in \mathbf{R}^{n\times n}$ are real, $\lambda_{max(min)}(M)$ is the maximum (minimum) eigenvalue of M.

$$V(x) = x^T P x\,, \tag{3.4}$$

as a candidate Lyapunov function for (2.19). Utilizing (3.1), it follows that along any solution of (2.19),

$$\frac{dV(x(t))}{dt} \le -2\,\|x(t)\|_Q^2 + 2a\|x(t)\|_Q + 2b\,.$$

Thus, $\frac{dV(x(t))}{dt} < 0$ for all t such that

$$\|x(t)\|_Q > a/2 + (a^2/4 + b)^{1/2}\,;$$

hence $\frac{dV(x(t))}{dt} < 0$ for all t satisfying

$$V(x(t)) > \underline{d}^2\,.$$

Standard arguments in Lyapunov theory complete the proof; see, e.g., [5].

Remark. If Assumption A2 is satisfied with

$$a = b = 0\,,$$

then the corresponding controller yields a feedback-controlled "rigid" model which is globally uniformly asymptotically stable about 0.

In order to obtain a controller which is also stabilizing for the flexible model, the following assumption is introduced.

Assumption A3. Assumption A2 is assured with a function p which, for some non-negative number k, satisfies

$$\|p(t,z)\|\,,\ \ \|\frac{\partial p}{\partial t}(t,z)\| \le k(1 + \|z\|)\,, \tag{3.5a}$$

$$\|\frac{\partial p}{\partial z}(t,z)\| \le k \tag{3.5b}$$

for all $t \in \mathbf{R}$, $z \in \mathbf{R}^l$.

A proposed feedback controller is any function p which assures satisfaction of Assumptions A2, A3.

4. ROBUSTNESS IN THE PRESENCE OF UNMODELLED FLEXIBILITIES

The following result assures us that a controller whose design is based on satisfying the requirements of Assumptions A2 and A3 for the "rigid" model will also "stabilize" the flexible model, provided Assumption A1 is satisfied and μ is sufficiently small.

Theorem 4.1. Consider an uncertain flexible model described by (2.12) or (2.17) where $M(\omega)$, K^o are symmetric and

$$M(\omega), C^o, K^o > 0 . \tag{4.1}$$

Suppose A1 is satisfied and the corresponding "rigid" model satisfies A2 and A3 with a controller p.

Then, there exists $\mu^* > 0$ such that if $\mu < \mu^*$, the feedback-controlled flexible model (2.20) asymptotically tracks 0 to within a bounded neighborhood.

Proof. See [8].

5. EXAMPLES OF PROPOSED CONTROLLERS

In this section, we consider a specific class of uncertain mechanical systems whose "rigid" models satisfy Assumptions A2, A3. For these systems, we exhibit "stabilizing" controllers which are robust in the presence of unmodelled flexibilities. Two main characterizations of the "rigid" models treated here are that the number of independent scalar control inputs is the same as the number of coordinates and the complete state is available for feedback.

5.1 A Specific Class of Uncertain "Rigid" Models

Consider an uncertain mechanical system whose "rigid" model is described by

$$M_{11}(\omega)\ddot{\phi} = U(t, \phi, \dot{\phi}, \omega) + Wu \tag{5.1a}$$

with measurement vector

$$z = [\phi^T \, \dot{\phi}^T]^T \tag{5.1b}$$

where $t \in \mathbf{R}$, $\phi \in \mathbf{R}^{\bar{N}}$, $u \in \mathbf{R}^{\bar{N}}$; the uncertain element ω belongs to a known set Ω; $M_{11}(\omega) \in \mathbf{R}^{\bar{N}\times\bar{N}}$ is symmetric; $W \in \mathbf{R}^{\bar{N}\times\bar{N}}$; and, for each $\omega \in \Omega$, the function $U(\cdot, \omega)$: $\mathbf{R}\times\mathbf{R}^{\bar{N}}\times\mathbf{R}^{\bar{N}} \to \mathbf{R}^{\bar{N}}$ is continuous.

The following assumption is satisfied.

Assumption B1.

(a) W is nonsingular.

(b) There exist real numbers $\underline{\beta}, \bar{\beta} > 0$ such that for all $\omega \in \Omega$

$$\lambda_{min}[M_{11}(\omega)] \geq \underline{\beta} , \tag{5.2a}$$

$$\lambda_{max}[M_{11}(\omega)] \leq \bar{\beta} , \tag{5.2b}$$

and

$$\|U(t, \phi, \dot{\phi}, \omega)\| \leq \bar{\beta}[1 + \|\phi\| + \|\dot{\phi}\|] \tag{5.2c}$$

for all $t \in \mathbf{R}$, $\phi \in \mathbf{R}^{\bar{N}}$, $\dot{\phi} \in \mathbf{R}^{\bar{N}}$.

To demonstrate that B1 implies A1-A2, we present some controllers which assure satisfaction of A1-A2; see [8].

5.2 Examples of Proposed Controllers

Choosing any nonsingular matrix $T \in \mathbb{R}^{\bar{N} \times \bar{N}}$ and defining

$$x \triangleq [\phi^T \dot{\phi}^T]^T$$

the "rigid" model can be described by (2.15) with

$$\bar{F}(t,x,u,\omega) = Ax + B[h(t,x,\omega) + G(\omega)T^T W u] \tag{5.3a}$$

$$\bar{d}(t,x,\omega) = x\,, \tag{5.3b}$$

where

$$A \triangleq \begin{bmatrix} 0 & I \\ 0 & 0 \end{bmatrix}, \quad B \triangleq \begin{bmatrix} 0 \\ T \end{bmatrix}, \tag{5.3c}$$

$$h(t,x,\omega) \triangleq T^{-1} M_{11}(\omega)^{-1} U(t,\phi,\dot{\phi},\omega)\,, \tag{5.3d}$$

$$G(\omega) \triangleq [T^T M_{11}(\omega) T]^{-1}\,. \tag{5.3e}$$

A proposed controller is any function $p\colon \mathbb{R} \times \mathbb{R}^{\bar{N}} \to \mathbb{R}^{\bar{N}}$ of the form

$$p(t,z) = (T^T W)^{-1} [p^o(t,z) + p^\varepsilon(z)] \tag{5.4}$$

where p^ε is specified below and p^o is any function satisfying requirements (3.5) of A3; p^o is chosen to reduce the magnitude of the uncertain term

$$e(t,x,\omega) \triangleq h(t,x,\omega) + G(\omega) p^o(t,x)\,. \tag{5.5}$$

5.2.1. Construction of p^ε. First choose any positive definite symmetric matrix $Q \in \mathbb{R}^{n \times n}$, $n \triangleq 2\bar{N}$, and any positive real number σ and solve the Riccati equation

$$PA + A^T P - 2\sigma PBB^T P + 2Q = 0 \tag{5.6}$$

for a positive definite symmetric $P \in \mathbb{R}^{n \times n}$; since (A, B) is controllable such a solution exists.

Choose any non-negative numbers γ, ρ, κ which, for all $\omega \in \Omega$, satisfy

$$\gamma \geq \lambda(\omega)[\bar{\sigma} + \frac{1}{4} \beta_1(\omega)^2]\,, \tag{5.7a}$$

$$\rho \geq \lambda(\omega) \beta_o(\omega)\,, \tag{5.7b}$$

$$\kappa \geq \beta_o(\omega)\,, \tag{5.7c}$$

where $\lambda(\omega)$, $\beta_o(\omega)$, $\beta_1(\omega)$, $\bar{\sigma}$, are chosen to satisfy

$$\lambda_{max}[T^T M_{11}(\omega) T] \leq \lambda(\omega)\,, \tag{5.7d}$$

$$\|e(t,x,\omega)\| \leq \beta_o(\omega) + \beta_1(\omega) \|x\|_Q\,, \tag{5.7e}$$

$$\left.\begin{array}{l}\bar{\sigma}\geq\sigma \text{ if } \beta_1(\omega)\equiv 0\\ \bar{\sigma}>\sigma \text{ if } \beta_1(\omega)\not\equiv 0\end{array}\right\} . \tag{5.7f}$$

Part (b) of B1 guarantees the existence of the above bounds.

Now, for any $\varepsilon > 0$, let $s: \mathbf{R}^{\bar{N}} \rightarrow \mathbf{R}^{\bar{N}}$ be any differentiable function with bounded derivative which satisfies

$$\|\eta\| s(\eta) = \|s(\eta)\| \eta , \tag{5.8a}$$

$$\|\eta\| \geq \varepsilon \Rightarrow \|s(\eta)\| \geq 1 - \|\eta\|^{-1}\varepsilon , \tag{5.8b}$$

for $\eta \in \mathbf{R}^{\bar{N}}$.

Then

$$p^{\varepsilon}(z) \triangleq -\gamma B^T P z - \rho s(\kappa B^T P z) . \tag{5.9}$$

As an example of a function satisfying the above requirements on s, consider

$$s(\eta) = (\|\eta\| + \varepsilon)^{-1} \eta . \tag{5.10}$$

6. AN ILLUSTRATIVE EXAMPLE

Consider a system consisting of two rotors B_1 and B_2 connected by a massless shaft B_3; see Figure 1.

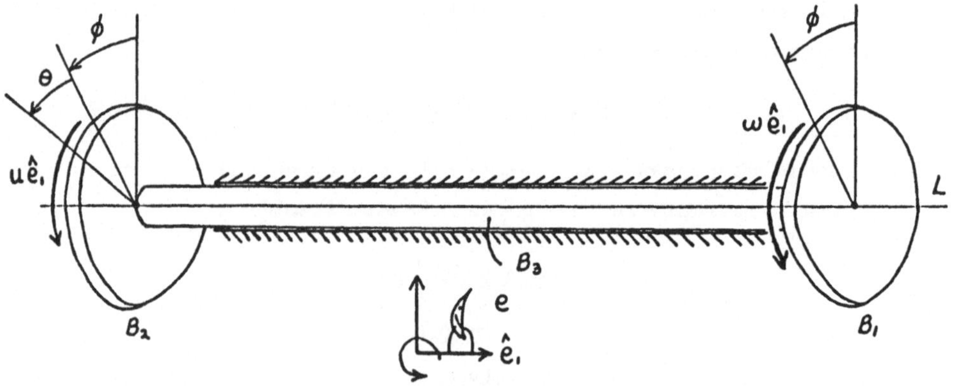

Figure 1. The system considered in the example.

Relative to inertial reference frame e, the system is constrained to rotate about a line L parallel to $\hat{e}_1$; $I_1, I_2 > 0$ are the moments of inertia of B_1, B_2 respectively, about L. Rotor B_2 is subject to a control moment $u(t)\hat{e}_1$. Rotor B_1 is subject to an unknown disturbance torque $\omega(t)\hat{e}_1$; the only information assumed available on ω is an upper bound on $|\omega(t)|$, i.e.,

$$|\omega(t)| \leq \beta \qquad \forall\ t \in \mathbb{R} \tag{6.1}$$

where β is known. The system configuration can be described by ϕ and θ, the angular displacements of B_1 relative to e and B_2 relative to B_1, respectively; thus

$$q \triangleq [\phi\ \theta]^T\ . \tag{6.2}$$

We shall treat B_3 as a neglected component; when it is rigid, $\theta = 0$.

The kinetic energy of the system is

$$\frac{1}{2}\,[I_T\dot{\phi}^2 + 2I_2\dot{\phi}\dot{\theta} + I_2\dot{\theta}^2]\ ,$$

where $I_T \triangleq I_1 + I_2$; hence,

$$M \triangleq \begin{bmatrix} I_T & I_2 \\ I_2 & I_2 \end{bmatrix}\ . \tag{6.3}$$

Utilizing Newtonian mechanics, the motion of the system can be described by

$$\begin{aligned} I_T\ddot{\phi} + I_2\ddot{\theta} &= u + \omega(t)\ , \\ I_2\ddot{\phi} + I_2\ddot{\theta} &= u + \lambda\ , \end{aligned} \tag{6.4}$$

where $-\lambda\hat{e}_1$ and $\lambda\hat{e}_1$ are the torques exerted by B_3 on B_1 and B_2, respectively.

As measurements available for feedback, we shall consider ϕ and $\dot{\phi} + \dot{\theta}$; hence

$$z \triangleq [\phi\ \ \dot{\phi} + \dot{\theta}]^T\ . \tag{6.5}$$

With B_3 rigid, the "rigid" model is

$$I_T\ddot{\phi} = u + \omega(t)\ , \tag{6.6a}$$

$$z = [\phi\ \ \dot{\phi}]^T\ . \tag{6.6b}$$

Modelling B_3 as a parallel combination of a linear torsional spring of spring constant μ^{-2} and a linear torsional damper of damping coefficient μ^{-1},

$$\lambda = -\mu^{-1}\dot{\theta} - \mu^{-2}\theta\ , \tag{6.7}$$

and the flexible model is given by

$$\begin{aligned} I_T\ddot{\phi} + I_2\ddot{\theta} &= u + \omega(t)\ , \\ I_2\ddot{\phi} + I_2\ddot{\theta} &= -\mu^{-1}\dot{\theta} - \mu^{-2}\theta + u\ , \end{aligned} \tag{6.8a}$$

$$z = [\phi\ \ \dot{\phi} + \dot{\theta}]^T\ . \tag{6.8b}$$

Clearly, (6.6) and (6.8) are in the form of (2.6) and (2.12) respectively.

We also note that the "rigid" model is an example of the type of system considered in Section 5; (6.6) is in the form of (5.1) with

$$M_{11}(\omega) = I_T ,$$

$$U(t, \phi, \dot{\phi}, \omega) = \omega(t) , \quad W = 1 .$$

Assumption B1 is satisfied; let

$$\underline{\beta} = I_T , \quad \bar{\beta} = \max\{I_T, \beta\} .$$

Choosing $T = 1$, $p^\circ = 0$ in (5.4), and utilizing (5.3), (5.5), one obtains

$$A = \begin{bmatrix} 0 & 1 \\ 0 & 0 \end{bmatrix} , \quad B = \begin{bmatrix} 0 \\ 1 \end{bmatrix} , \quad e(t, x, \omega) = I_T^{-1}\omega(t) ;$$

hence inequalities (5.7) can be satisfied by choosing γ, ρ, κ such that

$$\gamma \geq I_T \sigma , \quad \rho \geq \beta , \quad \kappa \geq I_T^{-1}\beta .$$

Now choose any $\alpha_1, \alpha_2 > 0$ which satisfy

$$\alpha_2^2 - \alpha_1 > 0$$

and let

$$Q = \begin{bmatrix} \alpha_1^2 & 0 \\ 0 & \alpha_2^2 - \alpha_1 \end{bmatrix} , \quad \sigma = 1 .$$

Solving Riccati equation (5.6) for $P > 0$ yields

$$B^T P = [\alpha_1 \ \alpha_2] .$$

Choosing any $\varepsilon > 0$ and utilizing (5.9), a "stabilizing" controller is given by

$$u(t) = p^\varepsilon(z(t)) ,$$

$$p^\varepsilon(z) = -\gamma(\alpha_1 z_1 + \alpha_2 z_2) - \rho s[\kappa(\alpha_1 z_1 + \alpha_2 z_2)] ,$$

where $s: \mathbf{R} \to \mathbf{R}$ is any differentiable function with bounded derivative which satisfies (5.8).

The significance of parameters α_1, α_2 is as follows. If

$$\omega(t) \equiv 0 , \quad \rho = 0 , \quad \gamma = I_T ,$$

the resulting, undisturbed, linear, feedback-controlled "rigid" model is given by

$$\ddot{\phi} + \alpha_2\dot{\phi} + \alpha_1 = 0 .$$

7. APPENDIX

Consider any system described by

$$\dot{y} = Y(t, y) \tag{7.1}$$

where $t \in \mathbb{R}$, $y \in \mathbb{R}^q$, $Y: \mathbb{R}\times\mathbb{R}^q \to \mathbb{R}^q$ and let B be a set containing $0 \in \mathbb{R}^q$.

Definition 7.1. System (7.1) *asymptotically B-tracks* 0 or *asymptotically tracks* 0 *to within B* iff it has the following properties.

(i) *Existence of solutions.* Given any $t_o \in \mathbb{R}$, $y_o \in \mathbb{R}^q$, there exists a solution $y(\cdot)$ of (7.1) with $y(t_o) = y_o$.

(ii) *Indefinite extension of solutions.* Every solution $y(\cdot)$: $[t_o, t_1) \to \mathbb{R}^q$ of (7.1) has an extension over $[t_o, \infty)$.

(iii) *Global uniform boundedness.* Given any bound $r \in \mathbb{R}_+$, there exists a bound $d(r) \in \mathbb{R}_+$ such that for any $t_o \in \mathbb{R}$ and any solution $y(\cdot)$ of (7.1),

$$\|y(t_o)\| \le r \;\Rightarrow\; \|y(t)\| \le d(r) \quad \forall\, t \ge t_o .$$

(iv) Given any neighborhood B_ε of B, there exists a neighborhood B_δ of 0 such that for any $t_o \in \mathbb{R}$ and any solution $y(\cdot)$ of (7.1),

$$y(t_o) \in B_\delta \;\Rightarrow\; y(t) \in B_\varepsilon \quad \forall\, t \ge t_o .$$

(v) *Global uniform attractivity of B*. Given any bound $r \in \mathbb{R}_+$ and any neighborhood B_ε of B, there exists $T(r, B_\varepsilon) \in \mathbb{R}_+$ such that for any $t_o \in \mathbb{R}$ and any solution $y(\cdot)$ of (7.1),

$$\|y(t_o)\| \le r \;\Rightarrow\; y(t) \in B_\varepsilon \quad \forall\, t \ge t_o + T(r, B_\varepsilon) .$$

Remark 7.1. If (7.1) satisfies the requirements of the above definition with $B = \{0\}$, then it is globally uniformly asymptotically stable about zero.

REFERENCES

[1] Ambrosino, G., G. Celentano, and F. Garofalo, Robust Model Tracking Control for a Class of Nonlinear Plants, *IEEE Trans. Automatic Control*, AC-30, 275, 1985.

[2] Ambrosino, G., G. Celentano, and F. Garofalo, Tracking Control of High-Performance Robots via Stabilizing Controllers for Uncertain Systems, *J. Optimiz. Theory Appl.*, 2, 239, 1986.

[3] Barmish, B. R., M. Corless, and G. Leitmann, A New Class of Stabilizing Controllers for Uncertain Dynamical Systems, *SIAM J. Contr. Optimiz.*, 21, 246, 1983.

[4] Chen, Y. H., Robust Control of Mechanical Manipulators, *J. Dynam. Syst. Meas. Control*, submitted.

[5] Corless, M., and G. Leitmann, Continuous State Feedback Guaranteeing Uniform Ultimate Boundedness for Uncertain Dynamic Systems, *IEEE Trans. Automatic Control*, **AC-26**, 1139, 1981.

[6] Corless, M., Modelling "Flexible Constraints" in Mechanical Systems, *Proc. 20th Midwestern Mechanics Conference*, Purdue University, West Lafayette, Indiana, 1987.

[7] Corless, M., Stability Robustness of Linear Feedback-Controlled Mechanical Systems in the Presence of a Class of Unmodelled Flexibilities, *Proc. 27th Conf. Decision Control*, Houston, Texas, 1988.

[8] Corless, M., Controllers for Uncertain Mechanical Systems with Robustness in the Presence of Unmodelled Flexibilities, in preparation.

[9] Corless, M. and G. Leitmann, Deterministic Control of Uncertain Systems: A Lyapunov Theory Approach, *Deterministic Nonlinear Control of Uncertain Systems: Variable Structure and Lyapunov Control*, (A. Zinober, ed.), IEE Publishers, to appear.

[10] Corless, M., and G. Leitmann, Controller Design for Uncertain Systems Via Lyapunov Functions, *Proc. American Control Conference*, Atlanta, Georgia, 1988.

[11] Corless, M., G. Leitmann and E. P. Ryan, Tracking in the Presence of Bounded Uncertainties, *Proc. 4th IMA Int. Conf. Control Theory*, Cambridge University, England.

[12] Gutman, S., Uncertain Dynamical Systems--Lyapunov Min-Max Approach, *IEEE Trans. Automatic Control*, **AC-24**, 437, 1979.

[13] Ha, I. J., and E. G. Gilbert, Robust Tracking in Nonlinear Systems, *IEEE Trans. Automatic Control*, **AC-32**, 763, 1987.

[14] Madani-Esfahani, S. M., R. A. DeCarlo, M. J. Corless, and S. H. Zak, On Deterministic Control of Uncertain Nonlinear Systems, *Proc. American Control Conf.*, Seattle, Washington, 1986.

[15] Ryan, E. P., G. Leitmann and M. Corless, Practical Stabilizability of Uncertain Dynamical Systems, Application to Robotic Tracking, *J. Optimiz. Theory Applic.*, **47**, 235, 1985.

[16] Spong, M. W., Modeling and Control of Elastic Joint Robots, *J. Dynam. Syst. Meas. Control*, **109**, 310, 1987.

OUTPUT FEEDBACK CONTROL OF UNCERTAIN SYSTEMS IN THE PRESENCE OF UNMODELED ACTUATOR AND SENSOR DYNAMICS

Stanislaw H. Żak and Mehrez Hached

School of Electrical Engineering
Purdue University
West Lafayette, IN 47907
USA

ABSTRACT

This paper analyzes the performance of output feedback controllers for a class of uncertain time-varying nonlinear systems in the presence of unmodeled actuator and sensor dynamics. In particular, on the basis of known nominal model and bounds on the uncertainties, and initially neglecting actuator and sensor dynamics, high-gain output feedback schemes are determined which force the output to track a given signal. Then, the effects of actuator and sensor dynamics are investigated on the performance of the tracking system.

KEY WORDS: Nonlinear systems, Output feedback, Uncertain systems, Singular perturbations.

1. INTRODUCTION

Recently, major progress has been made in the analysis and design of nonlinear control systems. Different approaches have been proposed (Utkin, [1], [2], Corless and Leitmann [17], Hunt et al. [5], Su et al. [6], Glad [22], [23], Bauman and Rugh [19], DeCarlo et al. [10], Isidori [15], Walcott and Żak [8], Steinberg and Corless [12]). An important property of control systems is their robustness, i.e. the ability of the system to retain certain performance measures in the presence of perturbations. Or in other words; "the ability of a control system to function even when the actual system differs from the model used for designing the controller" (Glad [22]). The system model used by the designer may differ from the controlled system because of model uncertainties or neglected high-frequency dynamics. Specifically, when devising a model of the plant, small time constants corresponding to actuator and/or sensor dynamics are neglected. Furthermore, it is often impossible to measure directly all the components of the state or output vectors. In order to restore them additional sensors are used which lead to motions different from the motions predicted by the plant model.

The problem of controlling a system in the presence of unmodeled actuator and sensor dynamics has received recently the attention of many researchers. In particular Bondarev et al. [7], and Żak et al. [25] studied the influence of neglected high-frequency dynamics on the variable structure control systems. Leitmann et al. [9] studied the robustness with respect to neglected actuator and sensor dynamics of state feedback controllers for uncertain systems. Glad [23] considered the sensitivity of the system to variations in gain at the input, corresponding to nonideal behavior of the actuators. The problem of the robustness of various output feedback control algorithms based on a reduced-order model with neglected high-frequency dynamics was investigated by O'Reilly [18] and Vostrikov et al. [24] using singular perturbation techniques.

The purpose of this paper is to analyze the effect of neglected high-frequency dynamics on various output feedback control designs for nonlinear uncertain systems. Our approach is inspired by Marino [4], Utkin [2], and Vostrikov et al. [24]. The tools we use in this paper are the high-gain output feedback and Lie derivatives.

The paper is organized as follows. Section 2 is devoted to the description of the class of nonlinear systems we consider along with the problem statement. The next section presents some background material and preliminary results. The following sections discuss different high-gain output feedback control schemes. Then the effects on the performance of the closed-loop system of unmodeled actuator and sensor dynamics are investigated. Finally, Section 6 contains concluding remarks.

2. PROBLEM STATEMENT

In this paper we consider a class of dynamical systems governed by the following equations

$$\left.\begin{aligned} \dot{x}(t) &= f(t,x) + G(t,x)\,[u(t) + \xi(t,x)] \\ y(t) &= h(x)\,, \end{aligned}\right\} \tag{2.1}$$

where $x \in \mathbb{R}^n$, $u \in \mathbb{R}^m$, $y \in \mathbb{R}^m$, and $\xi(\cdot)\ \mathbb{R}\times\mathbb{R}^n \to \mathbb{R}^m$ is the lumped uncertain element. We assume that the norm of the uncertain element is bounded by a known bounded nonnegative function; that is for all $(t,x) \in \mathbb{R}\times\mathbb{R}^n$

$$\|\xi(t,x)\| \leqq \rho(t,x)\,,$$

where $\rho(\cdot):\ \mathbb{R}\times\mathbb{R}^n \to \mathbb{R}_+$, and $\|\cdot\|$ is the Euclidean norm i.e., $\|x\| = (\sum_{i=1}^{n} |x_i|^2)^{1/2}$.

Note that the only information assumed about the uncertain vector is its maximum possible energy. If the uncertainties $\xi(t,x)$ enter structurally into the state equations as in (2.1) then we say that the matching condition is satisfied [17].

The function $f(\cdot)$ is a continuous single-valued vector-function and $G(\cdot)$ is a continuous single-valued matrix function with rank $G = m$. Furthermore, we require that $f(t,0) = 0$ for all t. The output vector function $h(\cdot)$ is continuously differentiable and

$h(0) = 0$.

In this paper we analyze two different output feedback control strategies. The first is the high-gain output feedback stabilization scheme. In the synthesis of this control law we utilize a nonlinear transformation which brings the original system into the "regular form" ([20]) from where the design is performed.

The aim of the second control law is to ensure the tracking property of the output of some given reference signal.

For both control strategies we will investigate the effects of the unmodeled actuator and sensor dynamics on the performance of the closed-loop systems.

3. PRELIMINARY RESULTS

LIE DERIVATIVES

Time-Invariant Lie Derivatives

Let $f: \mathbb{R}^n \longrightarrow \mathbb{R}^n$ and $g: \mathbb{R}^n \longrightarrow \mathbb{R}^n$ be C^∞ vector fields on $\mathbb{R}^n$. The Lie bracket is defined by

$$[f,g] \triangleq \frac{\partial f}{\partial x} g - \frac{\partial g}{\partial x} f ,$$

where $\frac{\partial f}{\partial x}$ and $\frac{\partial g}{\partial x}$ are the Jacobian matrices of f and g, respectively. Using an alternative notation, one can represent the Lie bracket as follows

$$[f,g] = (ad^1 f, g) .$$

Also, define

$$(ad^k f, g) = [f, (ad^{k-1} f, g)] ,$$

where, by definition

$$(ad^0 f, g) = g .$$

Next, consider a C^∞ function $h : \mathbb{R}^n \longrightarrow \mathbb{R}$. Let $dh = \nabla^T h$ be the derivative of h with respect to x, where ∇h is the gradient of h with respect to x. Then the Lie derivative of h with respect to f is defined by

$$L_f h = L_f(h) = \langle dh, f \rangle = \nabla^T h \cdot f .$$

The following notation is employed throughout this paper

$$L_f^0 h = h$$
$$L_f^1 h = L_f h$$
$$\vdots$$
$$L_f^k h = L_f(L_f^{k-1} h) .$$

The Lie derivative of dh with respect to the vector field f is defined by

$$L_f(dh) = \left(\frac{\partial (dh)^T}{\partial x} f \right)^T + (dh) \frac{\partial f}{\partial x} .$$

One may easily verify that these three Lie derivatives obey the following so-called Leibnitz formula

$$L_{[f,g]} h = \langle dh, [f,g] \rangle = L_g L_f h - L_f L_g h .$$

Furthermore, the following relation is valid

$$dL_f h = L_f(dh) .$$

Time-Varying Lie Derivatives

Suppose now f and g are C^∞ time-varying vector fields, i.e. $f(\cdot) : \mathbb{R} \times \mathbb{R}^n \rightarrow \mathbb{R}^n$, $g(\cdot) : \mathbb{R} \times \mathbb{R}^n \rightarrow \mathbb{R}^n$. Then the time-varying Lie bracket is defined by

$$(\Gamma^1 f, g) \triangleq (ad^1 f, g) - \frac{\partial g}{\partial t} ,$$

and

$$(\Gamma^k f, g) = (\Gamma^1 f, (\Gamma^{k-1} f, g)) ,$$

where

$$(\Gamma^0 f, g) \triangleq g .$$

Next consider a C^∞ function $h(\cdot) : \mathbb{R} \times \mathbb{R}^n \rightarrow \mathbb{R}$. Then the time-varying Lie derivative of h with respect to f is defined by

$$\mathscr{L}_f h = \mathscr{L}_f(h) \triangleq L_f h + \frac{\partial h}{\partial t} .$$

We define

$$\mathscr{L}_f^0 h \triangleq h ,$$
$$\mathscr{L}_f^k h \triangleq \mathscr{L}_f(\mathscr{L}_f^{k-1} h) = L_f(\mathscr{L}_f^{k-1} h) + \frac{\partial \mathscr{L}_f^{k-1} h}{\partial t} .$$

The time-varying Lie derivative of dh with respect to the time-varying vector field f is defined by

$$\mathscr{L}_f dh = f^T \left(\frac{\partial(dh)}{\partial x}\right) + (dh)\,\frac{\partial f}{\partial x} + \frac{\partial}{\partial t}\,(dh)\,.$$

Note that

$$d\mathscr{L}_f h = \mathscr{L}_f(dh)\,.$$

One may verify that the above defined time-varying Lie derivatives obey the following formula

$$\begin{aligned} &<dh,(\Gamma^1 f,g)> \\ &\quad = L_{(\Gamma^1 f,g)} h = L_g L_f h - L_f L_g h - L_{\frac{\partial g}{\partial t}} h \\ &\qquad\qquad = L_g \mathscr{L}_f h - \mathscr{L}_f L_g h\,. \end{aligned}$$

MARKOV PARAMETERS

The affine Markov parameters are defined as the elements of the matrix resulting from the product of the observability and controllability matrices of an affine nonlinear system described by the following equations

$$\left.\begin{aligned} &\dot{x} = f(t,x) + g_1(t,x)u_1 + \dots + g_m(t,x)u_m \\ &y = h(x) = [h_1(x)\,,\ \dots\,,\ h_p(x)]^T\,, \end{aligned}\right\} \tag{3.1}$$

where $f, g_1, \dots, g_m\ :\ \mathbb{R}\times\mathbb{R}^n \rightarrow \mathbb{R}^n$ and $h\ :\ \mathbb{R}^n \rightarrow \mathbb{R}^p$ are C^∞ vector fields.

The observability matrix of such a system is defined by the following (np)×n matrix

$$\mathcal{O} = \begin{bmatrix} dh_1 \\ \vdots \\ dh_p \\ \mathcal{L}_f(dh_1) \\ \vdots \\ \mathcal{L}_f(dh_p) \\ \vdots \\ \mathcal{L}_f^{n-1}(dh_1) \\ \vdots \\ \mathcal{L}_f^{n-1}(dh_p) \end{bmatrix}. \tag{3.2}$$

The controllability matrix is defined by the following n×(nm) matrix

$$\mathcal{C} = \left[g_1,\dots,g_m,(\Gamma^1 f,g_1),\dots,(\Gamma^1 f,g_m),\dots,(\Gamma^{n-1} f,g_1),\dots,(\Gamma^{n-1} f,g_m)\right]. \tag{3.3}$$

So the elements of the matrix $\mathcal{O}\mathcal{C}$ have the form

$$\begin{aligned}(\mathcal{L}_f^j(dh_\beta))\,(\Gamma^i f,g_\alpha) &= <\mathcal{L}_f^j(dh_\beta),\ (\Gamma^i f,g_\alpha)> \\ &= <d\mathcal{L}_f^j h_\beta\ ,\ (\Gamma^i f,g_\alpha)> \\ &= L_{(\Gamma^i f,g_\alpha)}\ \mathcal{L}_f^j h_\beta \end{aligned} \tag{3.4}$$

for i,j = 1,...,n−1, $\alpha = 1,\dots,m$, $\beta = 1,\dots,p$, and are referred to as the affine Markov parameters.

Theorem 3.1: If there exist constants c_k, $k = 0,1,\dots$ such that the Markov parameters satisfy

$$L_{(\Gamma^i f,g_\alpha)}\ \mathcal{L}_f^j h_\beta = c_k = c_{i+j}\ , \tag{3.5}$$

then

$$L_{(\Gamma^i f,g_\alpha)}\mathcal{L}_f^j h_\beta = L_{g_\alpha}\mathcal{L}_f^{i+j} h_\beta = \text{const} = c_{i+j}\ .$$

Proof: Repeated application of the definitions of Lie derivatives and condition (3.5) yields the following

$$\begin{aligned}L_{(\Gamma^i f,g_\alpha)}\mathcal{L}_f^j h_\beta &= <d\mathcal{L}_f^j h_\beta\ ,\ (\Gamma f,(\Gamma^{i-1} f,g_\alpha))> \\ &= L_{(\Gamma^{i-1} f,g_\alpha)}\mathcal{L}_f\mathcal{L}_f^j h_\beta - \mathcal{L}_f L_{(\Gamma^{i-1} f,g_\alpha)}\mathcal{L}_f^j h_\beta\end{aligned}$$

$$= L_{(\Gamma^{i-1}f, g_\alpha)} \mathscr{L}_f^{j+1} h_\beta - \mathscr{L}_f c_{i+j-1}$$

$$= L_{(\Gamma^{i-1}f, g_\alpha)} \mathscr{L}_f^{j+1} h_\beta \, .$$

Continuing in this manner we find that

$$L_{(\Gamma^i f, g_\alpha)} \mathscr{L}_f^j h_\beta = L_{g_\alpha} \mathscr{L}_f^{i+j} h_\beta = \text{const} = c_{i+j} \, .$$

□

For further information about Markov parameters for nonlinear time-invariant systems the reader is referred to [11] and [14].

Consider now a plant modeled by (3.1), where $p = m$, and the high gain control law

$$u = k\, s(x) \tag{3.6}$$

where $k > 0$ is a scalar and the function $s(\cdot) : \mathbb{R}^n \rightarrow \mathbb{R}^m$ is continuously differentiable. Assume that $\det SG \neq 0$, where

$$S = \frac{\partial s}{\partial x} \, , \quad \text{and} \quad G = [g_1, g_2, \ldots, g_m] \, .$$

Then we have

Theorem 3.2 ([2],[21]):

If

(i) the functions $f(t,x)$, $G(t,x)s(x)$, and $f_0 = f - G(SG)^{-1}Sf$ satisfy Lipschitz conditions for all x

(ii) the system

$$\frac{ds}{dt} = (SG)s$$

is uniformly exponentially stable, that is there exist positive $A \geqq 1$ and α such that

$$||s(x)|| < A||s(x(0))||e^{-\alpha t} \, ,$$

then for any positive Δ, and T there exists a positive k_0 such that

$$||s(x(t))|| < \Delta$$

for $k > k_0$ and $t_0 + t_1 < t < T$ on the solutions of (3.1) with the control $u = ks(x)$, and $\lim_{k \to \infty} t_1 = 0$.

4. THE OUTPUT REGULATION PROBLEM

Consider the nominal system, that is the system without uncertainty as described by

$$\left.\begin{aligned}\dot{x} &= f(t,x) + G(t,x)u \\ y &= h(x)\,.\end{aligned}\right\} \tag{4.1}$$

First we define the decoupling indices for the system (4.1). We consider each of the m output channels separately. So considering the first output channel we form the following row vector which we will call the decoupling vector for channel one

$$[L_{g_1}h_1,\ L_{g_2}h_1,\ldots,L_{g_m}h_1]\,. \tag{4.2}$$

If this row vector is not identically equal to zero, then we define the decoupling index of the first channel to be zero, or $d_1 = 0$.

However, if the row vector is identically equal to zero we proceed to form the following decoupling vector

$$[L_{g_1}\mathscr{L}_f h_1,\ L_{g_2}\mathscr{L}_f h_1\ ,\ \ldots\ ,\ L_{g_m}\mathscr{L}_f h_1]\,.$$

Again we determine if it is identically equal to zero, or not. If it is not we stop and define $d_1 = 1$. If it is zero we proceed further by forming

$$[L_{g_1}\mathscr{L}_f^2 h_1,\ L_{g_2}\mathscr{L}_f^2 h_1\ ,\ \ldots\ ,\ L_{g_m}\mathscr{L}_f^2 h_1]\,,$$

and so on.

So the decoupling index of channel 1, is equal to the smallest integer d_1 for which the decoupling vector,

$$[L_{g_1}\mathscr{L}_f^{d_1} h_1,\ L_{g_2}\mathscr{L}_f^{d_1} h_1\ ,\ \ldots\ ,\ L_{g_m}\mathscr{L}_f^{d_1} h_1]\,,$$

is not identically equal to zero.

Similar procedure for the other output channels yields a set of m parameters, d_i for $i = 1,2,\ldots m$.

The decoupling indices are an indication of what the lowest derivative of each output channel needs to be utilized for an output control to be effective. By taking the time derivative of the ith output channel we obtain

$$\dot{y}_i = \frac{\partial h_i}{\partial x}\dot{x} = \frac{\partial h_i}{\partial x}\,(f + g_1 u_1 + \ldots + g_m u_m)\,,$$

hence

$$\dot{y}_i = L_f h_i + [L_{g_1}h_i\ ,\ \ldots\ ,\ L_{g_m}h_i]\ u\,.$$

Thus if $[L_{g_1}h_i,\ L_{g_2}h_i\ ,\ \ldots\ ,\ L_{g_m}h_i] = [0]$ then u has no effect on the output y_i, so we need to form $\ddot{y}_i$ where

$$\ddot{y}_i = \frac{\partial \dot{y}_i}{\partial x}\dot{x} = \mathscr{L}_f^2 h_i + [L_{g_1}\mathscr{L}_f h_i \ , \ \dots \ , \ L_{g_m}\mathscr{L}_f h_i]u \ .$$

Again if $[L_{g_1}\mathscr{L}_f h_i, L_{g_2}\mathscr{L}_f h_i \ , \dots , L_{g_m}\mathscr{L}_f h_i] = [0]$, then u has no effect on the output and we need to take higher derivatives of y_i in a similar fashion as before.

Now that we have obtained the set of decoupling indices, we consider all the output channels together to form the following matrix

$$N = \begin{bmatrix} L_{g_1}\mathscr{L}_f^{d_1}h_1 & \dots & L_{g_m}\mathscr{L}_f^{d_1}h_1 \\ \vdots & & \vdots \\ L_{g_1}\mathscr{L}_f^{d_m}h_m & \dots & L_{g_m}\mathscr{L}_f^{d_m}h_m \end{bmatrix} . \tag{4.3}$$

We will assume that the matrix N is nonsingular and we will further assume that the Markov parameters of the system (4.1) are constant. Hence by the virtue of Theorem 3.1 the matrix N is constant.

With the N matrix constant and nonsingular, we proceed to construct a high-gain output control which will regulate the output to zero.

We will consider two cases. The first case is when all decoupling indices are equal to zero, and the second case when some, or all, decoupling indices are not equal to zero.

For a rigorous treatment of the decoupling problem for nonlinear time-invariant systems the reader is referred to [3], [14], [16].

Case 1: For this case the N matrix will have the following form

$$N = \begin{bmatrix} L_{g_1}h_1 & \dots & L_{g_m}h_1 \\ \vdots & & \vdots \\ L_{g_1}h_m & \dots & L_{g_m}h_m \end{bmatrix} = \frac{\partial h}{\partial x}G = HG \ , \tag{4.4}$$

where $\frac{\partial h}{\partial x}$ is the Jacobian matrix of h and $G = [g_1,\dots,g_m]$.

If we employ the following diffeomorphic state variable transformation

$$\left.\begin{aligned} \bar{x}_1 &= \phi_1(t,x) \\ \bar{x}_2 &= \phi_2(t,x) \\ &\vdots \\ \bar{x}_{n-m+1} &= h_1(x) \\ &\vdots \\ \bar{x}_n &= h_m(x)\,, \end{aligned}\right\} \tag{4.5}$$

where the ϕ_i's are chosen such that

$$L_{g_j}\phi_i = 0\,,\ \ j = 1,\ldots,m \ \text{ for all } \ i = 1,\ldots,n-m, \tag{4.6}$$

then the system (2.1) in the new coordinates will have the following form

$$\begin{bmatrix} \dot{\bar{x}}_1 \\ \vdots \\ \dot{\bar{x}}_{n-m} \end{bmatrix} = \begin{bmatrix} \mathcal{L}_f\phi_1 \\ \vdots \\ \mathcal{L}_f\phi_{n-m} \end{bmatrix} \tag{4.7a}$$

$$\begin{bmatrix} \dot{\bar{x}}_{n-m+1} \\ \vdots \\ \dot{\bar{x}}_n \end{bmatrix} = \begin{bmatrix} L_f h_1 \\ \vdots \\ L_f h_m \end{bmatrix} + \begin{bmatrix} L_{g_1}h_1 & \cdots & L_{g_m}h_1 \\ \vdots & & \vdots \\ L_{g_1}h_m & \cdots & L_{g_m}h_m \end{bmatrix}(u + \xi)\,. \tag{4.7b}$$

We will now employ the high gain output feedback control as given by

$$u = \frac{1}{\epsilon}\,K^o h(x)\,, \tag{4.8}$$

where ϵ is a small constant and K^o is an m×m constant matrix. Under the influence of this control, the system equations become

$$\begin{bmatrix} \dot{\bar{x}}_1 \\ \vdots \\ \dot{\bar{x}}_{n-m} \end{bmatrix} = \begin{bmatrix} \mathcal{L}_f\phi_1 \\ \vdots \\ \mathcal{L}_f\phi_{n-m} \end{bmatrix} \tag{4.9a}$$

$$\begin{bmatrix} \dot{\bar{x}}_{n-m+1} \\ \vdots \\ \dot{\bar{x}}_n \end{bmatrix} = \begin{bmatrix} L_f h_1 \\ \vdots \\ L_f h_m \end{bmatrix} + \frac{1}{\epsilon} N K^o h(x) + N\xi . \tag{4.9b}$$

We see that the application of this control decouples the system into the slow and fast subsystems. The dynamics of the slow subsystem are given by

$$\left.\begin{aligned} \begin{bmatrix} \dot{\bar{x}}_1 \\ \vdots \\ \dot{\bar{x}}_{n-m} \end{bmatrix} &= \begin{bmatrix} \mathcal{L}_f \phi_1 \\ \vdots \\ \mathcal{L}_f \phi_{n-m} \end{bmatrix} \\ y = h(x) &= \begin{bmatrix} \bar{x}_{n-m+1} \\ \vdots \\ \bar{x}_n \end{bmatrix} = 0 , \end{aligned}\right\} \tag{4.10}$$

whereas by invoking the following change in the time variable,

$$t = \epsilon \tau , \tag{4.11}$$

the equations describing the dynamics of the fast system are given by

$$\frac{d}{d\tau} \begin{bmatrix} \bar{x}_{n-m+1} \\ \vdots \\ \bar{x}_n \end{bmatrix} = \epsilon \begin{bmatrix} \mathcal{L}_f h_1 \\ \vdots \\ \mathcal{L}_f h_m \end{bmatrix} + N K^o h(x) + \epsilon N\xi , \tag{4.12}$$

and for sufficiently small ϵ, the above equations simplify to

$$\frac{d}{d\tau} \begin{bmatrix} \bar{x}_{n-m+1} \\ \vdots \\ \bar{x}_n \end{bmatrix} = N K^o h(x) = N K^o \begin{bmatrix} \bar{x}_{n-m+1} \\ \vdots \\ \bar{x}_n \end{bmatrix} . \tag{4.13}$$

Observing that the part of our transformation in (4.5) is $y = [\bar{x}_{n-m+1} , \dots , \bar{x}_n]^T$, we can rewrite the above equation as

$$\frac{dy}{d\tau} = N K^o y . \tag{4.14}$$

Note that by an appropriate choice of the matrix K^o the fast subsystem can be made uniformly exponentially stable. Thus if J is the required uniformly exponentially stable matrix then,

$$K^o = N^{-1} J \,, \tag{4.15}$$

and K^o can be evaluated since N is assumed to be nonsingular.

By invoking Theorem 3.2 we see that the stability of the fast subsystem will result in the trajectories of the system (4.7) converging to the Δ-vicinity of the manifold $y(x) = 0$. Thus the output is regulated to zero. Within the Δ-vicinity of the manifold, the system will be governed by equation (4.10) which represents the dynamics of the slow subsystem. From equation (4.10), we notice that we do not have any influence on the internal stability of the slow subsystem when the output is regulated to zero. We assume however that the slow subsystem is asymptotically stable. The stability of the slow subsystem is a structural property of the plant. This subject requires further research.

Although Theorem 3.2 was stated for nonlinear systems without uncertainties, it also applies to our particular case. This is because the uncertainties in the system (2.1) are bounded by a known bounded function.

Case 2: Let us first reorder the output channels so that they are ordered in ascending values of their decoupling indices. Thus y_1 is assigned to the channel with the smallest d_i, and y_m to the one with the largest d_i.

We then employ the following diffeomorphic state variable transformation

$$\bar{x} = \begin{bmatrix} \phi_1(t,x) \\ \phi_2(t,x) \\ \vdots \\ h_m(x) \\ \vdots \\ \mathcal{L}_f^{d_1-1} h_1 \\ \vdots \\ \mathcal{L}_f^{d_m-1} h_m \\ \mathcal{L}_f^{d_1} h_1 \\ \vdots \\ \mathcal{L}_f^{d_m} h_m \end{bmatrix}, \tag{4.16}$$

where the ϕ_i's are chosen such that

$$L_{g_j} \phi_i = 0 \quad j = 1, \ldots, m \,.$$

The system (2.1) in the new coordinates will have the following form

$$\left.\begin{aligned} \dot{\bar{x}}^1 &= f_1(t,\bar{x}) \\ \dot{\bar{x}}^2 &= f_2(t,\bar{x}) + N(u+\xi)\,, \end{aligned}\right\} \tag{4.17}$$

where $\bar{x}^1 \in \mathbb{R}^{n-m}$, $\bar{x}^2 \in \mathbb{R}^m$, and N is given by (4.3). The existence conditions of the transformation (4.16) can be deduced from the results of [5], [20], [26], [27].

Note that

$$\bar{x}_i^2 = y_i^{(d_i)} \quad , \; i = 1,\dots,m \,,$$

where $(\ )^{(j)}$ denotes the j-th derivative of ( ) with respect to t. The control law will have the form

$$u = \frac{1}{\epsilon} N^{-1} \begin{bmatrix} k_{1,1}y_1 + \dots + k_{1,d_1+1}y_1^{(d_1)} \\ \vdots \\ k_{m,1}y_m + \dots + k_{m,d_m+1}y_m^{(d_m)} \end{bmatrix}. \tag{4.18}$$

In the new coordinates the closed-loop system (4.17), (4.18) is decoupled into the slow and fast subsystems. The slow subsystem is governed by the equations

$$\left.\begin{aligned} \dot{\bar{x}}^1 &= f_1(t,\bar{x}) \\ y &= 0\,. \end{aligned}\right\} \tag{4.19}$$

As in the previous case, we have no influence on the stability of the slow subsystem. Therefore for the controller to be effective we have to assume that the system (2.1) without uncertainties is asymptotically stable when restricted to the manifold $y = 0$ which is equivalent to requirement that the system (4.19) is asymptotically stable.

As with regard to the fast subsystem we utilize a change in the time variable $t = \epsilon\,\tau$ to obtain

$$\begin{bmatrix} y_1^{(d_1+1)} \\ \vdots \\ y_m^{(d_m+1)} \end{bmatrix} = \epsilon(f_2 + N\xi) + \begin{bmatrix} k_{1,1}y_1 + \dots + k_{1,d_1+1}y_1^{(d_1)} \\ \vdots \\ k_{m,1}y_m + \dots + k_{m,d_m+1}y_m^{(d_m)} \end{bmatrix}. \tag{4.20}$$

If we now choose k_{ij} in such a way that the simplified fast subsystem

$$\begin{bmatrix} y_1^{(d_1+1)} \\ \vdots \\ y_m^{(d_m+1)} \end{bmatrix} = \begin{bmatrix} k_{1,1}y_1 + \dots + k_{1,d_1+1}y_1^{(d_1)} \\ \vdots \\ k_{m,1}y_m + \dots + k_{m,d_m+1}y_m^{(d_m)} \end{bmatrix}$$

is uniformly exponentially stable then by the virtue of Theorem (3.1) the closed-loop system is asymptotically stable.

The above output feedback stabilization schemes are quite restrictive. Their effectiveness depends on the stability of the nominal system ($\dot{x} = f + Gu$) when restricted to the manifold $y = h(x) = 0$. In the following section we provide a more effective control scheme. Before that however, we will analyze the effect of unmodeled actuator dynamics on the performance of the closed-loop system with the high-gain output feedbacks.

SYSTEMS WITH FAST UNMODELED MOTIONS

We now investigate the effects of the introduction of uncertain actuator dynamics on the performance of the system (2.1) with high gain output feedback controllers.

Case 1: We will assume that the actuator dynamics is modeled by the following equation

$$\mu_a \dot{r} = Lr + M\bar{u} \,, \quad u = Nr + R\bar{u} \,, \ \bar{u} = c\, K^o y \,, \tag{4.21}$$

where $r \in \mathbb{R}^q$, $q \geqq m$, L is a Hurwitz matrix, μ_a is a positive constant that reflects the "fastness" of the actuator, the matrices L, M, R, and N satisfy the condition $R - NL^{-1}M = I_m$, and $c = \dfrac{1}{\epsilon}$ is a large constant.

Proposition 4.1: If the matrix L is Hurwitz, (the fast subsystem described by (4.21) is exponentially stable) then as μ_a approaches 0, the motion of the slow subsystem is described by (2.1) with $u = \bar{u} = c\, K^o y$.

Proof: The fast subsystem is described by (4.21). Replacing $\bar{u}$ by its value yields

$$\mu_a \dot{r} = Lr + cMK^o y \,. \tag{4.22}$$

Let $\tau = \mu_a^{-1} t$, hence (4.22) becomes

$$\frac{dr}{d\tau} = Lr + cMK^o y \,. \tag{4.23}$$

Since L is a Hurwitz matrix, then as τ approaches infinity we have $y =$ constant and

$$\lim_{\tau \to \infty} r = -cL^{-1}MK^{o}y ,$$

hence

$$u = Nr + cRK^{o}y = [-NL^{-1}M + R]cK^{o}y = cK^{o}y . \tag{4.24}$$

The expression for u as per (4.24) can also be found by setting $\mu_a = 0$. Hence the slow subsystem is described by (2.1) and (4.24).

□

Case 2: If the actuator dynamics for this case is also described by (4.21) with

$$\bar{u} = cN^{-1} \begin{bmatrix} K_{1,1}y_1 + \dots + K_{1,d_1+1}y_1^{(d_1)} \\ \vdots \\ K_{m_1,}y_m + \dots + K_{m,d_m+1}y_m^{(d_m)} \end{bmatrix},$$

then using a similar argument as in the previous case we conclude that the slow subsystem is described by (2.1) with $u = \bar{u}$.

In conclusion, for a sufficiently fast actuator the proposed control schemes will stabilize the output.

5. THE TRACKING PROBLEM

Our goal now is to design a controller such that the output of the system (2.1) will track a given reference signal.

A sufficient condition for the output y to track the reference signal $\nu(t)$ is

$$\frac{d}{dt}[y - \nu(t)] = V[y - \nu(t)] \triangleq F(y, \nu(t)) , \tag{5.1}$$

where V is a Hurwitz matrix. If $\nu(t)$ = constant, then (5.1) becomes $\dot{y} = V[y - \nu]$.

We require that the closed-loop system (2.1) be asymptotically stable with respect to the time-varying manifold

$$\Omega = \{x : h(x(t)) - \nu(t) = y(t) - \nu(t) = 0\} .$$

The projection of the overall system on this manifold is

$$\dot{y}(t) - \dot{\nu}(t) = H\dot{x} - \dot{\nu}(t)$$

$$= Hf + HG(u + \xi) - \dot{\nu}(t) .$$

Using equation (5.1) and solving for u, we obtain the following control law

$$\tilde{u} = (HG)^{-1}[F(y, \nu(t)) - Hf + \dot{\nu}(t)] - \xi . \tag{5.2}$$

In order to implement the control law (5.2), we would have to have the exact knowledge of the uncertain vector $\xi(t,x)$. Hence this control strategy is impractical. In what follows we propose a practical control algorithm which approximates the controller (5.2).

Consider the following control strategy

$$u = K[V(y - \nu(t)) - (\dot{y} - \dot{\nu}(t))] \,, \tag{5.3}$$

where K is the matrix of gain coefficients, $K = cK^o$, and c is a scalar large factor. At the present time, we will assume that $\dot{y}$ can be measured exactly. Later, we will investigate the case in which $\dot{y}$ is measured by a sensor.

To analyze the behavior of the system (2.1) with the control law (5.3) in the presence of unmodeled actuator and sensor dynamics we will employ the arguments of Vostrikov et al. [24] used for systems without uncertainties.

Along the trajectories of the motion of the dynamical system (2.1), $\dot{y}$ is given by

$$\dot{y} = Hf(t,x) + HG(t,x)(u + \xi(t,x)) \,. \tag{5.4}$$

Proposition 5.1: If $\det(I + cHGK^o) \neq 0$, and $\det(HG) \neq 0$, then

(a) $\lim_{c\to\infty} \frac{d}{dt} [y - \nu(t)] = F(y,\nu(t))$,

(b) $\lim_{c\to\infty} u = (HG)^{-1} [F(y,\nu(t)) - Hf + \dot{\nu}(t) - HG\xi]$.

Proof: In what follows we shall utilize the arguments of Vostrikov et al. [24].

We first prove part (a). Recall that

$$\dot{y} = H\dot{x} \,,\; \dot{x} = f + G(K[F - (\dot{y} - \dot{\nu})] + \xi) \,,$$

thus, we have

$$\dot{y} = Hf + HGK(F - \dot{y} + \dot{\nu}) + HG\xi \,,$$

regrouping the $\dot{y}$ terms leads to

$$(I + HGK)\dot{y} = Hf + HGK(F + \dot{\nu}) + HG\xi \,,$$

Hence, for $K = cK^o$, we have

$$\dot{y} = (I + cHGK^o)^{-1}(Hf + HG\xi) + (I + cHGK^o)^{-1} cHGK^o(F + \dot{\nu}) \,.$$

Taking $\lim_{c\to\infty} \dot{y}$, the first term approaches zero, while the second term approaches $F + \dot{\nu}$, therefore

$$\lim_{c\to\infty} \dot{y} = F + \dot{\nu} \,.$$

We now prove part (b).

We have

$$u = K[F - (\dot{y} - \dot{\nu})] , \ \dot{y} = Hf + HG(u + \xi) ,$$

therefore

$$u = K[F - Hf - HG(u + \xi) + \dot{\nu}] .$$

Regrouping the u terms leads to

$$(I + KHG)u = K[F - Hf - HG\xi + \dot{\nu}] ,$$

hence, for $K = cK^{o}$, we have

$$u = (I + cK^{o}HG)^{-1}cK^{o} \ [F - Hf - HG\xi + \dot{\nu}] .$$

Thus

$$\lim_{c \to \infty} u = (HG)^{-1}[F - Hf - HG\xi + \dot{\nu}] .$$

□

SYSTEMS WITH FAST UNMODELED MOTIONS

We will now investigate the effects of the neglected actuator dynamics on the performance of the system (2.1) with the control law (5.3).

Suppose that the actuator dynamics is modeled by the following equation

$$\mu_a \dot{r} = Lr + M\bar{u} , \ u = Nr , \ \bar{u} = K(F - \dot{y} + \dot{\nu}) , \tag{5.5}$$

where $r \in \mathbb{R}^q$, $q \geqq m$, L is a Hurwitz matrix, μ_a is a positive constant that reflects the "fastness" of these dynamics, and the matrices L, M, N satisfy the condition $-NL^{-1}M = I$.

The system described by (2.1), and (5.5) may be studied by the methods of the theory of differential equations with small parameters in some of the derivatives [24].

For such systems, the overall motion can be decoupled into the fast and slow components [21] [24]. The method of decoupling motions is advantageous in systems involving high-gain feedback and/or singular perturbations. The main idea behind the theory is to decouple the system into two subsystems of lower dimensionality. The equations of the slow motions and the convergence conditions for the fast motions are examined in [21] and [24].

In the following proposition we investigate the effects of the actuator dynamics on the performance of the system (2.1).

Proposition 5.2: If the matrix $(L - MKHGN)$ is a Hurwitz matrix, then as $c \rightarrow \infty$ the motion of the slow subsystem will be described by (2.1) with $u = \tilde{u}$.

Proof: As $\mu_a \rightarrow 0$, the slow subsystem is described by the following equations

$$\dot{x} = f + G(u + \xi)\,, \quad u = \bar{u} = K(F - \dot{y} + \nu)\,.$$

We now examine the condition for the stability of the fast subsystem.

The fast subsystem is described by equation (5.5). Replacing $\bar{u}$ by its value yields

$$\mu_a \dot{r} = Lr + MK(F - Hf - HGNr - HG\xi + \nu)\,. \tag{5.6}$$

Let $\tau = \mu_a^{-1} t$, hence equation (5.6) becomes

$$\frac{dr}{d\tau} = (L - MKHGN)r + MK(F - Hf - HG\xi + \nu)\,, \tag{5.7}$$

where x = constant, t = constant.

If the matrix $(L - MKHGN)$ is Hurwitz, then

$$\lim_{\tau \rightarrow \infty} r = -(L - MKHGN)^{-1} MK(F - Hf - HG\xi + \nu)\,.$$

Applying twice the following matrix identity know as the matrix inversion lemma

$$(A_{11} + A_{12}A_{22}A_{21})^{-1} = A_{11}^{-1} - A_{11}^{-1}A_{12}(A_{21}A_{11}^{-1}A_{12} + A_{22}^{-1})^{-1}A_{21}A_{11}^{-1}\,,$$

and the condition $-NL^{-1}M = I_m$ we obtain

$$\begin{aligned} &\lim_{c \rightarrow \infty} N(L - MK^{\circ}(cHG)N)^{-1} cMK^{\circ} \\ &= \lim_{c \rightarrow \infty} cK^{\circ}[-I_m + (K^{\circ} + (cHG)^{-1})^{-1}K^{\circ}] \\ &= -(HG)^{-1}\,. \end{aligned} \tag{5.8}$$

Hence

$$\lim_{c \rightarrow \infty} u = \lim_{c \rightarrow \infty} Nr = (HG)^{-1}[F - Hf + \nu] - \xi\,.$$

□

INFLUENCE OF SENSOR DYNAMICS

To implement the control law (5.3), the vector $\dot{y}$ has to be measured by a sensor (approximate differentiator). Suppose that the approximate differentiator is modeled by the following equation

$$\left.\begin{aligned} \mu_s \dot{z} &= Az + Dh(x) \\ \hat{y} &= Pz\ , \end{aligned}\right\} \tag{5.9}$$

where $z \in \mathbb{R}^q$, $\hat{y} \in \mathbb{R}^m$, $q \geqq m$, and μ_s is a "small" parameter that reflects the "fastness" of the approximate differentiator, $\hat{y}$ is the estimate of y, A is a Hurwitz matrix, and the matrices P, A, and D satisfy the condition $-PA^{-1}D = I$. We shall also use $\dot{\hat{y}}$ instead of $\dot{y}$ in the control law (5.3). Therefore, we have

$$\dot{\hat{y}} = P\dot{z} = \mu_s^{-1}P(Az + Dh(x))\ .$$

Again, to examine the system (2.1) with the control strategy (5.3) and the approximate differentiator (5.9), we shall refer to the theory of decoupling motions ([21], [24]).

If we denote

$$s = Az + Dh(x) = \mu_s \dot{z}\ , \tag{5.10}$$

then the method of decoupling motions described in [21] is suitable for the resulting system. We now examine the condition for the convergence of the fast motions to the manifold $s = 0$.

The projection of the overall system on the manifold s is given by

$$\begin{aligned} \dot{s} &= A\dot{z} + D\dot{y} \\ &= A\dot{z} + DH\dot{x} \\ &= A\mu_s^{-1}s + DH[f + G(u + \xi)]\ . \end{aligned}$$

Replacing u by its value yields

$$\dot{s} = (A - DGHKP)\mu_s^{-1}s + DH(f + GKF + GK\nu + G\xi)\ . \tag{5.11}$$

If we now multiply both sides of the above equation by μ_s, and let $t = \mu_s\tau$, we get

$$\frac{ds}{d\tau} = \mu_s\frac{ds}{dt} = (A - DHGKP)s + \mu_s DH(f + GKF + G\xi + GK\nu)\ ,$$

where $x =$ constant, $t =$ constant. If the matrix $[A - DHGKP]$ is Hurwitz then

$$\lim_{\tau\to\infty} s = -\mu_s[A - DHGKP]^{-1}[DHf + DHGK(F + \nu) + DHG\xi]\ .$$

Using twice the matrix inversion lemma and the condition $-PA^{-1}D = I_m$ we obtain

$$\lim_{c\to\infty} (A - DHGKP)^{-1}$$

$$= \lim_{c\to\infty} P^{-1}P(A - D(cHG)K^{o}P)^{-1}$$

$$= \lim_{c\to\infty} P^{-1}[I - (K^{o} + (cHG)^{-1})^{-1}K^{o}]PA^{-1}$$

$$= + P^{-1}(HGK)^{-1}PA^{-1} .$$

Hence

$$\lim_{\substack{r\to\infty \\ c\to\infty}} s = \mu_s P^{-1}(F + \dot{\nu}) , \tag{5.12}$$

or $\mu_s^{-1}Ps = F + \dot{\nu}$, which gives

$$\dot{\hat{y}} - \dot{\nu} = F(\hat{y}, \nu) .$$

To derive the equation of the slow motions, we let μ_s equal to zero. Hence using equation (5.9) we get

$$z = - A^{-1}Dh(x) ,$$
$$\hat{y} = Pz = - PA^{-1}Dh(x) .$$

Using the fact that $- PA^{-1}D = I$, we obtain

$$\hat{y} = y .$$

Hence the equation of the slow motions is given by

$$\begin{aligned}\dot{x} &= f + G[K(F(\hat{y},\nu) - \dot{\hat{y}} + \dot{\nu}) + \xi] \\ &= f + G[u + \xi] ,\end{aligned}$$

where

$$u = K(F(\hat{y},\nu) - \dot{\hat{y}} + \dot{\nu}) = K(F(y,\nu) - \dot{y} + \dot{\nu}) .$$

Remark 5.1: Note that for large but finite values of the K-matrix, the value of the control signal u remains finite (as shown in Proposition 5.1 part (b)).

INFLUENCE OF NOISE

We now investigate the influence of noise on the behavior of the system (2.1) with the control law (5.3). Assume that the output vector y is corrupted by the continuously differentiable noise r(t), thus

$$\hat{y} = y + r(t) , \tag{5.13}$$

We now find values for $\dot{y}$, $\dot{\hat{y}}$, and u. We assume that $\det(I + HGK) \neq 0$.

(a) Recall

$$\dot{y} = H[f + G[K(F(\hat{y},\nu) - \dot{\hat{y}} + \dot{\nu}(t)) + \xi]] ,$$

using equation (5.13) we get

$$\dot{y} = H[f + G[K(F - \dot{y} - \dot{r} + \dot{\nu}) + \xi]] .$$

Solving for $\dot{y}$ we obtain

$$\dot{y} = (I + HGK)^{-1}[Hf + HGKF - HGK\dot{r} + HGK\dot{\nu} + HG\xi] . \tag{5.14}$$

(b) For the controller u,

$$\begin{aligned} u &= K(F - \dot{\hat{y}} + \dot{\nu}) \\ &= K(F - \dot{y} - \dot{r} + \dot{\nu}) , \end{aligned}$$

substituting $\dot{y} = Hf + HGu + HG\xi$, we get

$$u = K(F - Hf - HGu - HG\xi - \dot{r} + \dot{\nu}) ,$$

solving for u yields

$$u = (I + KHG)^{-1}K(F - Hf - \dot{r} - HG\xi + \dot{\nu}) . \tag{5.15}$$

(c) The derivative of the output vector with noise is

$$\dot{\hat{y}} = Hf + HGu + HG\xi ,$$

using $u = K(F - \dot{\hat{y}} + \dot{\nu})$ we obtain

$$\dot{\hat{y}} = Hf + HGK(F - \dot{\hat{y}} + \dot{\nu}) + HG\xi ,$$

solving for $\dot{\hat{y}}$ yields

$$\dot{\hat{y}} = (I + HGK)^{-1}(Hf + HGKF + HG\xi + HGK\dot{\nu}) . \tag{5.16}$$

In the limit the equations (5.14), (5.15), and (5.16) become

(i) $\dot{y} = F - \dot{r} + \dot{\nu}$,

(ii) $u = (HG)^{-1}[F - Hf - \dot{r} - HG\xi + \dot{\nu}]$,

(iii) $\dot{\hat{y}} - \dot{\nu} = F(\hat{y},\nu)$.

In part (i) above we can see that for an actual system, in the limiting case, the noise r(t) is fully "repeated" in the output. As for the controller u, apart from the "basic" control law $u = (HG)^{-1}(F - Hf - HG\xi + \dot{\nu})$, we have an additional component due to the additive noise.

6. CONCLUDING REMARKS

In this paper, we discussed the robustness of high-gain output feedback control designs for nonlinear time-varying uncertain models to unmodeled high-frequency dynamics. Our approach followed on the papers by Vostrikov et al. [24], and Utkin [2].

Two different control strategies were analyzed. The first one was concerned with the output regulation. To facilitate the synthesis we utilized a diffeomorphic state variable transformation of the given model into the regular form. This regular form was found very useful in the design. However the problem of constructing a transformation which brings the system into this form requires further investigation.

The aim of the second output feedback control design was to ensure the tracking by the output of a given reference signal. The proposed control algorithm involved the output vector derivative. Following Vostrikov et al. [24], we suggested a sensor estimating the output derivative. One may argue that using differentiating filters is impractical. However one has to recognize that the essential information about a given process has significant spectral components only at low frequencies [13 p. 227]. Hence if we use an approximate differentiator which is sufficiently fast then the system will hardly feel the difference between the ideal and approximate differentiators. Thus, this approximate differentiator acts as an ideal one and its gain levels off or decreases at higher frequencies. In this paper we attempted to prove that the approximate differentiator is a viable tool in the synthesis of control algorithms.

REFERENCES

[1] V. I. Utkin, *Sliding modes and their applications in variable structure systems,* Mir Publishers, Moscow, 1978.

[2] V. Utkin, "Application of equivalent control method to systems with large feedback gain," IEEE Trans. Automat. Contr., Vol. AC-23, No. 3, pp. 484-486, 1978.

[3] R. M. Hirschorn, "Invertibility of multivariable nonlinear control systems," IEEE Trans. Automat. Contr., Vol. AC-24, No. 6, pp. 855-865, 1979.

[4] R. Marino, "High-gain feedback in nonlinear control systems," Int. J. Contr., Vol. 42, No. 6, pp. 1369-1385, 1985.

[5] L. R. Hunt, R. Su, and G. Meyer, "Global transformations of nonlinear systems," IEEE Trans. Automat. Contr., Vol. AC-28, No. 1, pp. 24-31, Jan. 1983.

[6] R. Su, G. Meyer and L. R. Hunt, "Robustness in nonlinear control," in "Differential geometric control theory" Ed. R. W. Brockett et. al., pp. 316-337, Birkhäuser, Boston, 1983.

[7] A. G. Bondarev, S. A. Bondarev, N. E. Kostyleva, and V. I. Utkin, "Sliding modes in systems with asymptotic state observers," Automation and Remote Control, Vol. 46, No. 6, Part 1, pp. 679-684, June 1985.

[8] B. L. Walcott and S. H. Żak, "Output feedback control of nonlinear dynamical systems," Presented at the Int. Symp. on MTNS, Phoenix, Arizona, June 15-19, 1987.

[9] G. Leitmann, E. P. Ryan, and A. Steinberg, "Feedback control of uncertain systems: Robustness with respect to neglected actuator and sensor dynamics," Int. J. Contr., Vol. 43, No. 4, pp. 1243-1256, 1986.

[10] R. A. DeCarlo, S. H. Żak and G. P. Matthews, "Variable structure control of nonlinear multivariable systems: A tutorial," to appear, Proceedings of the IEEE, March 1988.

[11] L. R. Hunt, M. Luksic and R. Su, "Exact linearizations of input-output systems," Int. J. Control, Vol. 43, No. 1, pp. 247-255, 1986.

[12] A. Steinberg and M. Corless, "Output feedback stabilization of uncertain dynamical systems," IEEE Trans. Automat. Contr., Vol. AC-23, No. 10, pp. 1025-1027, 1985.

[13] H. M. Power and R. J. Simpson, "Introduction to dynamics and control," McGraw-Hill, England, 1978.

[14] J. Descusse and C. H. Moog, "Decoupling with dynamic compensation for strong invertible affine non-linear systems," Int. J. Control, Vol. 42, No. 6, pp. 1387-1398, 1985.

[15] A. Isidori, "The matching of a prescribed linear input-output behavior in a nonlinear system," IEEE Trans. Automat. Contr., Vol. AC-30, No. 3, pp. 258-265, 1985.

[16] L. J. Ha and E. G. Gilbert, "A complete characterization of decoupling control laws for a general class of nonlinear systems," IEEE Trans. Automat. Contr., Vol. AC-31, No. 9, pp. 823-830, 1986.

[17] M. J. Corless and G. Leitmann, "Continuous state feedback guaranteeing uniform ultimate boundedness for uncertain dynamic systems," IEEE Trans. Automat. Contr., Vol. AC-26, No. 5, pp. 1139-1144, 1981.

[18] J. O'Reilly, "Robustness of linear feedback control systems to unmodelled high-frequency dynamics," Int. J. Control, Vol. 14, No. 4, pp. 1077-1088, 1986.

[19] W. T. Bauman and W. J. Rugh, "Feedback control of nonlinear systems by extended linearization," IEEE Trans. Automat. Contr., Vol. AC-31, No. 1, pp. 40-46, Jan. 1986.

[20] A. G. Lukýanov and V. I. Utkin, "Method of reducing equations for dynamic systems to a regular form," Automation and Remote Control, No. 4, pp. 5-13, April 1981.

[21] V. I. Utkin and A. S.Vostrikov, "Control systems with decoupling motions," Proc. 7th Triennial World Congress of the IFAC, Vol. 2, pp. 967-973, Helsinki, Finland, June 12-16, 1978.

[22] S. T. Glad, "Robustness of nonlinear state feedback - A survey," Automatica, Vol. 23, No. 4, pp. 425-435, 1987.

[23] S. T. Glad, "On the gain margin of nonlinear and optimal regulators," IEEE Trans. Automat. Contr., Vol. AC-29, No. 7, pp. 615-620, July 1984.

[24] A. S. Vostrikov, V. I. Utkin, and G. A. Frantsuzova, "Systems with state vector derivative in the control," Automation and Remote Contr., Vol. 43, No. 3, pp. 283-286, 1982.

[25] S. H. Żak, J. D. Brehove, and M. J. Corless, "Control of uncertain systems with unmodeled actuator and sensor dynamics and incomplete state information," to appear; IEEE Trans. Systems, Man, and Cybernetics.

[26] R. Sommer, "Control design for multivariable non-linear time-varying systems," Int. J. Control, Vol. 31, No. 5, pp. 883-891, 1980.

[27] D. Bestle and M. Zeitz, "Canonical form observer design for non-linear time-variable systems," Int. J. Control, Vol. 38, No. 2, pp. 419-431, 1983.

ROBUST MODEL TRACKING FOR A CLASS OF SINGULARLY PERTURBED NONLINEAR SYSTEMS VIA COMPOSITE CONTROL

F. Garofalo and L. Glielmo

Dipartimento di Informatica e Sistemistica
Universita' degli Studi di Napoli

1. Introduction

Typical problems encountered in the design of a control system are the presence of parameter uncertainties and the coexistence of slow and fast dynamics in the plant to be controlled. When the uncertainties are described assigning their range of variation and these variations belongs to appropriate subspaces, the so called deterministic control of uncertain systems (Leitmann, 1980; Corless-Leitmann, 1981) represents an useful tool for the design of controllers capable of guaranteeing certain performance no matter what the realization of the uncertainties is. The rigorous treatment of systems with two-time scale behavior can be done utilizing singular perturbation theory (Kokotovic *et al.*; 1986). The simultaneous use of these two methods for the control of uncertain two-time scale systems has recently received some attention (see Leitmann (this volume) and its references).
In this paper we use a composite control technique in conjunction with the robust design of controllers for uncertain systems to synthesize a nonlinear controller which forces a class of two-time scale nonlinear system to follow a two-time scale linear reference model. The controllers that are used in the two phases of the design are obtained via a constructive use of Lyapunov functions (Kalman-Bertram, 1960). The same Lyapunov functions are successively combined (as suggested by Saberi-Khalil, 1984) for obtaining the proof of ultimate boundedness of the model tracking error.

2. Problem Statement

We consider a two-time scale nonlinear system described by the following equations

$$\dot{x}(t) = A_{11}(x(t))x(t)+A_{12}(x(t))z(t)+B_1(x(t))u(t)+a_1(x(t)); \tag{2.1a}$$

$$\mu\dot{z}(t) = A_{21}(x(t))x(t)+A_{22}(x(t))z(t)+B_2(x(t))u(t)+a_2(x(t)); \tag{2.1b}$$

$$x(t_0) = x_0; \tag{2.1c}$$

$$z(t_0) = z_0; \tag{2.1d}$$

where $x(t)\in R^n$, $z(t)\in R^m$ represent the state of the system, $u(t)\in R^p$ is the control input, $a_1(x(t))$ and $a_2(x(t))$ are nonlinear vectors, $\mu\in(0,\infty)$ is the singular perturbation parameter, and $A_{ij}(\cdot)$ and $B_i(\cdot)$, $i=1,2$, $j=1,2$ are matrices of appropriate dimensions.

The reference model specifying the state behavior expected from the controlled plant is described by the linear, time-invariant system

$$\dot{\hat{x}}(t) = \hat{A}_{11}\hat{x}(t)+\hat{A}_{12}\hat{z}(t)+\hat{B}_1\hat{u}(t); \tag{2.2a}$$

$$\mu\dot{\hat{z}}(t) = \hat{A}_{21}\hat{x}(t)+\hat{A}_{22}\hat{z}(t)+\hat{B}_2\hat{u}(t); \tag{2.2b}$$

$$\hat{x}(t_0) = \hat{x}_0; \tag{2.2c}$$

$$\hat{z}(t_0) = \hat{z}_0; \tag{2.2d}$$

where $\hat{x}(t)\in R^n$ and $\hat{z}(t)\in R^m$ is the state and $\hat{u}(t)\in R^p$ is a reference signal.

The following assumptions define the class of nonlinear plants considered here.

Assumption 1. There exist full rank matrices B_i, $i=1,2$ such that, for all $x\in R^n$, the following decomposition holds:

$$B_i(x) = \bar{B}_i + \bar{B}_i E_i(x), \qquad i=1,2,$$

$$a_i(x) = \bar{B}_i d_i(x), \qquad i=1,2 ,$$

where $E_i(\cdot)$ (resp. $d_i(\cdot)$) is a matrix (resp. a vector) of appropriate dimensions, continuously differentiable with respect to x.

The relationship between the system (2.1) and the reference model represented by equations (2.2) is precised by the following assumptions.

Assumption 2. For all $x \in R^n$ the following equalities hold

$$A_{ij}(x) - \hat{A}_{ij} = \bar{B}_i C_{ij}(x) , \qquad i,j = 1,2$$

$$\hat{B}_i = \bar{B}_i \hat{C}_i , \qquad i = 1,2$$

where $C_{ij}(x)$ are continuously differentiable matrices.

Moreover, the singularly perturbed model is assumed in *standard form*, i.e.,

Assumption 3. Matrix $\hat{A}_{22}$ is full rank.

Defining

$$\hat{A}_0 \triangleq \hat{A}_{11} - \hat{A}_{12}\hat{A}_{22}^{-1}\hat{A}_{21} \qquad (2.3)$$

we hypothesize that

Assumption 4. The pairs $(\hat{A}_0,\bar{B}_1)$ and $(\hat{A}_{22},\bar{B}_2)$ are controllable.

Assumption 5. The matrices $A_{ij}(x)$, $B_i(x)$, $a_i(x)$, for $i=1,2$ and $j=1,2$, are norm bounded in R^n. In particular we define

$$h_{ij} = \sup_{x\in R^n} \|C_{ij}(x)\|,$$

$$\kappa_i = \sup_{x\in R^n} \|E_i(x)\|,$$

$$\nu_i = \sup_{x\in R^n} \|d_i(x)\|,$$

Moreover $\kappa_i < 1$, $i=1,2$.

Finally we make the following

Assumption 6. The input reference signals $\hat{u}(\cdot)$ are such that there exist finite positive constants

$$k_s = \sup_{t\in[t_0,\infty)} \|\hat{u}_s(t)\|,$$

$$k_f = \sup_{t\in[t_0,\infty)} \|\hat{u}_f(t)\|,$$

where $\hat{u}_s(t)$ and $\hat{u}_f(t)$ represent the slow and the fast time scale components of $\hat{u}(t)$ and $\hat{u}(t)\overset{\Delta}{=}\hat{u}_s(t)+\hat{u}_f(t)$. Corresponding to these signals, there exists a positive constant $\bar{\mu}$ such that, for $\mu\in(0,\bar{\mu})$ the state variables of the reference model are uniformly bounded by known constants:

$$k_{\hat{x}} = \sup_{\substack{t\in[t_0,\infty)\\ \mu\in(0,\bar{\mu})}} \|\hat{x}(t)\|,$$

$$k_{\hat{z}} = \sup_{\substack{t \in [t_0, \infty) \\ \mu \in (0, \bar{\mu})}} \|\hat{z}(t)\| .$$

Remark 1. Assumption 1 is the so called "matching assumption" and defines the manner in which the nonlinearities enter the plant. The equalities in Assumption 1 and 2 are the so called "model matching conditions" and determine the class of model that can be tracked by the nonlinear system under consideration.

Remark 2. System (2.1) belongs to the class of singularly perturbed nonlinear system with slow nonlinearities considered by Chow-Kokotovic (1981). Note, however, that for design purposes, it is not strictly necessary to know the nonlinearities affecting the system but only a nominal linear behavior and an evaluation of the maximum deviation from this behavior as precised in Assumption 5. The composite control design for the practical stabilization of a similar class of plants is also considered by Garofalo (to appear).

The objective of the control is to synthesize a feedback control function guaranteeing that the plant tracks the model to within a bounded neighbourhood of the zero state tracking error.[1]
The procedure we propose for the synthesis of the controller is based on the separate design of controllers guaranteeing tracking of the slow approximation and of the the boundary layer approximation of the reference model. On the basis of these control laws the composite control is constructed which guarantees tracking of the model for sufficiently small values of the singular perturbation parameter μ.

---

[1] *A formal definition can be found in Corless (1987) or in Appendix 1.*

3. Slow Time Scale Control

Following Kokotovic *et al.* (1986) the slow approximation of the behavior of the reference model is obtained considering $\mu=0$ in (2.2b) and substituting the resulting value for variable z in (2.2a), obtaining

$$\dot{\hat{x}}_s(t) = \hat{A}_0\hat{x}_s(t)+\hat{B}_0\hat{u}_s(t), \tag{3.1}$$

where

$$\hat{B}_0 \triangleq \hat{B}_1 - \hat{A}_{12}\hat{A}_{22}^{-1}\hat{B}_2, \tag{3.2}$$

and the subscript s stands for slow time-scale approximation.
In order to design the controller for tracking the slow component (3.1) of the reference model, we need an approximation of system (2.1) in the slow time scale. To this end, we assume that z variable has a nominal behavior z_n which is exactly the one that $\hat{z}$ variable takes in the reference model, that is

$$\mu\dot{z}_n(t) = \hat{A}_{21}x(t)+\hat{A}_{22}z_n(t)+\hat{B}_2\hat{u}(t). \tag{3.3}$$

Correspondingly, the approximate model of slow dynamics neglects the nominal fast transients, i.e.,[2]

$$\dot{x}_s = A_{11}(x_s)x_s+A_{12}(x_s)\bar{z}_n+B_1(x_s)u_s+a_1(x_s), \tag{3.4a}$$

$$0 = \hat{A}_{21}x_s+\hat{A}_{22}\bar{z}_n+\hat{B}_2\hat{u}_s, \tag{3.4b}$$

---

[2]*Sometimes, when no confusion is likely to occur, we delete the time argument of the functions.*

which gives

$$\bar{z}_n = \hat{\Gamma}(x_s,\hat{u}_s) \triangleq -\hat{A}_{22}^{-1}(\hat{A}_{21}x_s + \hat{B}_2\hat{u}_s) \tag{3.5a}$$

and

$$\dot{x}_s = A_0(x_s)x_s + B_1(x_s)u_s - A_{12}(x_s)\hat{A}_{22}^{-1}\hat{B}_2\hat{u}_s + a_1(x_s), \tag{3.5b}$$

with

$$A_0(x_s) \triangleq A_{11}(x_s) - A_{12}(x_s)\hat{A}_{22}^{-1}\hat{A}_{21}. \tag{3.6}$$

Define now the slow time scale tracking error as

$$\xi_s \triangleq x_s - \hat{x}_s\ . \tag{3.7}$$

On the basis of (3.1), (3.2), (3.5) and (3.6) the slow time scale tracking error dynamics can be written as

$$\dot{\xi}_s = F_s\xi_s + \bar{B}_1u_s + \bar{B}_1E_1(x_s)u_s + \bar{B}_1[H_1(x_s)+K_s]\xi_s +$$
$$+ \bar{B}_1[H_1(x_s)\hat{x}_s - H_2(x_s)\hat{u}_s + d_1(x_s)], \tag{3.8}$$

where $F_s \triangleq \hat{A}_0 - \bar{B}_1K_s$, $K_s \in R^{pxn}$ is a matrix which makes matrix F_s asymptotically stable with specified eigenvalues (which is always possible by virtue of Assumption 4), and

$$H_1(x_s) \triangleq [C_{11}(x_s) - C_{12}(x_s)\hat{A}_{22}^{-1}\hat{A}_{21}], \tag{3.9a}$$

$$H_2(x_s) \triangleq [\hat{C}_1 + C_{12}(x_s)\hat{A}_{22}^{-1}\hat{B}_2]. \tag{3.9b}$$

From the knowledge of matrices $C_{11}(x)$ and $C_{12}(x)$ (given in Assumption 2), and matrices $\hat{A}_{22}$, $\hat{A}_{21}$ and $\hat{B}_2$, we can compute the following constants

$$k_{\xi_s} \overset{\Delta}{=} \sup_{x \in R^n} \|H_1(x)+K_s\|, \tag{3.10a}$$[3]

$$k_{d_1} \overset{\Delta}{=} \sup_{\substack{x \in R^n \\ t \in [t_0,\infty) \\ \mu \in (0,\bar{\mu})}} \|H\ (x)\hat{x}_s - H_2(x)\hat{u}_s + d_1(x)\|. \tag{3.10b}$$

Consider now the nonlinear feedback control law (Ambrosino-Celentano-Garofalo, 1985; Garofalo-Glielmo, to appear)

$$p_s(\xi_s) \overset{\Delta}{=} -\gamma_s \bar{B}_1^T P_s \xi_s, \tag{3.11a}$$

where P_s is the solution of the Lyapunov equation

$$F_s^T P_s + P_s F_s = -Q_s, \qquad Q_s \text{ positive definite}, \tag{3.11b}$$

and

$$\gamma_s = \gamma_s(\|\xi_s\|) \overset{\Delta}{=} \frac{\gamma_{s1} + \gamma_{s2}\|\xi_s\|}{\|B_1^T P_s \xi_s\| + \delta_s}, \quad \delta_s > 0. \tag{3.11c}$$

This feedback control has the tracking capabilities described in the next theorem.

Theorem 1. Consider the slow approximation (3.4) of system (2.1) subject to the feedback control law in (3.11). If constants γ_{si}, $i=1,2$, in (3.11c) are chosen so as to satisfy

$$\gamma_{s1} \geq \frac{k_{d_1}}{1-\kappa_1}, \tag{3.12a}$$

---

[3] *Notice that the suprema can always be replaced by upper bounds.*

$$\gamma_{s2} \geq \frac{k_{\xi_s}}{1-\kappa_1}\,, \tag{3.12b}$$

then system (3.5b) tracks the slow approximation (3.1) of the reference model (2.2) to within a spherical neighbourhood of $\xi_s=0$ whose radius can be made arbitrarily small by a suitable selections of constants γ_{si}, i=1,2, and/or of constant δ_s in (3.11c).

Proof. The proof of the theorem can be found in Appendix 2.

4. Fast Time Scale Control

The boundary layer approximation of the reference model (2.2) is given by (Kokotovic *et al.*, 1986)

$$\frac{d\hat{z}_f}{d\tau} = \hat{A}_{22}\hat{z}_f + \hat{B}_2\hat{u}_f, \tag{4.1}$$

where $\tau=t/\mu$, $\hat{u}_f$ represents the fast component of the reference signal, and

$$\hat{z}_f \triangleq \hat{z} - \hat{\Gamma}(\hat{x},\hat{u}_s) = \hat{z} + \hat{A}_{22}^{-1}(\hat{A}_{21}\hat{x}+\hat{B}_2\hat{u}_s). \tag{4.2}$$

The fast time scale approximation of system (2.1) is obtained substituting the slow control expression (3.11a) in equation (2.1b) and approximating variable $x_s(t)$ by $x(t)$ and $\hat{x}_s(t)$ by $\hat{x}(t)$. So doing we obtain

$$\mu\dot{z} = A_{21}(x)x-\gamma_s B_2(x)\bar{B}_1^T P_s\xi + A_{22}(x)z + B_2(x)u_f+a_2(x), \tag{4.3}$$

where u_f is the fast component of the control law and $\xi \triangleq x-\hat{x}$.

Defining

$$z_f \triangleq z - \hat{\Gamma}(x,\hat{u}_s) = z + \hat{A}_{22}^{-1}(\hat{A}_{21}x+\hat{B}_2\hat{u}_s), \tag{4.4}$$

the boundary layer model of the system can be written as

$$\frac{dz_f}{d\tau} = A_{22}(x)z_f+B_2(x)u_f-\gamma_s B_2(x)\bar{B}_1^T P_s \xi \\ +\bar{B}_2 G_1(x)x+\bar{B}_2 G_2(x)\hat{u}_s+\bar{B}_2 d_2(x), \tag{4.5}$$

with

$$G_1(x) \triangleq [C_{21}(x) - C_{22}(x)\hat{A}_{22}^{-1}\hat{A}_{21}], \tag{4.6a}$$

$$G_2(x) \triangleq -[\hat{C}_2+C_{22}(x)\hat{A}_{22}^{-1}\hat{B}_2]. \tag{4.6b}$$

The fast time scale tracking error can be defined as

$$\zeta_f \triangleq z_f-\hat{z}_f , \tag{4.7}$$

and, on the basis of (4.1) and (4.5), its dynamics can be written as

$$\frac{d\zeta_f}{d\tau} = F_f\zeta_f + \bar{B}_2u_f + \bar{B}_2E_2(x)u_f + \bar{B}_2[C_{22}(x) + K_f]\zeta_f+ \\ + \bar{B}_2[G_1(x) - \gamma_s(I_p + E_2(x))\bar{B}_1^T P_s]\xi \\ + \bar{B}_2[G_1(x)\hat{x} + C_{22}(x)\hat{z}_f + G_2(x)\hat{u}_s - \hat{C}_2\hat{u}_f + d_2(x)], \tag{4.8}$$

where $F_f \triangleq \hat{A}_{22}-\bar{B}_2K_f$ and $K_f \in R^{p\times m}$ is a matrix which makes matrix F_f asymptotically stable with specified eigenvalues (see Assumption 4).

On the basis of Assumptions 1, 2, 5 and 6, we can evaluate the finite constants

$$k_{d_2} \triangleq \sup_{\substack{x\in R^n \\ t\in[t_0,\infty) \\ \mu\in(0,\bar{\mu})}} \|G_1(x)\hat{x} + C_{22}(x)\hat{z}_f + G_2(x)\hat{u}_s - \hat{C}_2\hat{u}_f + d_2(x)\|, \tag{4.9a}$$

$$k_\zeta \triangleq \sup_{x\in R^n} \|C_{22}(x) + K_f\|, \tag{4.9b}$$

$$k_\xi \triangleq \sup_{x\in R^n} \|G_1(x) - \gamma_s(I_p + E_2(x))\bar{B}_1^T P_s\|, \tag{4.9c}$$

In the fast time scale the variables x and ξ can be considered constants, and the fast control law we propose for making the boundary layer system track the boundary layer reference model has the form

$$p_f(\zeta_f) \triangleq -\gamma_f \bar{B}_2^T P_f \zeta_f, \tag{4.10a}$$

where P_f is the solution of the Lyapunov equation

$$F_f^T P_f + P_f F_f = -Q_f, \qquad Q_f \text{ positive definite}, \tag{4.10b}$$

and

$$\gamma_f = \gamma_f(\|\zeta_f\|,\|\xi\|) \triangleq \frac{\gamma_{f1}+\gamma_{f2}\|\zeta_f\|+\gamma_{f3}\|\xi\|}{\|\bar{B}_2^T P_f \zeta_f\| + \delta_f}, \quad \delta_f>0. \tag{4.10c}$$

We can state the following

Theorem 2. Consider the boundary layer approximation (4.5) of system (2.1) subject to the feedback control law (4.10). If constants γ_{fi}, i=1,...,3 are chosen so as to satisfy

$$\gamma_{f1} \geq \frac{k_{d_2}}{1-\kappa_1}, \tag{4.11a}$$

$$\gamma_{f2} \geq \frac{k_{\zeta}}{1-\kappa_1}, \tag{4.11b}$$

$$\gamma_{f3} \geq \frac{k_{\xi}}{1-\kappa_1}, \tag{4.11c}$$

then system (4.5) tracks the boundary layer reference model (4.1) to within a spherical neigbourhood of $\zeta_f=0$ whose radius can be made arbitrarily small by a suitable selection of constants γ_{fi}, $i=1,\dots,3$ and/or constant δ_f in (4.10c).

Proof. The proof can be found in Appendix 2.

5. Guaranteed Performance of the Composite Control

The composite control is obtained as the sum of the slow and the fast control law with variable ζ_f replaced by $\zeta-[\hat{\Gamma}(x,\hat{u})+\hat{\Gamma}(\hat{x},\hat{u})] = \zeta + \hat{A}_{22}^{-1}\hat{A}_{21}\xi$, and ξ_s by its approximation ξ, obtaining

$$u_c = -\gamma_s\bar{B}_1^T P_s \xi - \gamma_f \bar{B}_2^T P_f \zeta - \gamma_f \bar{B}_2^T P_f \hat{A}_{22}^{-1}\hat{A}_{21}\xi, \tag{5.1}$$

where $\zeta \triangleq (z-\hat{z})$.

For this control law we can establish the following theorem.

Theorem 3. Consider system (2.1) subject to the control law (5.1). The closed loop system tracks the reference model to within a spherical neighbourhood of the zero state tracking error, if the following conditions are satisfied.

i) The constant γ_{s1} satisfies the inequality:

$$\gamma_{s1} \geq \frac{k_{d_1}'}{1 - \kappa_1}, \tag{5.2}$$

with

$$k_{d_1}' \triangleq \sup_{\substack{x \in R^n \\ t \in [t_0, \infty) \\ \mu \in (0, \bar{\mu})}} \| H(x)\hat{x} - H_2(x)\hat{u} + d_1(x) \|, \tag{5.3}$$

and the constant γ_{s2} satisfies the inequality (3.12b);

ii) the constants γ_{fj}, j=1,2,3 in the control law (5.1) are chosen so as to satisfy inequalities (4.11);

iii) constant γ_{f3}, besides satisfying (4.11c), satisfies

$$\gamma_{f3} < \frac{\lambda_{min}(Q_s)}{2\|P_s\| \sup_{x \in R^n} \|B_1(x)\|} \tag{5.4}$$

iv) the singular perturbation parameter is such that $0<\mu<\mu^*$ where μ^* is a constant whose value can be *a priori* computed.

Proof. The proof of Theorem 3 and the expression for the upper bound of paramerer μ are given in Appendix 3.

6. Conclusions

The robust model tracking control presented here is designed using the approach of deterministic control of uncertain systems, together with the composite control technique developed for singularly perturbed systems. This enables the designer to guarantee the model following within a spherical neighbourhood of the zero error, in the presence of "slow" nonlinearities. It must be pointed out that this technique does not require the knowledge of the form of the nonlinearities, but just the possible range of their variations.

Appendix 1

Some definitions and a useful lemma.

Consider the equation of a model tracking error dynamics in the form

$$\dot{\epsilon} = \varphi(\epsilon, t) \ , \ \epsilon(t_0) = \epsilon_0, \qquad \text{(A1.1)}$$

where $t \in R$, $\epsilon \in R^p$, and $\varphi: R^p x R \to R^p$ We say that the system tracks the reference model to within a spherical neighbourhood of radius R of $\epsilon = 0$ (indicated with B(R)) iff the following properties are satisfied:

i) *Existence of the solution*. Given any $(\epsilon_0, t_0) \in R^p x R$ there exists a solution $\epsilon(\cdot):[t_0, t_1) \to R^p$, $t_1 > t_0$ of (A1.1).

ii) *Indefinite extension of solution*. Every solution $\epsilon(\cdot):[t_0, t_1) \to R^p$ of (A1.1) has an extension over $[t_0, \infty)$.

iii) *Global uniform boundedness*. Given any bound $r \in R_+$, there exists a bound $d(r) \in R_+$ such that if $\epsilon(\cdot):[t_0, t_1) \to R^p$ is a solution of (A1.1) with $\|\epsilon_0\| \le r$, then $\|\epsilon(t)\| \le d(r)$ for all $t \in [t_0, t_1)$.

iv) *Local boundedness within* B(R). There exists a spherical neighbourhood $B(R_0)$ of $\epsilon = 0$ such that if $\epsilon(\cdot):[t_0, t_1) \to R^p$ is a solution of (A1.1) with $\epsilon_0 \in B(R_0)$ then $\epsilon(t) \in B(R)$ for all $t \in [t_0, t_1)$.

v) *Global uniform ultimate boundedness within* B(R). Given any bound $r \in R_+$ there exists $T(r) \in R_+$ such that if $\epsilon(\cdot):[t_0, t_1) \to R^p$ is a solution of (A1.1) with $\|\epsilon_0\| \le r$, then $\epsilon(t) \in B(R)$ for all $t \ge t_0 + T(r)$.

The listed properties of the solution $\epsilon(\cdot):[t_0, t_1) \to R^p$ can be stated with the aid of the following lemma (for the proof see Corless-Leitmann, 1981).

Lemma. Given system (A1.1) suppose $\varphi(0,t)=0$ for all $t\in R$. If there exists a C^1 function L defined on $\|\epsilon\|\geq s$ and $t\in R$, and if there exist class KR functions χ_1 and χ_2 and a class K function χ_3 such that

$$\chi_1(\|\epsilon\|)\leq L(\epsilon,t)\leq\chi_2(\|\epsilon\|), \tag{A1.2a}$$

$$\frac{\partial}{\partial t}L(\epsilon,t)+\nabla_\epsilon^T L(\epsilon,t) \leq -\chi_3(\|\epsilon\|), \tag{A1.2b}$$

then for all $\|\epsilon\|\geq s$ and $t\in R$ the system tracks the reference model to within any spherical neighbourhood $B(\bar{R})$ of $\epsilon=0$ with $\bar{R}>\chi_1^{-1}\circ\chi_2(s)$.

Appendix 2

Proofs of Theorems 1 and 2.

Consider as Lyapunov function candidate for system (3.8) with the feedback control (3.11) the following

$$V(\xi_s) \triangleq \xi_s^T P_s \xi_s. \tag{A2.1}$$

Evaluating the derivative along the solutions of the closed loop system by virtue of (3.9), (3.10), (3.11), (3.12) and Assumption 5, we have

$$\begin{aligned}(1/2)\dot{V}(\xi_s) = &-(1/2)\xi_s^T Q_s \xi_s - \gamma_s \xi_s^T P_s \bar{B}_1 \bar{B}_1^T P_s \xi_s \\ &- \gamma_s \xi_s^T P_s \bar{B}_1 E_1(x_s) \bar{B}_1^T P_s \xi_s + \xi_s^T P_s \bar{B}_1 H_1(x_s) x_s \\ &- \xi_s^T P_s \bar{B}_1 H_2(x_s)\hat{u}_s + \xi_s^T P_s \bar{B}_1 d_1(x_s) + \xi_s^T P_s \bar{B}_1 K_s \xi\end{aligned}$$

$$\begin{aligned}\leq -(1/2)\xi_s^T Q_s \xi_s - \gamma_s \|\bar{B}_1^T P_s \xi_s\|^2 (1-\kappa_1) + \|\bar{B}_1^T P_s \xi_s\| \|H_1(x_s)+K_s\| \|\xi_s\| \\ + \|\bar{B}_1^T P_s \xi_s\| \|H_1(x_s)\hat{x}_s - H_2(x_s)\hat{u}_s + d_1(x_s)\|\end{aligned}$$

$$\le -(1/2)\xi_s^T Q_s \xi_s - (\gamma_{s1}+\gamma_{s2}\|\xi_s\|)(\|\bar{B}_1^T P_s \xi_s\| - \delta_s)(1-\kappa_s)$$
$$+ \|\bar{B}_1^T P_s \xi_s\| \|H_1(x_s)+K_s\| \|\xi_s\|$$
$$+ \|\bar{B}_1^T P_s \xi_s\| \|H_1(x_s)\hat{x}_s - H_2(x_s)\hat{u}_s + d_1(x_s)\|$$

$$\le -(1/2)\xi_s^T Q_s \xi_s + k_{d_1}\delta_s + k_{\xi_s}\delta_s\|\xi_s\|$$
$$\le (1/2)[-v_1\|\xi_s\|^2 + v_2\|\xi_s\| + v_3] \qquad (A2.2)$$

where $v_1 \triangleq \lambda_{min}(Q_s)$, $v_2 \triangleq 2k_{\xi_s}\delta_s$, and $v_3 \triangleq 2k_{d_1}\delta_s$.

At this stage the application of the lemma reported in Appendix 1 proves the statement of the Theorem 1.

The proof of Theorem 2 proceeds exactly in the same way. We define as Lyapunov candidate for system (4.5) subject to the feedback control (4.10)

$$W(\zeta_f) \triangleq \zeta_f^T P_f \zeta_f \ . \qquad (A2.3)$$

The derivative along the solutions of the closed loop system, considering x constant in the fast time scale, can be proved to satisfy the following inequality

$$\dot{W}(\zeta_f) \le -w_1\|\zeta_f\|^2 + w_2\|\zeta_f\| + w_3 \qquad (A2.4)$$

with $w_1 \triangleq \lambda_{min}(Q_f)$, $w_2 \triangleq 2\delta_f k_\xi$, and $w_3 \triangleq w_3' + w_3''\|\xi\| \triangleq 2\delta_f k_{d_2} + 2\delta_f k_\xi \|\xi\|$.

Appendix 3

Proof of Theorem 3.

The proof of the theorem is based on the combined use of two Lyapunov functions, one for each component of the model reference tracking error.

For the first component we can write

$$\dot{\xi} = A_{11}(x)x+A_{12}(x)z-\gamma_s B_1(x)\bar{B}_1^T P_s\xi-\gamma_f B_1(x)\bar{B}_2^T P_f\zeta$$
$$+\ \gamma_f B_1(x)\bar{B}_2^T P_f\hat{A}_{22}^{-1}\hat{A}_{21}\xi+a_1(x)-\hat{A}_{11}\hat{x}-\hat{A}_{12}\hat{z}-\hat{B}_1\hat{u}$$
$$= \{F_s\xi-\gamma_s\bar{B}_1\bar{B}_1^T P_s\xi-\gamma_s\bar{B}_1 E_1(x)\bar{B}_1^T P_s\xi+\bar{B}_1[H_1(x)+K_s]\xi+$$
$$+\bar{B}_1[H_1(x)\hat{x}-H_2(x)\hat{u}+d_1(x)]\}\ -\gamma_f B_1(x)\bar{B}_2^T P_f[\zeta+\hat{A}_{22}^{-1}\hat{A}_{21}\xi]$$
$$+A_{12}(x)[\zeta+\hat{A}_{22}^{-1}\hat{A}_{21}\xi]+\bar{B}_1 C_{12}(x)[\hat{z}-\hat{\Gamma}(\hat{x},\hat{u})] \tag{A3.1}$$

The terms within braces are exactly the same as in the slow model (3.8), taking apart the substitution of $\hat{x}_s$ and $\hat{u}_s$ with $\hat{x}$ and $\hat{u}$. On the basis of Assumptions 5 and 6, and recalling (4.10c), it is possible to find constants α_i, $i=1,3$ and a_i, $i=1,3$ such that

$$\|-\gamma_f B_1(x)\bar{B}_2^T P_f[\zeta+\hat{A}_{22}^{-1}\hat{A}_{21}\xi]$$
$$+A_{12}(x)[\zeta+\hat{A}_{22}^{-1}\hat{A}_{21}\xi]-\bar{B}_1 C_{12}(x)[\hat{z}-\hat{\Gamma}(\hat{x},\hat{u})]\|$$
$$\le \alpha_1\|\xi\|+\alpha_2\|\zeta+\hat{A}_{22}^{-1}\hat{A}_{21}\xi\|+\alpha_3, \tag{A3.2a}$$

$$\|\dot{\xi}\| = \|A_{11}(x)x+A_{12}(x)z-\gamma_s B_1(x)\bar{B}_1^T P_s\xi-\gamma_f B_1(x)\bar{B}_2^T P_f\zeta$$
$$-\ \gamma_f B_1(x)\bar{B}_2^T P_f\hat{A}_{22}^{-1}\hat{A}_{21}\xi+a_1(x)-\hat{A}_{11}\hat{x}-\hat{A}_{12}\hat{z}-\hat{B}_1\hat{u}\|$$
$$\le a_1'\|\xi\|+a_2'\|\zeta+\hat{A}_{22}^{-1}\hat{A}_{21}\xi\|+a_3'. \tag{A3.2b}$$

For the second component of the model tracking error we simply rewrite equation (4.8) as

$$
\begin{aligned}
\mu\dot{\zeta} = & [F_f-\gamma_{f2}B_2(x)\bar{B}_2^T P_f](\zeta+\hat{A}_{22}^{-1}\hat{A}_{21}\xi) \\
& + \bar{B}_2[C_{22}(x)+K_f](\zeta+\hat{A}_{22}^{-1}\hat{A}_{21}\xi) \\
& + \bar{B}_2[G_1(x)-\gamma_s(I_p+E_2(x))\bar{B}_1^T P_s]\xi \\
& + \bar{B}_2[G_1(x)\hat{x}+C_{22}(x)\hat{z}_f+G_2(x)\hat{u}_s-\hat{C}_2\hat{u}_f+d_2(x)].
\end{aligned}
\tag{A3.3}
$$

Consider the function

$$
W(\xi,\zeta) \triangleq 1/2\ (\zeta+\hat{A}_{22}^{-1}\hat{A}_{21}\xi)^T P_f(\zeta+\hat{A}_{22}^{-1}\hat{A}_{21}\xi), \tag{A3.4}
$$

and evaluate the derivative along the solutions of the closed loop tracking error system (A3.1), (A3.3). One obtains

$$
\begin{aligned}
\dot{W}(\xi,\zeta) = & -(\zeta+\hat{A}_{22}^{-1}\hat{A}_{21}\xi)^T P_f\hat{A}_{22}^{-1}\hat{A}_{21}\dot{\xi} + (\zeta+\hat{A}_{22}^{-1}\hat{A}_{21}\xi)^T P_f\dot{\zeta} \\
\le & \|P_f\hat{A}_{22}^{-1}\hat{A}_{21}\|\|\zeta+\hat{A}_{22}^{-1}\hat{A}_{21}\xi\|[a_1'\|\xi\|+a_2'\|\zeta+\hat{A}_{22}^{-1}\hat{A}_{21}\xi\|+a_3'] \\
& + \frac{1}{\mu}[-w_1\|\zeta+\hat{A}_{22}^{-1}\hat{A}_{21}\xi\|^2 + w_2\|\zeta+\hat{A}_{22}^{-1}\hat{A}_{21}\xi\|] \\
\triangleq & -(\frac{w_1}{\mu} - a_2)\|\zeta+\hat{A}_{22}^{-1}\hat{A}_{21}\xi\|^2 + a_1\|\zeta+\hat{A}_{22}^{-1}\hat{A}_{21}\xi\|\|\xi\| \\
& +(a_3+\frac{w_2}{\mu})\|\zeta+\hat{A}_{22}^{-1}\hat{A}_{21}\xi\| + \frac{w_3'}{\mu} + \frac{w_3''}{\mu}\|\xi\|.
\end{aligned}
\tag{A3.5}
$$

Consider now the function

$$
V(\xi) \triangleq 1/2\ \xi^T P_s\xi, \tag{A3.6}
$$

and evaluate the time derivative along the solutions of the closed loop tracking error system (A3.1), (A3.3). In view of (A3.2a), and conditions (5.2) and (5.3), we have

$$\dot{V}(\xi) = -v_1\|\xi\|^2+v_2\|\xi\|+v_3+2\|P_s\|\|\xi\|[\alpha_1\|\xi\|+\alpha_2\|\zeta+\hat{A}_{22}^{-1}\hat{A}_{21}\xi\|+\alpha_3]$$

$$\triangleq -(v_1-2\alpha_1\|P_s\|)\|\xi\|^2 + b_1\|\xi\| + b_2\|\xi\|\|\zeta+\hat{A}_{22}^{-1}\hat{A}_{21}\xi\| + v_3. \tag{A3.7}$$

We can choose as Lyapunov candidate for the closed loop tracking error system (A3.1), (A3.3) the following

$$L(\xi,\zeta) \triangleq \begin{bmatrix} \xi \\ \zeta+\hat{A}_{22}^{-1}\hat{A}_{21}\xi \end{bmatrix}^T P(c) \begin{bmatrix} \xi \\ \zeta+\hat{A}_{22}^{-1}\hat{A}_{21}\xi \end{bmatrix}, \tag{A3.8}$$

where

$$P(c) \triangleq \begin{bmatrix} (1-c)P_s & 0 \\ 0 & cP_f \end{bmatrix}, \quad 0\le c\le 1. \tag{A3.9}$$

In view of (A3.5) and (A3.7) the time derivative of (A3.8) along the solutions of the closed loop tracking error system satisfies

$$\dot{L}(\xi,\zeta) \le - \begin{bmatrix} \|\xi\| \\ \|\zeta+\hat{A}_{22}^{-1}\hat{A}_{21}\xi\| \end{bmatrix}^T M(c) \begin{bmatrix} \|\xi\| \\ \|\zeta+\hat{A}_{22}^{-1}\hat{A}_{21}\xi\| \end{bmatrix}$$

$$+ m^T(c) \begin{bmatrix} \|\xi\| \\ \|\zeta+\hat{A}_{22}^{-1}\hat{A}_{21}\xi\| \end{bmatrix} + \bar{m}, \tag{A3.10}$$

where

$$M(c) \triangleq \begin{bmatrix} (1-c)(v_1-2\alpha_1\|P_s\|) & -1/2(ca_1+(1-c)b_2 \\ -1/2(ca_1+ (1-c)b_2 & c(\frac{w}{\mu}1- a_2) \end{bmatrix}, \tag{A3.11a}$$

$$m(c) \triangleq [(1-c)b_1+c\frac{w''_3}{\mu} \qquad c(a_3+\frac{w_2}{\mu})], \tag{A3.11b}$$

and

$$\bar{m} \triangleq v_3(1-c) + c\frac{w'_3}{\mu} . \tag{A3.11c}$$

Provided that $\alpha_1 < \frac{v_1}{2\|P_s\|}$ (which is guaranteed by condition (5.4)), the upper bound μ_p of parameter μ which guarantees the definite positivity of matrix M(c) is given by (see Saberi-Khalil, 1984)

$$\mu_p = \frac{(v_1-2\alpha_1\|P_s\|)w_1}{(v_1-2\alpha_1\|P_s\|)a_2 + a_1 b_2} \tag{A3.12}$$

Chosen $\mu^* \triangleq \min(\bar{\mu},\mu_p)$, for each $0<\mu<\mu^*$ the application of the lemma contained in Appendix 1 completes the proof of the Theorem.

References

G. AMBROSINO, G. CELENTANO, F. GAROFALO

(1985) Robust model tracking control for a class of nonlinear plants, IEEE Trans. Automat. Contr. AC-30, pp. 275-279.

J.H. CHOW, P.V. KOKOTOVIC

(1981) A two-stage Lyapunov-Bellman feedback design of a class of nonlinear systems, IEEE Trans on Automat. Control AC-26, pp. 656-663.

M. CORLESS

(1987) Robustness of a class of feedback-controlled uncertain nonlinear systems in the presence of singular perturbations, Proc. of the ACC Conf., Minneapolis, MI.

M. CORLESS, G. LEITMANN

(1981) Continuous state feedback guaranteeing uniform ultimate boundedness for uncertain dynamic systems, IEEE Trans. Automat. Contr. AC-26, pp.1139-1144.

F. GAROFALO

(to appear) Composite control of a singularly perturbed uncertain system with slow nonlinearities, Int. J. of Control.

F. GAROFALO, L. GLIELMO

(to appear) Nonlinear continuous feedback control for robust tracking, in "Deterministic Control of Uncertain Systems", IEE Press: London.

R.E. KALMAN, J.E. BERTRAM

(1960) Control system analysis and design via the second method of Lyapunov: Continuous-time systems , ASME Journal of Basic Engineering, pp. 371-393.

P.V. KOKOTOVIC, H.K. KHALIL, J. O'REILLY

(1986) "Singular Perturbation Methods in Control: Analysis and Design", Academic Press: London.

G. LEITMANN

(1980) Deterministic control of uncertain systems, Astronautica Acta 7, pp. 1457-1461.

(this volume) Controlling singularly perturbed uncertain dynamical systems.

A. SABERI, H. KHALIL

(1984) Quadratic-type Lyapunov functions for singularly perturbed systems, IEEE Trans. Automat. Contr. AC-29, pp. 542-550.

CONTROL OF UNCERTAIN DYNAMICAL SYSTEMS : SIMULTANEOUS STABILIZATION PROBLEMS

Bijoy K. GHOSH
Washington University
Saint-Louis, Missouri 63130, U.S.A.

In the last decade, significant progress have been witnessed in the design of a robust compensator for a family of multi input multi output systems. The main objective is to construct a dynamic compensator which simultaneously stabilizes a family of plants and satisfies various other design restrictions. The motivation is to extend various classically well-known compensator design methods for a single plant to a family of plants. Such a family of plants may occur as a result of parameter uncertainty or parameter variation in the plants and the goal is to construct a compensator which is insensitive to these parametric changes.

To begin with, we consider the "simultaneous stabilization problem" described as follows:

Given a r tuple $G_1,\dots,G_r$ of pxm proper transfer functions, does there exist a compensator $K(s)$ such that the closed loop systems $G_1[I + KG_1]^{-1},\dots, G_r[I + KG_r]^{-1}$ are internally stable?

This problem arises in reliable system design where $G_2,\dots, G_r$ represent a plant G_1 operating in various modes of failure and $K(s)$ is a non-switching stabilizing compensator. It also arises in the stability analysis and design of a plant which can be switched into various operating modes. It has been shown in [1] that

The integer max(m,p) is the critical number of plants below which the simultaneous stabilization problem is solvable almost always i.e. generically (in a suitable topology) by a compensator of McMillan degree q_0 where q_0 is the smallest integer satisfying

$$q_0[\max(m,p) + 1-r] \geq \sum_{i=1}^{r} n_i - \max(m,p) \qquad (1)$$

In the above formula, n_i is the McMillan degree of the plant G_i for $i=1,\ldots,r$ respectively. In fact, if $\min(m,p) = 1$ then the formula (1) also computes the minimum order of the generically stabilizing compensator. It may be remarked that the minimum order compensator problem is a classically unsolved problem and in [1] the problem is solved for the special case $\min(m,p) = 1$.

However, beyond saying that the simultaneous stabilization problem is solvable for certain classes, it is of great interest to parameterize all those cases where the problem is indeed solvable. Moreover, for ease of computation, such a parameterization has to be explicit. This question is parameterizing the set of r tuples of plants $(G_1',\ldots,G_r')$ is addressed in [2] and one of his main results is a considerable conceptual breakthrough, since to check simultaneous stabilizability using this result one only needs to know which path component $(G_1,\ldots, G_r)$ lies in; i.e. the problem is reduced to the problem of analyzing big pieces of the space of r tuples of systems rather than individual r-tuples. Similar results on simultaneous stabilization and pole assignment for a parameterized family of plants by a parameterized family of compensators is also obtained by Dr. Ghosh and is reported in [2]. To my knowledge, use of semialgebraic geometric methods for the purpose of parameterizing stabilizable or unstabilizable path components has been done for the first time in [2].

Considering more than $\max(m,p)$ plants for the purpose of simultaneous stabilization (is quite a reasonable objective in robust system design), but unfortunately in particular in [3] it is shown that, **"Pairs of simultaneously stabilizable single input single output plants of bounded McMillan degree may not have simultaneously stabilizing compensators of apriori bounded McMillan degree."**

It is shown by Dr. Ghosh in [3] that there exists a sequence of pairs of simultaneously stabilizable plants of degree one for which the minimum degree of the stabilizing compensator is arbitrarily large. A consequence of the above proposition is that a simultaneously stabilizing compensator cannot be constructed by solving a set of simultaneous equations or inequalities in the coefficients of a parameterized family of compensators of a given McMillan degree. Stated differently, if $r > \max(m,p)$, the classically known algebraic and semialgebraic geometric methods are inapplicable since the compensator space is not finite dimensional and in particular, any numerical computation of the associated compensator needs to use a more appropriate transcendental method proposed by Dr. Ghosh in [4]. Also in [4] a new 'partial pole placement' problem is proposed which arises from a more practical design requirement to place an arbitrary number of self conjugate poles in the closed loop while restricting the remaining poles in the region of stability. The following result is shown:

The problem of simultaneously stabilizing three single input single output plants chosen generically is equivalent to the problem of partially pole placing one single input single output plant by a stable minimum phase compensator.

Use and application of a stable, minimum phase compensator is introduced in [4] for the first time. Furthermorea folklore example

$$\frac{s-7}{s-4.6}, \quad \frac{s-2}{2s-2.6}, \quad \frac{s-6}{4.8s-24.6}$$

of a triplet of simultaneously unstabilizable plants that are stabilizable in pairs is constructed by Dr. Ghosh [4]. These results to multi input multi output problems are further generalized in [4] to show that

"If $r \min(m,p) \leq m+p$, the simultaneous partial pole assignment problem may be analyzed via interpolation methods and one obtains a semialgebraic parameterization of the partially pole assignable r-tuples of plants. If $r \min(m,p) > m+p$, the simultaneous partial pole assignment problem is to be analyzed via transcendental methods introduced in [4]."

The above result, therefore, characterizes the "degree of difficulty" and in particular asserts the existence of certain cases (say for example $m=p$, $r \geq 3$) when interpolation methods are inapplicable in the simultaneous stabilization problem.

We have seen so far that transcendental methods are useful when the degree of the compensators under consideration is not apriori bounded. Frequently in system identification and control, it is of interest to study a family of plants for which the McMillan degree is not fixed. In particular the degree may degenerate to a lower value. Thus rather than fixing the McMillan degree of a plant, it is useful to parameterize plants of McMillan degree $\leq n$ for some n. We ,therefore, pose the following question --

"Parameterize the set Ω_n of plants of degree $\leq n$ (possibly as a semialgebraic subset of an algebraic set) such that every p in Ω_n has an open neighborhood $N(p)$ of p in Ω_n such that $N(p)$ is simultaneously stabilizable by a compensator of degree $\leq q$ for some q."

Note that this question poses robust stabilization as a parameterization problem. In [5] an explicit parameterization of Ω_n is obtained as a subset of $\mathbb{RP}^{2n+1}$ for the single input single output systems and in particular we show that --

"Assume $m=p=1$, then Ω_n is a semialgebraic, open, connected and dense subset of $\mathbb{RP}^{2n+1}$."

More surprisingly we show that

"Ω_n is a trivial vector bundle over a circle. In particular Ω_n is diffeomorphic to $S^1 \times \mathbb{R}^{2n}$."

The space Ω_n has been parameterized for a multi input multi output plant in [6] as a vector bundle over a Grassmanian, a well known object in algebraic geometry. We argue that Ω_n and not rat n (the space of strictly proper single input single output transfer functions of a given degree) or $\sum_{m,p}^{n}$ (the space of pxm transfer functions of degree n) is a more natural space for system identification and control. Various properties of this space has been reported in [8].

The geometry of Ω_n is useful in the study of a structured family of plants wherein the degree is apriori bounded. In practice, however, one is also interested in the study of a family of plants possibly with some unmodelled dynamics. For example, under the presence of a high frequency "parasitics" it is unreasonable to assume that the McMillan degree of a family of plants is bounded by n. In [6] we, therefore, construct the space Ω_∞ as a direct limit of the spaces $\Omega_1 \subset \Omega_2 \subset \ldots$ where Ω_∞ is a subspace of $\mathbb{R}^\infty$. Of course two points in Ω_∞ can model the same dynamical system and one therefore considers the quotient space $\tilde{\Omega}_\infty$ where two points in Ω_∞ are equivalent if they correspond to the same dynamical system. Various properties of $\tilde{\Omega}_\infty$ are being studied. In particular, we show that in $\tilde{\Omega}_\infty$ there exists arbitrary small open neighborhood N with the following property--

There exists a sequence $\xi_0, \xi_1, \ldots$ of plants in N such that the minimum degree of the stabilizing dynamic compensator for the plants corresponding to $\xi_0, \xi_1, \ldots$ increases arbitrarily.

This fact in particular implies that

"There exists $p \in \tilde{\mathbb{R}}^\infty$ such that every open neighborhood N of p in $\tilde{\Omega}_\infty$ cannot be stabilized even by an adaptive controller of arbitrary large degree q."

Thus we obtain a major limitation of the adaptive controllers that are currently of interest in system theory, viz. open neighborhoods of points in $\tilde{\Omega}_\infty$ that cannot be robustly stabilizable even by an adaptive controller. The structure of $\tilde{\Omega}_\infty$ also enables us to define a hybrid family of plants, (i.e. a family of plants with structured and unstructured uncertainty). In particular in [6] we characterize (for the first time in the literature) hybrid families of plants that can be stabilized simultaneously by an adaptive controller.

The proposed hybrid parameterization has many advantages over the currently existing graph parameterization due to Vidyasagar. In fact the hybrid parameterization is graded by the degree of the dynamical systems and each one of the graded space is diffeomorphic to an Euclidean space if the plant is strictly proper. The Euclidean structure is of particular importance in system identification. Furthermore, the sequence of plants for example

$$\delta_n(s) = \frac{s^n}{s^{n+1} + \frac{1}{n+2}}$$

converges to $\frac{1}{s}$ as $n \to \infty$ in the graph-topology. Thus in graph parameterization, arbitrary close to a plant of a given degree there exists plants of arbitrary large degree which is clearly a deficiency from the point of view of robustness and obtaining an apriori bound on the complexity of the compensators. Hybrid parameterization does not suffer from these disadvantages and therefore appears to be a good parameterization for system identification and adaptive control.

In [7] we study the problem of simultaneous stabilization of a family F of plants described as follows --

$$F \triangleq \{g(s):\ g(s) = [\sum_{i=0}^{n-1} a_i s^i] \,/\, [\sum_{i=0}^{n-1} b_i s^i + s^n],$$
$$a_i \epsilon [\alpha_i, \beta_i],\ b_i \epsilon [\gamma_i, \delta_i],\ \alpha_i \le \beta_i$$
$$\gamma_i \le \delta_i,\ i=0,\dots, n-1,\ \deg g(s) = n\}$$

We prove the following rather surprising result

"A necessary and sufficient condition that every plant in F is simultaneously stabilizable by a feedback gain k is that eight plants in F (suitably chosen) is simultaneously stabilizable by a feedback gain k."

We find the above result quite surprising. Indeed it asserts the existence of a suitable family of uncountably many plants, stabilizability of which can be asserted via the simultaneous stabilization problem of a finite number of plants. This we consider is a major conceptual breakthrough.

The main idea of the preceding paragraph can be generalized to include dynamic compensation as well. In fact one can obtain a sufficient condition which can be made asymptotically necessary by increasing the computational complexity of the algorithm. Roughly speaking one therefore concludes the existence of a computational technique to construct a robust compensator which can be asymptotically improved by considering increased computational load. This in my view is a computational breakthrough and in particular such a sequence of algorithms did not exist in the literature previously.

For the purpose of constructing a compensator with an apriori bounded McMillan degree it is important to consider to following problem.

"Given a family F of linear dynamical systems that can be stabilized simultaneously by a fixed non-switching compensator. Does there exist an apriori bound on the degree of the compensator which simultaneously stabilizes F."

In general the above problem is unsolved. However for a 1 parameter family of plant we have a surprising result: Let $x_1(s)/y_1(s)$ and $x_2(s)/y_2(s)$ be a pair of proper but not strictly proper plants. Consider a 1 parameter family F of plants described as follows

$$F = \{g_\lambda(s): \ g_\lambda(s) = [\lambda x_1 + (1-\lambda)x_2]/[\lambda y_1 + (1-\lambda)y_2]$$
$$\lambda \in [0, 1], \ \deg g_\lambda(s) \le n \ \forall \lambda\}.$$

Let $a_1, \ldots, a_t$ denote the zeros of $x_1y_2 - x_2y_1$ in the open left half of the complex plane. Let

$b_j = x_2/x_1(a_i)$ if the multiplicity of a_j as a common zero of x_1, x_2 is $\le$ multiplicity of a_i as a common zero of y_1, y_2

$= y_2/y_1(a_i)$ otherwise.

for $i=1, \ldots, t$. Let $s_i = (a_i-1)/(a_i+1)$ and $z_i = (\sqrt{b_i}-1)/(\sqrt{b_i}-1)$ where the branch cut for the square root is taken to be the non-positive real axis. Furthermore let k be the largest real number such that

$$[1 - k^2 z_i z_j]/[1 - s_i s_j]_{i,\ j=1}^{t}$$

is non-negative definite. The main result is now described as follows

"The following three statements are equivalent.

1. F is simultaneously stabilizable by some dynamic compensator.
2. F is simultaneously stabilizable by some dynamic compensator of degree $\leq 3n-2$.
3. $k > 1$

We find that the above result is quite surprising. In fact, where as the conjecture - "pairs of simultaneoulsy stabilizable plants of bounded McMillan degree have simultaneously stabilizing compensators of bounded McMillan degree" - is false, the conjecture that "simultaneously stabilizable linear 1-parameter family of plants of bounded McMillan degree have simultaneously stabilizing compensators of bounded McMillan degree" is indeed true. Of course it is unknown if similar results would continue to be true for multiparameter family of plants. It appears however, in view of the above result, that the problem of stabilizing a discrete r-tuple of plants (in particular a pair of plants) simultaneously is a much harder problem to solve compared to simultaneously stabilizing a continuous family of plants. This fact indeed appears to be quite contrary to our original expectation - in fact the problem of simultaneous stabilization of a pair of plants was originally used with an idea of simplifying the robust stabilization problem of a family of plants.

In order to arbitrary tune the closed loop frequencies of a plant, it is necessary to consider the simultaneous pole assignment problem. In [6] we analyze the pole placement problem as an intersection problem and apply Schubert enumerative calculus to compute (under appropriate cases) the number of complex dynamic compensators that would place the closed loop poles of a set of r-plants in a given set of self-conjugate complex numbers. We compactify the space of compensators and define a set of points known as 'base locus' and a set of points known as 'critical points.' Roughly speaking, we assert in [6] that a compensator has to avoid the base locus and the critical points for otherwise the closed loop response of the control system would either be sensitive or would fail to be robust with respect to changes in the parameters. An explicit parameterization of these points also open up some new restrictions in the compensator design problem previously unknown in system theory.

To summarize, we maintain that the use of semialgebraic geometric, algebraic geometric and transcendental methods are three distinct foundational techniques that have been applied in robust system design. Extensions of these methods to parameterization, design, identification problems, and adaptive control would be useful and are currently being explored. These techniques are also being extended to nonlinear and time varying systems.

References:

[1] B. K. Ghosh and C. I. Byrnes, "Simultaneous Stabilization and Simultaneous Pole-placement by Non-switching Dynamic Compensation," IEEE Transactions on Automatic Control, Vol. AC-28, No. 6, June 1983, pp. 735-741.

[2] B. K. Ghosh, "An Approach to Simultaneous System Design, Part I: Semialgebraic Geometric Methods," SIAM J. on Control and Optimization, May, 1986.

[3] B. K. Ghosh, "Simultaneous Partial Pole Placement--A New Approach to Multimode System Design," IEEE Trans. on Automatic Control, May, 1986.

[4] B. K. Ghosh, "Transcendental and Interpolation Methods in Simultaneous Stabilization and Simultaneous Partial Pole Placement Problem," Accepted by SIAM J. on Control and Optimization.

[5] B. K. Ghosh and W. P. Dayawansa, "A hybrid parameterization of linear single input single output systems," accepted by Systems and Control Letters.

[6] B. K. Ghosh, "An approach to simultaneous System Design, Part II: Dynamic Compensation by Algebraic Geometric Methods," (To appear in SIAM J. of Control and Optimization.)

[7] B. K. Ghosh, "Some New Results on the Simultaneous Stabilizability of a Family of Single Input, Single Output Systems," Systems and Control Letters, 6, (1985), pp. 39-45.

[8] B. K. Ghosh and W. P. Dayawansa, "An approach to linear system identification, I Differential geometric methods in hybrid parameterization problems," submitted to SIAM J. on Control and Optimization.

A NEW APPROACH TO THE MODELLING UNCERTAINTY PROBLEM OF SYSTEMS DESCRIBED IN STATE SPACE FORM

David Bensoussan
Ecole de technologie supérieure
Université du Québec
4750, rue Henri-Julien
Case postale 1000, Succursale E
Montréal, Québec, CANADA
H2T 1R0

ABSTRACT

Modelling of systems is generally done by frequency response methods or state variable methods. It is our object to show how frequency domain robustness results can be extrapolated to their state space counterpart. Using properties of input-output relations of systems and different compatible norms we will show how a corresponding frequency response robustness result can be applied. The method can be used to solve a certain class of non linear equations. It can apply to the control of non linear multivariable systems in order to better stability, sensitivity as well as decentralized control results. It can also apply to assess the state feedback, the output feedback and the observer with regard to the robustnees problem.

1. INTRODUCTION

Multivariable control theory evolved in the sixties, using the state variable approach. This approach together with growing computer technology gave rise to tremendous research. Interesting results on system stability, controlability, observability, reachability and detectability were developed. This was a sharp contrast to the single input-single output frequency response approach involving polynomial approaches, Nyquist criterium, and root locus methods.

However, many of the answers given by state space methods lack the suppleness of multivariable methods as they apply to well defined models with no modelling uncertainty. Adaptive control is a partial response for the modelling uncertainty problem as far as parametric uncertainty is concerned. Clearly, in any state space representation (A, B, C, D), there is no way to predict the behaviour of eigenvalues whenever the matrix representation is modified to $(A+\Delta A,B,C,D)$. On the other hand, frequency response methods apply better to the uncertainty problem: in the case of a single input single output Nyquist diagram for instance, a Nyquist plot could be replaced by some Nyquist band representing the modelling uncertainty at each frequency.

Multivariable frequency response methods such as the inverse Nyquist area [1], the multivariable Nyquist criterium [2], and the multivariable root locus [3] are concerned mainly with system stability. However, the input output approach to systems [4,5,6,7,8] which apply to any normed algebraic representation of systems fit particularily to the frequency response setting. Such an approach allows us to handle the problem of modelling uncertainty. It is our purpose to show how multivariable frequency response uncertainty methods can be extrapolated to the multivariable state space uncertain models case.

2. Mathematical notations

We shall consider frequency responses defined in Hardy spaces, namely space $H^n_{\infty e}$, H^n_∞, and H^n_2 . $H^n_{\infty e}$ is the space of n x n matrices of frequency responses in $H^1_{\infty e}$, i.e. frequency responses which are holomorphic and bounded in some right half plane $Re(s)>\sigma_u > 0$, $\sigma_u > 0$. H^n_∞ is the space of n x n matrices of elements in H^n_∞, i.e. frequency responses which are holomorphic and bounded in the open right half plane $Re(s) > 0$. H^n_2 is the space of n- tuple vectors whose elements $u_i = 1,2,..n$ are in H^1_2, i.e. frequency responses which are holomorphic and bounded in $Re(s) > 0$. Frequency responses in H^n_2 will be normed as follows:

$$\|u_2\| = \sum_{i=1}^{n} \| u_i\|_2^2$$

$$\|u_i\|_2 = \int_{-\infty}^{\infty} | u_i(j\omega)|^2 d\omega \qquad i = 1,2,...n$$

We underline the fact that the H_2 norm is equivalent to the L_2 norm, i.e.

$$\|u_i\|_2 = \int_{-\infty}^{\infty} | u_i(t)|^2 dt$$

Functions T in H^n_∞ are normed as follows

$$\|T\|_\infty = \sup_{u\neq 0} \frac{\| Tu \|2}{\| u \|_2}$$

$$\|T\|_\infty = \sup_{\omega} \bar{\sigma} \{T(j\omega)\}$$

where $\bar{\sigma}(.)$ represents the maximal singular value of the matrix $T(j\omega)$.

We introduce the matrix $G\{T\}$ whose elements are $\|T_{ij}\|_\infty$ and a new norm $g(T) = \bar{\sigma} \{G(T)\}$.

It has been shown [9] that such a norm is compatible with the H_∞ norm, i.e.

$$\|T\|_\infty \leq g(T) \leq n \|T\|_\infty$$

We shall extend this new norm to any matrix A in which an induced L_2 norm is introduced for each of its elements a_{ij}, i.e.

$$g(A) = \bar{\sigma}\{G(A)\}$$

3. Multivariable frequency response uncertainties

Given an n x n matrix of frequency responses P(s) in $H^n_{\infty e}$, we shall consider the unity feedback system with a series compensator C(s) in H^n_∞ (figure 1). We shall assume that the matrix uncertainty $\Delta P(s)$ is in H^n_∞ and that the feedback system is altered by the following additive H_2 perturbations: input plant perturbation w(s), external output perturbation d(s), and sensor noise perturbation w(s).

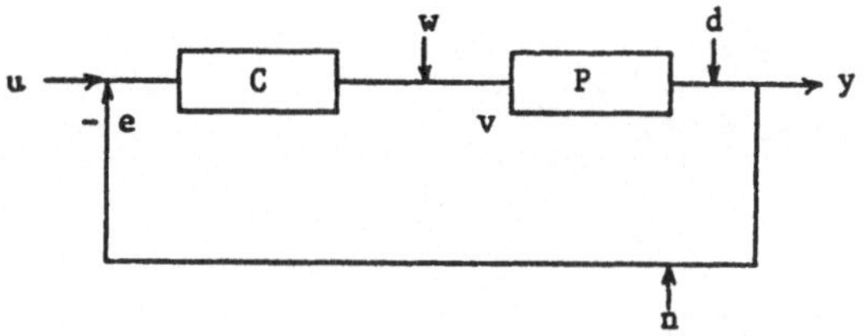

FIGURE 1

For simplicity, we omit the (s) and deduce:

$$\begin{aligned} v &= Ce + w \\ y &= Pv + d \qquad (1) \\ e &= u - y - n \end{aligned}$$

$$\begin{bmatrix} y \\ E \\ v \end{bmatrix} = \begin{bmatrix} (PC)(I+PC)^{-1} & +(I+PC)^{-1} & -PC(I+PC)^{-1} & PC(I+PC)^{-1} \\ -(I+PC)^{-1} & -(I+PC)^{-1} & -(I+PC)^{-1} & -P(I+PC)^{-1} \\ -C(I+PC)^{-1} & -C(I+PC)^{-1} & -C(I+PC)^{-1} & I-CP(I+PC)^{-1} \end{bmatrix} \begin{bmatrix} u \\ d \\ w \end{bmatrix} \qquad (2)$$

Transmissions from e and v to u, n and w are related to the "internal stability" of the system, while transmission from y to d refer to the output disturbance effects. Whenever P is replaced by P + ΔP where ΔP represents the modelling uncertainty, quantitative measures of feedback quality can be performed, with respect to external disturbances and plant uncertainty.

We shall focus our attention solely on the effect of feedback on plant uncertainty ΔP.

3.1 Stability

Stability concerns itself with the holomorphicity and the boundedness of the transmissions (2).

Our first concern is to ensure the stability of these transmissions whenever P is replaced by $P' = P + \Delta P$. Small gain and positivity conditions permit us to derive stability criteria. Defining

$$\begin{aligned} e &= (I + PC)^{-1}u = Eu \\ y &= PC\,(I + PC)^{-1}u = Tu \end{aligned} \qquad (3)$$

and

$$\begin{aligned} e' &= (I + P'C)^{-1}u = E'u \\ y &= P'C\,(I + P'C)^{-1}u = T'u \end{aligned} \qquad (4)$$

it can be shown that

$$E' = E - E\Delta PCE\,[I + \Delta PCE]^{-1} \qquad (5)$$

Whenever the original nominal transmissions (3) are stable, E' will be stable if $\|\Delta PCE\|<1$. Moreover, T' will be automatically stable as $E' + T' = I$. Similarily, whenever $(I + PC)$ or $(I + P'C)$ is η positive definite, it follows that the transmissions E' and T' are stable. η positivity refers to positivity within a finite margin η.

3.2 Sensitivity

It is not sufficient to require that the transmissions E or E' be stable, i.e. holomorphic and bounded in the right half plane. Indeed E represents the transmission between output and output additive external disturbances d and it is desirable to reduce the effect of such disturbances.

Norms restricted on a given bandwidth $|\omega|<\omega_1$, such as maximal singular values over a given bandwidth $\|(.)\|$ can be introduced. We therefore require $\|E\|_{\omega 1} < \varepsilon$ and $\|E'\|_{\omega 1} < \varepsilon'$ where ε, ε' are positive constants which are less than unity.

Assuming the nominal plant satisfies $\|E\|_{\omega 1} < \varepsilon$, designing the compensator C such that $\|\Delta PCE\|_{\omega 1} < \alpha$ where $\alpha \leq \{(\varepsilon-\varepsilon')/\varepsilon\} < 1$ ensures that $\|E'\|_{\omega 1} < \{\varepsilon/(1-\alpha)\} = \varepsilon'$.

Such a design has been proven to be feasible [10,11].

3.3 Decentralized control

The problem of decentralized control is to attain some diagonal closed loop transmission, i.e. each input of the feedback controls one ouptut of the feedback system regardless of the cross-coupling of the plant.

A model of this situation involves a multivariable plant P with n input-output pairs (v_i, y_i, $i = 1, 2 \ldots n$) and a controller C which has access to both inputs and outputs of P. (Fig. 1) In others words, C is a n x n matrix compensator. Assumming (I + PC) has an inverse, our aim is to diagonalize the closed loop transmission $T = PC\,(I + PC)^{-1}$.

Using the relation $E + T = I$, we deduce

$$\|T\|_{\omega_1} \leq 1 + \|E\|_{\omega_1} < 1 + \varepsilon$$
$$\|T'\|_{\omega_1} \leq 1 + \|E'\|_{\omega_1} < 1 + \varepsilon'$$

i.e. sensitivity reduction is linked to decentralized control over a frequency band of interest.

4. Multivariable state space uncertainty

Given the state space system representation

$$\dot{x} = A\,x + B\,u$$
$$y = C\,x + D\,u$$

We are interested in studying the system performance whenever systems dynamics change to $A' = A + \Delta A$, i.e.

$$\dot{x}' = A'\,x' + B\,u$$

$$y' = C\,x' + D\,u$$

Any change in the matrix A involves changes in the eigenvalues defined by $(sI-A) = 0$ for the nominal system and $(sI - A') = 0$ for the perturbed system. However, we cannot predict in which "direction" these eigenvalues are changing.

One way of checking the stability of the perturbed system is to study the class of disturbances $\Delta A = A' - A$ which preserve the stability of the matrix $(sI - A')$. Assuming for example, that $(sI-A)$ is invertible in H_∞^n, the condition $\|\Delta A(sI-A)^{-1}\| < 1$ would ensure the stability of the perturbed system.

4.1 Correspondance between multivariable frequency responses and state space models

Given state space models (A, B, C, D) and (A', B, C, D), it is possible to assign to each model multivariable frequency responses T(s) and T'(s) such that

$$\begin{aligned} T(s) &= C\,(sI - A)^{-1}\,B + D \\ T'(s) &= C\,(sI - A')\,B + D \end{aligned} \tag{5}$$

These relations apply also to matrix entries, i.e.

$$\begin{aligned} t_{i,j}(s) &= \{C\,(sI - A)^{-1}\,B + D\}_{i,j} \quad i,j = 1, 2, ..n \\ t'_{i,j}(s) &= \{C(sI - A')\,B + D\}_{i,j} \quad i,j = 1,2,..n \end{aligned} \tag{6}$$

Assuming the entries of the matrix $\{a_{ij}\}$ change to $\{a'_{ij}\} = \{a_{ij} + \delta a_{ij}\}$, it is possible to relate changes ΔA to changes in the individual entries using the norm $g(\Delta A)$.

Assuming $(sI - A)$ and $(sI - A - \Delta A)$ are invertible, and that $g[(sI - A)^{-1}\,\Delta A] < 1$ it can be shown that

$$g[T'(s) - T(s)] \leq \frac{g[C(sI - A)^{-1}\ B]\ g(\Delta A)\ g[C(sI - A)^{-1}\ ]}{1 - g[(sI - A)^{-1}]\ g(\Delta A)} \tag{7}$$

The upper bound for $g\,[T'(s) - T(s)]$ as a function of $g\,(\Delta A)$ is represented in figure 2. If such a bound can be related to one of the many robustness results developed in the frequency response framework [11,13,15,16, 17,18], we may then consider their use in the state space framework. Any change in the matrix A will be normed by $g\,(\Delta A)$ and it will be possible to verify that the corresponding frequency response bound $g[T'(s) - T(s)]$ respects some frequency response robustness performance result.

Similarily, any frequency response robustness results can be found to hold on some state space distributed model $A + \Delta A$. The bound on the norm of ΔA can be deduced graphically and be related to variations of the norms of the entries of the matrix ΔA.

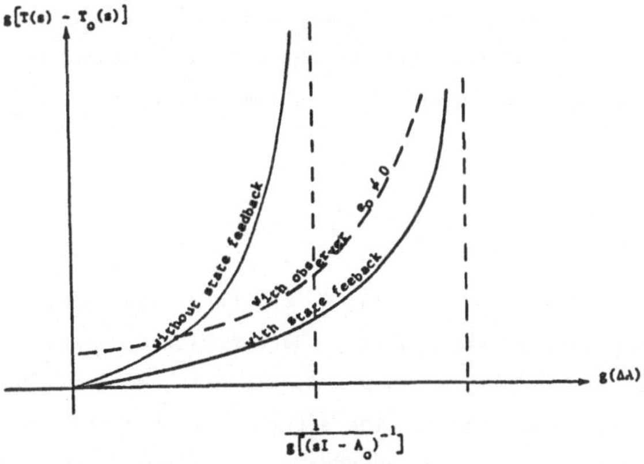

FIGURE 2

4.2 State Feedback, output feedback and observers

Assuming a state space model (A, B, C, D) in which the pair (A, B) is controllable, it has been shown [21] that a stabilizing state feedback involving a matrix compensation K leads to a robustness bound assymptotic to the critical value $g(\Delta A) = 1/\{g(sI - A + BK)^{-1}\}$. Similarily, assuming the pair (A, B) controllable and the pair (A, C) is observable, a stabilizing output feedback controller involving a matrix conpensation H leads to a robustness bound assymptotic to the critical value $g(\Delta A) = 1/[g(sI - BHC - A)^{-1}\}$. These two robustness bounds show that high gain state feedback or high gain output feedback reduce the output variations due to plant perturbation.

Moreover, it can be shown that in the case of a system with observer gain G and control gain K, conresponing to a stable system model

$$\begin{bmatrix} \dot{x} \\ \dot{e} \end{bmatrix} = \begin{bmatrix} A + BK & -BK \\ 0 & A + GC \end{bmatrix} \begin{bmatrix} x \\ e \end{bmatrix} \tag{8}$$

where e is the error signal existing between the real state and the estimated state, uncertainty in the plant dynamics ΔA leads to a robustness bound which is an increasing functions of a (ΔA) and which is also proportionnal to the initial value of the error signal, i.e. the error in the estimation of the initial state. Such a conclusion can be drawn after rewriting (8) as

$$\begin{aligned} x &= -(sI - A - \Delta A - BK)^{-1} BKe \\ e &= (sI - A - \Delta A - GC)^{-1} e_o \end{aligned} \tag{9}$$

The robustness uncertainty due to the choice of initial conditions can be optimized by choosing the initial value of the estimated state to be $C^{+}y_{o}$ where y_{o} is the initial measurement value and C^{+} is the pseudo inverse of C[22].

5. Conclusion

The correspondances shown in section 4 can be extrapolated to multivariable results involving robustness, sensitivity, decentralized control and applied to perturbed state space models. These models are derived from differential equations with varying coefficients. The effect of such variations (which can be non linear provided that they can be adequatly normed, such as in the case of conic/sector non linearities [19,20,9]) on the stability of the solutions may be deduced from the bounds of the frequency response counterparts.

Acknowledgement : The Author wishes to thank the National Research Council of Canada which supported this research.

[1] M.H. Rosenbrock, "Design of Multivariable Control Systems Using the Inverse Nyquist Array", IEEE Procedings, vol. 116, no. 11, November 1969.

[2] J.F. Barman and J. Katzenelson, "The Generalized Nyquist Type Stability Criterion for Multivariable Feedback Systems", International Journal of Control, vol. 20, pp. 593-622, 1974.

[3] A.G.J. MacFarlane and I. Postlewaite, "The Generalized Nyquist Stability Criterion and Multivariable Root Loci", International Journal of Control, vol. 25, pp. 81-127, 1977.

[4] G. Zames, "Feedback and Optimal Sensitivity: Model Reference Transformations, Multiplicative Seminorms, and Approximate Inverses", IEEE Trans on Automatic Control, Vol. AC-26, No.3, pp. 301-320, 1981.

[5] G. Zames and B.A. Francis, "Feedback, Minimax Sensitivity, and Optimal Robustness", IEEE Trans. on Automatic Control, Vol. AC-28, No. 5, pp. 585-601, May 1983.

[6] B.A. Francis and G. Zames, "On H_{∞} Optimal Sensitivity Theory for SISO Feedback Systems", IEEE Tran. on Automatic Control, Vol. AC-29, No. 1, pp. 9-16, January 1984.

[7] B.A. Francis, J.W. Helton and G. Zames, "H_{∞} Optimal Feedback Controllers for Linear Multivariable Systems", IEEE Transactions on Automatic Control, Vol. AC-29, No. 10, pp.888-900, October 1984.

[8] C. Foias, A. Tannenbaum and G. Zames, "Weighted Sensitivity Minimization for Delay Systems", Proceedings of 24th Conference on Decision and Control, Ft. Lauderdale, Florida, pp. 244-249, December 1985.

[9] D. Bensoussan, "Decentralized Control In Non Linear Systems", Proceeding of the 25th IEEE Conference on Detection and Control, Athens, Greece, pp. 865-867, December 10-12th 1986.

[10] Bensoussan, David, "Sensitivity Reduction in Single Input Single Output Systems", International Journal of Control, vol. 39, no. 2, pp. 321-335, 1984.

[11] D. Bensoussan, "Decentralized Control and Sensibility Reduction in Weakly Coupled Systems", International Journal of Control, Vol. 40, No. 6, pp. 1099-1118, 1984.

[12] G.Zames and D. Bensoussan, "Multivariable Feedback, Sensitivity, and Decentralized Control", IEEE Trans. on Automatic Control, Vol. AC-28, No. 11, pp. 1030-1035, November 1983.

[13] D. Bensoussan, "Commande décentralisée de systèmes multivariables avec incertitude de modélisation", AMSE Conf. on Modeling and Simulation, June 17-29, 1984.

[14] D. Bensoussan, "Robustesse et commande décentralisée des systèmes multivariables non linéaires", Proceedings of the Seventh International Conference on Analysis and Optimization of Systems. INRIA, Springer Verlag, Volume 83, pp. 630-645.

[15] J.S. Freudenberg and D.P. Looze. "Right half Plane Poles and Zeros and Design Trade offs in Feedback Systems." IEEE Transactions on Automatic Control, AC-30, p. 555, 1985.

[16] S.D. O'Young and B.A. Francis,"Optimal Performance and Robust Stabilisation", Automatica, To appear.

[17] S.D. O'Young and B.A. Francis,"Sensitivity Trade-offs for Multivariable Plants", IEEE Transactions on Automatic Control. AC-30. p. 625, 1985.

[18] H. Kwakernaak, "Minimax Frequency Domain Performance and Robustness Optimization of Linear Feedback Systems", IEEE Transactions on Automatic Control, vol. AC-30, No. 10, October 1985, pp.994-1004.

[19] G.Zames, "On the Input Output Stability of Time Varying Non Linear Systems. Part I", IEEE Trans on Automatic Control, Vol. AC-11, No. 2, pp. 228-238, April 1966.

[20] G. Zames, "On the Input Output Stability of Time Varying Non Linear Systems - Part II", IEEE Trans on Automatic Control, Vol. AC-11, No. 3, pp. 465-476, July 1966.

[21] Bensoussan, David, "On the Equivalence between State Space and Frequency Response Models : A Missing Link for the Study of the Robustness Problem", Proceedings of the 26th IEEE Conference on Decision and Control, Los Angeles, California, December 9-11, 1987

[22] G. A. Johnson, Appendix of the paper "Linear Adaptive Control Via Disturbance Accomodation, Some Case Studies", Proceedings of the American Control Conference, Washingston, vol. 1, p.546, June 1986.

NONSMOOTH SOLUTIONS OF HAMILTON-JACOBI-BELLMAN EQUATION

Halina Frankowska
Centre National de la Recherche Scientifique, CEREMADE,
Université PARIS-IX Dauphine, Paris, France *and*
International Institute for Applied Systems Analysis, Laxenburg, Austria

Introduction

We are concerned here with the Hamilton-Jacobi equation

$$-\frac{\partial V}{\partial t} + H(t\,,x\,,-\frac{\partial V}{\partial x}) = 0\,,\quad V(1\,,\cdot) = g \tag{1}$$

arising in optimal control.

It is well known that in general it does not admit classical (C^1) solutions, even when the data are smooth. This led to several weaker notions of solutions (see for instance [5]-[9], [12], [17] and bibliographies contained therein). We do not have the ambition here to provide the reader with a complete overview of existing definitions of solutions, but we shall compare few of them having their origins in Nonsmooth Analysis.

The importance of HJB (Hamilton-Jacobi-Bellman) equation for the investigation of properties of dynamical systems and, in particular, of control systems was recognized a long time ago.

One may see that the value function of an optimal control problem verifies (1) whenever it is smooth. This allows in some (very restrictive) cases to obtain a short proof of Pontriagin's necessary conditions for optimality, to prove some sufficient conditions for optimality and to construct optimal feedbacks [11].

Let us emphasize that the value function arising in control theory is nondecreasing along all trajectories of the system and is constant along optimal trajectories. This leads to a verification technique in optimal control (see [19] for a complete discussion and references). However, "computing" the value function from its definition is a very difficult task.

On the other hand if we are able to find a solution of (1) having the cornerstone properties of the value function, then we may hope to use it for the same purposes.

One can seek for instance to define the solution of HJB equation in such a way that the value function is the unique solution to (1). The notion of viscosity solution introduced in [8], [9], [17] (see also [10] for bibliographical comments) fulfills that objective, but only partially: the uniqueness results are proved up to now only for continuous solutions on open sets. Although a large class of free end point optimal control problems have a locally Lipschitz value function, it is well known that for the target problem and, more generally, for problems with state constraints, the value function is not continuous and very strong controllability conditions are needed to prove its continuity. Controllability conditions exclude however from consideration a large number of control problems.

A different approach was developed in [5]-[7], where a solution of HJB equation associated to the target problem:

$$\text{minimize } g(x(1))$$
$$\dot{x} = f(t, x, u(t)),\ u(t) \in U \text{ is measurable} \tag{2}$$
$$x(0) = x_0,\ x(1) \in K$$

was defined in such a way that it is locally Lipschitz and nondecreasing along trajectories of the control system (2). Naturally this led to apply the verification technique to functions which may be different from the value function. Such a generalized solution is not uniquely defined, but it allows to consider a broader class of control problems, to adapt the verification technique to problems with discontinuous value function and to get some necessary and sufficient conditions for optimality ([6], [7]).

Another important property of smooth value function is the possibility to construct optimal feedback laws. When the value function is the unique locally Lipschitz viscosity solution of (1) it allows as well to associate feedback laws to the solution of HJB equation. This feature fails whenever the value function is discontinuous. Generalized solutions of (1) defined in [6] enjoy more regularity than the value function does and for the very same reason are only far relatives of this last one:

It is impossible to associate the optimal feedback law with an arbitrary generalized solution of (1). A counter example was constructed in [4].

One would wonder which way to choose. Clearly, we cannot expect from the solution of HJB equation to be unique, locally Lipschitz and at the same time to be equal to the value function.

In this paper we show that a necessary and sufficient condition for a function $V : [0,1] \times \mathbf{R}^n \longrightarrow \mathbf{R} \bigcup \{\pm\infty\}$ to be nondecreasing along all trajectories of the control system (2) is:

$$\sup_{u \in U} D_{\uparrow}(-V)(t, x)(1, f(t, x, u)) \leq 0 \tag{3}$$

where $D_{\uparrow}(-V)(t, x)$ is the contingent epiderivative of the function $-V$ at (t, x) (see Section 2 for precise definitions).

Such a necessary and sufficient condition leads to a verification technique. Then we investigate necessary and sufficient conditions for a function $V : [0,1] \times \mathbf{R}^n \longrightarrow \mathbf{R} \bigcup \{\pm\infty\}$ to have the following property:

for all $(t, \xi) \in [0, 1] \times \mathbf{R}^n$ with $V(t, \xi) \neq \pm \infty$ there exists a solution $\bar{x} : [t, 1] \longrightarrow \mathbf{R}^n$ of the control system

$$\begin{cases} x' &= f(s, x, u(s)), \ u(s) \in U \text{ is measurable} \\ x(t) &= \xi \end{cases}$$

such that the function $s \longrightarrow V(s, \bar{x}(s))$ is nonincreasing.

In this way we obtain the second contingent inequality

$$\inf_{u \in U} D_{\uparrow} V(t, x)(1, f(t, x, u)) \leq 0 \tag{4}$$

We also observe that a function V verifying (3), (4) and such that $V(1, \cdot) = g(\cdot)$ is a viscosity solution to (1).

In this way solutions of contingent inequalities (3), (4) form a subset of viscosity solutions containing the value function. Let us emphasize that the value function of the target problem is equal to $+\infty$ at every point (t_0, x_0) for which there is no solution x of (2) satisfying $x(t_0) = x_0$, $x(1) \in K$. This creates an additional difficulty to define the solution of (1) properly.

In the example given in Section 5 we show why it is rather hopeless to expect uniqueness in the discontinuous case. *In a nutshell, the function V verifying (3), (4) allows to narrow the class of admissible controls and to get some information about optimal feedback laws and above all, by accepting as solutions extended functions, can be used in control problems with state constraints.*

The outline of the paper is as follows. In Sections 1, 2 we derive contingent inequalities (3) and (4). Section 3 is devoted to the optimal feedback laws. In Section 4 we prove that solutions of (3), (4) are viscosity solutions of (1) and discuss their relationship. Section 5 compares viscosity solution of HJB equation and solutions in the sense of Clarke. We do not provide here complete proof of many results. They may be found in [12], [13].

1. Monotone behaviour of V.

We consider a dynamical system described by a differential inclusion.

Let F be a set-valued map from $[0, 1] \times \mathbf{R}^n$ to $\mathbf{R}^n$. We associate with it the differential inclusion

$$x' \in F(t, x) \tag{1.1}$$

A function $x \in W^{1,1}(t, T)$, $T \geq t$ (the Sobolev space) is called a trajectory of the differential inclusion (1.1) if for almost all $s \in [t, T]$, $x'(s) \in F(s, x(s))$. The set of all trajectories of (1.1) defined on the time interval $[t, T]$ and starting at ξ, $(x(t) = \xi)$ is denoted by $S_{[t, T]}(\xi)$.

Let B denote the closed unit ball in $\mathbf{R}^n$. Throughout the whole paper we assume that for all $(t, x) \in [0, 1] \times \mathbf{R}^n$

(H_1) $F(t, x)$ is a nonempty compact set,

that for all $x \in \mathbf{R}^n$

(H_2) $F(\cdot, x)$ is continuous on $[0, 1]$

and

(H_3) F is locally Lipschitzian in x, in the sense that for every $(t_0, x_0) \in [0, 1] \times \mathbf{R}^n$ there exists a neighborhood $\mathbf{N}$ in $[0, 1] \times \mathbf{R}^n$ and a constant L such that for all $(t, x), (t, y) \in \mathbf{N}$

$$F(t, x) \subset F(t, y) + L \| x - y \| B \tag{1.2}$$

Differential inclusion (1.1) is a convenient expression of laws governing a dynamical system. Many systems arising in control theory may be reduced to (1.1). For instance consider the closed loop control system

$$x'(t) = f(t, x(t), u(t)), \ t \in [0, 1] \tag{1.3}$$

$$u(t) \in U(x(t)) \tag{1.4}$$

where $f : [0, 1] \times \mathbf{R}^n \times \mathbf{R}^m \longrightarrow \mathbf{R}^n$ is a continuous function and $U : \mathbf{R}^n \rightrightarrows \mathbf{R}^m$ is a continuous control map with nonempty compact images. Admissible controls are measurable functions on $[0, 1]$ satisfying (1.4).

For all $(t, x) \in \mathbf{R} \times \mathbf{R}^n$ set

$$F(t, x) = \{f(t, x, u) : u \in U(x)\} \tag{1.5}$$

Clearly every trajectory of (1.3), (1.4) is a trajectory of the differential inclusion (1.1) with F defined as in (1.5). Conversely, with every trajectory $x \in S_{[0, 1]}$ of differential inclusion (1.1) we can associate a measurable selection $u(t) \in U(x(t))$ such that (1.3) holds true almost everywhere in $[0, 1]$. This follows from Lusin's theorem exactly by the same arguments as in [1, p. 91] (see also [6, pp. 111-112]).

The implicit control system

$$\begin{cases} f(t, x(t), x'(t), u(t)) = 0, & t \in [0, 1] \\ u(t) \in U \end{cases} \tag{1.6}$$

where $f : [0, 1] \times \mathbf{R}^n \times \mathbf{R}^n \times \mathbf{R}^m \longrightarrow \mathbf{R}^k$ is continuous and U is a compact metric space may be also reduced to (1.1) by setting

$$F(t, x) = \{v : 0 \in f(t, x, v, U)\} \tag{1.7}$$

In [14] it was shown that solutions of (1.6) and (1.7) do coincide. For further discussion and applications of differential inclusions in control theory see bibliographies contained in [12], [14].

For all $t \in \mathbf{R}$, $T \geq t$ and $\xi \in \mathbf{R}^n$ set

$$R(T, t)\, \xi = \{x(T) : x \in S_{[t, T]}(\xi)\} \tag{1.8}$$

This is the so-called reachable set of (1.1) from (t, ξ) at time T.

When F is sufficiently regular the set $co\, F(t, \xi)$ is the infinitesimal generator of the semigroup $R(\cdot, t)\, \xi$:

Theorem 1.1. [12]. Assume that the assumptions (H_1) - (H_3) are verified. Then for every $(t_0, \xi_0) \in [0, 1[\times \mathbf{R}^n$ and all (t, ξ) near (t_0, ξ_0) and small $h > 0$

$$R(t + h, t)\, \xi = \xi + h\, coF(t_0, \xi_0) + o(t, \xi, h) \tag{1.9}$$

where

$$\lim_{\substack{(t,\xi) \longrightarrow (t_0,\xi_0) \\ h \longrightarrow 0+}} \| o(t, \xi, h) \| / h = 0$$

Remark. Equality (1.9) means that

$$R(t + h, t)\, \xi \subset \xi + h\, coF(t_0, \xi_0) + \| o(t, \xi, h) \|\, B \text{ and}$$

$$\xi + h\, coF(t_0, \xi_0) \subset R(t + h, t)\, \xi + \| o(t, \xi, h) \|\, B$$

Definition 1.2 (contingent epiderivative). Let X be a subset of $\mathbf{R}^m$, $\varphi : X \longrightarrow \mathbf{R} \bigcup \{\pm \infty\}$ be a given function and $x_0 \in X$ be such that $\varphi(x_0) \neq \pm \infty$. The contingent epiderivative of φ at x_0 is the function $D_\uparrow \varphi(x_0) : \mathbf{R}^m \longrightarrow \mathbf{R} \bigcup \{\pm \infty\}$ defined by: for all $u \in \mathbf{R}^m$

$$D_\uparrow \varphi(x_0)\, u = \liminf_{\substack{(u', h) \longrightarrow (u, 0+) \\ x_0 + hu' \in X}} [\, \varphi(x_0 + hu') - \varphi(x_0)\,] / h$$

The contingent epiderivative $D_\uparrow \varphi\, (x_0)\, u$ is defined only for those $u \in \mathbf{R}^m$ for which there exists at least one sequence $(u_n, h_n) \longrightarrow (u, 0+)$ satisfying $x_0 + h_n u_n \in X$.

The epigraph of $D_\uparrow \varphi(x_0)$ is equal to the contingent cone to the epigraph of φ at $(x_0, \varphi(x_0))$. If for all $u \in \mathbf{R}^m$, $D_\uparrow \varphi(x_0) u > -\infty$ then $D_\uparrow \varphi(x_0)$ is positively homogeneous and lower semicontinuous (see [2, Chapter 7]).

Theorem 1.3. Let $V : [0, 1] \times \mathbf{R}^n \longrightarrow \mathbf{R} \bigcup \{\pm \infty\}$ be a given function and assume that for every trajectory $x \in W^{1,1}(t, T)$, $T > t$ of (1.1) the function $[t, T] \ni s \longrightarrow V(s, x(s))$ is nondecreasing. Then for all $t \in [0, 1[$, $x \in \mathbf{R}^n$ satisfying $V(t, x) \neq \pm \infty$

$$\sup_{u \in coF(t, x)} D_\uparrow(-V)(t, x)(1, u) \leq 0 \tag{1.10}$$

Proof. Fix $t \in [0, 1[$, $x \in \mathbf{R}^n$ with $V(t, x) \neq \pm \infty$ and $u \in coF(t, x)$. By Theorem 1.1 there exist $u_h \in coF(t, x)$ such that $\lim_{h \longrightarrow 0+} u_h = u$ and $x + hu_h \in R(t + h, t)x$. By the assumption on V, $V(t + h, x + hu_h) \geq V(t, x)$. Hence

$$D_\uparrow(-V)(t, x)(1, u) \leq \liminf_{h \longrightarrow 0+} \frac{V(t, x) - V(t + h, x + hu_h)}{h} \leq 0$$

Since $u \in coF(t, x)$ is arbitrary we end the proof. □

To prove that (1.10) is as well a sufficient condition for monotonicity of V along trajectories of (1.1) it is necessary to require some regularity of V and F.

Let Dom V denote the domain of V, i.e. the set

$$\text{Dom } V := \{(t, x) \in [0, 1] \times \mathbf{R}^n : V(t, x) \neq \pm\infty\}.$$

For a set $L \subset \mathbf{R}^m$ and $x \in L$ we denote by $T_L(x)$ the contingent cone to L at x, i.e.

$$T_L(x) = \{v \in \mathbf{R}^m : \liminf_{h \to 0+} \frac{\text{dist}(x + hv, L)}{h} = 0\}$$

The following result is a slight generalization of [12, Theorem 2.1].

Theorem 1.4 Let $L \subset [0, 1] \times \mathbf{R}^n$ be a closed set and $V : L \longrightarrow \mathbf{R} \bigcup \{\pm\infty\}$ be an upper semicontinuous function such that

$$\begin{cases} \forall (t, x) \in \text{Dom } V, t < 1 \text{ we have } (1, F(t, x)) \subset T_L(t, x) \ \& \\ \sup_{u \in F(t, x)} D_\uparrow(-V)(t, x)(1, u) \leq 0 \end{cases} \tag{1.11}$$

Assume that F is locally Lipschitz in both variables. Then for every trajectory $x \in S_{[t,1]}$ satisfying graph $x \subset$ Dom V, the function $s \longrightarrow V(s, x(s))$ is nondecreasing on $[t, 1]$.

Proof. Consider the closed set $K = \text{epi}(-V)$. By [2, p.418], epi $D_\uparrow(-V)(s, x)$ is equal to the contingent cone $T_K(s, x, -V(s, x))$, and, by (1.11), for all $s \in [0, 1[$, $(s, x) \in$ Dom V, $q \geq -V(s, x)$

$$(1, F(s, x), 0) \subset T_K(s, x, -V(s, x)) \subset T_K(s, x, q) \tag{1.12}$$

Fix a trajectory $x \in S_{[t, 1]}$, $t \in [0, 1]$ such that $(s, x(s)) \in$ Dom V and consider the function $g : [t, 1] \longrightarrow \mathbf{R}_+$ defined by

$$g(s) = \text{dist}_K(s, x(s), -V(t, x(t)))$$

Observe that $g(t) = 0$.

Step 1. We claim that $g \equiv 0$ on $[t,1]$. Indeed assume for a while that for some $T \in [t, 1]$, $g(T) > 0$. For all $s \in [t, 1]$, let $\pi(s) \in K$ be such that

$$g(s) = \| (s, x(s), -V(t, x(t))) - \pi(s) \| \tag{1.13}$$

By continuity of g there exist $t \leq t_0 < t_1 \leq T$ such that $g(t_0) = 0$, $g > 0$ on $]t_0, t_1]$ and for all $s \in]t_0, t_1]$

$$\pi(s) = (\bar{s}, \bar{y}, q)) \text{ for some } \bar{s} \in [0, 1[, \| \bar{y} - x(s) \| \leq 1, q \geq -V(\bar{s}, \bar{y}). \tag{1.14}$$

To end the proof of Step 1 we verify that $g = 0$ on $[t_0,t_1]$. Indeed g being a Lipschitzian function, by Gronwall's inequality, it is enough to show that for a constant $\bar{L}$>0

$$g'(s) \leq \bar{L} g(s) \ \ a.\ e. \text{ in } [t_0 , t_1] \tag{1.15}$$

Let $\bar{L}$ be the Lipschitz constant of F on the set $\{(s , x(s) + B) : s \in [0 , 1]\}$. By the Rademacher theorem g is differentiable almost everywhere. Let $s \in [t_0 , t_1]$ be a point where the derivatives $g'(s)$ and $x'(s) \in F(x(s))$ do exist. Since $x(s + h) = x(s) + hx'(s) + o(h)$, applying the inequality of [1, p.202] we obtain that $g'(s) \leq \text{dist}((1 , x'(s) , 0) , T_K(\pi(s)))$. Thus by (1.14), (1.12) for some $\bar{s} \in [0 , 1[$, $\bar{y} \in x(s) + B$

$$g'(s) \leq \text{dist}((1 , x'(s) , 0) , (1 , F(\bar{s} , \bar{y}) , 0)) \leq \bar{L}(\| \bar{s} - s \|^2 + \| x(s) - \bar{y} \|^2)^{1/2} \leq \bar{L} g(s)$$

and (1.15) follows. □

When the function V considered in Theorem 1.4 is locally Lipschitz then the assumption (1.11) may be relaxed.

Theorem 1.5 Let $L \subset [0 , 1] \times \mathbf{R}^n$ and $V : L \longrightarrow \mathbf{R}$ be a locally Lipschitz function satisfying

$$\forall\, u \in F(t , x) \text{ with } (1 , u) \in T_L(t , x) ,\ D_{\uparrow}(-V)(t , x)(1 , u) \leq 0$$

Then for every trajectory $x : [t , 1] \longrightarrow L$ of (1.1), the function $s \longrightarrow V(s , x(s))$ is nondecreasing on $[t , 1]$.

Proof Fix $x \in S_{[t , 1]}$ with graph $x \subset L$. The function $s \longrightarrow \varphi(s) := -V(s , x(s))$ is absolutely continuous. Let $s \in]\, t , 1\, [$ be such that $\varphi'(s)$ and $x'(s) \in F(s , x(s))$ do exist and $h_i \longrightarrow 0+$ be such that

$$D_{\uparrow}(-V)(s , x(s))(1 , x'(s)) = \lim_{i \longrightarrow \infty} \frac{-V(s + h_i , x(s + h_i)) + V(s , x(s))}{h_i}$$

Then $\varphi'(s) \leq 0$ and, consequently, φ is nonincreasing.

We study next the target problem of optimal control: Let $x_0 \in \mathbf{R}^n$, the subset $K \subset \mathbf{R}^n$ and the function $g : \mathbf{R}^n \longrightarrow \mathbf{R}$ be given.

Consider the problem:

$$\text{minimize } \{g(x(1)) : x \in S_{[0 , 1]}(x_0) , x(1) \in K\} \tag{1.16}$$

The set K is the target of the problem (1.16).

Theorem 1.6 Let $L \subset [0 , 1] \times \mathbf{R}^n$, $V : L \longrightarrow \mathbf{R} \bigcup \{\pm \infty\}$, $F : [0 , 1] \times \mathbf{R}^n \longrightarrow \mathbf{R}^n$ satisfy all the assumptions of Theorem 1.4 or 1.5 and the boundary condition

$$V(1 , x) = \begin{cases} g(x) \text{ when } x \in K \\ +\infty \text{ otherwise} \end{cases} \tag{1.17}$$

Further assume that for every $x \in S_{[0 , 1]}$ with $x(1) \in K$ we have graph $x \subset \text{Dom}\, V$. If $\bar{x} \in S_{[0 , 1]}$ is such that $\bar{x}(0) = x_0$ and $V(t , \bar{x}(t)) \equiv \text{const} \neq \pm \infty$, then $\bar{x}$ is an optimal solution of the problem (1.16).

Proof. Observe that $\bar{x}(1) \in K$. By Theorem 1.4 or 1.5 for every trajectory $x \in S_{[0,1]}(x_0)$ satisfying $x(1) \in K$:

$$V(0, x_0) \leq V(1, x(1)) = g(x(1)) .$$

Since

$$V(0, x_0) = V(0, \bar{x}(0)) = V(1, \bar{x}(1)) = g(\bar{x}(1)) \leq g(x(1))$$

the result follows. □

Remark The dynamical programming approach associates with the target problem its value function

$$V(t, \xi) = \inf \{g(x(1)) : x \in S_{[t,1]}(\xi), x(1) \in K\}$$

where $V(t, \xi) = +\infty$ when there is no $x \in S_{[t,1]}(\xi)$ verifying $x(1) \in K$. Clearly V is nondecreasing along trajectories of (1.1), hence it verifies the inequality (1.10).

2. Existence of "constant" trajectories.

Let $V : [0, 1] \times \mathbf{R}^n \longrightarrow \mathbf{R} \cup \{\pm\infty\}$ be a given function. In this section we investigate sufficient conditions for: $\forall (t, \xi) \in \operatorname{Dom} V$ there exists a trajectory $x \in S_{[t,1]}(\xi)$ verifying graph $x \subset \operatorname{Dom} V$ and such that the function $s \longrightarrow V(s, x(s))$ is nonincreasing on $[t, 1]$. This condition together with the results of Section 1 will infer that $V(s, x(s)) \equiv$ const along at least one trajectory of (1.1) defined on $[t, 1]$ with $x(t) = \xi$.

Theorem 2.1 Assume that for some $(t, \xi) \in \operatorname{Dom} V$, $\tau > 0$, there exists a trajectory $x \in W^{1,1}(t, t+\tau)$ of (1.1) with $x(t) = \xi$ and such that the function $s \longrightarrow V(s, x(s))$ is nonincreasing. Then

$$\exists\, u \in coF(t, \xi) \text{ such that } D_{\uparrow} V(t, \xi)(1, u) \leq 0 \tag{2.1}$$

Proof. By Theorem 1.1 for all small $s > 0$

$$x(t+s) \in \xi + s\, coF(t, \xi) + o(s)$$

where $\lim_{s \longrightarrow 0+} \frac{o(s)}{s} = 0$. Let $s_i \longrightarrow 0+$, $u \in coF(t, x)$ be such that $\lim_{i \longrightarrow \infty} \frac{x(t+s_i)-\xi}{s_i} = u$.

Then

$$\liminf_{h \longrightarrow 0+,\, u' \longrightarrow u} \frac{V(t+h, \xi+hu') - V(t, \xi)}{h} \leq \liminf_{i \longrightarrow \infty} \frac{V(t+s_i, x(t+s_i)) - V(t, \xi)}{s_i} \leq 0$$

This ends the proof. □

To get the statement opposite to Theorem 2.1, we have to require more regularity of V.

From now until the end of the section we assume that for some $a > 0$ and all $t \in [0, 1]$, $x \in \mathbf{R}^n$

$$\sup \{ \|u\| : u \in F(t, x)\} \le a(\|x\| + 1) \tag{2.2}$$

and that F has convex images.

Theorem 2.2. Let $P : [0, 1] \rightrightarrows \mathbf{R}^n$ be a set valued map with nonempty images and closed graph and let $V : \text{graph } P \longrightarrow \mathbf{R}$ be a continuous function satisfying the inequality:

for all $t \in [0, 1]$, $x \in P(t)$ there exists $u \in F(t, x)$ such that $D_\uparrow V(t, x)(1, u) \le 0$. (2.3)

Then for every $(t, \xi) \in \text{graph } P$ we can find $z \in S_{[t, 1]}(\xi)$ such that the function $s \longrightarrow V(s, z(s))$ is nonincreasing on $[t, 1]$.

Proof.

Step 1. Fix $(t, \xi) \in \text{graph } P$ with $t < 1$, $z_0 \ge V(t, \xi)$ and let $h > 0$ be such that $t + h \le 1$. Set $K = \text{epi } V$,

$$F_1(s, x) = \{1\} \times F(t + s, x) \times \{0\}$$

$$\hat{F}(s, x) = \begin{cases} F_1(s, x) \text{ for } s < h \\ \overline{co}(F_1(h, x) \cup \{0\}) \text{ for } s = h \end{cases}$$

Then $\hat{F}$ is upper semicontinuous on $[0, h] \times \mathbf{R}^n$.

Consider the viability problem

$$\begin{cases} y' \in \hat{F}(s, y), y(s) \in K \\ y(0) = (t, \xi, z_0) \end{cases} \tag{2,4}$$

Fix $0 \le s \le h$, $(t + s, x, z) \in K$ and let $u \in F(t + s, x)$ be such $D_\uparrow V(t + s, x)(1, u) \le 0$. Then $(1, u, D_\uparrow V(t + s, x)(1, u)) \in T_K(t + s, x, z)$ and we proved that $\hat{F}(s, x) \cap T_K(t + s, x, z) \ne \emptyset$.

For $s = h$ we have $0 \in \hat{F}(s, x)$ for all x. Thus $\hat{F}$ verifies the viability condition on K. By the Haddad viability theorem ([1]) there exists $0 < T \le h$ and a trajectory $y : [0, T[\longrightarrow \mathbf{R}^n$ of the differential inclusion

$$y' \in \hat{F}(s, y), y(0) = (t, \xi, z_0)$$

satisfying $y([0, T[) \subset K$. Since $y'(\cdot)$ is essentially bounded (thanks to (2.2), definition of $\hat{F}$ and the Gronwall lemma), y may be extended on the whole interval $[0, T]$.

Using that K is closed we obtain that $y([0, T]) \subset K$. Since $(t, \xi, z_0) \in K$ is arbitrary, using the assumption (2.2) we prove that y may be extended on the whole time interval $[0, h]$. Let $\omega(s) \in \mathbf{R}^n$, $y_0(s) \in \mathbf{R}$ be such that $y(s) = (t + s, \omega(s), y_0(s))$. By definition of $\hat{F}$, $y_0(s) \equiv V(t, \xi)$. Setting $x(t + s) = \omega(s)$ we obtain that $x \in S_{[t, t+h]}(\xi)$ and

$$V(t+h\,,\,x(t+h)) \le V(t\,,\,\xi) = y_0(h)$$

Step 2. Fix $(t\,,\,\xi) \in P\,,\, t < 1\,,\, t = t_0 < \ldots < t_i < t_{i+1} < \ldots < t_k = 1$. Set $x(t_0) = \xi$. By Step 1 there exists $x_i \in S_{[t_i\,,\,t_{i+1}]}\,(x_i(t_i))$ such that for all $i \ge 0$

$$V(t_{i+1}\,,\,x_i(t_{i+1})) \le V(t_i\,,\,x_i(t_i))$$

Hence there exists $x \in S_{[t\,,\,1]}(\xi)$ such that $\forall\, i = 0\,,\ldots,\,k-1\,,\; V(t_{i+1}\,,\,x(t_{i+1})) \le V(t_i\,,\,x(t_i))$.

Step 3. Let $\{t_i\}_{i\ge 0}$ be a dense subset of $[t\,,\,1]$. Fix $j \ge 0$. By Step 2 there exists $x_j \in S_{[t\,,\,1]}(\xi)$ such that for all $i\,,\,r \in \{0\,,\ldots,\,j\}$

$$t_i \le t_r \Longrightarrow V(t_r\,,\,x_j(t_r)) \le V(t_i\,,\,x_j(t_i))$$

Since F has convex images and satisfies the growth condition (2.2), the sequence $\{x_j\}$ has a subsequence converging uniformly to some $z \in S_{[t\,,\,1]}(\xi)$. Then for all $i\,,\,r \ge 0$

$$t_i \le t_r \Longrightarrow V(t_r\,,\,z(t_r)) \le V(t_i\,,\,z(t_i))$$

Fix $t \le \tau < s \le 1$ and let $\{t_{i_j}\}\,,\,\{t_{i_r}\}$ be subsequences converging to τ and s respectively. Then for all large j and r

$$V(t_{i_r}\,,\,z(t_{i_r})) \le V(t_{i_j}\,,\,z(t_{i_j}))$$

Using continuity of V and taking the limit in the above inequality we get $V(\tau\,,\quad z(\tau)) \le V(t\,,\,z(t))$. Hence V is nonincreasing along z. □

Theorems 1.4 and 2.2 yield

Corollary 2.3 Let V and P be as in Theorem 2.2. Further assume that for every $(t\,,\,x) \in \text{graph}\,P\,,\, t < 1$ we have $(1\,,\,F(t\,,\,x)) \subset T_{\text{graph}\,P}(t\,,\,x)$ and

$$\sup_{u \in F(t\,,\,x)} D_\uparrow(-V)(t\,,\,x)(1\,,\,u) \le 0$$

If F is locally Lipschitz in both variables then for every $(t\,,\,\xi) \in \text{graph}\,P$ there exists $z \in S_{[t\,,\,1]}(\xi)$ such that $V(s\,,\,z(s)) = \text{const}$ on $[t\,,\,1]$.

From Theorems 2.2 and 1.5 we also deduce

Corollary 2.4 Let V and P be as in Theorem 2.2 and assume that V is locally Lipschitz on graph P and for all

$$(t\,,\,x) \in \text{graph}\,P\,,\, \forall\, u \in F(t\,,\,x) \text{ with } (1\,,\,u) \in T_{\text{graph}\,P}\,(t\,,\,x)\,,$$

$$D_\uparrow(-V)(t\,,\,x)(1\,,\,u) \le 0$$

Then for every $(t\,,\,\xi) \in \text{graph}\,P$ there exists $z \in S_{[t\,,\,1]}(\xi)$ such that $V(s\,,\,z(s)) = \text{const}$ on $[t\,,\,1]$.

Theorem 2.5 Let $\Omega \subset \mathbf{R}^n$ be an open set and $V : [0, 1] \times \Omega \longrightarrow \mathbf{R}$ be a continuous function such that for all $t \in [0, 1]$, $x \in \partial \Omega$

$$\lim_{x' \longrightarrow_\Omega x} V(t, x') = +\infty$$

Further assume that for every $t \in [0, 1[$, $x \in \Omega$

$$\exists\, u \in F(t, x) \text{ such that } -M(\|x\| + 1) \leq D_\uparrow V(t, x)(1, u) \leq 0$$

where M does not depend on (t, x). Then for every $(t, \xi) \in [0, 1] \times \Omega$ there exists $z \in S_{[t, 1]}(\xi)$ such that the function $s \longrightarrow V(s, z(s))$ is nonincreasing on $[t, 1]$.

The above result and Theorem 1.4 imply

Corollary 2.6. Under all assumptions of Theorem 2.5 assume that F is locally Lipschitz in both variables and for every $t \in [0, 1[$, $x \in \Omega$

$$\sup_{u \in F(t, x)} D_\uparrow(-V)(t, x)(1, u) \leq 0$$

Then for every $(t, \xi) \in [0, 1] \times \Omega$ there exists $z \in S_{[t, 1]}(\xi)$ such that $V(s, z(s)) = \text{const}$ on $[t, 1]$.

Proof of Theorem 2.5. Set $K = \text{graph } V$, $F_1(s, x) = \{1\} \times F(t + s, x) \times [M(\|x\| + 1), 0]$, define $\hat{F}$ in the same way as before. Then K is closed. Fix $(t, \xi) \in [0, 1[\times \Omega$. By the same arguments we obtain the existence of a trajectory $y : [0, T] \longrightarrow \mathbf{R}^n$ of (2.4) satisfying $y([0, T]) \subset K$. Let $z(s)$, $y_0(s)$ be such that $y(s) = (t + s, z(s), y_0(s))$.

Then $y_0(T) \neq +\infty$ and thus $z(T) \in \Omega$. Using the same arguments as above we prove the existence of a trajectory $\bar{x} : [t, 1] \longrightarrow \mathbf{R}^n$ of (1.1) such that $s \longrightarrow V(s, \bar{x}(s))$ is a nonincreasing function.

3. Optimal Feedback

Observe that if V and z are as in Corollary 2.3 (or 2.4, or 2.6) then for all $s \in [t, 1]$ and all small $h > 0$, $V(s + h, z(s + h)) = V(s, z(s))$. Thus $D_\uparrow V(s, z(s))\,(1, z'(s)) \leq 0$ whenever the derivative $z'(s)$ does exist. For all $(s, x) \in \mathbf{R} \times \mathbf{R}^n$ set

$$G(s, x) = \{u \in F(s, x) : D_\uparrow V(s, x)(1, u) \leq 0\} \tag{3.1}$$

and consider the differential inclusion

$$x' \in G(s, x) \tag{3.2}$$

Under all assumptions of Corollary 2.3 (or 2.4, or 2.6) for every $(t, \xi) \in [0, 1] \times \mathbf{R}^n$ there exists a solution z of (3.2) with $z(t) = \xi$ and such that $V(s, z(s)) = \text{const}$ on $[t, 1]$. A natural question arises if for every trajectory $z \in W^{1,1}(t, 1)$ of (3.2) the function $s \longrightarrow V(s, z(s)) \equiv \text{const}$ on $[t, 1]$.

Definition 3.1. The set-valued map $G : [0, 1] \times \mathbf{R}^n \longrightarrow \mathbf{R}^n$ is called an optimal feedback low associated to V if

(a) $\forall (t, \xi) \in Dom\ V$ there exists a solution x of (3.2) defined on $[t, 1]$ such that $x(t) = \xi$

(b) For every solution $x \in W^{1,1}(t, 1)$ of (3.2), the function $s \longrightarrow V(s, x(s)) \equiv$ const on $[t, 1]$.

Theorem 3.2. Let $P : [0, 1] \longrightarrow \mathbf{R}^n$ be a set-valued map with nonempty images and closed graph and $V :$ graph $P \longrightarrow \mathbf{R}$ be a locally Lipschitz function satisfying (2.3) and such that

$$\forall (t, x) \in \text{graph } P, \forall u \in F(t, x) \text{ with } (1, u) \in T_{\text{graph } P}(t, x),$$

$$D_{\uparrow}(-V)(t, x)(1, u) \leq 0$$

If F has convex images and verifies the growth condition (2.2) then the map G defined by (3.1) is an optimal feedback low.

Proof. By Corollary 2.4 the requirement (a) of Definition 3.1 is satisfied. Pick a solution $z \in W^{1,1}(t, 1)$ of (3.2). Then the function $s \longrightarrow V(s, z(s))$ is absolutely continuous.

Let s be so that $\varphi'(s)$, $z'(s) \in F(s, z(s))$ do exist and $h_i \longrightarrow 0+$ be such that

$$D_{\uparrow} V(s, z(s))(1, z'(s)) = \lim_{i \longrightarrow 0+} \frac{V(s + h_i, z(s + h_i)) - V(s, z(s))}{h_i}$$

Thus $\varphi'(s) = D_{\uparrow} V(s, z(s))(1, z'(s)) \leq 0$ and φ is nonincreasing. By Theorem 1.5, φ is also nondecreasing. Thus $\varphi =$ const. □

Exactly the same arguments yield

Theorem 3.3. Let Ω be an open subset of $\mathbf{R}^n$ and $V : [0, 1] \times \Omega \longrightarrow \mathbf{R}$ be a locally Lipschitz function satisfying all the assumptions of Corollary 2.6. If F has convex images and satisfies (2.2) then the map G defined by (3.1) is an optimal feedback low.

Remark: In the above theorem we may avoid the assumption of Lipschitz continuity of V with respect to t, if instead we assume that for every $(t_0, x_0) \in Dom\ V$ there exist $L > 0$, $\epsilon > 0$ such that for all $(t, x) \in [t_0 - \epsilon, t_0 + \epsilon] \bigcap [0, 1] \times (x_0 + \epsilon B)$, $V(t, \cdot)$ is L-Lipschitz on $\{x : (t, x) \in \text{graph } P \bigcap \{t\} \times (x_0 + \epsilon B)\}$ and restrict our attention only to those solutions $x \in W^{1,1}(t, 1)$ of (3.2) for which $s \longrightarrow V(s, x(s))$ is absolutely continuous.

We apply the above results to the target problem considered at the end of the first section.

Theorem 3.4 Let V be as in Theorem 3.2 or as in Theorem 3.3 and verifies the boundary condition (1.17). Then every trajectory $z \in W^{1,1}(0, 1)$ of (3.2) (with G defined by (3.1)) satisfying $z(0) = x_0$ is an optimal solution of the target problem. Moreover if F has convex images and verifies (2.2), then (3.2) has at least one solution starting at x_0.

4. Viscosity solutions of Hamilton-Jacobi-Bellman equation and contingent inequalities.

We associate with F its Hamiltonian H defined by

$$\forall (t , x) \in [0 , 1] \times \mathbf{R}^n , \forall q \in \mathbf{R}^n , H(t , x , q) = \sup \{ <q , e> : e \in F(t , x) \} \tag{4.1}$$

Our aim is to show that for any open set $\Omega \subset [0 , 1] \times \mathbf{R}^n$ every solution of the problem

$$\left.\begin{array}{l} \inf\limits_{u \in coF(t , x)} D_{\uparrow} V(t , x) (1 , u) \leq 0 , (t , x) \in \Omega \\ \sup\limits_{u \in coF(t , x)} D_{\uparrow} (-V) (t , x) (1 , u) \leq 0 , (t , x) \in \Omega \end{array}\right\} \tag{4.2}$$

is the viscosity solution of Hamilton-Jacobi equation

$$-\frac{\partial}{\partial t} V + H \left(t , x , -\frac{\partial}{\partial x} V\right) = 0 , \ (t , x) \in \Omega \tag{4.3}$$

(see Crandall-Lions [8], and Crandall-Evans-Lions [9]). Some related results can be found in [18]. We recall first

Definition 4.1 (super- and subdifferentials). Let $\Omega \subset \mathbf{R}^m$ be an open set, φ be a function from Ω to $\mathbf{R}$ and $x_0 \in \Omega$.

The *superdifferential* of φ at x_0 is the set

$$\partial_+ \varphi(x_0) = \{p : \limsup_{x \to x_0} [\varphi(x) - \varphi(x_0) - < p , x - x_0 >] / \| x - x_0 \| \leq 0\} .$$

The *subdifferential* of φ at x_0 is the set

$$\partial_- \varphi(x_0) = \{p : \liminf_{x \to x_0} [\varphi(x) - \varphi(x_0) - < p , x - x_0 >] / \| x - x_0 \| \geq 0\} .$$

The super and subdifferentials are closed, possibly empty, convex sets.

Definition 4.2 (viscosity solution). A function $V : \Omega \longrightarrow \mathbf{R}$ is called a viscosity solution of the equation (4.3) if for every $(t , x) \in \Omega$ we have

a) for all $p = (p_0 ,\ldots, p_n) \in \partial_+ V(t , x)$

$$-p_0 + H(t , x , -(p_1 ,\ldots, p_n)) \leq 0$$

(viscosity subsolution).

b) for all $p = (p_0 ,\ldots, p_n) \in \partial_- V(t , x)$

$$-p_0 + H(t , x ,-(p_1 ,\ldots, p_n)) \geq 0 .$$

(viscosity supersolution).

Lemma 4.3. [12] Let Ω be an open set and $\varphi : \Omega \longrightarrow \mathbf{R}$. Then

$$\partial_- \varphi(x_0) = \{p : \forall r \in \mathbf{R}^m , D_+ \varphi (x_0) r \geq < p , r >\} \tag{4.4}$$

$$\partial_+ \varphi(x_0) = \{p : \forall r \in \mathbf{R}^m , D_+ (-\varphi) (x_0) r \geq < -p , r >\} = -\partial_- (-\varphi)(x_0) \tag{4.5}$$

Theorem 4.4. If a function $V : \Omega \longrightarrow \mathbf{R}$ verifies relations (4.2), then V is a viscosity solution to Hamilton-Jacobi equation (4.3).

Proof. By (4.4), (4.2) for all $(p\ ,\ q) \in \partial_-\ V(t\ ,\ x) \subset \mathbf{R} \times \mathbf{R}^n$

$$p - H(t\ ,\ x\ ,\ -\ q) = \inf_{u\ \in\ coF(t\ ,\ x)} (p + < q\ ,\ u >) \leq \inf_{u\ \in\ coF(t\ ,\ x)} D_\uparrow\ V(t\ ,\ x)\ (1\ ,\ u) \leq 0$$

Thus for all $(p\ ,\ q) \in \partial_-\ V(t\ ,\ x)$

$$-p + H(t\ ,\ x\ ,\ -\ q) \geq 0 \tag{4.6}$$

On the other hand by (4.5), (4.2) for all $(p\ ,\ q) \in \partial_+\ V(t\ ,\ x)$

$$-p + H(t\ ,\ x\ ,\ -\ q) = \sup_{u\ \in\ coF(t\ ,\ x)} (-p - < q\ ,\ u >) \leq \sup_{u\ \in\ coF(t\ ,\ x)} D_\uparrow(-\ V)\ (t\ ,\ x)\ (1\ ,\ u) \leq 0\ .$$

The very definition of viscosity solution ends the proof. □

Hence solutions of (4.2) form a subset of viscosity solutions of (4.3). Since contingent derivative is not convex, in general a viscosity solution of (4.3) does not verify (4.2). For this reason results of Sections 1,2,3 do not apply to an arbitrary viscosity solution.

For the target problem considered in Section 1, the value function is nondecreasing along trajectories of (1.1) and is constant along optimal trajectories. Hence results of Section 3 may be applied to the value function when it verifies additional regularity requirements. However to find the value function may be a difficult task.

If the HJB equation (4.3) (1.17) has a unique solution V and the value function is a solution of (4.3) then the results of Section 3 may be applied to V when it is regular enough.

However uniqueness theorems for viscosity solutions concerns (up to now) only continuous solutions (see [8]-[10]). When the value function is discontinuous, we have to take into consideration contingent inequalities from Sections 1,2. When the viscosity solution verifies in addition these contingent inequalities, then results of Section 3 may be applied.

We provide next an example of the target problem having a Lipschitz viscosity solution different from the value function. It is obtained thanks to an appropriate choice of the domain of definition.

Example Let U denote the closed unit ball in $\mathbf{R}^2$ and consider the set-valued map $F : \mathbf{R} \times \mathbf{R}^2 \longrightarrow \mathbf{R}^2$ defined by $F(t\ ,\ x) = U$. We consider the point target $K = (1\ ,\ 0)$ and the function $g \equiv 0$. The reachable set at time 1 of the inclusion

$$x' \in F(t\ ,\ x)\ ,\ x(0) = x_0 \tag{4.7}$$

is equal to

$$R(1) = x_0 + U$$

It is easy to see that

$$V(t\ ,\ x) = \begin{cases} 0 & \text{if } \|\ x - (1\ ,\ 0)\ \| \leq 1 - t \\ +\ \infty & \text{otherwise} \end{cases}$$

Set

$$W(t, x_1, x_2) = \begin{cases} 0 & \text{if } | x - 1 | \le 1 - t, | x_2 | \le 1 - t \\ + \infty & \text{otherwise} \end{cases}$$

We first observe that $V \neq W$ and that $V(1, \cdot) = W(1, \cdot)$, i.e. V and W verify the same boundary condition. On the other hand for all $(t, x) \in \text{Int Dom } W$, $\nabla W(t, x) = 0$. Hence W is a solution of the Hamilton-Jacobi equation (4.3) on Int Dom W.

Even if we apply the definition of viscosity solution on Dom W from [10], it is still possible to check that W solves (4.3) on Dom W. Thus the Hamilton-Jacobi-Bellman equation (4.3) has at least two solutions verifying the same boundary condition

$$V(1, x) = \begin{cases} 0 & \text{if } x = (1, 0) \\ + \infty & \text{otherwise} \end{cases}$$

Observe that $V = W$ on Dom V and one would wonder if the above negative statements would be improved if we restricted our attention only to Dom V. However for an arbitrary nonlinear control system (1.1), the set Dom V may be as difficult to find as the function V itself (the global controllability on nonlinear systems remains an unsolved problem!). Therefore such improvement would be only an illusory one. On the other hand the map W does not verify the second contingent inequality (2.1) for (t, ξ) from the boundary of Dom W with $0 < t < 1$. In this way the function W may be excluded from consideration.

5. Clarke's verification technique.

In [6], [7] a different approach to the target problem was developed.

Definition 5.1. Let $\varphi : \mathbf{R}^m \longrightarrow \mathbf{R}$ be a locally Lipschitzian at x function. The epiderivative $\varphi^0(x) : \mathbf{R}^m \longrightarrow \mathbf{R}$ is defined by: for all $u \in \mathbf{R}^m$

$$\varphi^0(x)u = \limsup_{\substack{x' \longrightarrow x \\ h \longrightarrow 0+}} [\varphi (x' + hu) - \varphi(x')] / h$$

The generalized gradient $\partial \varphi(x)$ is given by

$$\partial \varphi(x) = \{p \in \mathbf{R}^m : \forall u \in \mathbf{R}^m, < p, u > \le \varphi^0(x)u\}$$

Observe that $\varphi^0(x) \ge D_\uparrow \varphi(x)$ and therefore

$$\partial_- \varphi(x) \subset \partial \varphi(x), \partial_+(-\varphi)(x) \subset - \partial \varphi(x) = \partial(-\varphi)(x).$$

Let $\Omega \subset \mathbf{R} \times \mathbf{R}^n$ be an open set. A locally Lipschitz function $V : \Omega \longrightarrow \mathbf{R}$ is called generalized solution of Hamilton-Jacobi equation (4.3) if for all $(t, x) \in \Omega$

$$\sup_{(p, q) \in \partial V(t, x)} \left\{ -p + H(t, x, - q) \right\} = 0$$

It was shown in [7] that if $z \in W^{1,1}(t\ ,\ 1)$ is a trajectory of (1.1) such that for some $\epsilon > 0$ and all $t < s < 1$, $(s\ ,\ z(s) + \epsilon\ B) \subset \Omega$, then $s \longrightarrow V(s\ ,\ z(s))$ is nondecreasing on $[t\ ,\ 1]$. This led to a verification technique.

To make a comparison with the results of Section 1, observe that for all $(t\ ,\ x) \in \Omega$

$$\sup_{(p\ ,\ q) \in \partial V(t\ ,\ x)} \ \sup_{u \in coF(t\ ,\ x)} -p + < -\ q\ ,\ u >$$

$$= \sup_{u \in coF(t\ ,\ x)} \ \sup_{(p\ ,\ q) \in \partial(-V)(t\ ,\ x)} -p + < -\ q\ ,\ u >$$

$$= \sup_{u \in coF(t\ ,\ x)} (-V)^o\ (t\ ,\ x)(1\ ,\ u) \leq 0$$

Hence

$$\sup_{u \in coF(t\ ,\ x)} D_{\uparrow}\ (-V)\ (t\ ,\ x)(1\ ,\ u) \leq 0 \text{ and,}$$

consequently, V verifies the contingent inequality (1.10). Hence replacing V by its restriction to $\{(s\ ,\ z(s) + \epsilon\ \mathring{B}) \colon s \in [t\ ,\ 1]\}$ and using Theorem 1.5 we prove that V is nondecreasing along z.

We provide next a comparison of viscosity and generalized solutions of (4.3).

Theorem 5.1 A locally Lipschitz function $V : \Omega \longrightarrow \mathbf{R}$ is a viscosity solution of (4.3) if and only if V is a generalized solution and

$$\forall (p\ ,\ q) \in \partial_-\ V(t\ ,\ x)\ , \quad -p + H(t\ ,\ x\ ,\ -q) = 0$$

Proof By the definition of Hamiltonian H, for every $t \in [0\ ,\ 1]\ ,\ x \in \mathbf{R}^n\ ,\ H(t\ ,\ x\ ,\ \cdot)$ is convex. By the assumption on F , H is continuous. Thus [13, Theorem 2.3] ends the proof. □

References

[1] Aubin, J.-P. and A. Cellina (1984). *Differential Inclusions.* Springer-Verlag.

[2] Aubin, J.-P. and I. Ekeland (1984). *Applied Nonlinear Analysis.* Wiley-Interscience.

[3] Bellman, R.E. (1957). *Dynamic Programming.* Princeton University Press.

[4] Berkowitz, L. (to appear)

[5] Clarke, F.H. (1982). The applicability of the Hamilton-Jacobi verification technique, Proceedings IFIP, New York, September 1981, in System Modeling and Optimization 38, Springer-Verlag, pp. 88-94.

[6] Clarke, F.H. (1983). *Optimization and Nonsmooth Analysis.* Wiley- Interscience.

[7] Clarke, F.H. and R.B. Vinter (1983). Local optimality conditions and Lipschitzian solutions to the Hamilton-Jacobi equation, SIAM J. of Control and Optimization, 21(6), pp. 856-870.

[8] Crandall, M.G. and P.L. Lions (1983). Viscosity solutions of Hamilton-Jacobi equations, Trans. Amer. Math. Soc., 277, pp. 1-42.

[9] Crandall, M.G., L.C. Evans and P.L. Lions (1984). Some properties of viscosity solutions of Hamilton-Jacobi equation, Trans. Amer. Math. Soc., 282(2), pp. 487-502.

[10] Ishii, H. (1988). Representation of solutions of Hamilton-Jacobi equations, Nonlinear Analysis 12, pp. 121-146.

[11] Fleming, W.H. and R.W. Rishel (1975). *Deterministic and Stochastic Optimal Control.* Springer-Verlag.

[12] Frankowska, H. (1988). Optimal trajectories associated to a solution of contingent Hamilton-Jacobi equation. Working Paper 87-069, IIASA, to appear in Applied Mathematics and Optimization.

[13] Frankowska, H. (1989). Hamilton-Jacobi Equations: viscosity solutions and generalized gradients. J. Math. Analysis and Appl. (to appear)

[14] Frankowska, H. (to appear). Set-valued analysis and some control problems, Proceedings of Conference 30 Years of Modern Control Theory, E. Roxin, (editor), June 1988, Marcel Dekker.

[15] Fleming, W.H. and R.W. Rishel (1975). *Deterministic and Stochastic Optimal Control.* Springer-Verlag.

[16] Lions, P.L. (1983). Existence results for first order Hamilton-Jacobi equations, Ric. Mat. 32, pp. 3-23.

[17] Lions, P.L. (1982). *Generalized solutions of Hamilton-Jacobi Equations*, Pitman, Boston.

[18] Lions, P.L., P.E. Souganidis (1985). Differential games, optimal control and directional derivatives of viscosity solutions of Bellman's and Isaaks' Equations, SIAM J. of Control and Optimization, Vol. 23, no. 4.

[19] Zeidler, E. (1984). *Nonlinear Functional Analysis and its Applications.* Vol. III, Springer-Verlag.

SYSTEMES STOCHASTIQUES ET QUANTIQUES

STOCHASTIC AND QUANTUM SYSTEMS

STOCHASTIC QUANTIZATION[1]

Sanjoy K. Mitter
Department of Electrical Engineering and Computer Science
and
Laboratory for Information and Decision Systems
Massachusetts Institute of Technology
Cambridge, MA 02139
U.S.A.

1. INTRODUCTION

In recent work [1] we have studied stochastic differential equations related to the free field and $(\varphi^4)_2$-fields in finite volume following the earlier work of Jona-Lasinio and Mitter [2]. In [3] we have studied Lattice approximations to these stochastic differential equations and proved a limit theorem when the lattice spacing goes to zero. We now describe the nature of the results we have obtained.

Let $\Lambda\subset\mathbf{R}^2$ be a finite open rectangle and S' denote $\mathcal{D}'(\Lambda)$ the space of distributions on Λ and let $\mathcal{S}'$ denote the space of tempered distributions on Λ. Let $C_i = (-\Delta+I)^{-1}$, i = 1,2 with Dirichlet (resp. free) boundary conditions on Λ. C_i, i = 1,2 are covariance operators and for C a covariance operator let $C(\cdot,\cdot)$ denote its integral kernel, C^α its αth operator power and let μ_C denote the centered Gaussian measure with variance operator C. Consider the following S'-valued stochastic differential equation

$$\begin{cases} d\varphi(t) = -\dfrac{1}{2}\, C_1^{-\varepsilon}\varphi(t)dt + dw(t) \\ \\ \varphi(0) = \phi \in S', \;\; 0<\varepsilon<1 \end{cases} \tag{1.1}$$

where W(t) is a Wiener process with covariance $C_1^{1-\varepsilon}$. It is not difficult to prove that this equation has a unique solution and has a path continuous version as an $H^{-\alpha}$-valued process on $(0,\infty)$. Moreover $\varphi(\cdot)$ is ergodic and has μ_{C_1} as it unique invariant measure. The same claims can be made with C_1 replaced by C_2. This procedure of creating a stochastic differential equation with unique invariant measure a desired invariant measure is termed stochastic quantization. It is worth observing that the random field $\varphi(t)$ for each t is a Markov random field and satisfies the Osterwalder-Schrader axioms. A proof of this will follow from that of Nelson [4]. Note that we cannot take ε=0 in equation (2.1), since the transition probabilities $p(t;\varphi,.)$ of the process φ for different t's are no longer mutually absolutely continuous, a fact needed to prove ergodicity of the process $\varphi(\cdot)$. The case ε=1 is excluded since W(t) is then no longer a genuine Wiener process.

Since the process $\varphi(\cdot)$ is ergodic with unique invariant measure μ_{C_1}, correlation functions

[1]This research was been supported by the Air Force Office of Scientific Research grant AFOSR-85-0227 and the Army Research Office under grant DAAL03-86-K-0171 through the Center for Intelligent Control Systems.

$E_{\mu_{C_1}}(\tilde{\varphi}(x_1)\ldots\tilde{\varphi}(x_n))$, $((\tilde{\varphi})$ denotes the gaussian random field with covariance $\mu_{C_1})$ can be computed by exchanging time and space averages. This is the basic idea behind Monte Carlo calculations of statistics of the random field.

We study this differential equation in a space of distributions since the invariant measure μ_{C_1} can only be supported in some space of distributions. This is a consequence of the Minlos Theorem. It can be shown that the measure μ_{C_1} is supported in the space $H^{-1}(\Lambda)$, the dual of the Sobolev space $H^1(\Lambda)$.

In [1] and [3], we have also studied the infinite-dimensional non-linear stochastic differential equation

$$d\varphi(t) = -\frac{1}{2}(C^{-\varepsilon}\varphi(t) + C^{1-\varepsilon}: \varphi(t)^3:)dt + dw(t) \tag{1.2}$$

with $\varphi(0)$ having initial law μ given by:

$$\left.\begin{aligned} \frac{d\mu}{d\mu_{C_1}} &= \exp\left(-\frac{1}{4}\int_\Lambda :\phi^4:dx\right)\Big/Z \\ Z &= \int_\Lambda \exp\left(-\frac{1}{4}\int :\varphi^4:dx\right) d\mu_{C_1}(\varphi) \end{aligned}\right\} \tag{1.3}$$

In the above $:\varphi(t)^3:$ denotes Wick-ordering with respect to μ_{C_1} and has the explicit definition:

$$:\varphi(t)^3: = \varphi^3(t) - 3\left(E_{\mu_{C_1}}\varphi(t)^2\right)\varphi(t) \tag{1.4}$$

and is well-defined as an element of $L^2(d\mu_{C_1})$. Similarly $:\varphi^4:$ denotes WIck-ordering with respect to μ_{C_1} and the integral $\int_\Lambda :\varphi^4:dx$ is well-defined as an element of $L^2(d\mu_C)$ via an appropriate limiting procedure. The fact that μ is a well-defined probabability measure is a consequence of Nelson's estimate [4].

The difficulty of studying equation (1.2) is that since the non-linear drift term $:\varphi(t)^3:$ is only defined in some limiting sense we cannot interpret it in the Ito snese and hence we have to interpret it in a weak sense. In [1] it is shown that the new measure P_0 defined by

$$\begin{aligned} \frac{dP_0}{dP} = \exp\Big(\frac{1}{2}\int_0^T <:\varphi^3(s):, dw(s)> - \frac{1}{8}\int_0^T <:\varphi^3(s):, C^{1-\varepsilon}: \varphi^3(s):>_{\dot{H}^{\alpha}, H^{\alpha}} ds \\ + \frac{1}{4}\int_\Lambda :\varphi^4(0): dx\Big)\Big/Z \end{aligned} \tag{1.5}$$

where Z is a normalizing constant, is a well-defined probability measure. The proof uses both estimates from quantum field theory and probabilistic arguments (in particular Novikov's criterion for an exponential super-martingale to be a martingale).

In [1] a limit theorem at the process level when $\Lambda \uparrow \mathbf{R}^2$ is also proved.

2. STOCHASTIC QUANTIZATION AND IMAGE ANALYSIS

Our interest in these problems arose from problems of image analysis. To see this note that the measure μ corresponds to Hamiltonian

$$H = \int_{\Lambda} \left[\|\nabla\varphi\|^2 + m_0^2\,\varphi^2 + \lambda{:}\varphi^4{:} \right] dx \tag{2.1}$$

where m_0 is the bare mass and λ the coupling constant (taken both to be 1 in the previous section). Corresponding to the Hamiltonian we can construct the limit Gibbs measure in the sense of Sinai (cf. [5] and [4]).

Consider the following problems of Image Analysis.

Problem I.

Let $\Omega \subset \mathbf{R}^2$ be an open bounded set and let $\psi \in L^\infty(\Omega)$ be given. We think of ψ as an observed noisy image. We wish to construct an estimate $\varphi \in H^1(\Omega)$ such that

$$J(\varphi) = \int_{\Omega} |\psi - \varphi|^2 dx + \int_{\Omega} \|\nabla\varphi\|^2 dx$$

is minimized.

It is natural to think of $J(\varphi)$ as a conditional Hamiltonian $H_0(\varphi|\psi)$ and construct a conditional measure $\mu(\varphi|\psi)$ by making appropriate probabilistic hypotheses on ψ (for example by associating an Hamiltonian for ψ). To construct estimates we would have to compute statistics corresponding to the measure $\mu(\varphi|\psi)$ and this would be done using the ideas of stochastic quantization for both φ and ψ. A start towards doing this has been made in [6].

Problem II.

Let $\Omega \subset \mathbf{R}^2$, be bounded and open and let $\psi \in L^\infty(\Omega)$. Consider the following variational problem. Minimize

$$J(\varphi, \Gamma) = \int_{\Omega} |\psi - \varphi|^2 dx + \int_{\Omega \setminus \Gamma} |\nabla\varphi|^2 dx + \mathcal{H}^1(\Gamma),$$

where Γ is a closed set with $\Gamma \subset \bar{\Omega}$ and $\mathcal{H}^1(\Gamma)$ denotes the one-dimensional Hausdorff measure. The interpretation of this functional is that we want to find an estimate $(\hat{\varphi}, \hat{\Gamma})$ of the observed noisy image ψ which preserves the discontinuities of the image, there are not too many discontinuities and $\hat{\Gamma}$ is an estimate of the discontinuities. It can be shown that a minimizing solution $(\hat{\varphi}, \hat{\Gamma})$ exists [7], [8]. A detailed study of the first variation of J has been done in [9].

It is not clear how to give a probabilistic interpretation to this problem. However, if we consider a lattice analog, then we can give a probabilistic interpretation by constructing a measure on the lattice $Z^2x(Z^2)^*$, where $(Z^2)^*$ denotes the dual lattice. This was one of the motivations for our work reported in [3]. For details of this problem in a discrete space setting, see our paper [10] and the references cited there.

3. RENORMALIZATION GROUP METHODS AND A BELLMAN EQUATION

The main purpose of this section is to describe the renormalization group method of K.G. Wilson for U-V cut-off removal as formulated by P.K. Mitter [11, 12]. A certain infinite-dimensional Hamilton-Jacobi-Bellman equation arises in this context which has a natural control-theoretic interpretation.

Consider the linear parabolic equation in R^n x(0,T]

$$\left.\begin{aligned} dp^\varepsilon(x,t) &= L_\varepsilon^* p^\varepsilon(x,t) + \frac{1}{\varepsilon} V(x,t)p^\varepsilon(x,t) \\ p^\varepsilon(x,0) &= p_0^\varepsilon(x) = K_\varepsilon \exp\left(-\frac{1}{\varepsilon} S_0(x)\right) \end{aligned}\right\} \tag{3.1}$$

Here $\varepsilon > 0$, $S_0(x) > 0$, $\lim_{\varepsilon\to 0} \varepsilon \ell n\, K_\varepsilon = 0$ and L_ε^* is the formal adjoint of the diffusion operator

$$L_\varepsilon = \frac{\varepsilon}{2}\sum_{i=1}^{n} \frac{\partial^2}{\partial x_i^2} + \sum_{i=1}^{n} f_i(x)\frac{\partial}{\partial x_i} \tag{3.2}$$

We assume that f is a C^∞-function with bounded derivatives upto order 3, -V is a C^∞-function which is bounded below by zero.

Following, for example, Fleming-Mitter [13], introduce the logaarithmic transformation

$$S^\varepsilon(x,t) = -\varepsilon \ln p^\varepsilon(x,t). \tag{3.3}$$

Then $S^\varepsilon(x,t)$ satisfies the Bellman-Hamilton-Jacobi equation

$$\left.\begin{aligned} &\frac{\partial}{\partial t} S^\varepsilon(x,t) - \frac{\varepsilon}{2}\Delta S^\varepsilon(x,t) + H^\varepsilon(x,t,\nabla S(x,t)) = 0 \\ &S^\varepsilon(x,0) = -\varepsilon \ln p_0^\varepsilon(x), \end{aligned}\right\} \tag{3.4}$$

and $H^\varepsilon(x,t,p) = p'f(x) + \frac{1}{2}\|p\|^2 - V(x,t)$.

Formally, letting $\varepsilon \to 0$, we obtain the Hamilton-Jacobi equation

$$\left.\begin{aligned} &\frac{\partial}{\partial t} \tilde{S}(x,t) + H(x,t,\nabla\tilde{S}(x,t)) = 0, \\ &S(x,0) = S_0(x) . \end{aligned}\right\} \tag{3.5}$$

One can prove that $\lim_{\varepsilon \to 0} \varepsilon \ln p^{\varepsilon}(x,t) = -J(x,t)$ on compact subsets of $R^n x[0,T]$, where $J(x,t)$ is the value function of a deterministic optimal control problem:

Minimize

$$J(t; x_0, u) = S_0(x_0) + \frac{1}{2}\int_0^t \|u(s)\|^2 ds \tag{3.6}$$

subject to

$$\left.\begin{aligned} \frac{dx}{ds} &= f(x(s)) + u(s) \\ x(0) &= x_0. \end{aligned}\right\} \tag{3.7}$$

Let $U_{x,t} = \{(x_0, u) | x_u(0) = x_0, x_u(t) = x, u \in L^2(0,t;R^n)\}$, and

$$J(x,t) = \text{Inf}[J(t; x_0,u) | (x_0, u) \in U_{x,t}].$$

Then finally J satisfies (3.5). Note that this is a minimum energy optimum control problem. In a similar manner, $S^{\varepsilon}(x,t)$ has the interpretation of a value function for a Markovian stochastic opitmal control problem [12].

We now return to the ideas of section 1. We consider the random field $\phi(x)$ on R^d, $d>2$ with measure μ_C. The covariance C has a kernel $C(x-y)$ given by the formula (in terms of Fourier transforms)

$$C(x-y) = \frac{1}{(2\pi)^d}\int d^d\omega \, \frac{1}{\omega^2} \, e^{i\omega.(x-y)}$$

(the covariance operator is $(-\Delta)^{-1}$ in contrast to the covariance operator $(-\Delta+I)^{-1}$ in Section 1). Let the measure μ_{C_κ} be defined by giving the kernel

$$C_\kappa(x-y) = \int \frac{d^d\omega}{(2\pi)^d} \frac{e^{-\frac{\omega^2}{\kappa^2}}}{\omega^2} e^{i\omega.(x-y)}$$

A computation gives the scaling properties

$$C_\kappa(x-y) = \kappa^{d-2} C_1(\kappa(x-y)), \tag{3.8}$$

and if ϕ denotes the random field with measure μ_{C_κ} given by covariance C_κ and Φ denotes the random field with measure μ_{C_1} given by covariance C_1, then

$$\phi(x) = \kappa^{\frac{d-2}{2}} \Phi(\kappa x). \tag{3.9}$$

The measure μ_{C_κ} is supported on smooth functions. By virtue of the above

$$E_{\mu_{C_\kappa}}(\phi(x_1)...\phi(x_n)) = \kappa^{n(\frac{d-2}{2})} E_{\mu_{C_1}}(\Phi(\kappa x_1)...\Phi(\kappa x_n)) \tag{3.10}$$

The problem of studying the behaviour of the n-point correlations for fixed $x_1,...,x_n$ as $\kappa \to \infty$ is equivalent to studying the long distance (infinite volume limit) problem at a fixed cut-off.

Let $V_0(\phi)$ be an even polynomial and consider the new measure with interaction V_0

$$d\mu_\kappa = d\mu_{C_\kappa} \exp(-V_0(\phi)) \tag{3.11}$$

and the corresponding characteristic function

$$Z_\kappa(f) = \int d\mu_\kappa \exp(i\phi(f)) \tag{3.12}$$

There are two steps in the renormalization group method.

Step 1 (Scaling)

From (3.9),

$$d\mu_\kappa(\Phi) = d\mu_{C_1}(\Phi) \exp\left(-V_0\left(\kappa^{\frac{d-2}{2}} \Phi(\cdot)\right)\right). \tag{3.13}$$

Set

$$V_0\left(\kappa^{\frac{d-2}{2}} \Phi(\cdot)\right) = \mathcal{V}_0^{(\kappa)}(\Phi(\cdot)).$$

Then

$$Z_\kappa(f) = \int d\mu_{C_1}(\Phi)\exp\left(-\mathcal{V}_0^{(\kappa)}(\Phi(\cdot)) + \Phi(f_\kappa)\right) \tag{3.14}$$

where $f_\kappa(x) = \kappa^{\frac{d-2}{2}-d} f(\kappa^{-1}x)$.

Step 2. Lowering the Cut-Off.

Consider the transformation

$$1 \to e^{-t}.1, \quad t \in R_+ .$$

We know,

$C_1(x-y) = \int \frac{d^d\omega}{(2\pi)^d} \frac{e^{(-\frac{\omega^2}{1^2})}}{\omega^2} e^{i\omega.(x-y)}$, and hence

$$C_{e^{-t}.1}(x-y) = \int \frac{d^d\omega}{(2\pi)^d} \frac{e^{e^{(\frac{-\omega^2}{-2t})}}}{\omega^2} e^{i\omega.(x-y)} \tag{3.15}$$

Now $C_1 > C_{e^{-t}.1}$ as operators.

Let $C_1 = C_{e^{-t}.1} + C_t^{(h)}$. (3.16)

In the above C_1 is the covariance of the field Φ at unit cut-off, $C_{e^{-t}.1}$ the covariance corresponding to the lowered cut-off and $C_t^{(h)}$ the covariance corresponding to a fluctuating field.

From the (3.16) we have the decomposition $\Phi = \phi^{(1)} + \zeta$, ζ denoting the fluctuating field and $\phi^{(1)}$ and ζ are independent Gaussian field. The covariance kernel of $C_t^{(h)}$ has exponential decay as $|x-y| \to \infty$.

We now integrate out the fluctuating field and scale back.

$$\int d\mu_{C_1}(\Phi) \exp(-\mathcal{V}_0^{(\kappa)}(\Phi)) = \int d\mu_{C_{e^{-t}.1}}(\phi^{(1)}) d\mu_{C_t}(h) \exp(-\mathcal{V}_0^{(\kappa)}(\phi_1+\zeta))$$

$$= \int d\mu_{C_1}(\Phi) d\mu_{C_t^{(h)}}(\zeta) \exp\left[-\mathcal{V}_0^{(\kappa)}\left(e^{-\frac{d-2}{2}t}\Phi(e^{-t}.) + \zeta\right)\right].$$

The renormalization group transformation is defined by

$$\exp(-\mathcal{V}_t^{(\kappa)}(\Phi)) = \int d\mu_{C_t^{(h)}}(\zeta) \exp\left[-\mathcal{V}_0^{(\kappa)}\left(e^{-\frac{d-2}{2}t}\Phi(e^{-t}.) + \zeta\right)\right] \tag{3.17}$$

which sends

$$\mathcal{V}_0^{(\kappa)} \to \mathcal{V}_t^{(\kappa)}.$$

$\mathcal{V}_t^{(\kappa)}$ is called the effective potential.

A computation shows that $\mathcal{V}_t$ (dropping the superscript κ) satisfies the infinite-dimensional Bellman-Hamilton-Jacobi equation

$$\frac{\partial \mathcal{V}_t}{\partial t} = -\int d^dx \left(\left[\frac{d-2}{2} + x.\nabla_x\right]\Phi(x)\right) \frac{\delta \mathcal{V}_t}{\delta \Phi(x)}$$

$$- \int d^dx \,.\, d^dy\, K(x-y) \left[-\frac{\delta^2 \mathcal{V}_t}{\delta\Phi(x)\delta\Phi(y)} + \frac{\delta \mathcal{V}_t}{\delta\Phi(x)} \frac{\delta \mathcal{V}_t}{\delta\Phi(y)}\right] \tag{3.18}$$

where

$$K(x-y) = \int \frac{d^d\omega}{(2\pi)^d} e^{-i\omega.(x-y)} e^{-\omega^2}.$$

$\mathcal{V}_0^{(K)}$ will have parameters which will have to be fixed so that we start at a critical surface. Studying the fixed point of the renormalization group transformation is equivalent to studying the asymptotic behavior of the equation (3.18) (at least in the small region).

Equation (3.18) has a stochastic control interpretation as suggested earlier in the section, and $\mathcal{V}_t^{(K)}$ has the interpretation of a Bellman Value function. The machinery of non-linear semigroups may be useful for this purpose.

4. NEW PROBLEMS

We would like to suggest that the ideas of the renormalization group method as exposed in the previous section could be generalized to yield a dynamic renormalization group method which would be relevant to problems of stochastic quantization. A program for this is described below.

We consider the stochastic differential equation (1.1). The solution of this equation for each t gives us a Gaussian measure in path space. This path space Gaussian measure plays the role of the measure μ_C of section 3. Cut-offs can be introduced for this measure and scaling properties analogous to (3.8) and (3.9) obtained. Note that this Gaussian measure can be obtained via a Girsanov Transformation of Wiener measure. The interaction measure is now introduced by a second Girsanov transformation as in (1.5). The proposal is to proceed as in Section 3 where the renormalization group transformation is now a transformation of Girsanov functionals thereby creating an effective Girsanov functional. The details of this will be presented elsewhere.

REFERENCES

[1] V.S. Borkar, R.T. Chari and S.K. Mitter, Stochastic Quantization of Field Theory in Finite and Infinite Volume, to appear in J. of Functional Analysis, 1988.

[2] G. Jona-Lasinio and P.K. Mitter, On the Stochastic Quantization of Field Theory, Comm. Math. Phys. 101 (1985), 409-436.

[3] V.S. Borkar and S.K. Mitter, Lattice Approximation in the Stochastic Quantization of Fields, Proceedings Meeting on Stochastic Partial Differential Equations and Applications II, Trento Italy, February 1-6, 1988.

[4] E. Nelson, Probability Theory and Euclidean Field Theory, in Constructive Quantum Field Theory, eds. G. Velo and A.S. Wightman, Springer-Verlag, New York-Berlin, 1973.

[5] Ya. G. Sinai, Theory of Phase Transitions: Rigorous Results, Pergamon Press, 1982.

[6] A. Dembo and O. Zeitouni, Maximum A-Posteriori Estimation of Random Fields - Part I: Elliptic Gaussian Fields Observed Via a Noisy Nonlinear Channel, submitted to the Journal of Multivariate Analysis.

[7] L. Ambrosio, Variational Problems in SBV, Center for Intelligent Control Systems report CICS-P-86.

[8] T. Richardson, Existence Result for a Variational Problem Arising in Computer Vision, Center for Intelligent Control Systems report, CICS-P-63.

[9] D. Mumford and J. Shah, Optimal Approximations by Piecewise Smooth Functions and Associated Variational Problems, submitted to Comm. in Pure and Applied Math.

[10] S.K. Mitter, Estimation Theory and Statistical Physics, in Lecture Notes in Mathematics, Vol. 1203, Springer-Verlag, New York-Berlin, 1986.

[11] P.K. Mitter and T.R. Ramdas, Continuous Wilson renormalization group and the 2-D O(n) non-linear σ-model, to appear in Proceedings of 1987 Cargese Summer School on "Non-perturbative Quantum Field Theory".

[12] P.K. Mitter, Lectures on the Renormalization Group Method delivered at the Laboratory for Information and Decision Systems, M.I.T., May 1988.

[13] W.H. Fleming and S.K. Mitter, Optimal Control and Non-Linear Filtering for Nondegenerate Diffusion Processes, Stochastic 8 (1982), 63-77.

LOCAL CONTROLLABILITY OF GENERALIZED QUANTUM MECHANICAL SYSTEMS

T.J. Tarn[+], John W. Clark[++] and Garng M. Huang[+++]

+ Department of Systems Science and Mathematics, and Center of Robotics and Automation, Washington University, St. Louis, Missouri 63130, U.S.A.

++ Department of Physics, and McDonnell Center for the Space Sciences, Washington University, St. Louis, Missouri 63130, U.S.A.

+++ Department of Electrical Engineering, Texas A & M University, College Station, Texas 77843, U.S.A.

ABSTRACT

The concept of local controllability is investigated for non-relativistic quantum systems. Sufficient conditions will be sought such that the solution of the controlled Schrodinger equation can be guided, over a short time interval, to any chosen point in a suitably prescribed neighborhood of the solution in the absence of control. Evolution equations which are linear in the controls but nonlinear in the quantum state ψ are considered. Our formulation and analysis will (for the most part) run parallel to those of Hermes.

I. INTRODUCTION

In recent years, there has been a growing interest in the system theoretic problems of filtering and control of quantum mechanical systems. Several noteworthy efforts exist: (i) Tarn, Huang and Clark [1] and van der Schaft [2] have explored the formal basis for the modelling of quantum mechanical control systems. (ii) Clark, Tarn and their associates [3-6] have obtained results on quantum nondemolition filtering problem. (iii) Belavkin [7] has investigated the measurement and control problem in quantum dynamical systems. (iv) Pierce, Dahleh and Rabitz [8] have studied the optimal control problem of quantum mechanical systems. (v) Butkovskiy and collaborators have discussed the control of quantum objects in broad terms and have set forth general conditions for controllability of pure quantum states [9-11].

To the authors' knowledge very little has been published in the way of mathematically definitive results on the controllability of quantum systems. In [12] the authors are able to establish a series of global controllability conditions for the Schrodinger equation which is linear in state and linear in the external controls by extending the geometric approach as implemented by Sussmann and Jurdjevic [13,14], Krener [15], Brockett [16], Kunita [17] and others.

In the present contribution, we shall consider evolution equations which are linear in the controls but nonlinear in the quantum state; in this case the work of Hermes [18] is extended to obtain conditions for local controllability along an unguided reference solution.

II. PROBLEM FORMULATION WITH NONLINEAR GENERATORS

In adapting Hermes' work [18] to our ends, it is convenient to think in terms of the x representation [19]. Thus the state vector $\xi \in \mathcal{H}$ will be represented by the wave function $\xi(x) \in L_2(R^n)$, where $x \in R^n$ stands (ordinarily) for the set of spatial coordinate variables associated with the quantum system. (More generally, x may stand for any complete set of compatible variables [19] built from the position and momentum variables. Spin and other internal degrees of freedom can be incorporated by essentially trivial modifications.) Now, let us define a class of operators H in $\mathcal{H}$ which are supposed to be skew-Hermitian (norm preserving) and time independent and to have, in the x representation, the mode of action

$$(H\xi)(x) \equiv H\xi|_x = \sum_{\lambda=1}^{p} f_{\lambda,1}((H_{\lambda,1}\xi)(x)) \ldots f_{\lambda,q}((H_{\lambda,q}\xi)(x)). \qquad (1)$$

Here, p, q are some integers, the $H_{\lambda,\mu}$ ($\lambda = 1,\ldots,p$; $\mu = 1,\ldots,q$) are closed, skew-Hermitian linear operators acting in $\mathcal{H}$, and the mappings $f_{\lambda,\mu}$: $C^1 \to C^1$ are real analytic. (By the last requirement we mean that $f_{\lambda,\mu}(w)$ is a real analytic function of its argument w, this argument in itself being generally complex, $w \in C^1$. Also, in expression (1), $f_{\lambda,\mu}(w) f_{\lambda',\mu'}(w')$ is to be interpreted as the usual product of complex functions.) Throughout the current section, the generators $H_0,\ldots,H_r$ entering the "controlled Schrödinger equation" will be assumed to be of this more general form. Thus, while $H_0,\ldots,H_r$ are still taken skew-Hermitian, they need not be linear--although the linear case is certainly included.

We shall further assume that a unique local solution exists for the initial value problem

$$\frac{d}{dt}\psi_t = \left[H_0 + \sum_{\ell=1}^{r} u_\ell(t) H_\ell\right]\psi_t, \qquad \psi_{t=0} = \phi \in \mathcal{H}, \qquad (2)$$

posed by the Schrödinger equation so generalized, the admissible controls u_ℓ now being real, analytic, bounded functions of t. To establish that this is a viable assumption, we note that it is automatically fulfilled within the framework of [12], provided ϕ belongs to the analytic domain D_ω; moreover, in Ref. 20 it has been shown to be valid for a certain relevant class of partial differential equations. On the other hand the formulation of general conditions on $H_0 + \sum u_\ell H_\ell$ for the existence of a unique local solution of (2) awaits further mathematical developments.

Our next task is to specify the Lie bracket appropriate to the (generally) infinite-dimensional, (generally) nonlinear control problem (2), wherein the H_k, $k=0,\ldots,r$ are of type (1). First, we appeal to the chain rule to define a sort of derivative operator, DH, corresponding to an operator H of that type:

$$((DH\xi)(x))\zeta(x) = \sum_{\lambda=1}^{p}\sum_{\mu=1}^{q} f_{\lambda,1}((H_{\lambda,1}\xi)(x))\ldots f_{\lambda,\mu-1}((H_{\lambda,\mu-1}\xi)(x))$$

$$\cdot f'_{\lambda,\mu}((H_{\lambda,\mu}\xi)(x))f_{\lambda,\mu+1}((H_{\lambda,\mu+1}\xi)(x))\ldots f_{\lambda,q}((H_{\lambda,q}\xi)(x))(H_{\lambda,\mu}\zeta)(x), \quad (3)$$

where $\zeta \in H$ and $f'(w)$ is the derivative of $f(w)$ with respect to its argument. The Lie bracket of two operators H, K of the indicated class is then specified by

$$([H,K]\xi)(x) \equiv [H,K]\xi|_x = \{(DH\xi)(x)\}(K\xi)(x) - \{(DK\xi)(x)\}(H\xi)(x), \quad (4)$$

to apply $\forall\ \xi \in H$ and $\forall\ x$. Again we shall employ the notation $ad_H K = [H,K]$, $ad_H^{\nu+1} K = [H, ad^{\nu} K]$, $\nu = 1, 2,\ldots$; also, $ad_H^0 K \equiv K$. The prescription (4) for the Lie product is obviously consistent with that of [12], for, if H and K are linear, $[H,K] = HK - KH$ as in [12].

Remark 1. The above definitions and specifications are tenable even if H and $H_{\lambda,\mu}$ of (1) are not skew-Hermitian (or even if skew-Hermiticity is not a meaningful concept). As is well known, skew-Hermiticity of the generators of time displacement is an indispensible requirement in conventional quantum theory, where it is necessary for the probability interpretation of ψ_t. On the other hand, there are circumstances in which one may be led to drop this requirement, namely, (i) in approximate treatments of the Schrödinger equation designed to yield simple pictures of complicated phenomena involving many degrees of freedom, and (ii) in radical revisions of conventional quantum theory aimed at a more fundamental description of the microscopic world. The optical model of nuclear reactions, [21] wherein a complex potential is introduced to simulate the effects of inelastic processes, is a good example of circumstance (i), while the hadronic theory proposed by Santilli [22] suffices to illustrate possibility (ii). Obviously, in the latter context new interpretations as well as a new formal apparatus (see, e.g., Ref. 23) must accompany the enlarged mathematical framework.

Remark 2. The message of this comment is similar to that of Remark 1, except that the subject is nonlinearity of the generators $H_0, \ldots, H_r$ rather than violation of their skew-Hermiticity. Conventional quantum mechanics is necessarily a linear theory, in that the superposition principle is an essential property. Specifically, linearity of the operators $H_0, \ldots, H_r$ is required to maintain this property. But again one might agree, either (i) in the framework of approximation methods, or (ii) in fundamental extensions of quantum theory, to sacrifice linearity. The Hartree-Fock approximation [19,21] of atomic and nuclear physics furnishes a prominent example of a nonlinear approximation to the conventional quantum description. On the other side of the coin, nonlinear quantum theories at the first-principles level have been considered by a number of authors; for example, Wigner [24] has suggested that a resolution of the mysteries associated with "collapse of the wave packet" might be sought in terms of such a theory. [25]

III. GENERALIZED DECOMPOSITION THEOREM

Consider the system (2), wherein it is assumed that $\phi \in D \equiv \bigcap_{k=0}^{r} \mathrm{dom}\, H_k \neq$ null set. Let $V_t(\phi) \in D$ denote the solution (evaluated at time t) of the associated reference problem

$$\frac{d}{dt}\eta_t = H_0\eta_t \quad , \quad \eta_0 = \phi. \tag{5}$$

This problem corresponds to free evolution of the quantum system, the external controls being turned off; accordingly $\eta_t = V_t(\phi)$ will be referred to as the homogeneous reference solution. Treating ϕ, rewritten ζ, as an arbitrary element of the allowed domain D, we obtain a mapping $\zeta \to V_t(\zeta)$, which in general defines a nonlinear operator. (We note that in the special case that the generator H_0 is linear, $V_t(\zeta)$, which traces an integral curve of the vector field H_0, serves to define a linear evolution operator V_t. However, in the nonlinear setting of the present analysis, we are strictly not allowed to divorce operator from operand, since an operator of class (1) generally depends on the point of H at which it acts.) The differential of the mapping $\zeta \to V_t(\zeta)$, to be denoted $DV_t(\zeta)$, is also (generally) a nonlinear operator. One may loosely interpret $DV_t(\zeta)$ as the derivative of the object $V_t(\zeta)$, a state vector, with respect to its argument, which is again a state vector. By $DV_t(\zeta)|_x$ we will mean the differential of the (wave function) $\to$ (wave function) map $\zeta(x) \equiv \zeta|_x \to V_t(\zeta)|_x$.

Definition 1. A complex-valued function g: $t \to g(t) = g_1(t) + ig_2(t)$ is said to be complex analytic in the variable t, where $t \in R^1$, if the functions g_1 and g_2 are real analytic in t.

Theorem 1. (Generalized Decomposition Theorem (cf. Refs. 18,26)). Let ζ be an arbitrary element of the common domain D of the operators $H_0, \ldots, H_r$, and suppose that (i) the maps $t \to V_t(\zeta)|_x$ and $t \to DV_t(\zeta)|_x$ are complex analytic in t for all x and (ii) the differential $DV_t(\zeta)$ converges in the strong operator topology to the identity operator id, as $t \to 0^+$. Then, a sufficient condition for

$V_t(W_t(\phi))$ to provide a solution of the <u>controlled</u> dynamical problem (2), is that $W_t(\phi)$ satisfy

$$\frac{d}{dt}\xi_t = \sum_{\nu=0}^{\infty} \frac{(-t)^\nu}{\nu!}\left[ad^\nu_{H_0} \sum_{\ell=1}^{r} u_\ell H_\ell\right]\xi_t,\ \xi_0 = \phi \in D. \tag{6}$$

If $DV_t(\zeta)$ is one-to-one, the stated condition is also necessary.

<u>Proof</u>. A necessary and sufficient condition for $V_t(W_t(\phi))$ to be a solution of (2), given that $V_0(W_0(\phi)) = W_0(\phi) = \phi$, is $H_0V_t(W_t(\phi)) + \sum_{\ell=1}^{r} u_\ell H_\ell V_t(W_t(\phi))$

$$= \frac{d}{dt} V_t(W_t(\phi)) = \frac{\partial}{\partial t} V_t(\zeta)|_{\zeta=W_t(\phi)} + DV_t(\phi)\frac{d}{dt} W_t(\phi) \quad . \tag{7}$$

Since by definition $V_t(\zeta)$ must satisfy the differential equation $\partial V_t(\zeta)/\partial t = H_0V_t(\zeta)$, where ζ may be regarded as an <u>independent variable</u> so far as the time derivative is concerned, the initial terms in the first and last members of (7) cancel. Thus condition (7) may be distilled to

$$DV_t(\phi)\frac{d}{dt} W_t(\phi) = \left[\sum_{\ell=1}^{r} u_\ell(t)H_\ell\right]V_t(W_t(\phi)) \ . \tag{8}$$

The crucial step is to prove that, for all ζ and for all x,

$$DV_t(\zeta) \sum_{\nu=0}^{\infty} \frac{(-t)^\nu}{\nu!}\left[ad^\nu_{H_0} \sum_{\ell=1}^{r} u_\ell(t)H_\ell\right]\zeta|_x = \left[\sum_{\ell=1}^{r} u_\ell(t)H_\ell\right] V_t(\zeta)|_x. \tag{9}$$

Once property (9) is established, the theorem is in hand; for if $W_t(\phi)$ satisfies (6), it will then follow from the sufficiency of (8) that $V_t(W_t(\phi))$ solves problem (2).

In order to establish (9), we examine the quantity

$$g_\ell(t;H_\ell)|_x = DV_t(\zeta) \sum_{\nu=0}^{\infty} \frac{(-t)^\nu}{\nu!}\left[ad^\nu_{H_0} H_\ell\right]\zeta|_x - H_\ell V_t(\zeta)|_x. \tag{10}$$

With ζ an element of the allowed domain, the maps $t \to V_t(\zeta)|_x$ and $t \to H_\ell V_t(\zeta)|_x$ are complex analytic by our hypotheses, as is the map $t \to DV_t(\zeta)|_x$. Consequently, the right-hand side of (10) is complex analytic in t, for all ζ and for all x. Therefore it is legitimate to evaluate $g_\ell(t;H_\ell)|_x$ be means of its Taylor expansion in t.

To begin with, we know $g_\ell(0;H_\ell)|_x = 0$, because $DV_t(\zeta) \to id$ in the strong operator topology as $t \to 0^+$, and $V_0(\zeta) = \zeta$. Next, consider that

$$\frac{d}{dt}DV_t(\zeta) = D\frac{\partial}{\partial t} V_t(\zeta) = D_\zeta[H_0V_t(\zeta)] = D_\zeta[H_0(V_t(\zeta))]$$
$$= \{DH_0(V_t(\zeta))\}\{DV_t(\zeta)\} \quad .$$

(The differentials in the first line are all with respect to ζ, as is indicated explicitly in places where confusion might arise. The differential $D_\zeta[H_0(V_t(\zeta))]$ is computed as the product of the differential of the mapping $V_t(\zeta) \to H_0(V_t(\zeta))$ and the differential of the initial mapping $\zeta \to V_t(\zeta)$.) In similar vein,

$$\frac{d}{dt}[H_\ell V_t(\zeta)] = \frac{d}{dt}[H_\ell(V_t(\zeta))] = DH_\ell(V_t(\zeta))H_0V_t(\zeta) \quad .$$

Using these last two relations, we may obtain (with the dot indicating time derivative)

$$\begin{aligned}
\dot{g}_\ell(t;H_\ell) &= DH_0(V_t(\zeta))DV_t(\zeta)\sum_{\nu=0}^{\infty}\frac{(-t)^\nu}{\nu!}[ad^\nu_{H_0}H_\ell]\zeta \\
&\quad - DV_t(\zeta)\sum_{\nu=0}^{\infty}\frac{(-t)^\nu}{\nu!}[ad^{\nu+1}_{H_0}H_\ell]\zeta - \frac{d}{dt}[H_\ell V_t(\zeta)] \\
&= DH_0(V_t(\zeta))[DV_t(\zeta)\sum_{\nu=0}^{\infty}\frac{(-t)^\nu}{\nu!}[ad^\nu_{H_0}H_\ell]\zeta - H_\ell V_t(\zeta)] \\
&\quad + [DH_0(V_t(\zeta))H_\ell V_t(\zeta) - DH_\ell(V_t(\zeta))H_0V_t(\zeta)] \\
&\quad - DV_t(\zeta)\sum_{\nu=0}^{\infty}\frac{(-t)^\nu}{\nu!}[ad^{\nu+1}_{H_0}H_\ell]\zeta \\
&= DH_0(V_t(\zeta))g_\ell(t;H_\ell) - g_\ell(t;ad_{H_0}H_\ell) \quad . \qquad (11)
\end{aligned}$$

But we know, from previous argument or its extension, that $g_\ell(t;H_\ell)|_x$ and $g_\ell(t;ad_{H_0}H_\ell)|_x$ tend to zero as $t \to 0^+$; it follows that $\dot{g}_\ell(0;H_\ell)|_x = 0$ for all ζ and for all x.

The pattern is now set for an inductive construction of successive time derivatives of $g(t;H_\ell)$. In particular, based on the above results we may form

$$\begin{aligned}
\ddot{g}_\ell(t;H_\ell) &= \frac{d}{dt}[DH_0(V_t(\zeta))g_\ell(t;H_\ell)] + DH_0(V_t(\zeta))\dot{g}_\ell(t;H_\ell) \\
&\quad - DH_0(V_t(\zeta))\dot{g}_\ell(t;ad_{H_0}H_\ell) + g_\ell(t;ad^2_{H_0}H_\ell) \quad ,
\end{aligned}$$

and it follows that $\ddot{g}_\ell(t;H_\ell)|_x \to 0$ as $t \to 0^+$. Continuing the process indefinitely, we arrive at the result that at $t = 0$ all the time derivatives of $g_\ell(t;H_\ell)|_x$ vanish, to arbitrarily high order. Thus $g_\ell(t;H_\ell)|_x$ is identically 0, for all ζ, for all x, i.e.,

$$DV_t(\zeta)\sum_{\nu=0}^{\infty}\frac{(-t)^\nu}{\nu!}[ad^\nu_{H_0}H_\ell]|_x = H_\ell V_t(\zeta)|_x \ ,$$

$\ell = 1, \ldots, r$. The desired property (9) ensues upon multiplying this equality by $u_\ell(t)$ and summing over ℓ.

Corollary 1. Same as Theorem 1, except that "complex analytic" is everywhere to be replaced by "real analytic". (See Ref. 18)

Proof. Direct observation.

IV. LOCAL CONTROLLABILITY ALONG A REFERENCE HOMOGENEOUS SOLUTION

Definition 2. The system (2) is said to be locally controllable along the solution $\eta_t = V_t(\phi)$ of the control-free problem (5) on the manifold $M \subset H$ if, for small $t > 0$, there exists a set of $u_\ell(t)$, $\ell = 1, \ldots, r$, such that the solution ψ_t of (2) can be controlled to a neighborhood of η_t on M. The precise meaning of the last phrase is that ψ_t can be steered into any direction of the tangent space TM_{η_t} of M at the point $\eta_t = V_t(\phi) \in M$, for all $\phi \in M$.

We may now formulate the central result of this section.

Theorem 2. Assume that the homogeneous solution of system (2), i.e., the solution $\eta_t = V_t(\phi)$ of the uncontrolled system (5), satisfies the hypotheses (i) and (ii) of Theorem 1 for ζ (and specifically ϕ) on a finite-dimensional submanifold M, $M \subset D \subset H$, $\dim M = m$. Assume further that there exist integers $\nu_{\ell j_\ell}$ (with $\ell = 1, \ldots, r$ and $j_\ell = 1, \ldots, k_\ell < \infty$, and $0 \le \nu_{\ell 1} < \nu_{\ell 2} < \ldots < \nu_{\ell k_\ell}$) such that the set $\{[\mathrm{ad}_{H_0}^{\nu_{\ell j_\ell}} H_\ell]\phi\}$ spans the tangent space TM_{η_t} of M at $\eta_t = V_t(\phi)$ for all $\phi \in M$. It follows that system (2) is locally controllable along η_t on M. (Cf. Theorem 2, Ref. 18.)

Proof. If the functions $u_{\ell j_\ell}(t)$, where $\ell = 1, \ldots, r$ and $j_\ell = 1, \ldots, k_\ell$, qualify as admissible controls (real, analytic, bounded functions of t), then so do the finite linear combinations

$$u_\ell^{a_\ell}(t) = \sum_{j_\ell=1}^{k_\ell} a_{\ell j_\ell} u_{\ell j_\ell}(t), \qquad \ell = 1, \ldots, r ,$$

wherein the real coefficients $a_{\ell 1}, \ldots, a_{\ell k_\ell}$, are chosen (for convenience) to obey

$$\sum_{j_\ell=1}^{k_\ell} |a_{\ell j_\ell}| = 1 \quad , \ell = 1, \ldots, r .$$

Let us abbreviate the set $\{a_{\ell j_\ell}\}$ simply as $\underline{a}$. By generalized decomposition in the multi-input, complex case of the preceding subsection (i.e., by virtue of Theorem 1), the solution of problem (2), with the $u_\ell^{a_\ell}$ as controls, is given by $\psi_t^a = V_t(W_t^a(\phi))$. The solution $\xi_t^a = W_t^a(\phi)$ of the boundary value problem (6), restated for the controls $u_\ell^{a_\ell}$, evidently obeys the integral equation

$$W_t^a(\phi) = \phi + \sum_{\ell=1}^{r} \int_0^t u_\ell^{a_\ell}(t) \sum_{\nu=0}^{\infty} \frac{(-s)^\nu}{\nu!} \, ds \left[\mathrm{ad}_{H_0}^\nu H_\ell\right] W_t^a(\phi) \quad .$$

Thus

$$\frac{\partial}{\partial a_{\ell j_\ell}} V_t(W_t^a(\phi)) \,|_{a=0} = DV_t(W_t^a(\phi)) \frac{\partial}{\partial a_{\ell j_\ell}} W_t^a(\phi) \,|_{a=0}$$

$$= \sum_{\nu=0}^{\infty} \left(\int_0^t u_{\ell j_\ell}(s) \frac{(-s)^\nu}{\nu!} ds\right) DV_t(\phi)\left[ad_{H_0}^\nu H_\ell\right]\phi , \tag{12}$$

where $a=0$ means all of the $a_{\ell j_\ell}$ are zero. By assumption, we can find a set of integral (or zero) powers $\nu_{\ell j_\ell}$, where $\ell = 1, \ldots, r$, $j_\ell = 1, \ldots, k_\ell$, $0 \le \nu_{\ell 1} < \nu_{\ell 2} < \ldots < \nu_{\ell k_\ell}$, and $\nu_{max} = \max(\nu_{\ell j_\ell}) < \infty$, such that the set

$\{[ad_{H_0}^{\nu_{\ell 1}} H_\ell]\phi, \ldots, [ad_{H_0}^{\nu_{\ell k_\ell}} H_\ell]\phi, \ell = 1, \ldots, r\}$ spans TM_{η_t}. Then, since (also by assumption) $DV_t(\phi) \to id$ strongly as $t \to 0^+$, there must exist a time $t_1 > 0$ such that the set

$$\left\{DV_t(\phi)\left[ad_{H_0}^{\nu_{\ell 1}} H_\ell\right]\phi, \ldots, DV_t(\phi)\left[ad_{H_0}^{\nu_{\ell k_\ell}} H_\ell\right]\phi\right\}$$

spans TM_{η_t}, over the time interval $0 \le t \le t_1$.

We now proceed to make a judicious choice of the original functions $u_{\ell j_\ell}(t)$ involved in (12). One can realize admissible controls $\tilde{u}_{\ell j_\ell}(t)$ obeying the conditions

$$\int_0^{t_1} \tilde{u}_{\ell j_\ell}(s) \frac{(-s)^\nu}{\nu!} ds = \begin{cases} 0, & \text{for } \nu \ne \nu_{\ell j_\ell}, \quad 0 \le \nu \le \nu_{max} + 1 , \\ c_{\ell j_\ell} \ne 0 , & \text{for } \nu = \nu_{\ell j_\ell} , \end{cases} \tag{13}$$

where $\ell = 1, \ldots, r$, $j_\ell = 1, \ldots, k_\ell$, and the $c_{\ell j_\ell}$ are real constants. The connection between the $\tilde{u}_{\ell j_\ell}$ and the $u_{\ell j_\ell}$ will be specified shortly. The power ν being integral, inversion of relations (13) is in effect just a classical finite-moments problem. (Note that in the upper range $\nu > \nu_{max} + 1$, we have

$$\int_0^{t_1} \tilde{u}_{\ell j_\ell}(s) \frac{(-s)^\nu}{\nu!} ds = O\left(t_1^{\nu_{max} + 3}\right) ,$$

since $|\tilde{u}_{\ell j_\ell}|$ is by assumption bounded. This implies that the higher moments not specified by (13) will be negligible.)

With t in the interval $[0,t_1]$, we now carry out the change of variable $s = t_1 h/t$ in the integral on the left of (13):

$$\int_0^{t_1} \tilde{u}_{\ell j_\ell}(s) \frac{(-s)^\nu}{\nu!} ds = \left(\frac{t_1}{t}\right)^{\nu+1} \int_0^t u_{\ell j_\ell}(t_1 h/t) \frac{(-h)^\nu}{\nu!} dh .$$

Hence

$$\int_0^t \tilde{u}_{\ell j_\ell}(t_1 h/t) \frac{(-h)^\nu}{\nu!} dh = \begin{cases} 0, & \text{for } \nu \neq \nu_{\ell j_\ell}, \quad 0 \le \nu \le \nu_{max} + 1 , \\ \left(\frac{t}{t_1}\right)^{\nu+1} c_{\ell j_\ell} , & \text{for } \nu = \nu_{\ell j_\ell} , \\ 0\left(t^{\nu_{max} + 3}\right) , & \text{for } \nu > \nu_{max} + 1 . \end{cases}$$

Setting $u_{\ell j_\ell}(s) = \tilde{u}_{\ell j_\ell}(t_1 s/t)$ in (12), we arrive finally at the result

$$\frac{\partial}{\partial a_{\ell j_\ell}} V_t(W_t^a(\phi)) \,|_{a=0} = c_{\ell j_\ell}\left(\frac{t}{t_1}\right)^{\nu_{\ell j_\ell} + 1} DV_t(\phi) \left[ad_{H_0}^{\nu_{\ell j_\ell}} H_\ell\right]\phi + 0 \left(t^{\nu_{max} + 2}\right) ,$$

where, for $t < t_1$, the last term can be neglected, t_1 being small. Consequently the set $\{\partial V_t(W_t^a(\phi))/\partial a_{\ell j_\ell}, \ \ell = 1, \ldots, r, \ j_\ell = 1, \ldots, k_\ell\}$ spans TM_{η_t} for t in the interval $[0,t_1]$, where t_1 has been chosen above. This means that we have been able to choose the controls so that, for small $t > 0$, the state defined by system (2) can be steered into any direction of the tangent space on M at the point $\eta_t = V_t(\phi)$. Then by definition the system is locally controllable along the reference solution $V_t(\phi)$, for all $\phi \in M$.

Remark 3. Theorems 1 and 2 remain true as stated if the H_k, $k = 0, \ldots, r$, are not skew-Hermitian.

Example 1. The theorems of the present paper are aimed at an infinite-dimensional space of quantum states. However, the results obtained herein are still valid (with trivial alterations) for a finite-dimensional state space. As pointed out in Remark 3, from a mathematical standpoint we may also dispense with the assumption that the generators $H_0, \ldots, H_r$ are skew-Hermitian.

For example, consider a nonlinear control system on R^m, $m < \infty$, defined by

$$\frac{d}{dt} \chi(t) = A(\chi(t)) + u(t)B(\chi(t)) \quad , \ \chi(0) = \chi_0 , \tag{14}$$

where A and B are real analytic vector fields corresponding to nonlinear operators of the sort introduced in Section II. Then, as argued in Ref. 18, a sufficient condition for local controllability along the homogeneous ($u \equiv 0$) solution of (14) is $\text{rank}\{[ad_A^\nu B]\chi_0, \ \nu = 0,1,2,\ldots,\infty\} = m$. This is precisely the condition which would enter the finite-state-space version of Theorem 2. Problem (14) does not strictly refer to a quantum-mechanical system; its study is, nevertheless, illuminating.

While surely of high interest, the identification and analysis of "non-trivial" examples of the utility of Theorem 2, meaning examples concerned with novel quantum control systems characterized by nonlinear generators, exceeds the scope of the present work.

V. SUMMARY AND OUTLOOK

It has been our aim to augment the foundation for the concept of controllability of quantum-mechanical systems [12]. In the generalized, nonlinear formulation of the quantum control problem, we were able to determine conditions for the property of local controllability along a homogeneous (i.e., control-free) solution, without having to refer to the existence of an analytic domain which was assumed in the global analysis of [12]. (Our treatment of this case amounts to an extension of Hermes' work [18] to a multi-input, complex-state problem.) From the results obtained herein on the controllability of the solution of nonlinear Schrodinger equations, one may regain, upon appropriate specialization or adaptation, certain well-known systems-theoretic results in finite-dimensional state space (see, in particular, Refs. 13-18).

Clearly, only a modest beginning has been made toward achieving the larger goal of a comprehensive theory of quantum control. The following problems, among others, await concerted effort:

(i) Adaptation of the notions of observability, identification, realization, and feedback to the quantum context [27].

(ii) Study of a controlled version of the Schrödinger equation for the time evolution of the density operator, [19] so as to extend control theory to the realm of quantum statistical mechanics.

It is evident that powerful mathematical techniques must be invoked to carry through this program; moreover, one must confront the profound conceptual obstacles intrinsic to the quantum measurement process [25,28,29].

ACKNOWLEDGMENTS

This work was supported in part by the National Science Foundation under Grant Nos. DMR-8519077 and ECS-8515899.

REFERENCES

[1] Tarn, T.J., G.M. Huang and J.W. Clark: Modelling of quantum mechanical control systems, Mathematical Modelling, 1 (1980), 109-121.

[2] van der Schaft, Aryan J.: Hamiltonian and quantum mechanical control systems, in: Proceedings of the 4th International Seminar on Mathematical Theory of Dynamical Systems and Microphysics, Udine (Ed. A. Blaquiere, S. Diner and G. Lochak), Springer-Verlag, Wien-New York 1987.

[3] Clark, J.W. and T.J. Tarn: Quantum nondemolition filtering, in: Proceedings of the 4th International Seminar on Mathematical Theory of Dynamical Systems and Microphysics, Udine(Ed. A. Blaquiere, S. Diner and G. Lochak), Springer-Verlag, Wien-New York 1987.

[4] Tarn, T.J., J.W. Clark, C.K. Ong and G.M. Huang: Continuous-time quantum mechanical filter, in: Proceedings of the Joint Workshop on Feedback and Synthesis of Linear and Nonlinear Systems, Bielefeld and Rome (Ed. D. Hinrichsen and A. Isidori), Springer-Verlag, Berlin 1982.

[5] Clark, J.W., C.K. Ong, T.J. Tarn and G.M. Huang: Quantum nondemolition filters, Mathematical Systems Theory, 18(1985), 33-35.

[6] Ong, C.K., G.M. Huang, T.J. Tarn and J.W. Clark: Invertibility of quantum-mechanical control systems, Mathematical Systems Theory, 17(1984), 335-350.

[7] Belavkin, Viacheslav: Non-demolition measurement and control in quantum dynamical systems, in : Proceedings of the 4th International Seminar on Mathematical Theory of Dynamical Systems and Microphysics, Udine (Ed. A. Blaquiere, S. Diner and G. Lochak), Springer-Verlag, Wien-New York 1987.

[8] Peirce, A.P., M.A. Dahleh and H. Rabitz: Optimal control of quantum mechanical systems: existence, numerical approximations, and applications, to appear in: Proceedings of the IEEE International Conference: Control 88, University of Oxford, UK, April (1988).

[9] Butkovskiy, A.G. and Yu. I. Samoilenko, Control of quantum systems, automation and remote control, No. 4, April (1979), 485-502; Control of quantum systems, automation and remote control, No. 5 May (1979), 629-645.

[10] Butkovskiy, A.G. and Ye. I. Pustil'nykova: The method of seeking finite control for quantum mechanical processes, in: Proceedings of the 4th International Seminar on Mathematical Theory of Dynamical Systems and Microphysics, Udine (Ed. A. Blaquiere, S. Diner and G. Lochak), Springer-Verlag, Wien-New York 1987.

[11] Butkovskiy, A.G. and Yu. I. Samoilenko: Controllability of quantum mechanical systems, Dokl, Akad. Nauk SSSR 250, 51 (1980) [Sov, Phys, Dokl, 25, 22 (1980)].

[12] Huang, G.M., T.J. Tarn and J.W. Clark: On the controllability of quantum mechanical systems, J. Math. Phys. 24 (1983) 2608-2618.

[13] Sussmann, H. and V. Jurdjevic: Controllability of non-linear systems, Journal of Differential Equations, 12 (1962), 95-116.

[14] Jurdjevic, V. and H. Sussmann: Control systems on lie groups, Journal of Differential Equations, 12 (1972), 313-329.

[15] Krener, Arthur J.: A Generalization of chow's theorem and the bang-bang theorem to nonlinear control problems, SIAM Journal of Control, Vol. 12, No. 1, Feb. (1974).

[16] Brockett, Roger W.: Nonlinear systems and differential geometry: Proceedings of IEEE, Vol. 64, No. 1, Jan (1976).

[17] Kunita, Hiroshi: On the controllability of nonlinear systems, with applications of polynomial systems, Applied Mathematics and Optimization (1976), 89-99.

[18] Hermes, H.: Local controllability of observables in finite and infinite dimensional nonlinear control systems, Applied Mathematics and Optimization, 5, (1979), 117-125.

[19] Messiah, A.: Quantum mechanics, Vols. I and II, Wiley, New York (1961).

[20] Beals, R. and C. Feffermann: On Local solvability of linear partial differential equations, Annals of Mathematics, 97 (1973), 483-498.

[21] Brown, G.E.: Unified theory of nuclear models and forces, North-Holland, Amsterdam, (1971).

[22] Santilli, R.M.: Need of subjecting to an experimental verification the validity within a hadron of Einstein's special relativity and Pauli's exclusion principle, Hadronic Journal, 1, (1978), 574-901.

[23] Abraham, R., and J.E. Marsden: Foundations of mechanics, 2nd ed. Benjamin, Reading, (1978).

[24] Wigner, E.P.: The Scientist speculates, I.J. Good, Ed. W. Heinemann, London, (1961).

[25] d'Espagnat, B.: Conceptual foundations of quantum mechanics, Benjamin, Reading, (1976).

[26] Hermes, H.: Controllability of nonlinear delay differential equations, Nonlinear Analysis, Theory, Methods and Applications, 3 (1979).

[27] Kailath, T.: Linear systems, Prentice-Hall, Inc., Englewood Cliffs, (1980).

[28] Helstrom, C.W.: Quantum detection and estimation theory, Academic Press, New Yok, (1976).

[29] Ilic, D.: D. Sc. Dissertation, Washington University (1978), unpublished.

CALCULUS OF VARIATIONS AND QUANTUM PROBABILITY

J.C. Zambrini(*)
Mathematics Institute, University of Warwick
Coventry CV4 7AL, England

Key words: Quantum physics, Feynman's path integrals, heat equation, Bernstein processes, Calculus of variations, Euclidean Quantum Mechanics.

Contents

(*) Partly supported by SERC Grant No. GRD23404.

§1 Feynman's approach revisited

Feynman left us, recently, and we still have to face the challenge to understand the deep meaning of his original approach to quantum physics [1]. No serious theoretical physicist doubts that, in a certain way, Feynman was profoundly right, but one has to admit that this way is still unknown to us, at least for those of us who think that a good physical theory should be mathematically consistent.

It has been thought for a long time that *the* rigorous version of Feynman's path integral approach is Kac's one, involving Wiener integral and the theory of Brownian motion [2]. Although it is rigorous, there are suspicions that this is not what we need for quantum physics. For example, the (Euclidean) program of constructive field theory, founded on Feynman-Kac formula, did not succeed in producing realistic quantum field theories [3]. On the other hand, it is certainly true that to deal with the heat equation (i.e. the "imaginary time" Schrödinger equation)

$$-\hbar\frac{\partial\theta^*}{\partial t} = H\theta^* \tag{1.1}$$

in $L^2(\mathbb{R}^3)$, for instance, with H the Hamiltonian observable of the system, simplifies considerably the analysis. This transfer principle from the Schrödinger equation

$$i\hbar\frac{\partial\Psi}{\partial t} = H\Psi \tag{1.2}$$

to the heat equation (1.1) is called the Euclidean point of view. It has proved to be technically very useful in both non-relativistic and relativistic quantum physics [3], but the theoretical reasons for this efficiency are lacking.

Another, completely different, Euclidean starting point for quantum physics has been introduced recently [4-5]. It involves a new class of (well defined) diffusion processes, the Bernstein processes, whose properties differ notably from the properties of the stochastic processes associated with Feynman-Kac formula. In particular, these processes have dynamical characteristics very close to the (formal) diffusions underlying Feynman's path integral, including the time reversibility.

It is not the first time that time-symmetric diffusion processes are used in relation with foundational questions of quantum mechanics: this has been done twenty years ago by E. Nelson [7]. The basic point of his theory ("Stochastic Mechanics") was to deal directly with the Schrödinger equation (1.2). The theory that we are advocating here ("Euclidean Quantum Mechanics" [5]), although it uses some common technical tools, is different and much easier to link to Feynman's original strategy. In particular, it shed some new light on Feynman's version of Least action principle, in a generalization of classical calculus of variation which may be of some intrinsic interest for the specialists of control theory.

It should also be noticed that Euclidean Quantum Mechanics (EQM) has been inspired by a forgotten idea of E. Schrödinger [8].

In the present contribution, we are going to focus on the case of a unit mass and charge particle in an electromagnetic field, namely a self adjoint Hamiltonian for (1.2) of the form

$$H = -\frac{\hbar^2}{2}[\nabla - i\hbar^{-1}\vec{a}]^2 + \varphi \tag{1.3}$$

on $L^2(\mathbb{R}^3)$, where $\vec{a} = \vec{a}(x)$ is the vector potential and $\varphi = \varphi(x)$ is the scalar potential such that the magnetic and electric fields $\vec{H}$ and $\vec{E}$ satisfy

$$\begin{cases} \vec{H} = \text{rot } \vec{a} \\ \vec{E} = -\nabla\varphi \end{cases} \tag{1.4}$$

Afterwards, the arrows are reserved for electromagnetic data. First, we are going to assume the existence of a certain $\mathbb{R}^3$-valued diffusion process Z_t, t in $I = [-\frac{T}{2}, \frac{T}{2}]$ and to describe its properties. We shall discuss later the question of its existence.

§2. Action functional and dynamics

Let us denote by $h(x,t-s,y)$ the integral kernel of $e^{-(t-s)H}$, as an operator in $L^2(\mathbb{R}^3)$. Even when the scalar potential φ is zero, the Hamiltonian H of (1.3) is such that $e^{-(t-s)H}$ is not positivity preserving (it is not even reality preserving).

In a Euclidean approach, however, the quantum mechanical momentum $P = -i\nabla$ should become $-\nabla$ since the time parameter τ of the Schrödinger equation is analytically continued, $\tau \to -it$. It is therefore natural to consider, as Euclidean version of (1.3),

$$H_A = -\frac{\hbar^2}{2}[\nabla + \hbar^{-1}\vec{A}]^2 + \varphi \tag{1.3'}$$

where $\vec{A}$ corresponds to $-i\vec{a}$.

Let us consider a positive (smooth) solution $\mathcal{O}^*$ of Eq. (1.1) for the Hamiltonian H_A. Let us define

$$I_*(x,t) = -h \log \mathcal{O}^*(x,t) \tag{2.1}$$

It solves the following nonlinear partial differential equation

$$\frac{\partial I_*}{\partial t} - \frac{\hbar}{2}\Delta I_* + (\nabla I_* - \vec{A})\nabla I_* = \tfrac{1}{2}|\nabla I_* - \vec{A}|^2 + \varphi + (\nabla I_* - \vec{A}).\vec{A} - \frac{\hbar}{2}\nabla.\vec{A} \tag{2.2}$$

Comparing the left hand side with the backward infinitesimal generator of an $\mathbb{R}^3$-valued diffusion process Z_t with diffusion coefficient $\hbar$ and (backward) drift $B_* = \nabla I_* - \vec{A}$, namely the second order differential operator D_* defined for any smooth $f: \mathbb{R}^3 \times \mathbb{R} \to \mathbb{R}^3$ by

$$D_* f(x,t) = \left(\frac{\partial}{\partial t} + B_*\nabla - \frac{\hbar}{2}\nabla\right) f(x,t) \tag{2.3}$$

we observe that (2.2) reduces to

$$D_*I_*(z,(t),t) = \tfrac{1}{2}|D_*Z(t)|^2 + \varphi(Z(t)) + \vec{A}.D_*Z(t) - \frac{\hbar}{2}\nabla.\vec{A} \tag{2.2'}$$

Now consider (formally) the classical limit $h = 0$ of this expression. Then $t \to Z(t)$ has a continuous strong derivative, and $D_*f(Z(t),t)$ reduces to $\frac{d}{dt}f(Z(t),t)$. In particular, the r.h.s. of (2.2') is the (Euclidean) Lagrangian of the system and, therefore, the l.h.s. is the derivative of the (Euclidean) Hamilton's principal function. For $h \neq 0$, we call I_*, defined by (2.1), the backward Hamiltonian principal function (or simply Action function) associated with the starting Hamiltonian (1.3'). We also define the Lagrangian L of the system by the right hand side of (2.2'), namely

$$L(Z(t),D_*Z(t)) \equiv \tfrac{1}{2}|D_*Z(t)|^2 + \varphi(Z(t)) + \vec{A}.D_*Z(t) - \frac{\hbar}{2}\nabla.\vec{A} \tag{2.4}$$

Let us show that these definitions are dynamically consistent. Taking the gradient of (2.2) and using the vector identity $\nabla(\vec{a}.\vec{b}) = \vec{b}\nabla\vec{a} + \vec{a}\nabla\vec{b} + \vec{a} \times \mathrm{rot}\,\vec{b} + \vec{b} \times \mathrm{rot}\,\vec{a}$, we obtain

$$D_*D_*Z(t) = \nabla\varphi + D_*Z(t) \times \mathrm{rot}\,\vec{A} - \frac{\hbar}{2}\,\mathrm{rot}\,\mathrm{rot}\,\vec{A} \tag{2.5}$$

This is clearly a generalization of the classical Newton equation for the Lorentz force acting on the charged particle. The change of sign of the force is natural in a Euclidean description.

The presence of h in the definition of the Lagrangian (2.4) may seem alarming, but it is not. The conditional expectation of (2.2'), given the future position $Z(t) = x$, and a time integration, yields

$$I_*(x,t) - E_{x,t}[I_*(Z(-\tfrac{T}{2}), -\tfrac{T}{2})] =$$

$$E_{x,t}\int_{-T/2}^{t} \{\tfrac{1}{2}|D_*Z(\tau)|^2 + \varphi(Z(\tau))\}d\tau + E_{x,t}\int_{-T/2}^{t} \{\vec{A}.D_*Z(\tau) - \frac{\hbar}{2}\nabla.\vec{A}\}d\tau \tag{2.6}$$

But, according to Itō's calculus [14], the second term of the right hand side reduces to

$$E_{x,t}\int_{-T/2}^{t}\{\vec{A}.dZ - \frac{\hbar}{2}\int_{-T/2}^{t}\nabla.\vec{A}\}d\tau$$

$$= E_{x,t}\int_{-T/2}^{t}\vec{A}\circ dZ$$

where $\int\vec{A}.dZ$ and $\int\vec{A}\circ dZ$ denote respectively the backward Itō's integral with respect to a decreasing filtration $\mathcal{F}_t$ (i.e. the sigma algebra generated by Z(u), u ≥ t) and the symmetric Fisk-Stratonovich integral. Therefore the action function can also be expressed as the very classical looking path integral

$$I_*(x,t) = E_{x,t}\int_{-T/2}^{t}\{\tfrac{1}{2}|D_*Z(\tau)|^2 + \varphi(Z(\tau))\}d\tau + E_{x,t}\int_{-T/2}^{t}\vec{A}\circ dZ + E_{x,t}[I_*(Z(-\tfrac{T}{2}), -\tfrac{T}{2})] \qquad (2.7)$$

Given (2.4), it is natural to call Generalized Potential $U = U(Z,D_*Z)$ the function

$$U(Z,D_*Z) \equiv \varphi(Z) + \vec{A}.D_*Z - \frac{h}{2}\nabla.\vec{A} \qquad (2.8)$$

Observing that

$$D_*\left(\frac{\partial U}{\partial D_*Z}\right) - \frac{\partial L}{\partial Z} = -(\nabla\varphi + D_*Z \times \text{rot}\,\vec{A} - \frac{\hbar}{2}\,\text{rot rot}\,\vec{A}) \qquad (2.9)$$

the Lagrangian (2.4) reduces to

$$L(Z,D_*Z) = T(D_*Z) + U(Z,D_*Z) \qquad (2.10)$$

for the Kinetic Energy $T(D_*Z) \equiv \frac{1}{2}|D_*Z|^2$.

It follows from Eqs. (2.5) and (2.10) that a (stochastic) Euler–Lagrange equation holds

$$D_*\left(\frac{\partial L}{\partial D_* Z}\right) - \frac{\partial L}{\partial Z} = 0. \tag{2.11}$$

This method of derivation uses only the decreasing filtration $\mathcal{F}_t$ and therefore breaks the natural time symmetry of quantum theory. But let us observe that we can do the same using an incresing filtration $\mathcal{P}_t$, the forward analogue ot (2.3) (for B the forward drift)

$$Df(x,t) = \left(\frac{\partial}{\partial t} + B.\nabla + \frac{\hbar}{2}\Delta\right) f(x,t)$$

and starting from a positive solution of the backward heat equation $\hbar \frac{\partial \mathcal{O}}{\partial t} = H_A \mathcal{O}$ (well defined in §5). Then, the analogue of Eq. (2.5) becomes

$$DDZ(t) = \nabla\varphi + DZ(t) \times \text{rot}\, \vec{A} + \frac{\hbar}{2} \text{rot rot}\, \vec{A}.$$

In particular, if a unique diffusion Z_t adapted simultaneously to $\mathcal{F}_t$ and $\mathcal{P}_t$ exists, it solves the time symmetric Newton equation

$$\tfrac{1}{2}\left(DDZ(t) + D_* D_* Z(t)\right) = \nabla\varphi + \tfrac{1}{2}\left(DZ + D_* Z\right) \times \text{rot}\, \vec{A}$$

involving exclusively a natural generalization of the Lorentz force. The §4 will elaborate the meaning of this remark.

On the other hand, the stochastic differential of the starting action function I_* is also defined by

$$dI_* = \nabla I_* \circ dZ + \frac{\partial I_*}{\partial t} dt \tag{2.12}$$

The comparison of this expression with its classical counterpart (when $\hbar = 0$) suggests to define the (backward) Momentum and Energy of the system respectively by

$$p_* \equiv \frac{\partial L}{\partial D_* Z} = \nabla I_* \tag{2.13}$$

and

$$\varepsilon_* \equiv \frac{\partial I_*}{\partial t} \tag{2.14}$$

A straightforward calculation using (2.2') shows that

$$D_* \varepsilon_*(Z(t),t) = 0 \tag{2.15}$$

In other words, ε_* is an $\mathcal{F}_t$-martingale, a natural generalization of the classical conservation of energy in our context.

A very important property of Lagrangian mechanics is the gauge invariance. Suppose that, to the starting Lagrangian (for example the one of (2.7)), is added the derivative of a smooth function F in such a way that

$$\tilde{I}_*(x,t) = E_{x,t} \int_{-T/2}^{t} L(Z,D_*Z)ds + E_{x,t} \int_{-T/2}^{t} D_*F(Z,s)ds + E_{x,t}\, \tilde{I}_*(Z(-\tfrac{T}{2}), -\tfrac{T}{2}) \tag{2.16}$$

or equivalently $\tilde{L}(Z,D_*Z,S) = L(Z,D_*Z) + D_*F(Z,S)$. The explicit form of the supplementary integrand, according to (2.3), is

$$D_*F = \frac{\partial F}{\partial S} + D_*Z.\nabla F - \frac{h}{2}\Delta F$$

It is immediate to verify that, for such a "Lagrangian" D_*F, the Euler–Lagrange equation (2.11) reduces to an identity. This means that, as in classical mechanics, the Euler–Lagrange equation

(2.11) are invariant under the gauge transformation $L \rightarrow L + D_* F$.

Clearly, (2.16) corresponds to the following relation between actions,

$$\tilde{I}_*(x,t) = I_*(x,t) + F(x,t) \tag{2.16'}$$

This means that, under the same transformation, the Momentum and Energy (2.13) and (2.14) become

$$\tilde{p}_* = p_* + \nabla F \tag{2.13'}$$

$$\tilde{\varepsilon}_* = \varepsilon_* + \frac{\partial F}{\partial t} \tag{2.14'}$$

as in classical mechanics (up to an "Euclidean" sign).

On the other hand, by (2.1), $\tilde{I}_*(x,t) = -h \log \tilde{\mathcal{O}}^*(x,t)$ and (2.16') gives the simple relation

$$\tilde{\mathcal{O}}_* = \mathcal{O}_*(x,t)\, e^{-F(x,t)} \tag{2.17}$$

It follows immediately that $\tilde{\mathcal{O}}_*$ solves the heat equation

$$-\hbar \frac{\partial \tilde{\mathcal{O}}_*}{\partial t} = -\frac{\hbar^2}{2}(\nabla + \hbar^{-1}\vec{A} - \nabla F)^2 \tilde{\mathcal{O}}_* + (\varphi + \frac{\partial F}{\partial t})\tilde{\mathcal{O}}_* \tag{2.18}$$

if $\mathcal{O}_*$ solves (1.1) for the Hamiltonian H_A. This is clearly the Euclidean version of a local gauge transformation in the Schrödinger equation (1.2)

The construction if valid for any regular positive solution of (1.1), in particular for its integral kernel. In this case, the similarity with Feynman path integral is very striking.

The necessity to use a symmetric integral to describe vector potential has been shown by Feynman is his original paper (1948). If our postulated diffusion process really exists, it has the dynamical properties we need to construct a natural extension of classical mechanics compatible with Feynman's ideas.

§3. Calculus of variation

The validity of (2.11) suggests the existence of a variational calculus in which the action functionals are of the form (2.7). For $\vec{A} = 0$, $\varphi = V$ the following result holds:

Theorem

Let L the (Euclidean) Lagrangian $L(q,\dot{q}) = \frac{1}{2}|\dot{q}|^2 + V(q)$. A smooth diffusion process Z(s), solution of the Newton equation

$$D_*D_*Z(s) = \nabla V(Z(s)) \qquad -\frac{T}{2} < s < t$$

minimizes the action functional

$$J[Z(\cdot)] = E_{x,t}\int_{-T/2}^{t} L(Z(s),D_*Z(s))ds + E_t I_*(Z(-\tfrac{T}{2}), -\tfrac{T}{2})$$

on the set of neighbouring processes Z^ε such that $D_*Z^\varepsilon(-\frac{T}{2}) = \nabla I_*(Z^\varepsilon(-\frac{T}{2}), -\frac{T}{2})$ and $Z^\varepsilon(t) = x$.

The relevant family of neighbouring diffusions contains $\mathbb{R}^3$-valued $\mathcal{F}_S$-semimartingales of the form

$$\begin{cases} d_*Z^\varepsilon(s) = B_*^\varepsilon(Z^\varepsilon(s),s)ds + d_*W_*(s) & -\frac{T}{2} < s < t \\ Z^\varepsilon(t) = y \end{cases} \tag{3.1}$$

for $W_*(s)$ an $\mathcal{F}_s$-martingale, and perturbed drifts defined by

$$B_*^{\varepsilon}(x,s) = -h\left(\frac{\nabla\theta^*}{\theta^*} + \varepsilon\nabla g_*\right)(x,s) \tag{3.2}$$

where g_* is smooth, but arbitrary, and ε is a small parameter. When $\varepsilon = 0$, the process Z^{ε} is the one of §2 (for $\vec{A} = 0$). The proof is elementary, starting from (2.1), $I_*(Z^{\varepsilon}(s),s) = -h \log \theta^*(Z^{\varepsilon}(s),s)$, using the Theorem of variation of a stochastic differential equation with respect to a parameter, and the inequality

$$D_* I_*(Z^{\varepsilon}(s),s) \leq \tfrac{1}{2}|B_*^{\varepsilon}|^2(Z^{\varepsilon}(s),s) + V(Z^{\varepsilon}(s)) \tag{3.3}$$

with equality iff $\varepsilon = 0$.

The point is that the stochastic Hamilton-Jacobi equation (2.2) is an equation of Dynamic Programming familiar in Optimal Stochastic Control. Eq. (2.1) is a logarithmic transformation in the sense of W.H. Fleming (Cf. [9], and these proceedings). The idea of a stochastic calculus of variations of this kind for quantum mechanics (in the context of Nelson theory) is due to K. Yasue [10]. It gave rise to a lot of interesting works (Cf. references in [7] (1985) and [15], for example).

§4. Existence of diffusions

The odd point of the strategy suggested in §2 is that it is time assymmetric. Clearly, if a diffusion process can be used to give a meaning to Feynman's path integral approach of quantum mechanics, it should be a time symmetric diffusion process. Since we are dealing with the heat equation (1.1), the situation seems hopeless.

The solution of the puzzle was suggested by Schrödinger 50 years ago [8] and developed in [4, 5 and 6]. We summarize here this solution since it cannot be regarded, as yet, as common knowledge. We say that a Hamiltonian H is in the Schrödinger class if $h(s,x,t,y) = h(x,t-s,y) =$ kernel $\{e^{-(t-s)H}\}$ is jointly continuous in x, y, $t-s$, and strictly nonnegative.

We observe that

$$h(s,x,t,\xi,v,y) = \frac{h(s,x,t,\xi)h(t,\xi,v,y)}{h(s,x,v,y)} \qquad -\frac{T}{2}<s<t<v<\frac{T}{2} \quad x,\xi,y \text{ in } \mathbb{R}^3 \tag{4.1}$$

is the density of a probability measure with respect to $d\xi$, called a **Bernstein transition** [4-5].

A stochastic process $Z_t : \Omega \to \mathbb{R}^3$ is a Bernstein process iff for any bounded Borel measurable g,

$$E[g(Z_t) \mid \mathcal{P}_s \cup \mathcal{F}_u] = E[g(Z_t) \mid Z_s, Z_u] \quad \forall s<t<v \text{ in } I = \left[-\frac{T}{2}, \frac{T}{2}\right] \tag{4.2}$$

where $\mathcal{P}_s$ is the increasing filtration till s (the "past") and $\mathcal{F}_u$ is the decreasing filtration from u (the "future"). This is, in modern terms, the local Markov property, perfectly time symmetric.

The usual construction of diffusion processes starts from the data of an initial probability, a (forward) transition probability and reconstruct the finite dimensional distributions according to the picture:

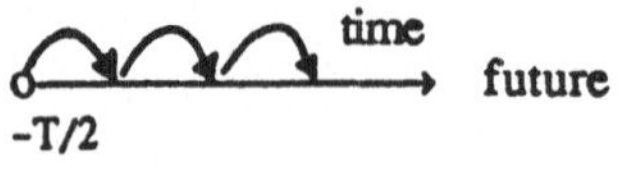

It is also possible to do it from a final probability and a (backward) transition probability:

The following Theorem, due to Jamison [11], shows that a joint probability measure and a Bernstein transition determine a Bernstein process on $I = \left[-\frac{T}{2}, \frac{T}{2}\right]$ according to

Theorem

Let $H(s,x,t,B,u,y) = \int_B h(s,x,t,d\xi,u,y)$ be a Bernstein transition, $-\frac{T}{2} < s < t < u < \frac{T}{2}$, m be a probability measure on the Borel sigma algebra $\mathcal{B}(\mathbb{R}^3) \times \mathcal{B}(\mathbb{R}^3)$. Then there is a unique probability measure P_m such that with respect to (Ω, σ_I, P_m) (where σ_I is the sigma-algebra generated by Z_t, t in I) Z_t is an $\mathbb{R}^3$-valued Bernstein process and

a) $P_m(Z_{-T/2} \in B_s, Z_{T/2} \in B_E) = m(B_s \times B_E)$, B_s, B_E in $\mathcal{B}(\mathbb{R}^3)$

b) $P_m(Z_t \in B \mid Z_s, Z_u) = H(s, Z_s, t, B, u, Z_u)$, B in $\mathcal{B}(\mathbb{R}^3)$

c) $P_m(Z_{-T/2} \in B_s, Z_{t_1} \in B_1, \dots, Z_{t_n} \in B_n, Z_{T/2} \in B_E) =$

$$\int_{B_s \times B_E} dm(x,y) \int_{B_1} H(-T/2, x, t_1, dx_1, T/2, y) \dots \int_{B_n} H(t_{n-1}, x_{n-1}, t_n, dx_n, T/2, y) .$$

The proof can be found in [11] (cf. also [4]). Its key idea is that, for fixed initial position, a Bernstein transition is a backward Markovian transition probability and, for a fixed final position it is a forward one.

The resulting Z_t, t in I, is obviously not Markovian in general. To get a Bernstein Markovian one and only one choice of joint probability m is possible, namely

$$m \equiv M(B_s \times B_E) = \int_{B_s \times B_E} \mathcal{O}^*_{-T/2}(x) h(x,T,y)\, \mathcal{O}_{T/2}(y) dx dy \tag{4.3}$$

for any (unspecified) bounded positive measurable $\mathcal{O}^*_{-T/2}$, $\mathcal{O}_{T/2}$. With this choice, the finite dimensional distributions of Z_t reduce to

$$P_M(dx_1, t_1, \dots, dx_n, t_n) = \int_{\mathbb{R}^3 \times \mathbb{R}^3} \mathcal{O}^*_{-T/2}(x) h(x, t_1 + T/2, dx_1) \dots h(dx_n, T/2 - t_n, y) \mathcal{O}_{T/2}(y) dx dy. \tag{4.4}$$

for $\frac{T}{2} < t_1 \leq t_2 \leq \cdots \leq \frac{T}{2}$.

The data that we are given, actually, are a pair of initial and final probability densities $p_{-T/2}(x)dx$ and $p_{T/2}(y)dy$. So the marginals of the joint probability M give a system of non-linear functional equations for $\theta^*_{-T/2}$ and $\theta_{T/2}$, discovered by Schrödinger [8]:

$$\begin{cases} \theta^*_{-T/2}(x) \int_{\mathbb{R}^3} h(x,T,y)\, \theta_{T/2}(y)dy = p_{-T/2}(x) \\ \theta_{T/2}(y) \int_{\mathbb{R}^3} \theta^*_{-T/2}(x)\, h(x,T,y)dx = p_{T/2}(y) \end{cases} \tag{4.5}$$

The existence theorem for the measure of the Markovian Bernstein process Z_t, t in I, uses a general result of Beurling [12] Cf. also [4-5]:

Theorem

For H in the Schrödinger class and $p_{-T/2}$, $p_{T/2}$ two strictly positive probability densities, then positive (not necessarily integrable) solutions $\{\theta^*_{-T/2}, \theta_{T/2}\}$ of (4.5) exist and are unique.

The process Z_t is then entirely determined, for t in I. It is a diffusion process without killing. For the Hamiltonian (1.3') its (backward) drift and diffusion matrix are, respectively, given by

$$\begin{aligned} B_*(Z(t),t) &= (\nabla I_* - \vec{A})\,(Z(t),t) \\ C_*(Z(t),t) &= \hbar\, \mathbb{1} \end{aligned} \tag{4.6}$$

where $\mathbb{1}$ is the 3×3 identity matrix, as predicted in §2.

The Bernstein diffusion Z_t is time symmetric. For each t in I, its probability density is given by

$$p(x,t)dx = \theta\theta^*(x,t)dx \tag{4.7}$$

where θ^* solves the initial value problem (1.1) for $\theta^*(x, -T/2) = \theta^*_{-T/2}$ and θ solves the final value problem

$$\frac{\partial\theta}{\partial t} = H\theta$$

for $\theta(x, T/2) = \theta_{T/2}(x)$,and the pair $\theta^*_{-T/2}, \theta_{T/2}$ is a solution of the Schrödinger systems (4.5). For a more accurate summary of the construction cf [16].

§5. Hilbert space approach.

The sense in which the new resulting theory can be regarded as an Euclidean version of quantum mechanics (EQM) is clarified by the following analytical description. (Cf. [6].)

For H bounded below, the functional calculus enables us to take the analytical continuation of a solution of the initial value problem for the Schrödinger equation (1.2) (for $h \equiv 1$). Then

$$\theta^*_\chi(x,t) = \left(\int_{-\infty}^{\infty} e^{-t\lambda}\, dE^H(\lambda)\chi\right)(x)\,, \tag{4.8}$$

where $\{E^H(\lambda)\}$ is the spectral family of H, solves the initial value problem for (1.1) in $L^2(\mathbb{R}^3)$.

Suppose that χ is an analytic vector for H with convergence radius T/2. Then

$$\theta_\chi(x,t) = \left(\int_{-\infty}^{\infty} e^{t\lambda}\, dE^H(\lambda)\right)(x) \tag{4.9}$$

is also well defined, for t in $I = [-T/2, T/2]$.

Observing that

$$\int_{\mathbb{R}^3} \mathcal{O}_{\bar{\chi}} \mathcal{O}^*_{\chi}(x,t)dx = \|\chi\|_2^2 \, , \tag{4.10}$$

one defines a dense linear subspace of $L^2(\mathbb{R}^3)$ by

$$\tilde{\mathcal{V}}^*_t = \{\mathcal{O}^*_{\chi}(t) = e^{-tH}\chi, \chi \text{ in } \mathcal{D}(e^{(T/2)H})\} \tag{4.11}$$

where $\mathcal{D}(e^{(T/2)H})$ denotes the dense set of analytical vectors for H with convergence radius T/2.

Define $U_t^{-1} : \tilde{\mathcal{V}}^*_t \to D(e^{(T/2)H})$ by $\mathcal{O}^*_{\chi}(t) \to \chi$ and a scalar product in $\tilde{\mathcal{V}}^*_t$ by

$$\begin{aligned}(\mathcal{O}^*_{\chi_1}(t) \,|\, \mathcal{O}^*_{\chi_2}(t))_t &= \langle U_t^{-1}\mathcal{O}^*_{\chi_1}(t) \,|\, U_t^{-1}\mathcal{O}^*_{\chi_2}(t)\rangle_2 \\ &= \langle \chi_1 \,|\, \chi_2 \rangle_2 \, .\end{aligned} \tag{4.12}$$

The completion of $\tilde{\mathcal{V}}^*_t$ with respect to $(\cdot\,|\,\cdot)_t$ is an Hilbert space denoted by $\tilde{\mathcal{V}}^*_t$, and called the forward Hilbert space. $(\tilde{\mathcal{V}}^*_t, (\cdot\,|\,\cdot)_t)$ can, actually, be identified with $(L^2, \langle\cdot\,|\,\cdot\rangle_2)$ because U_t^{-1} has a unitary extension from $\mathcal{V}^*_t$ onto L^2. In particular, each quantum mechanical operator on $L^2(\mathbb{R}^3)$ has its Euclidean analogue: if $A : \mathcal{D}(A) \to L^2(\mathbb{R}^3)$ is an observable, then

$$A_{-t}^F = U_t A U_t^{-1} = e^{-tH} A e^{tH} : U_t \mathcal{D}(A) \subset \mathcal{V}^*_t \to \mathcal{V}^*_t \tag{4.13}$$

defines the forward operator associated with A.

So, under proper restrictions on the domains, the Euclidean version of Heisenberg equation of motion is valid:

$$-\frac{d}{dt}A_t^F = [A,H]_t^F .\tag{4.14}$$

For example, the simplest Hamiltonian (1.3') with $\vec{A} = 0$ corresponds here to

$$H = -\tfrac{1}{2}P^2 + \varphi(Q)\tag{4.15}$$

(since iP is the quantum mechanical momentum) and then

$$\frac{dQ_t^F}{dt} = P_t^F , \quad \frac{dP_t^F}{dt} = \nabla\varphi(Q_t^F)\tag{4.16}$$

are the (Euclidean) Hamilton equations.

There is a natural probabilistic interpretation of these Euclidean observables, namely for χ', χ regular enough and positive such that, according to (4.7), $\eta_{\chi'}\eta^*_{\chi}(x,t)\,dx$ is the probability density of a Bernstein diffusion $Z(t)$, $t \in I$,

$$\begin{aligned}(\mathcal{O}^*\chi'(t) \mid Q\mathcal{O}^*\chi(t))_t &= (\mathcal{O}^*\chi'(t) \mid U_t Q_t^F U_t^{-1}\mathcal{O}^*\chi(t))_t \\ &= \langle \chi' \mid Q_t^F \chi\rangle_2 \\ &= E[Z(t)]\end{aligned}\tag{4.17}$$

and similarly for the other observables.

Some results of §2 become easier to interpret. For example, (2.13) means that

$$p_* = \nabla I_* = -\frac{\nabla\theta^*}{\theta^*}\tag{4.18}$$

$$= \frac{P\theta^*}{\theta^*}$$

where P is the Euclidean momentum. If we regard the l.h.s. of (4.18) as a random variable, using (4.7),

$$E[p_*] = \int \theta P\theta^* \, dx = (\theta^*(t) \,|\, P\theta^*(t))_t \ .$$

The same is true for the other observables.

Euclidean Quantum Mechanics has a dynamical structure which is notably different from Nelson's Stochastic Mechanics. For example, if one defines a constant of motion A_t^F by $\frac{d}{dt} A_t^F = 0$ as suggested by (4.14), one shows easily that the associated random variable $a_* = a_*(Z(t),t)$, t in I, is an $\mathcal{F}_t$-martingale, i.e. that $D_* a_*(Z(t),t) = 0$. (4.19)

This is an important aspect of the probabilistic structure of EQM, which is without equivalent in Nelson's theory.

§6. A brief history of Euclidean Quantum Mechanics.

The basic idea of EQM is due to E Schrödinger [8]. Then S. Bernstein, R. Fortet, A. Beurling and B. Jamison contributed to the mathematical clarification of Schrödinger's intuition : cf [11], [12] and references therein. The relation with quantum dynamics has been discovered in [4]. It was initially motivated by open problems in Nelson's Stochastic Mechanics. Then, the Euclidean approach has been developed on its own [5], under the name of "Euclidean Quantum Mechanics", and can be regarded as an alternative starting point for a Euclidean program of field theory. In [6] the Hilbert space approach of the theory is elaborated.

Acknowledgments.

It is a pleasure to thank the organizers of the third Bellman Continuum International Workshop, and in particular Professor A. Blaquiere, for the remarkable organization of the meeting and its stimulating atmosphere. Also, I cannot forget Alice and Peta, at Warwick, who showed unstinting patience during the past two years.

References.

[1] R.P. Feynman, A.R. Hibbs, *Quantum Mechanics and Path Integrals,* McGraw-Hill, New York (1965).

[2] M. Kac, "On some connections between probability theory and differential and integral equation", *Proc. of 2nd Berkeley Symp. on Probability and Statistics,* J. Neyman Ed., Univ. of California Press, Berkeley (1951).

[3] J. Glimm, A. Jaffe, *Quantum Physics, a Functional integral point of view,* Springer, New York (1981); B. Simon, *Functional Integration and Quantum Physics,* Academic Press, New York (1979).

[4] J.C. Zambrini, *J. Math. Physics,* 27, No.9, 2311 (1986).

[5] J.C. Zambrini, *Phys. Review A,* Vol.35, No.9 (1987).

[6] S. Albeverio, K. Yasue and J.C. Zambrini, "Euclidean Quantum Mechanics: analytical approach", to appear.

[7] E. Nelson, *Dynamical Theories of Brownian Motion,* Princeton University Press (1967); *Quantum Fluctuations,* Princeton Univ. Press, Princeton (1985).

[8] E. Schrödinger, *Ann. Inst. Henri Poincaré* 2,269 (Cf. p.296) (1932).

[9] W.H. Fleming, *Appl. Math. Optim.* 4, 329 (1978).

[10] K. Yasue, *J. Math. Phys.* 22, 1010 (1981); *J. of Funct. Analysis,* 41, 327 (1981).

[11] B. Jamison, Z. Wahrscheinlich. *Gebiete* 30, 65 (1974).

[12] A. Beurling, *Ann. Math.* 72, 189 (1960).

[13] E. Nelson, *Ann. Math.* 70, 572 (1959).

[14] K. Itô, *Appl. Math.* Opt. 1, 374 (1975).

[15] P. Blanchard, P. Combe and W. Zheng, *Mathematical and Physical Aspects of Stochastic Mechanics,* Lect. Notes in Phys. 281, Springer Verlag, Berlin (1987).

[16] J.C. Zambrini, "New probabilistic approach to the classical heat equation", in *Proc. of Swansea Meeting,* Aug. 4-8, 1986, Ed. A. Truman and I.M. Davies, Lect. Notes in Math. no. 1325, Springer-Verlag, Berlin (1988).

FROM TWO STOCHASTIC OPTIMAL CONTROL PROBLEMS TO THE SCHRODINGER EQUATION

K. Kime
Department of Mathematics and Statistics
Case Western Reserve University
Cleveland, Ohio 44106 (USA)

A. Blaquiere
Universite Paris 7, Laboratoire d'Automatique Theorique
Paris (FRANCE)

1. Introduction

In recent years, interest has developed in the connections between stochastic control theory, dynamic programming and quantum mechanics [1-4, 7, 12, 13] and (related) variational approaches [9, 11, 14, 15] to Nelson's stochastic mechanics [10]. In this paper, we will start by considering two stochastic optimal control problems, one "forward" in time, one "backward" in time. We show that, if there are solutions to the extended Hamilton-Jacobi equations associated with the control problems, then there is a solution of a Schrödinger equation and conversely, if there is a sufficiently well-behaved solution to a Schrödinger equation, there are solutions to a pair of extended H-J equations. We note connections with Nelson's stochastic mechanics. The H-J equations are equivalent to a pair of inhomogeneous "backward" and "forward" heat equations via a well-known exponential transformation. One may thus pass from these to a Schrödinger equation (and back).

2. Definitions and Notations

We assume a given underlying probability space (Ω,F,P). E^n denotes n-dimensional Euclidean space, (t_0,t_1) an interval in E^1. S denotes $(t_0,t_1) \times E^n$; $\bar{S} = [t_0,t_1] \times E^n$. Definitions of stochastic process, Brownian motion will be taken from [6] as will other elements of our framework which will be noted below.

A solution of a stochastic differential equation

$$d\xi = b(t,\xi(t))dt + \sigma(t,\xi(t))dw \tag{2.1}$$

with initial data $\xi(s) = x$ is to be interpreted as in [6] as a solution of the integral equation

$$\xi(t) = \xi(s) + \int_s^t b(r,\xi(r))dr + \int_s^t \sigma(r,\xi(r))dw(r) \tag{2.2}$$

Here, w is standard Brownian motion of dimension n. With the vector notation $\xi=(\xi_1,\ldots\xi_n)$, $b=(b_1\ldots b_n)$, we have

$$d\xi_i = b_i(t,\xi(t))dt + \sum_{\ell=1}^{n} \sigma_{i\ell}(t,\xi(t))dw_\ell \qquad i=1,\ldots n$$

The notation $C_p^{1,2}(S)$ denotes the class of functions ψ in $C^{1,2}(S)$ (meaning C^1 in t, C^2 in x) which satisfy $|\psi(t,x)| \le D(1+|x|^k)$ for some constants D,k, when $(t,x) \in S$.

3. Two Stochastic Optimal Control Problems

We consider first a "forward" stochastic optimal control problem, Problem 1, in 3.1, then the symmetric "backward" problem, Problem 2, in 3.2. The controls v and $\bar{v}$ will take values in E^n.

3.1 Problem 1

Consider the stochastic differential equation

$$d\xi = v(t,\xi(t))dt + \sigma dw \qquad (3.1.1)$$

with initial data $\xi(s) = x \in E^n$, at time $s \in (t_0,t_1)$. Here, w is a standard n-dimensional Brownian motion, and

$$\sigma_{ij} = \sqrt{2D}\,\delta_{ij}$$

where δ is the Kronecker delta, and D is a positive constant. We assume that v belongs to a class of admissible control functions defined as follows:

Definition 3.1.A [6]. A feedback control law v (the term feedback refers to the fact that the control is a function of the state $\xi(t)$) is admissible if v is a Borel measurable function from $\bar{S}$ into E^n, such that

(a) For each (s,x), $t_0 \le s \le t_1$, there exists a Brownian motion w such that (3.1.1) with initial data $\xi(s) = x$ has a solution ξ, unique in probability law; and

(b) For each $k > 0$, $E_{sx}|\xi(t)|^k$ is bounded for $s \le t \le t_1$, and

$$E_{sx}\int_s^{t_1} |v(t,\xi(t))|^k\, dt < \infty$$

(the bound may depend on (s,x)). The subscript sx refers to the fact that $\xi(s) = x$.

Either of the following conditions are sufficient for the admissibility of v:

(i) For some constant M_1, $|v(t,y)| \le M_1(1+|y|)$ for all $(t,y) \in \bar{S}$. Moreover, for any bounded Borel set $B \subset E^n$ and $t_0 < t' < t_1$, there exists a constant K_1 such that, for all $x,y \in B$ and $t_0 \le t \le t'$,

$$|v(t,x) - v(t,y)| \le K_1 |x-y|$$

(K_1 may depend on B,t'; and both M_1, K_1 may depend on v).

(ii) v satisfies a Lipschitz condition on $\bar{S}$. Further, if (i) or (ii) holds, the Brownian motion w can be specified in advance, which is the case in Problem 1.

Now, for $(t,x) \in \bar{S}$ and $\upsilon \in E^n$, let

$$L(t,x,\upsilon) = \frac{1}{2} m\upsilon^2 + Q(t,x) \tag{3.1.2}$$

where Q is continuous on $\bar{S}$, and let $W_1 : E^n \to R_+$ (R_+ denoting non-negative real numbers) be continuous and assume

$$\begin{aligned} |Q(t,x)| &\le C(1 + |x|)^k \\ W_1(x) &\le C(1 + |x|)^k \end{aligned} \tag{3.1.3}$$

for some constants C,k.

We define a <u>cost function</u>

$$J(s,x,v) = E_{sx} \left\{ \int_s^{t_1} L(t,\xi(t),v(t,\xi(t)))dt + W_1(\xi(t_1)) \right\} .$$

The conditions on Q and W_1 ensure that J is finite.

Now let the optimal control problem be as follows: Find an admissible feedback control v^*, among all admissible feedback controls, which minimizes $J(s,x,v)$. The following Verification Theorem gives sufficient conditions for the existence of a minimizing v^*.

<u>Theorem 3.1.B [6].</u> Let $W(s,x)$ be a solution of the dynamic programming equation

$$0 = \frac{\partial W}{\partial s} + \min_{\upsilon \in E^n} \left[D\Delta W + \sum_{i=1}^{n} \upsilon_i \frac{\partial W}{\partial x_i} + \frac{1}{2} m\upsilon^2 + Q(s,x) \right] \tag{3.1.4}$$

$$(s,x) \in S \quad ,$$

with boundary data

$$W(t_1,x) = W_1(x), \qquad x \in E^n, \tag{3.1.5}$$

such that W is in $C_p^{1,2}(S)$ and continuous on $\bar{S}$. Then,

(a) $W(s,x) \leq J(s,x,v)$ for any admissible feedback control v and any initial data $(s,x) \in S$.

(b) If v^* is an admissible feedback control such that

$$D\Delta W + \sum_{i=1}^{n} v_i^*(s,x) \frac{\partial W}{\partial x_i} + \frac{1}{2} m(v^*(s,x))^2 + Q(s,x)$$

$$= \min_{\upsilon \in E^n} \left[D\Delta W + \sum_{i=1}^{n} \upsilon_i \frac{\partial W}{\partial x_i} + \frac{1}{2} m\upsilon^2 + Q(s,x) \right] \tag{3.1.4}$$

for all $(s,x) \in S$, then $W(s,x) = J(s,x,v^*)$ for all $(s,x) \in S$. Thus, v^* is optimal.

Now let us assume that there exists a W satisfying the hypotheses of the Verification Theorem, and an optimal control v^*. Then, since the controls take values in E^n, which is open

$$mv^* = - \text{grad } W \text{ for all } (s,x) \in S \tag{3.1.6}$$

and

$$\frac{\partial W}{\partial s} = - D\Delta W + \frac{1}{2m} (\text{grad } W)^2 - Q \tag{3.1.7}$$

for all $(s,x) \in S$. Equation (3.1.7) is analogous to the Hamilton-Jacobi equation of classical mechanics; we shall refer to it as an extended Hamilton-Jacobi equation.

3.2 Problem 2

Now let us introduce another type of admissibility for a feedback control function as follows:

Definition 3.2.A A feedback control law $\bar{v}$ is <u>backward admissible</u> if $\bar{v}$ is such that

$\bar{v}(\tau,x) = - \hat{v}(t_0+t_1-\tau,x)$ for all $(\tau,x) \in \bar{S}$, and

$\hat{v}$ is an admissible feedback control law.

We consider the stochastic differential equation

$$d\eta = \bar{v}(\tau,\eta(\tau))d\tau + \sqrt{2D}d\bar{w} \tag{3.2.1}$$

where $\bar{v}$ is a backward admissible feedback control law, and

$$\bar{w}(\tau) = w(t_0+t_1-\tau).$$

We say that η is a solution to (3.2.1) with terminal data $\eta(\sigma) = y \in E^n$, with $t_0 \le \tau < \sigma \le t_1$, if η satisfies the integral equation

$$\eta(\tau) = \eta(\sigma) - \int_\tau^\sigma \bar{v}(r,\eta(r))dr - \int_\tau^\sigma \sqrt{2D}d\bar{w}(r). \tag{3.2.2}$$

By making the change of variable

$$\tau = t_0 + t_1 - t,$$
$$\sigma = t_0 + t_1 - s,$$
$$\ell = t_0 + t_1 - r$$

(3.2.2) becomes

$$\eta(t_0+t_1-t) = \eta(t_0+t_1-s) - \int_t^s \bar{v}(t_0+t_1-\ell,\ \eta(t_0+t_1-\ell))(-d\ell) - \int_t^s \sqrt{2D}\ d\bar{w}(t_0+t_1-\ell) \quad . \tag{3.2.3}$$

Define $\hat{\eta}(\ell) = \eta(t_0+t_1-\ell)$. Now (3.2.3) becomes

$$\hat{\eta}(t) = \hat{\eta}(s) + \int_s^t \hat{v}(\ell,\hat{\eta}(\ell))\ d\ell + \int_s^t \sqrt{2D}\ d\bar{w}(\ell) \quad , \tag{3.2.4}$$

and we have

$$\hat{\eta}(s) = y \quad . \tag{3.2.5}$$

We now define

$$\hat{J}(s,y,\hat{v}) = E_{sy}\left\{\int_s^{t_1}[\tfrac{m}{2}\left(\hat{v}(\ell,\hat{\eta}(\ell))\right)^2 + \hat{Q}(\ell,\hat{\eta}(\ell))]d\ell + \bar{W}_0(\hat{\eta}(t_1))\right\}$$
$$= E_{\sigma y}\left\{\int_{t_0}^{\sigma}[\tfrac{m}{2}\left(\bar{v}(r,\eta(r))\right)^2 + Q(r,\eta(r))]dr + \bar{W}_0(\eta(t_0))\right\}$$
$$= \bar{J}(\sigma,y,\bar{v}) \quad . \tag{3.2.6}$$

Here Q is the same as in Problem 1, $\bar{W}_0: E^n \to R_+$ is continuous and $\bar{W}_0(y) \le C(1 + |y|)^k$, (C,k as in (3.1.3)). Thus, $\hat{Q}(\ell,\hat{\eta}(\ell)) = Q(t_0+t_1-\ell,\ \eta(t_0+t_1-\ell))$.

We now consider, as in Problem 1, the problem of minimizing (3.2.6). For given terminal data $y \in E^n$ at time $\sigma \in (t_0,t_1]$, we shall say that $\bar{v}^*$ is <u>backward optimal</u> if

$\bar{v}^*$ is backward admissible, and

$$\bar{J}(\sigma,y,\bar{v}) \geq \bar{J}(\sigma,y,\bar{v}^*)$$

for all backward admissible $\bar{v}$.

In view of (3.2.4) - (3.2.6), we have the following version of the Verification Theorem:

Theorem 3.2.B Let $\hat{W}$ be a solution of the dynamic programming equation

$$0 = \frac{\partial \hat{W}}{\partial s} + \min_{\upsilon \in E^n} \left[D\Delta\hat{W} + \sum_{i=1}^{n} \hat{\upsilon}_i \frac{\partial \hat{W}}{\partial y_i} + \frac{1}{2} m\hat{\upsilon}^2 + \hat{Q}(s,y)\right] \tag{3.2.7}$$

$$(s,y) \in S \quad ,$$

with $\hat{W}(t_1,y) = \bar{W}_0(y)$, $y \in E^n$, such that $\hat{W}$ is in $C_p^{1,2}(S)$ and continuous on $\bar{S}$. Then:

(a) $\hat{W}(s,y) \leq \hat{J}(s,y,\hat{v})$ for any admissible feedback control $\hat{v}$ and any initial data $(s,y) \in S$.

(b) If $\hat{v}^*$ is an admissible feedback control such that

$$\begin{aligned} &D\Delta\hat{W} + \sum_{i=1}^{n} \hat{v}_i^*(s,y)\frac{\partial \bar{W}}{\partial y_i} + \frac{1}{2} m(\hat{v}^*(s,y))^2 + \hat{Q}(s,y) = \\ &\min_{\hat{\upsilon} \in E^n} \left[D\Delta\hat{W} + \sum_{i=1}^{n} \hat{\upsilon}_i \frac{\partial \hat{W}}{\partial y_i} + \frac{1}{2} m\hat{\upsilon}^2 + \hat{Q}(s,y)\right] \end{aligned} \tag{3.2.8}$$

for all $(s,y) \in S$, then $\hat{W}(s,y) = \hat{J}(s,y,\hat{v}^*)$ for all $(s,y) \in S$; $\hat{v}^*$ is optimal

Now suppose there exists a function $\hat{W}$ satisfying these hypotheses, and an optimal control $\hat{v}^*$. Define

$$\bar{W}(\sigma,y) = \hat{W}(t_0+t_1-\sigma,y) \ , \ t_0 < \sigma \leq t_1 \ .$$

Then $\bar{W}(t_0,y) = \hat{W}(t_1,y)$ and $\dfrac{\partial \bar{W}}{\partial \sigma} = -\dfrac{\partial \hat{W}}{\partial s}$.

We define

$$\bar{v}^*(\sigma,y) = \bar{v}^*(t_0+t_1-s,y) = -\hat{v}^*(s,y).$$

Now we have

$$0 = -\frac{\partial \bar{W}}{\partial \sigma} + D\Delta\bar{W} - \sum_{i=1}^{n} (\bar{v}_i^*(\sigma,y))\frac{\partial \bar{W}}{\partial y_i} + \frac{1}{2} m(\bar{v}^*(\sigma,y))^2 + Q(\sigma,y) \tag{3.2.9}$$

and, as in Problem 1,

$$m\bar{v}^* = \text{grad } \bar{W} \tag{3.2.10}$$

$$\frac{\partial \bar{W}}{\partial \sigma} = D\Delta\bar{W} - \frac{1}{2m}(\text{grad } \bar{W})^2 + Q \qquad \text{on } S. \tag{3.2.11}$$

We have let

$$\hat{v}^*(s,y) = -\bar{v}^*(t_0+t_1-s,y), \qquad y \in E^n.$$

From (3.2.6) we have

$$\bar{J}(\sigma,y,\bar{v}^*) = \hat{J}(s,y,\hat{v}^*). \tag{3.2.12}$$

If the Verification Theorem 3.2.B is satisfied, then $\bar{v}^*$ is optimal; that is

$$\hat{J}(s,y,\hat{v}) \geq \hat{J}(s,y,\hat{v}^*). \tag{3.2.13}$$

From (3.2.6) and (3.2.11), (3.2.12) implies

$$\bar{J}(\sigma,y,\bar{v}) \geq \bar{J}(\sigma,y,\bar{v}^*). \tag{3.2.14}$$

for all backward admissible $\bar{v}$.

Therefore, if $\hat{v}^*$ is an optimal control in the sense of Theorem 3.2.A, then $\bar{v}^*$ is a backward optimal control for Problem 2, and the converse is also true.

4. Extended Hamilton-Jacobi Equations, the Schrödinger Equation and Inhomogeneous Backward and Forward Heat Equations.

4.1 Extended Hamilton-Jacobi Equations and the Schrödinger Equation

We have seen, that if there exist W, $\bar{W}$, v^*, $\bar{v}^*$ satisfying the conditions of the Verification Theorems, then W is a solution of the equation

$$\frac{\partial G}{\partial t}(t,x) - \frac{1}{2m}(\text{grad } G(t,x))^2 + D\Delta G(t,x) + Q(t,x) = 0 \tag{4.1.1}$$

$$(t,x) \in S$$

with

$$G(t_1,x) = W_1(x), \tag{4.1.2}$$

and $\bar{W}$ is a solution of the equation

$$\frac{\partial \bar{G}}{\partial t}(t,x) + \frac{1}{2m} (\text{grad } \bar{G}(t,x))^2 - D\Delta\bar{G}(t,x) - Q(t,x) = 0 \qquad (4.1.3)$$

$$(t,x) \in S$$

with

$$\bar{G}(t_0,x) = \bar{W}_0(x). \qquad (4.1.4)$$

We now show that, when there are solutions G, $\bar{G}$ of (4.1.1), (4.1.3), then there are solutions of a Schrödinger equation. From now on D shall denote $\hbar/2m$.

Proceeding as in [4], with $G^* = \frac{\bar{G}+G}{2}$, $H^* = \frac{\bar{G}-G}{2}$, we have

$$\frac{\partial}{\partial t} (G^*-H^*) - \frac{1}{2m} (\text{grad}(G^*-H^*))^2 + D\Delta(G^*-H^*) + Q = 0 \qquad (4.1.5)$$

$$\frac{\partial}{\partial t} (G^*+H^*) + \frac{1}{2m} (\text{grad}(G^*+H^*))^2 + D\Delta(G^*+H^*) - Q = 0 \qquad (4.1.6)$$

Adding and subtracting (4.1.5), (4.1.6) gives

$$\frac{\partial H^*}{\partial t} + \frac{1}{2m} (\text{grad } H^*)^2 + \frac{1}{2m} (\text{grad } G^*)^2 - D\Delta G^* - Q = 0 \qquad (4.1.7)$$

$$\frac{\partial G^*}{\partial t} + \frac{1}{m} \text{grad } H^* \text{ grad } G^* - D\Delta H^* = 0 \qquad (4.1.8)$$

Equations (4.1.7), (4.1.8) are equations (19), (20), of [4], except for the potential Q which was taken to be zero in [4].

At this stage, we make the following observation: if we define

$$\tilde{Q} = \frac{1}{m} (\text{grad } G^*)^2 - 2D\Delta G^* - Q \qquad (4.1.9)$$

then (4.1.7) becomes

$$\frac{\partial H^*}{\partial t} + \frac{1}{2m} (\text{grad } H^*)^2 - \frac{1}{2m} (\text{grad } G^*)^2 + D\Delta G^* + \tilde{Q} = 0 \qquad (4.1.10)$$

(4.1.8) is unchanged:

$$\frac{\partial G^*}{\partial t} + \frac{1}{m} \text{grad } H^* \text{ grad } G^* - D\Delta H^* = 0 \qquad (4.1.8)$$

If we now multiply (4.1.10) by i, and subtract (4.1.8), we obtain

$$\frac{\partial}{\partial t}(-G^*+iH^*) = -D\Delta H^* + \frac{1}{m}\text{ grad } H^* \text{ grad } G^* + \frac{i}{2m}(\text{grad } G^*)^2$$

$$- \frac{i}{2m}(\text{grad } H^*)^2 - iD\Delta G^* - i\tilde{Q}$$

or

$$\frac{\partial}{\partial t}(-G^*+iH^*) = iD\Delta(-G^*+iH^*) + \frac{i}{2m}(\text{grad}(-G^*+iH^*))^2 - i\tilde{Q} \qquad (4.1.11)$$

Straightforward differentiation gives us

Proposition 4.1.A. If G, $\bar{G}$ are solutions to (4.1.1), (4.1.3), then

$$\psi = \exp\left(\frac{-G^*+iH^*}{\hbar}\right) \qquad (4.1.12)$$

is a solution to

$$i\hbar\frac{\partial\psi}{\partial t} = \frac{-\hbar^2}{2m}\Delta\psi + \left(\frac{(\text{grad } G^*)^2}{m} - \frac{\hbar\Delta G^*}{m} - Q\right)\psi. \qquad (4.1.13)$$

Conversely, suppose we start with the Schrödinger equation

$$i\hbar\frac{\partial\hat{\psi}}{\partial t} = \frac{-\hbar^2}{2m}\Delta\hat{\psi} - P\hat{\psi} \qquad (t,x) \in \bar{S} \qquad (4.1.14)$$

with given potential P. Assume there is a solution $\hat{\psi}$ of (4.1.14), $\hat{\psi} \neq 0$, all (t,x), with

$$\hat{\psi} = \exp\left(\frac{-M+iN}{\hbar}\right) \qquad (4.1.15)$$

and suppose that M and N are $C^{1,2}$ functions on $\bar{S}$. Running the above arguments backwards, we see

$$\frac{\partial N}{\partial t} + \frac{1}{2m}(\text{grad } N)^2 - \frac{1}{2m}(\text{grad } M)^2 + D\Delta M - P = 0 \qquad (4.1.16)$$

$$\frac{\partial M}{\partial t} + \frac{1}{m}\text{ grad } N \text{ grad } M - D\Delta N = 0 \qquad (4.1.17)$$

The passage from (4.1.14) to the pair of equations (4.1.16), (4.1.17) was used by Louis de Broglie for introducing his "theorie du guidage" (see [5]; equations (4.1.16), (4.1.17) are the so-called equations (J) and (C) of Louis de Broglie). Together with this pair of equations he defined the quantum potential Q_p by

$$Q_p = D\Delta M - \frac{1}{2m} (\text{grad } M)^2 \tag{4.1.18}$$

The purpose of the definition (4.1.18) was to reduce equation (4.1.16) to the form

$$\frac{\partial N}{\partial t} + \frac{1}{2m} (\text{grad } N)^2 + Q_p - P = 0 \tag{4.1.19}$$

which is the Hamilton-Jacobi equation of classical mechanics for the motion of a mass-point in the potential $P - Q_p$. As the reader may anticipate, if we next introduce the "modified potential" $\hat{Q}$ by

$$\hat{Q} = P - 2Q_p = P - 2D\Delta M + \frac{(\text{grad } M)^2}{m} \quad , \tag{4.1.20}$$

then

$$\frac{\partial}{\partial t}(N+M) + \frac{1}{2m} (\text{grad } (N+M))^2 - D\Delta(N+M) - \hat{Q} = 0 \tag{4.1.21}$$

$$\frac{\partial}{\partial t}(M-N) - \frac{1}{2m} (\text{grad } (M-N))^2 + D\Delta(M-N) + \hat{Q} = 0 \tag{4.1.22}$$

Thus we have

<u>Proposition 4.1.B.</u> If

$$\hat{\psi} = \exp\left(\frac{-M+iN}{\hbar}\right)$$

is a solution as above to

$$i\hbar \frac{\partial \psi}{\partial t} = \frac{-\hbar^2}{2m} \Delta\hat{\psi} - P\hat{\psi} \quad , \tag{4.2.12}$$

then (M-N) is a solution to

$$\frac{\partial G}{\partial t} - \frac{1}{2m} (\text{grad } G)^2 + D\Delta G + \hat{Q} = 0 \quad , \tag{4.1.23}$$

and (N+M) is a solution to

$$\frac{\partial \bar{G}}{\partial t} + \frac{1}{2m} (\text{grad } \bar{G})^2 - D\Delta\bar{G} - \hat{Q} = 0 \quad . \tag{4.1.24}$$

Equations (4.1.23, (4.1.24) are the equations (4.1.1), (4.1.3) with Q replaced by Q, which is given by (4.1.20) (note that $\hat{Q}$ is specified once P is given, and N and M subsequently determined).

Remark 4.1.C

Nelson's Stochastic Mechanics

In [10], Nelson considered a Markov process x(t) with forward and backward drifts b and b_* and diffusion coefficient $\hbar/2m$ as a model for the motion of a particle of mass m subjected to an external force F. He showed, under assumptions, that if one defines

$$u = \frac{1}{2}\left(b - b_*\right)$$
$$z = \frac{1}{2}\left(b + b_*\right) \quad ,$$

then u and z satisfy

$$\frac{\partial u}{\partial t} = \frac{\hbar}{2m}\,\mathrm{grad}(\mathrm{div}\, z) - \mathrm{grad}(z \cdot u) \tag{4.1.25}$$

$$\frac{\partial z}{\partial t} = F/m - (z\cdot\nabla)z + (u\cdot\nabla)u + \frac{\hbar}{2m}\,\Delta u \tag{4.1.26}$$

Then, using the fact that in the derivation of the above equations u was shown to be a gradient

$$u = \hbar/m \;\mathrm{grad}\; R$$

where $R = \frac{1}{2}\log\rho$, the probability density of x(t), and assuming that z is a gradient also,

$$z = \hbar/m \;\mathrm{grad}\; S,$$

it is shown that, in the case $F = -\mathrm{grad}\, V$,

$$\psi = \exp(R+iS)$$

satisfies the Schrödinger equation

$$i\hbar\,\frac{\partial\psi}{\partial t} = \frac{-\hbar^2}{2m}\,\Delta\psi + V\psi.$$

Conversely, one may start with any (normalized) solution ψ to a Schrödinger equation, writing

$$\psi = \exp(R+iS)$$

and

$$u = \frac{\hbar\;\mathrm{grad}\;R}{m} \quad , \qquad z = \frac{\hbar\;\mathrm{grad}\;S}{m} \quad ,$$
$$b = z+u \quad , \qquad b_* = z-u \quad , \qquad \rho = |\psi|^2 \quad .$$

The Markov process with forward and backward drifts b and b_* and diffusion coefficient $\hbar/2m$ has probability density ρ.

Now, if there exist W, $\bar{W}$, v^*, $\bar{v}^*$ satisfying the conditions of the Verification Theorem, then, letting

$$W^* = \frac{\bar{W}+W}{2} \quad , \quad V^* = \frac{\bar{W}-W}{2} \; ,$$

W^* and V^* are solutions to (4.1.8), (4.1.10), i.e.,

$$\frac{\partial W^*}{\partial t} + \frac{1}{m} \text{ grad } V^* \text{ grad } W^* - D\Delta V^* = 0 \tag{4.1.27}$$

$$\frac{\partial V^*}{\partial t} + \frac{1}{2m} (\text{grad } V^*)^2 - \frac{1}{2m} (\text{grad } W^*)^2 + D\Delta W^* + \tilde{Q} = 0, \tag{4.1.18}$$

$$\tilde{Q} = \frac{(\text{grad } W^*)^2}{m} - \frac{\hbar}{m} \Delta W^* - Q \; .$$

Then

$$\psi = \exp\left(\frac{-W^*+iV^*}{\hbar}\right)$$

is a solution to

$$i\hbar \frac{\partial \psi}{\partial t} = \frac{-\hbar^2}{2m} \Delta\psi + \left[\frac{(\text{grad } W^*)^2}{m} - \frac{\hbar}{m} \Delta W^* - Q\right] \psi$$

Letting

$$u = \frac{-\text{grad } W^*}{m} \quad , \quad z = \frac{\text{grad } V^*}{m} \quad ,$$

$$b = z+u \quad , \quad b_* = z-u \quad ,$$

we have

$$b = \frac{-\text{grad } W}{m} = v^*$$

$$b_* = \frac{\text{grad } \bar{W}}{m} = \bar{v}^* \; .$$

This gives an interpretation of the forward and backward drifts of the Markov process in Nelson's stochastic mechanics as the optimal controls v^* and $\bar{v}^*$. Also $\rho = |\psi|^2 = \exp(-2W^*/\hbar)$ (ψ normalized if necessary).

Taking gradients of (4.1.27), (4.2.28) gives

$$\frac{\partial u}{\partial t} = -D\Delta z - \text{grad}(z \cdot u) \tag{4.1.29}$$

$$\frac{\partial z}{\partial t} = \frac{-\text{grad } \tilde{Q}}{m} - \frac{1}{2} \text{ grad } z^2 + \frac{1}{2} \text{ grad } u^2 + D\Delta u \tag{4.1.30}$$

if W^*, V^* are $C^{1,3}$ functions, and Q is C^1 in x. As $D = \hbar/2m$ and u and z are gradients, these are the same as (4.1.25) and (4.1.26), with $F = -\text{grad}\,\tilde{Q}$.

4.2 Inhomogeneous "backward and forward" heat equations

Now, if we make the exponential transformation

$$\phi(t,x) = \exp(-G(t,x)/\hbar) \tag{4.2.1}$$

in equation (4.1.1), we have

$$\frac{\partial\phi}{\partial t} = -D\Delta\phi + \frac{Q\phi}{\hbar} \tag{4.2.2}$$

with

$$\phi(t_1,x) = \exp\left(\frac{-G(t_1,x)}{\hbar}\right) \quad .$$

Similarly, if

$$\bar{\phi}(t,x) = \exp\left(\frac{-\bar{G}(t,x)}{\hbar}\right) \tag{4.2.3}$$

is put in (4.1.3), we have

$$\frac{\partial\bar{\phi}}{\partial t} = -D\Delta\bar{\phi} + \frac{Q\bar{\phi}}{\hbar} \tag{4.2.4}$$

with

$$\bar{\phi}(t_0,x) = \exp\left(\frac{-\bar{G}(t_0,x)}{\hbar}\right)$$

Thus from Proposition (4.1.B) and the above transformation we have the following

Fact I. If $\hat{\psi}$ given by

$$\hat{\psi} = \exp\left(\frac{-M+iN}{\hbar}\right)$$

is a solution to (4.1.14), as in Proposition 4.1.B, then

i) $\phi = \exp\left(\frac{-(M-N)}{\hbar}\right)$ is a solution of

$$\frac{\partial\phi}{\partial t} = -D\Delta\phi + \frac{\hat{Q}\phi}{\hbar} \tag{4.2.5}$$

ii) $\bar{\phi} = \exp\left[\frac{-(M+N)}{\hbar}\right]$ is a solution of

$$\frac{\partial\bar{\phi}}{\partial t} = D\Delta\bar{\phi} - \frac{\hat{Q}\bar{\phi}}{\hbar} \tag{4.2.6}$$

iii) The square of the modulus of $\hat{\psi}(t,x)$ is given by

$$|\hat{\psi}(t,x)|^2 = \exp(-2M/\hbar) = \phi(t,x)\ \bar{\phi}(t,x) = \phi^*(t,x)$$

iv) ϕ^* is a solution of the Fokker-Planck equation

$$\frac{\partial\phi^*}{\partial t} = -\sum_{i=1}^{n} \frac{\partial}{\partial x_i}(v_i(t,x)\phi^*) + D\Delta\phi^* \tag{4.2.7}$$

where

$$v_i(t,x) = \frac{2D}{\phi(t,x)}\ \frac{\partial\phi(t,x)}{\partial x_i} \qquad i = 1,\ldots n. \tag{4.2.8}$$

$$(t,x) \in \bar{S}.$$

Conversely, let $\tilde{\bar{\phi}}$, $\tilde{\phi}$ be the solutions of the equations

$$\frac{\partial\tilde{\bar{\phi}}}{\partial t} = D\Delta\tilde{\bar{\phi}} - \frac{R}{\hbar}\tilde{\bar{\phi}} \quad \text{in} \quad (t_0,t_1)\times E^n \tag{4.2.9}$$

$$\frac{\partial\tilde{\phi}}{\partial t} = -D\Delta\tilde{\phi} + \frac{R}{\hbar}\tilde{\phi} \quad \text{in} \quad (t_0,t_1)\times E^n\ , \tag{4.2.10}$$

satisfying conditions

$$\tilde{\bar{\phi}}(t_0,\cdot) = \bar{\phi}_0 \tag{4.2.11}$$

$$\tilde{\phi}(t_1,\cdot) = \phi_1\ , \tag{4.2.12}$$

where R is bounded, continuous on $\bar{S}$ and satisfies a Hölder condition with respect to x, and $\bar{\phi}_0$ and ϕ_1 are non-negative, continuous, and bounded functions on E^n. (see [8] for existence and uniqueness of $\tilde{\bar{\phi}}$, $\tilde{\phi}$). It may be seen, [8], that

$$\tilde{\bar{\phi}}(t,x) > 0, \quad \text{and } \tilde{\phi}(t,x) > 0 \quad \text{in } S$$

provided that neither $\bar{\phi}_0$ nor ϕ_1 vanishes identically. Now, defining $\tilde{\bar{W}}$, $\tilde{W}$ by

$$\tilde{\bar{\phi}} = \exp\left(\frac{-\tilde{\bar{W}}}{\hbar}\right) \tag{4.2.13}$$

$$\tilde{\phi} = \exp\left(\frac{-\tilde{W}}{\hbar}\right) \tag{4.2.14}$$

we see that $\tilde{W}$ is a solution of

$$\frac{\partial G}{\partial t} - \frac{1}{2m}(\text{grad } G)^2 + D\Delta G + R = 0 \tag{4.2.15}$$

with

$$G(t_1,x) = -\hbar \log \phi_1 \tag{4.2.16}$$

and $\tilde{\bar{W}}$ is a solution of

$$\frac{\partial \bar{G}}{\partial t} + \frac{1}{2m}(\text{grad } \bar{G})^2 - D\Delta G - R = 0 \tag{4.2.17}$$

with

$$\bar{G}(t_0,x) = -\hbar \log \bar{\phi}_0 \tag{4.2.18}$$

Thus, from Proposition 4.1.A and the above arguments, we have

Fact II. If $\tilde{\bar{\phi}}$, $\tilde{\phi}$ are solutions to the Cauchy problems (4.2.9), (4.2.11) and (4.2.10), (4.2.12), then

$$\tilde{\psi} = \exp\left(\frac{-(\tilde{\bar{W}}+\tilde{W}) + i(\tilde{\bar{W}}-\tilde{W})}{2\hbar}\right) = \exp\left(\frac{-\tilde{W}^*+i\tilde{V}^*}{\hbar}\right)$$

where

$$\tilde{W}^* = \frac{\tilde{\bar{W}}+\tilde{W}}{2}, \quad \tilde{V}^* = \frac{\tilde{\bar{W}}-\tilde{W}}{2},$$

satisfies

$$i\hbar\frac{\partial \tilde{\psi}}{\partial t} = \frac{-\hbar^2}{2m}\Delta\psi + \left(\frac{\left[\text{grad}\left((\tilde{\bar{W}}+\tilde{W})/2\right)\right]^2}{m} - \hbar\Delta\left[\frac{\tilde{\bar{W}}+\tilde{W}}{2m}\right] - R\right)\tilde{\psi} \tag{4.2.19}$$

Note:

a) the solution $\tilde{\psi}$ of the Schrödinger equation (4.2.19) depends, like $\tilde{\bar{W}}$ and $\tilde{W}$, on the initial and terminal data of the Cauchy problems.

b)

$$|\tilde{\psi}(t,x)|^2 = \exp\left(\frac{-\tilde{\bar{W}}(t,x) + \tilde{W}(t,x)}{2mD}\right) = \tilde{\bar{\phi}}(t,x)\,\tilde{\phi}(t,x)$$
$$= \tilde{\phi}^*(t,x)\ . \qquad (4.2.20)$$

Fact I is obtained in the proof of Theorem 4.3 of [15]; Fact II is more or less implicit in Theorem 4.4 and Corollary 4.4.1 of [15], however, the arguments here give Fact II more directly.

<u>Example</u> Homogeneous "backward and forward" heat equations, $n = 1$

The solution of

$$\frac{\partial\tilde{\phi}}{\partial t} = -D\Delta\tilde{\phi} \quad \text{on } [t_0, t_1] \text{ where } 0 < t_0 < t_1 < T, \qquad (4.2.21)$$

$$\tilde{\phi}_1(x) = \frac{1}{\sqrt{4\pi D(T-t_1)}} \exp\left(\frac{-x^2}{4D(T-t_1)}\right) \qquad (4.2.22)$$

is known to be

$$\tilde{\phi}(t,x) = \frac{1}{\sqrt{4\pi D(T-t)}} \exp\left(\frac{-x^2}{4D(T-t)}\right) \qquad t_0 \le t \le t_1 \qquad (4.2.23)$$

Similarly, the solution of

$$\frac{\partial\tilde{\bar{\phi}}}{\partial t} = D\,\frac{\partial^2\tilde{\bar{\phi}}}{\partial x^2} \qquad (4.2.24)$$

$$\tilde{\bar{\phi}}_0(x) = \frac{1}{\sqrt{4\pi D t_0}} \exp\left(\frac{-x^2}{4Dt_0}\right) \qquad (4.2.25)$$

is known to be

$$\tilde{\bar{\phi}}(t,x) = \frac{1}{\sqrt{4\pi D t}} \exp\left(\frac{-x^2}{4Dt}\right). \qquad (4.2.26)$$

Then

$$\tilde{W}(t,x) = -2mD \log \tilde{\phi}(t,x)$$
$$= \frac{m}{2}\,\frac{x^2}{T-t} + mD\log(T-t) + mD\log 4\pi D \qquad (4.2.27)$$

$$\tilde{\bar{W}}(t,x) = \frac{m}{2}\,\frac{x^2}{t} + mD\log t + mD\log 4\pi D \quad . \qquad (4.2.28)$$

Now,

$$\tilde{W}^*(t,x) = \frac{(\tilde{\bar{W}}+\tilde{W})}{2}(t,x) = \frac{mx^2}{4}\left[\frac{1}{T-t} + \frac{1}{t}\right] + \frac{mD}{2}\log(t(T-t)) + mD\log 4\pi D \;,$$

$$\tilde{V}^*(t,x) = \frac{(\tilde{\bar{W}}-\tilde{W})}{2}(t,x) = \frac{mx^2}{4}\left[\frac{1}{t} - \frac{1}{T-t}\right] + \frac{mD}{2}\,(\log t - \log(T-t)) \;,$$

$$\nabla\tilde{W}^*(t,x) = \frac{mx}{2}\left[\frac{T}{t(T-t)}\right] \qquad \Delta\tilde{W}^*(t,x) = \frac{mT}{2t(T-t)} \;,$$

$$\nabla\tilde{V}^*(t,x) = \frac{mx}{2}\left[\frac{T-2t}{t(T-t)}\right] \qquad \Delta\tilde{V}^*(t,x) = \frac{m(T-2t)}{2t(T-t)} \;.$$

Thus, by Prop. 4.2.B,

$$\tilde{\psi} = \exp\left\{ \frac{-\left[\frac{mx^2}{4}\left[\frac{1}{T-t} + \frac{1}{t}\right] + \frac{mD}{2}\log(t(T-t)) + mD\log 4\pi D\right]}{\hbar} + i\,\frac{\left[\frac{mx^2}{4}\left[\frac{T-2t}{t(T-t)}\right] + \frac{mD}{2}\log\left[\frac{t}{T-t}\right]\right]}{\hbar} \right\} \qquad (4.2.29)$$

satisfies

$$i\hbar\frac{\partial\tilde{\psi}}{\partial t} = \frac{-\hbar^2}{2m}\Delta\tilde{\psi} + \left[\frac{mx^2}{4}\left(\frac{T}{t(T-t)}\right)^2 - \frac{\hbar T}{t(T-t)}\right]\tilde{\psi}. \qquad (4.2.30)$$

<u>References</u>

1. A. Blaquiere, Liens entre la theorie geometrique des processus optimaux et la mecanique ondulatoire, C.R. Acad. Sc. Paris, Serie A., Vol. 262 (1966), pp. 539-595.

2. A. Blaquiere, Interpretation d'un coefficient de diffusion complexe en mecanique ondulatoire, C.R. Acad. Sc. Paris, Serie A, Vol 268 (1969), pp. 1304-1306.

3. A. Blaquiere, System Theory: A new approach to wave mechanics, J. Optim. Thy. Appl., 32, 4 (1980), pp. 463-478.

4. A. Blaquiere and A. Marzollo, An alternative approach to wave mechanics of a particle at the non-relativistic approximation, Information, Complexity and Control in Quantum Physics, Proc. of the 4th International Seminar on Mathematical Theory of Dynamical Systems and Microphysics, Udine, 1985, Springer-Verlag, Wien, 1987.

5. De Broglie, L., Une tentative d'interpretation causale et nonlineaire de la mecanique ondulatoire, Gauthier-Villars, Paris, 1956.

6. W. Fleming and R. Rishel, Deterministic and Stochastic Optimal Control, Springer-Verlag, Berlin, 1975.

7. F. Guerra and L. Morato, Quantization of dynamical systems and stochastic control theory, Physical Review D, 27, 8 (1983), pp. 1774-1786.

8. A.M. Il'in, A.S. Kalashnikov, O.A. Oleinik, Linear Equations of a Second Order of Parabolic Type, Russian Mathematical Surveys, Vol. 17, Macmillan and Co., Ltd., London, 1962.

9. S. Mitter, Non-linear Filtering and Stochastic Mechanics, Stochastic Systems: The Mathematics of Filtering and Identification with Applications, Proc. NATO Advanced Study Institute, Les Arcs, Savoie, France 1980, Reidel, Dordrecht, 1981.

10. E. Nelson, Derivation of the Schrödinger Equation from Newtonian Mechanics, Physical Review, 150, 4 (1966), pp. 1079-1085.

11. E. Nelson, Quantum Fluctuations, Princeton U.P., Princeton, 1985.

12. L. Papiez, Stochastic optimal control and quantum mechanics, J. Math. Phys., 23, 6 (1982), pp. 1017-1019.

13. K. Yasue, Quantum mechanics and stochastic control theory, J. Math. Phys., 22, 5 (1981), pp. 1010-1020.

14. K. Yasue, Stochastic Calculus of Variations, J. Func. Analysis, 41 (1981), pp. 327-340.

15. J.C. Zambrini, Variational processes and stochastic versions of mechanics, J. Math. Phys., 27, 9 (1986), pp. 2307-2330.

APPENDIX. AN EXISTENCE THEOREM.

Theorem 1.

If $W_1 : E^n \to R_+$, continuous on E^n, satisfies

$$W_1(x) \leq C(1+x^2)$$

for some C, and if Q, continuous on $\overline{S}$, satisfies

(A'1) $\quad 0 < \frac{Q(s,x)}{2mD} \leq K$ for some K, and for all $(s,x) \in \overline{S}$,

(A'2) $\quad$ Q is uniformly Hölder continuous in (s,x) in compact subsets of $\overline{S}$,

then there exists W satisfying the hypotheses of the Verification Theorem 3.1.B.

First note that the exponential transformation has the property stated in

Remark 1.

The following conditions are equivalent

(i) F is a non negative continuous function on $\overline{S}$, and $F(t,x) \leq A(1+x^2)$ for some A, and for all $(t,x) \in \overline{S}$,

(ii) f is a continuous function on $\overline{S}$ which satisfies

$$1 \geq f(t,x) \geq M \exp(-\alpha x^2), \quad (t,x) \in \overline{S}, \tag{1}$$

for some constants M, α, with $1 \geq M > 0$, $\alpha \geq 0$, and

$$f = \exp\left(-\frac{F}{2mD}\right).$$

Proof of Remark 1.

(i) $\Rightarrow$ (ii)

F continuous on $\overline{S}$ implies that f is continuous on $\overline{S}$. Further

$$0 \leq F(t,x) \leq A(1+x^2), \quad (t,x) \in \overline{S},$$

implies that

$$1 \geq f(t,x) \geq \exp\left(-\frac{A}{2mD}\right)\exp\left(-\frac{A}{2mD}x^2\right), \quad (t,x) \in \overline{S}.$$

We let $M = \exp(-A/2mD)$, $\alpha = A/2mD$.

Since $A \geq 0$, we have $1 \geq M > 0$, $\alpha \geq 0$. Therefore (i) $\Rightarrow$ (ii) is established.

(ii) $\Rightarrow$ (i)

From condition (1) it follows that

$$f(t,x) > 0 \text{ for all } (t,x) \in \overline{S}.$$

Then, f continuous on $\overline{S}$ implies that $F \triangleq -2mD \operatorname{Log} f$ is continuous on $\overline{S}$. Further

$$1 \geq f(t,x) \geq M \exp(-\alpha x^2), \quad (t,x) \in \overline{S}$$

implies that

$$0 \leq F(t,x) \leq 2mD\,\alpha x^2 - 2mD \operatorname{Log} M, \quad (t,x) \in \overline{S}$$

with $-2mD \operatorname{Log} M \geq 0$ since $1 \geq M > 0$.

Letting $A \triangleq \operatorname{Sup}(2mD\alpha, -2mD \operatorname{Log} M)$ we have

$$0 \leq F(t,x) \leq A(1+x^2), \quad (t,x) \in \overline{S},$$

so that (ii) $\Rightarrow$ (i) is established.

Proof of Theorem 1.

Let $W_1 : E^n \to R_+$, continuous on E^n, satisfy

$$W_1(x) \leq C(1+x^2) \tag{2}$$

for some C, and let Q, continuous on $\overline{S}$, satisfy (A'1) and (A'2).

Consider the Cauchy problem

$$\frac{\partial\tilde{\rho}}{\partial s} = -D\Delta\tilde{\rho} + \frac{Q}{2mD}\,\tilde{\rho} \quad \text{in } [t_o,t_1)\times E^n \tag{3}$$

$$\tilde{\rho}(t_1,x) = \rho_1(x) \triangleq \exp\left(-\frac{W_1(x)}{2mD}\right). \tag{4}$$

From the Theorem of Il'in (8), it follows that there exists a unique continuous and bounded solution $\tilde{\rho}$ to that Cauchy problem, and further

$$\tilde{\rho}(s,x) > 0 \quad \text{for} \quad (s,x) \in \overline{S}. \tag{5}$$

Define $\tilde{W}$ by

$$\tilde{\rho}(s,x) = \exp\left(-\frac{\tilde{W}(s,x)}{2mD}\right), \quad (s,x) \in \overline{S}. \tag{6}$$

Substituting in (3) we obtain

$$\frac{\partial\tilde{W}}{\partial s} - \frac{1}{2m}(\operatorname{grad}\tilde{W})^2 + D\Delta\tilde{W} + Q = 0. \tag{7}$$

Now, let

$$\tilde{v} \triangleq -\frac{1}{m}\operatorname{grad}\tilde{W} \quad \text{for all} \quad (s,x) \in S. \tag{8}$$

Since

$$\frac{\partial\tilde{W}}{\partial s} + \sum_{i=1}^{n} v_i \frac{\partial\tilde{W}}{\partial x_i} + D\Delta\tilde{W} + \frac{1}{2}mv^2 + Q(s,x) \tag{9}$$

as a function of v on E^n, has a unique minimum for $v = \tilde{v}(s,x)$, and since from (7)

(8), (9), the value of this minimum is

$$\frac{\partial\tilde{W}}{\partial s} - \frac{1}{2m}(\text{grad }\tilde{W})^2 + D\Delta\tilde{W} + Q = 0 ,$$

we conclude that :

$\tilde{W}(s,x)$ is a solution of the dynamic programming equation

$$0 = \frac{\partial\tilde{W}}{\partial s} + \min_{\nu\in E^n}\left[D\Delta\tilde{W} + \sum_{i=1}^{n}\nu_i\frac{\partial\tilde{W}}{\partial x_i} + \frac{1}{2}m\nu^2 + Q(s,x)\right], \qquad (s,x)\in S$$

with boundary data

$$\tilde{W}(t_1,x) = W_1(x), \qquad x\in E^n .$$

Further, since $\tilde{\rho}$ is in $C^{1,2}(S)$, strictly positive and continuous on $\bar{S}$, $\tilde{W}$ is in $C^{1,2}(S)$ and continuous on $\bar{S}$.

What remains to be proved is that $\tilde{W}$ satisfies a polynomial growth condition. For that purpose, we will use an expression of the solution to a Cauchy problem given in Theorem 5.3, p.148, of A. Friedman†.

By (A'1), (A'2) together with the fact that ρ_1 is continuous and bounded, the assumptions of that theorem are satisfied in the case of the Cauchy problem (3), (4), so that we have

$$\tilde{\rho}(s,x) = \int_\Omega \rho(\tilde{\xi}_{sx}(t_1))\exp\left(-\int_s^{t_1}\frac{Q(\sigma,\tilde{\xi}_{sx}(\sigma))}{2mD}\,d\sigma\right)d\tilde{P}(\omega) \qquad (10)$$

with $\tilde{\xi}_{sx}$ solution of

$$d\tilde{\xi} = \sqrt{2D}\,d\tilde{w} \quad \text{on } [s,t_1] ,$$
$$\tilde{\xi}(s) = x,$$

where $\tilde{w}$ is a standard n-dimensional Brownian motion with respect to some probability space $(\Omega, F, \tilde{P})$.

Since, by (A'1), $0 < (Q(s,x)/2mD) < K$ for some K and for all $(s,x)\in\bar{S}$, we have

$$1 > \exp\left(-\int_s^{t_1}\frac{Q(\sigma,\tilde{\xi}_{sx}(\sigma))}{2mD}\,d\sigma\right) > \exp(-K(t_1-s)) > \exp(-K(t_1-t_o)). \qquad (11)$$

† A. Friedman, Stochastic Differential Equations and Applications, Vol. I, Academic Press, New York, 1975.

Therefore, from (10), (11)

$$\int_\Omega \rho(\tilde{\xi}_{sx}(t_1))d\tilde{P}(\omega) \geq \tilde{\rho}(s,x) \geq B \int_\Omega \rho(\tilde{\xi}_{sx}(t_1))d\tilde{P}(\omega), \qquad (12)$$

$$(s,x) \in \overline{S}$$

with $0 < B \triangleq \exp(-K(t_1-t_o)) \leq 1$.

Now, by the transformation formula for integrals (see [6]), (12) rewrites

$$\int_{E^n} k(s,x;t_1,y)\rho_1(y)dy \geq \tilde{\rho}(s,x) \geq B \int_{E^n} k(s,x;t_1,y)\rho_1(y)dy \qquad (13)$$

$$(s,x) \in \overline{S}, \quad 0 < B \leq 1,$$

where k is the transition density of Brownian motion.

Further, by (2), Remark 1 and (4)

$$1 \geq \rho_1(y) \geq M \exp(-\alpha y^2) , \qquad (14)$$

for some constants M, α, with $1 \geq M > 0$, $\alpha \geq 0$.

At last let us prove that (13), (14) imply that

$$1 \geq \tilde{\rho}(s,x) \geq N \exp(-\beta x^2), \quad (t,x) \in \overline{S} , \qquad (15)$$

for some constants N, β, with $1 \geq N > 0$, $\beta \geq 0$.

Remember that

$$k(s,x;t,y) = [4\pi D(t-s)]^{-n/2} \exp\left[-\frac{1}{4D} \frac{(y-x)^2}{t-s}\right] .$$

The left-hand inequality in (15) is a direct consequence of the left-hand inequalities in (13) and (14), since

$$\int_E k(s,x;t_1,y)\rho_1(y)dy \leq \int_{E^n} k(s,x;t_1,y)dy = 1 .$$

Note that it implies that $\tilde{W}(s,x) \geq 0$ for all $(s,x) \in \overline{S}$.

The right-hand inequality in (15) is a consequence of the right-hand inequalities in (13) and (14). Suppose $\alpha > 0$ and define $T > t_1$ by

$$\alpha = \frac{1}{4D} \frac{1}{T-t_1} ,$$

so that

$$M \exp(-\alpha y^2) = \frac{M}{[4\pi D(T-t_1)]^{-n/2}} k(t_1,y;T,0) .$$

Then, from the right-hand inequalities in (13), (14) we deduce

$$\tilde{\rho}(s,x) \geq \frac{BM}{[4\pi D(T-t_1)]^{-n/2}} \int_E k(s,x;t_1,y)k(t_1,y;T,0)dy \; .$$

By Chapman-Kolmogorov formula, this is

$$\frac{BM}{[4\pi D(T-t_1)]^{-n/2}} \; k(s,x;T,0).$$

Further

$$k(s,x;T,0) = [4\pi D(T-s)]^{-n/2} \exp(-\frac{1}{4D} \frac{x^2}{T-s}) \geq$$

$$\geq [4\pi D(T-t_o)]^{-n/2} \exp\,(-\frac{1}{4D} \frac{x^2}{T-t_1}) \; .$$

Therefore, the right-hand inequality in (15) is proved with

$$N = BM \left[\frac{T-t_1}{T-t_o}\right]^{n/2}, \qquad \beta = \frac{1}{4D} \frac{1}{T-t_1} = \alpha \; .$$

One can readily verify that $1 \geq N > 0$, and we already have $\beta > 0$.

For $\alpha = 0$, (15) is trivial with $N = BM$, $\beta = 0$.

At last, by (6), (15), Remark 1, and by the fact that $\tilde{\rho} \in C^{1,2}(S)$ and is continuous on $\overline{S}$:

$$\tilde{W} \in C_2^{1,2}(S)$$

and is continuous on $\overline{S}$, which concludes the proof of Theorem 1.

GIRSANOV TRANSFORMATION AND TWO STOCHASTIC OPTIMAL CONTROL PROBLEMS. THE SCHRÖDINGER SYSTEM AND RELATED CONTROLLABILITY RESULTS.

A. Blaquière
Université Paris 7, Laboratoire d'Automatique Théorique
Paris (France)

1. INTRODUCTION.

At the begining of Quantum mechanics, two theoretical frames in good agreement with experimental data emerged, namely the ones attached to the names of Heisenberg and Schrödinger. During the past say about twenty years, taking advantage of the evolution of mathematics in some domains as semi-groups, stochastic processes, stochastic control, other theoretical frames have been proposed (13-16) among which the most widely known is probably the one of Nelson. The fact is that, at the present time, we are faced with two kinds of theories deeply different : the older ones which deal with wave functions, wave equations and all the machinery associated with them and the more recent ones which deal with stochastic differential equations and parabolic equations. This has reopened an old debate since, among the theoretical physicists, there is a robust tradition of resistance to the interpretation of quantum phenomena in terms of classical diffusion processes.

One reason for physicists to reject the interpretation of quantum phenomena in terms of classical diffusion processes is that diffusion processes are unable to account for interference phenomena. This is partly true and partly wrong.

This is true in the sense that the theory of diffusion processes is based on a "superposition principle", whose one expression is Kac's formula (another one is the equation of Chapman-Kolmogorov) which does not account for interference phenomena. In contrast, the theory of wave equations such as the Schrödinger equation is based on another "superposition principle", whose one expression is Feynman's formula (another one is the principle of Huygens-Fresnel) which does account for interference phenomena.

This is wrong in the sense that, if one uses the theory of *control* of diffusion processes one is able to account for "interference patterns" without introducing any "interference principle" as an ingredient. Nelson's stochastic mechanics - and our Proposition 5.3.D - shows that, starting from a given probability distribution at the initial time, one can control the motion of a random particle in such a way that the terminal probability distribution coïncides with any preassigned "interference pattern" (under proper mathematical conditions indeed). The counterpart of such a description is that the value of the control at a state x and at a time

$t < t_1$ will depend in general on an information given at time t_1, say *in the future* of t. This is a typical situation in optimal control theory and, more generally in the theory of the decision. One can easily understand that it may hurt physicists, however we must point out that it is not Quantum mechanics which is implicated, since the laws of Quantum mechanics are invariant under time reversal, but the question of *causality* at the microphysics scale (which we will not discuss).

Even if the discussions concerning the interpretation of the basic concepts of Quantum mechanics may seem futile, the fall out of the resurgence of an interpretation relying on stochastic processes is a partial answer to an important question : *Can one control a Schrödinger equation in order to obtain a wave-function having prescribed properties at the terminal time, starting from some known initial data ?*

We have based our proofs of the main theorems of this paper (Propositions 5.2.B, 5.3.D and 6.B) on properties of the *Schrödinger system* (Sec.5), a result originated in a paper of Schrödinger (8) published in 1932. Mathematical development of Schrödinger's idea is due to Bernstein (9) and, later, to a few other authors, in particular to Beurling (10) and Jamison (11). More recently, Zambrini (16) has returned to Jamison's main theorems and has discussed some of their connections with Physics. So doing, he has found a number of formulas and results previously obtained in Ref. (14). However, whereas Ref. (14) starts from two stochastic optimal control problems, the work of Zambrini starts from the two parabolic equations already used by Schrödinger, namely the equation of heat transfer and its adjoint. The paper of Zambrini much clarifies the connections between the results of Ref.(14) and the theory of reciprocal processes of Jamison. It also clarifies the relations between this framework and the version created by Nelson.

In Sec.3, we refer to Problems 1 and 2 of Ref. (20) and we recall the results obtained by performing the exponential change of variables on the generalized Hamilton-Jacobi equations associated with them.

In Sec.4, we show how, starting from the generalized Hamilton-Jacobi equations for these problems, one can obtain Kac's formula through the use of Cameron-Martin-Girsanov transformation. This method is related to one developed by Mitter in Ref. (7). We derive a version of Kac's formula which enables comparison with Feynman's formula and, in a way, clarifies a point of divergence between the two sets of ideas we started with at the begining of this introduction.

2. BASIC DEFINITIONS, NOTATIONS AND FACTS.

Our definitions of a *stochastic process*, a *n-dimensional Brownian motion* ($n \geq 1$), a *solution to a system of stochastic differential equations*, a *Markov process* are taken from (1).

Unless otherwise specified, (Ω, F, P) will denote a given underlying probability space, E^n a n-dimensional Euclidean space, $B(E^n)$ (or B, when no confusion is possible) the σ-algebra of Borel subsets of E^n, $[t_o, t_1]$ a compact interval of E^1.

Let the strip $S \triangleq (t_o, t_1) \times E^n$, and define

$$\mathcal{D}_s f(s,x) \triangleq \sum_{i,j=1}^{n} a_{ij}(s,x) \frac{\partial^2 f(s,x)}{\partial x_i \partial x_j} + \sum_{i=1}^{n} b_i(s,x) \frac{\partial f(s,x)}{\partial x_i} - c(s,x) f(s,x) \qquad (2.1)$$

in E^n.

Our definition of a *fundamental solution* $Y : (s,x;t,y) \mapsto Y(s,x;t,y)$ of the differential equation

$$\frac{\partial f}{\partial s} + \mathcal{D}_s f = 0 \qquad (2.2)$$

in the strip $\overline{S}$ is taken from (5) (see also the Appendix of (4)).

The following theorem gives sufficient conditions for the existence and uniqueness of a fundamental solution.

Theorem 2.A (4). Let the differential operator (2.1) satisfy the following conditions :

(i) The coefficients a_{ij}, b_i, $(i,j = 1,\ldots n)$ and c are bounded and continuous on $\overline{S}$ and satisfy a Hölder condition with respect to x :

$$\left.\begin{aligned} |a_{ij}(s,x') - a_{ij}(s,x)| &< K|x' - x|^{\lambda} \\ |b_i(s,x') - b_i(s,x)| &< K|x' - x|^{\lambda} \\ |c(s,x') - c(s,x)| &< K|x' - x|^{\lambda} \end{aligned}\right\} \quad i,j = 1,\ldots n, \qquad \lambda > 0 .$$

(ii) The coefficients a_{ij} satisfy a Hölder condition with respect to s :

$$|a_{ij}(s',x) - a_{ij}(s,x)| < K|s' - s|^{\lambda} .$$

(iii) There exists a constant $\gamma > 0$ such that for all $(s,x) \in \overline{S}$ and any collection of real numbers $\lambda_1, \ldots \lambda_n$:

$$\sum_{i,j=1}^{n} a_{ij}(s,x)\lambda_i\lambda_j \geq \gamma \sum_{i=1}^{n} \lambda_i^2 .$$

Then equation (2.2) has a fundamental solution Y and this solution is unique. The solution has the following properties :

(iv) For any $t_o < s < t < t_1$,

$$Y(s,x,t,y) > 0 .$$

(v) The following inequalities are satisfied :

$$Y(s,x;t,y) \leq M(t-s)^{-n/2} \exp\left(-\alpha \frac{|y-x|^2}{t-s}\right)$$

$$\left|\frac{\partial Y(s,x;t,y)}{\partial x_i}\right| \leq M(t-s)^{-(n+1)/2} \exp\left(-\alpha \frac{|y-x|^2}{t-s}\right)$$

$$\left|\frac{\partial^2 Y(s,x;t,y)}{\partial x_i \partial x_j}\right| \leq M(t-s)^{-(n/2)-1} \exp\left(-\alpha \frac{|y-x|^2}{t-s}\right)$$

$$\left|\frac{\partial Y(s,x;t,y)}{\partial s}\right| \leq M(t-s)^{-(n/2)-1} \exp\left(-\alpha \frac{|y-x|^2}{t-s}\right)$$

where M and α are some positive constants.

(vi) If, in $\bar{S}$, the derivatives

$$\frac{\partial a_{ij}}{\partial x_j}, \quad \frac{\partial^2 a_{ij}}{\partial x_i \partial x_j}, \quad \frac{\partial b_i}{\partial x_i}, \qquad (i,j = 1,\ldots n)$$

exist, are bounded and continuous, and satisfy a Hölder condition in x, then $Y(s,x;t,y)$ as a function of t,y satisfies the equation :

$$\frac{\partial Y}{\partial t} = \sum_{i,j=1}^{n} \frac{\partial^2 (a_{ij}(t,y)Y)}{\partial y_i \, \partial y_j} - \sum_{i=1}^{n} \frac{\partial (b_i(t,y)Y)}{\partial y_i} - c(t,y)Y . \tag{2.3}$$

The proof that the two fundamental solutions Y and $\bar{Y}$ of (2.2) and (2.3), respectively, are equal is given in I1' in [5].

We now consider again the parabolic equation (2.2) with the terminal condition

$$f(t_1,y) = f_1(y) \quad \text{on} \quad E^n . \tag{2.4}$$

A solution to the classical Cauchy problem for (2.2), (2.4) is defined to be a function f on $\bar{S}$ which is continuous in $\bar{S}$, has continuous derivatives $\frac{\partial f}{\partial x_i}$, $\frac{\partial f}{\partial s}$, $\frac{\partial^2 f}{\partial x_i \partial x_j}$, $(i,j = 1,\ldots n)$, in $[t_o,t_1)\times E^n$, satisfies (2.2) in $[t_o,t_1)\times E^n$ and the terminal condition (2.4) where f_1 is a given *continuous* function.

We have the following theorem which, but for a change of variable, is Theorem 2 of Ref.[5] :

Theorem 2.B [5]. In $[t_o,t_1)\times E^n$ there exists a unique bounded solution of the Cauchy problem for (2.2),(2.4) if the following conditions hold :

(i) The coefficients a_{ij}, b_i, c are bounded and continuous in $\bar{S}$ and satisfy conditions (i), (ii), (iii) of Theorem 2.A, and

(ii) f_1 is continuous and bounded on E^n .

The solution is given by

$$f(s,x) = \int_{E^n} Y(s,x;t_1,y)f_1(y)dy \quad \text{on} \quad [t_o,t_1)\times E^n \; . \tag{2.5}$$

Note : Il'in actually considers the equation

$$\frac{\partial g}{\partial \sigma} = \mathcal{D}_\sigma g \tag{2.6}$$

with initial data

$$g(t_o,x) = g_o(x) \quad \text{on} \quad E^n \; , \tag{2.7}$$

and states that g is given by

$$g(\sigma,y) = \int_{E^n} \overline{Z}(t_o,x;\sigma,y)g_o(x)dx \; , \tag{2.8}$$

where $\overline{Z}$ is the fundamental solution of (2.6), and $\overline{Z}(t_o,x;\sigma,y)$ is equal to $Z(\sigma,y;t_o,x)$ of Il'in. To pass from (2.6), (2.7) to (2.2), (2.4) we make the change of variable $\sigma = t_o+t_1-s$.

From Theorem 2.B, and the corresponding theorem of Il'in, it is easy to see that, under the assumptions of these theorems :

$$f(s,x) = \int_{E^n} Y(s,x;t,y)f(t,y)dy, \qquad t_o \leq s < t \leq t_1 \; , \tag{2.9}$$

and

$$g(\sigma,y) = \int_{E^n} \overline{Z}(\tau,x;\sigma,y)g(\tau,x)dx \; , \qquad t_o \leq \tau < \sigma \leq t_1 \; . \tag{2.10}$$

3. EXPONENTIAL TRANSFORMATION.

We refer to Problems 1 and 2 of Ref. [20]. Keeping the same notation, we will assume that there exist solutions W, $\overline{W}$, v*, $\overline{v}$*, satisfying the hypotheses of the Verification Theorems. Thus

$$\frac{\partial W}{\partial s} = - D\Delta W + \frac{1}{2m}(\text{grad } W)^2 - Q, \tag{3.1}$$

$$mv^* = -\text{grad } W^{\dagger}, \tag{3.2}$$

for all $(s,x) \in S$; and

$$\frac{\partial \overline{W}}{\partial \sigma} = D\Delta\overline{W} - \frac{1}{2m}(\text{grad } \overline{W})^2 + Q, \tag{3.3}$$

$$m\overline{v}^* = \text{grad } \overline{W}, \tag{3.4}$$

for all $(\sigma,y) \in S$.

We now perform the exponential change of variable on (3.1), (3.2) ; that is, we let

---

† grad $\triangleq$ $(\partial/\partial x_1 , \ldots, \partial/\partial x_n)$.

$$\rho(s,x) \triangleq \exp\left(-\frac{W(s,x)}{2mD}\right), \qquad (s,x) \in \bar{S}\ . \tag{3.5}$$

(ρ can be defined on $\bar{S}$ since W is continuous on $\bar{S}$) .

Thus

$$\rho(t_1,x) = \exp\left(-\frac{W_1(x)}{2mD}\right) \triangleq \rho_1(x)\ . \tag{3.6}$$

Equation (3.1) becomes

$$\frac{\partial\rho}{\partial s} = -\ D\Delta\rho + \frac{Q}{2mD}\,\rho\ , \quad (s,x) \in S\ , \tag{3.7}$$

and, from (3.2) we have

$$v_i^*(s,x) = \frac{2D}{\rho(s,x)}\ \frac{\partial\rho(s,x)}{\partial x_i}\ , \qquad (s,x) \in S\ . \tag{3.8}$$

Further

$$v_i^*(t_1,x) = \frac{2D}{\rho_1(x)}\ \frac{\partial\rho_1(x)}{\partial x_i} \tag{3.9}$$

if the partial derivatives exist.

Finding a solution to the equation

$$\frac{\partial\tilde{\rho}}{\partial s} = -\ D\Delta\tilde{\rho} + \frac{Q}{2mD}\,\tilde{\rho}\ , \qquad (s,x) \in S\ , \tag{3.10}$$

with continuous and bounded terminal data $\tilde{\rho}_1(x)$, constitutes a Cauchy problem. We have the following

Proposition 3.1.C. If Q satisfies

(A1) Q is bounded in $\bar{S}$,

and

(A2) $|Q(s,x') - Q(s,x)| \leq M|x' - x|^{\lambda}$, $\quad x',x \in E^n$,

for some positive constants M, λ, then there exists a unique continuous and bounded solution $\tilde{\rho}$ to the Cauchy problem (3.10) with

$$\tilde{\rho}(t_1,x) = \tilde{\rho}_1(x).$$

This follows from a direct application of Theorem 2.B.

Thus, if we let p denote the fundamental solution of (3.10), then

$$\tilde{\rho}(s,x) = \int_{E^n} p(s,x;t,y)\,\tilde{\rho}\,(t,y)dy, \qquad s < t \leq t_1\ . \tag{3.11}$$

Since W_1 is continuous and non-negative, ρ_1 given by (3.6) is continuous and bounded in E^n. It then follows from (3.6) and (3.7) that if (A1), (A2) hold, then

$$\rho(s,x) = \int_{E^n} p(s,x;t,y)\,\rho\,(t,y)dy, \qquad s < t \leq t_1\ ; \tag{3.12}$$

that is

$$\exp\left(-\frac{W(s,x)}{2mD}\right) = \int_{E^n} p(s,x;t,y)\exp\left(-\frac{W(t,y)}{2mD}\right)dy, \quad s < t \leq t_1 . \tag{3.13}$$

Likewise, we make the exponential change of variable

$$\bar{\rho}(\sigma,y) \triangleq \exp\left(-\frac{\bar{W}(\sigma,y)}{2mD}\right), \quad (\sigma,y) \in \bar{S}, \tag{3.14}$$

with

$$\bar{\rho}_o(y) \triangleq \bar{\rho}(t_o,y) = \exp\left(-\frac{\bar{W}_o(y)}{2mD}\right). \tag{3.15}$$

Equation (3.3) becomes

$$\frac{\partial\bar{\rho}}{\partial\sigma} = D\Delta\bar{\rho} - \frac{Q}{2mD}\bar{\rho}, \quad (\sigma,y) \in S, \tag{3.16}$$

and, from (3.4) we have

$$\bar{v}_i^*(\sigma,y) = -\frac{2D}{\bar{\rho}(\sigma,y)}\frac{\partial\bar{\rho}(\sigma,y)}{\partial y_i}, \quad (\sigma,y) \in S. \tag{3.17}$$

Further

$$\bar{v}_i^*(t_o,y) = -\frac{2D}{\bar{\rho}_o(y)}\frac{\partial\bar{\rho}_o(y)}{\partial y_i} \tag{3.18}$$

if the partial derivatives exist.

Again from Theorem 2.B, assuming (A1), (A2) hold, we have the existence and uniqueness of a continuous and bounded solution $\tilde{\rho}$ to the Cauchy problem

$$\frac{\partial\tilde{\rho}}{\partial\sigma} = D\Delta\tilde{\rho} - \frac{Q}{2mD}\tilde{\rho}, \quad (\sigma,y) \in S, \tag{3.19}$$

with continuous and bounded initial data

$$\tilde{\rho}(t_o,y) = \tilde{\rho}_o(y), \quad y \in E^n. \tag{3.20}$$

Further, if we let $\bar{p}$ denote the fundamental solution of (3.19), then

$$\tilde{\rho}(\sigma,y) = \int_{E^n} \bar{p}(\tau,x;\sigma,y)\tilde{\rho}(\tau,x)dx, \quad t_o \leq \tau < \sigma. \tag{3.21}$$

From the remark following Theorem 2.A, $\bar{p} = p$. Thus

$$\tilde{\rho}(\sigma,y) = \int_{E^n} p(\tau,x;\sigma,y)\tilde{\rho}(\tau,x)dx, \quad t_o \leq \tau < \sigma. \tag{3.22}$$

Since $\bar{W}_o$ is continuous and non-negative, $\bar{\rho}_o$ given by (3.15) is continuous and bounded in E^n. It then follows from (3.15) and (3.16) that if (A1), (A2) hold, then

$$\bar{\rho}(\sigma,y) = \int_{E^n} p(\tau,x;\sigma,y)\bar{\rho}(\tau,x)dx, \quad t_o \leq \tau < \sigma ; \tag{3.23}$$

that is

$$\exp\left(-\frac{\overline{W}(\sigma,y)}{2mD}\right) = \int_{E^n} p(\tau,x;\sigma,y)\exp\left(-\frac{\overline{W}(\tau,x)}{2mD}\right)dx, \quad t_o \leq \tau < \sigma . \qquad (3.24)$$

4. SEMIGROUPS DEFINED BY KAC'S FORMULA .

4.1. Cameron-Martin-Girsanov Transformation.

It is usual (see for instance [6]) to associate with the parabolic equation (2.2) the system of stochastic differential equations

$$d\tilde{\xi} = \tilde{\beta}(t)dt + \gamma(t)d\tilde{w} \qquad (4.1.1)$$

with $\tilde{\beta}(t) \triangleq (b_1(t,\tilde{\xi}(t)),\dots b_n(t,\tilde{\xi}(t)))$, $\gamma^2(t) = 2(a_{ij}(t,\tilde{\xi}(t)))_{ij=1}^n$.

$\tilde{w}$ is a standard n-dimensional Brownian motion with respect to some probability measure $\tilde{P}$

In the case of equation (3.7) where

$$\mathcal{D}_s = D\sum_{i=1}^n \frac{\partial^2}{\partial x_i^2} - \frac{Q}{2mD} ,$$

(4.1.1) reads

$$d\tilde{\xi} = \sqrt{2D}\, d\tilde{w} . \qquad (4.1.2)$$

Now, in Problem 1, we are faced with two systems of stochastic differential equations, namely (4.1.2) and

$$d\xi = v^*(t,\xi(t))dt + \sqrt{2D}\, dw . \qquad (4.1.3)$$

A relation between them is provided by the Cameron-Martin-Girsanov transformation.

A similar discussion holds in the case of Problem 2.

We have assumed that there exists a W satisfying the hypotheses of the Verification Theorem, and an optimal control v*, and that Q is continuous and bounded in $\overline{S}$. Further let W satisfy

(B1) W is of class $C^{1,2}$ (see [1]) ; that is, the partial derivatives W_t, W_{x_i}, $W_{x_i x_j}$ are continuous on $\overline{S}$.

Let us apply the Ito stochastic differential rule [1] to $W(t,\tilde{\xi}(t))$ where $\tilde{\xi}$ satisfies (4.1.2). Then

$$W(t,\tilde{\xi}(t)) - W(s,\tilde{\xi}(s)) = \int_s^t \frac{\partial W}{\partial t}(\sigma,\tilde{\xi}(\sigma))d\sigma +$$

$$+ \sum_{i=1}^n \int_s^t \frac{\partial W}{\partial x_i}(\sigma,\tilde{\xi}(\sigma))d\tilde{\xi}_i(\sigma) + D\sum_{i=1}^n \int_s^t \frac{\partial^2 W}{\partial x_i^2}(\sigma,\tilde{\xi}(\sigma))d\sigma . \qquad (4.1.4)$$

Otherwise we deduce from (3.1)

$$\int_s^t Q(\sigma,\tilde{\xi}(\sigma))d\sigma = -\int_s^t \frac{\partial W}{\partial t}(\sigma,\tilde{\xi}(\sigma))d\sigma + \frac{1}{2m}\sum_{i=1}^{n}\int_s^t\left(\frac{\partial W}{\partial x_i}(\sigma,\tilde{\xi}(\sigma))\right)^2 d\sigma$$

$$- D\sum_{i=1}^{n}\int_s^t \frac{\partial^2 W}{\partial x_i^2}(\sigma,\tilde{\xi}(\sigma))d\sigma \ . \tag{4.1.5}$$

Note that, in view of (B1) and the continuity of Q on $\overline{S}$, (3.1) holds for all $(s,x)\in\overline{S}$, so that, in both relations (4.1.4), (4.1.5) we have $t_o < s < t < t_1$.

By adding (4.1.4) and (4.1.5) we obtain

$$\int_s^t Q(\sigma,\tilde{\xi}(\sigma))d\sigma + W(t,\tilde{\xi}(t)) - W(s,\tilde{\xi}(s)) =$$

$$= \sum_{i=1}^{n}\int_s^t \frac{\partial W}{\partial x_i}(\sigma,\tilde{\xi}(\sigma))d\tilde{\xi}_i(\sigma) + \frac{1}{2m}\sum_{i=1}^{n}\int_s^t\left(\frac{\partial W}{\partial x_i}(\sigma,\tilde{\xi}(\sigma))\right)^2 d\sigma \tag{4.1.6}$$

from which we readily deduce, by taking the exponentials of both sides after division by $-2mD$,

$$\exp \zeta_s^t(\theta) = \frac{\exp\left(-\frac{W(t,\tilde{\xi}(t))}{2mD}\right)}{\exp\left(-\frac{W(s,\tilde{\xi}(s))}{2mD}\right)}\ \exp\left(-\int_s^t \frac{Q(\sigma,\tilde{\xi}(\sigma))}{2mD}\, d\sigma\right) \tag{4.1.7}$$

with

$$\zeta_s^t(\theta) = \int_s^t \theta(\sigma)d\tilde{w}(\sigma) - \frac{1}{2}\int_s^t |\theta(\sigma)|^2 d\sigma \ , \text{ and} \tag{4.1.8}$$

$$\theta(t) = \frac{1}{\sqrt{2D}}\, v^*(t,\tilde{\xi}(t)) = -\frac{1}{m\sqrt{2D}}\,\text{grad}\ W(t,\tilde{\xi}(t)) \ . \tag{4.1.9}$$

Now assuming

(B2) $|v^*(t,x)| < M$ for some M, $(t,x)\in\overline{S}$,

we are ready to apply

Theorem 4.1.A (1). Let P be absolutely continuous with respect to $\tilde{P}$, with

$$P(d\omega) = \exp \zeta_s^T(\theta)\,\tilde{P}(d\omega)\ , \quad t_o < s < t < T < t_1 \ .$$

Then

(a) $P(\Omega) = 1$ (hence (Ω,F,P) is a probability space).

(b) Let $w(t) = \tilde{w}(t) - \int_s^t \theta(\sigma)d\sigma$. Then w is a standard n-dimensional Brownian motion with respect to P.

(c) Let $\beta = \tilde{\beta} + \gamma\theta$. Then

$$\xi(t) - \xi(s) = \int_s^t \beta(\sigma)d\sigma + \int_s^t \gamma(\sigma)dw(\sigma) \ .$$

Here ξ and $\tilde{\xi}$ have the same sample functions, but are considered as stochastic processes with respect to different probability measures P and $\tilde{P}$.

From (4.1.2) we have

$$\tilde{\xi}(t) - \tilde{\xi}(s) = \sqrt{2D}\,(\tilde{w}(t) - \tilde{w}(s)) \tag{4.1.10}$$

Then, from (b) of Theorem 4.1.A with (4.1.9) and (4.1.10) we deduce

$$w(t) = \frac{1}{\sqrt{2D}}(\tilde{\xi}(t) - \tilde{\xi}(s)) + \tilde{w}(s) - \int_s^t \frac{1}{\sqrt{2D}} v^*(\sigma,\tilde{\xi}(\sigma))d\sigma \tag{4.1.11}$$

and

$$w(s) = \tilde{w}(s) \tag{4.1.12}$$

At last (4.1.11) and (4.1.12) result in

$$\tilde{\xi}(t) = \tilde{\xi}(s) + \int_s^t v^*(\sigma,\tilde{\xi}(\sigma))d\sigma + \sqrt{2D}\,(w(t) - w(s)) \; .$$

This is what condition (c) of the Theorem states since, here, $\tilde{\beta} = 0$, $\gamma = \sqrt{2D}\, I$ where I is the n×n unity matrix, and

$$\beta = v^*(t,\tilde{\xi}(t)) \; .$$

In other words, provided that $\tilde{w}$ and w are related by the condition (b) of Theorem 4.1.A, and provided that the initial data satisfy $\tilde{\xi}(s,\omega) = \xi(s,\omega)$, for $\omega \in \Omega$ a.s.[†],then the solutions $\tilde{\xi}$ and ξ of (4.1.2) and (4.1.3) with these initial data, respectively, are stochastic processes on $(\Omega,F,\tilde{P})$ and (Ω,F,P), respectively, *which have the same sample functions*. Note that, here, the Brownian motion w is not given in advance (in contrast to Sect.3). It arises in the course of changing the probability measure.

4.2. Kac's Formula.

Now let the initial data associated with (4.1.2) be $\tilde{\xi}(s) = x \in E^n$, at time $s \in [t_0,t_1)$, and denote by $\tilde{\xi}_{sx}$ the corresponding solution. Then, in formula (4.1.7) we have $W(s,\tilde{\xi}(s)) = W(s,x)$. From (4.1.7) and (a) of Theorem 4.1.A we obtain

$$\exp(-\frac{W(s,x)}{2mD}) = \int_\Omega \exp\left(-\frac{W(T,\tilde{\xi}_{sx}(T))}{2mD}\right)\exp\left(-\int_s^T \frac{Q(\sigma,\tilde{\xi}_{sx}(\sigma))}{2mD}\, d\sigma\right)d\tilde{P}(\omega), \tag{4.1.13}$$

$$t_0 < s < T < t_1 \; .$$

(4.1.13) is Kac's formula. This formula turns out to be valid for a much larger class of potentials (see [7]). Here we have proved (4.1.13) under the assumptions (A1), (B1), (B2).

† P or $\tilde{P}$ - almost surely since $P \prec \tilde{P}$ and $\tilde{P} \prec P$.

In a restricted case, we shall give formula (4.1.13) another form, sometimes more suggestive. Now, in addition to the fact that Q is continuous and bounded in $\overline{S}$ (see (A1)) let Q satisfy

(B3) $$Q(t,x) > 0 \qquad \text{for all } (t,x) \in \overline{S}\ .$$

In view of the continuity of $\tilde{\xi}_{sx}$ on $[s,t_1]$ for $\tilde{P}$-almost every $\omega \in \Omega$, and of the continuity of Q, we have

$$\exp\left(-\frac{1}{2mD}\int_s^{s+h} Q(\sigma,\tilde{\xi}_{sx}(\sigma))d\sigma\right) = 1 - \frac{Q(s,x)}{2mD}\,h + o(h,\omega) \tag{4.1.14}$$

$\tilde{P}$-almost surely, for $h>0$ sufficiently small, where $\lim_{h\downarrow o} \frac{|o(h,\omega)|}{h} = 0$ for $\omega \in \Omega$.

Let us first prove that

$$\lim_{h\downarrow o} \int_\Omega \frac{|o(h,\omega)|}{h}\, d\tilde{P}(\omega) = 0\ . \tag{4.1.15}$$

Since, from (A1) and (B3), $0 \le (Q(t,x)/2mD) \le K$ for some K, and for all $(t,x) \in \overline{S}$, we have

$$0 \le \frac{1}{h}\left\{1-\exp\left(-\frac{1}{2mD}\int_s^{s+h} Q(\sigma,\tilde{\xi}_{sx}(\sigma))d\sigma\right)\right\} \le \frac{1}{h}\{1-\exp(-Kh)\} \le K \tag{4.1.16}$$

From (4.1.14) and (4.1.16) it follows that

$$-K \le \frac{o(h,\omega)}{h} \le K\ ,$$

so that $(|o(h,\omega)|/h) \le K$, and (4.1.15) follows from the Lebesgue dominated convergence theorem.

Now, from (4.1.13) and (4.1.14) it follows that

$$\exp\left(-\frac{W(s,x)}{2mD}\right) = \left(1-\frac{Q(s,x)}{2mD}\,h\right)\int_\Omega \exp\left(-\frac{W(s+h,\tilde{\xi}_{sx}(s+h))}{2mD}\right) d\tilde{P}(\omega) + \alpha(h) \tag{4.1.17}$$

with

$$\alpha(h) \triangleq \int_\Omega \exp\left(-\frac{W(s+h,\tilde{\xi}_{sx}(s+h))}{2mD}\right) o(h,\omega)\, d\tilde{P}(\omega)\ .$$

By the transformation formula for integrals (see (1)), the first term in the right-hand side of (4.1.17) rewrites

$$\left(1-\frac{Q(s,x)}{2mD}\,h\right)\int_{E^n} \exp\left(-\frac{W(s+h,y)}{2mD}\right) k(s,x;s+h,y)dy$$

where k is the transition density of Brownian motion†.

---

† Note that $\tilde{P}\{\tilde{\xi}_{sx}(t) \in B\} = \int_{\{\tilde{\xi}_{sx}(t)\in B\}} d\tilde{P}(\omega) = \int_B dP^*(y) = \int_B k(s,x;t,y)dy, \quad B \in \mathcal{B}\ .$

As concerns the second term, since (B3) implies that L, given by (3.1.2) of Ref. (20) is non-negative, and since W_1 is non-negative, the cost function

$$J(s,x,v) = E_{sx}\left\{\int_s^{t_1} L(t,\xi(t),v(t,\xi(t))dt + W_1(\xi(t_1))\right\}$$

is non-negative and so is the function W. Accordingly $\rho(t,x) \triangleq \exp(-W(t,x)/2mD)$ is bounded on $\overline{S}$, so that

$$\frac{|\alpha(h)|}{h} < \frac{A}{h}\int_\Omega |o(h,\omega)|d\widetilde{P}(\omega) \qquad \text{for some A,}$$

and consequently, in view of (4.1.15)

$$\lim_{h\downarrow o} \frac{|\alpha(h)|}{h} = 0 .$$

Therefore, if (A1), (B1), (B2) and (B3) hold, then the Kac formula can be written

$$\exp\left(-\frac{W(s,x)}{2mD}\right) = \int_{E^n} \exp\left(-\frac{W(s+h,y)}{2mD}\right)\left(1 - \frac{Q(s,x)}{2mD}h + \widetilde{o}(h)\right) k(s,x;s+h,y)dy \tag{4.1.18}$$

$$t_o \le s < s+h \le t_1 ,$$

where $\lim_{h\downarrow o} \frac{|\widetilde{o}(h)|}{h} = 0$,

or, equivalently, by writing k(s,x;s+h,y) explicitely and replacing

$1- \frac{Q(s,x)}{2mD}h + \widetilde{o}(h)$ by $\exp\{-\frac{1}{2mD}(Q(s,x)h + o'(h))\}$,

$$\exp(-\frac{W(s,x)}{2mD}) =$$
$$= (4\pi Dh)^{-n/2}\int_{E^n} \exp\left(-\frac{W(s+h,y)}{2mD}\right)\exp\left(-\frac{1}{2mD}[\frac{1}{2}m\frac{(y-x)^2}{h} + Q(s,x)h + o'(h)]\right)dy \tag{4.1.19}$$

$$t_o \le s < s+h \le t_1 .$$

where $\lim_{h\downarrow o} \frac{|o'(h)|}{h} = 0$.

A similar discussion, starting from Problem 2, leads to

$$\exp(-\frac{\overline{W}(t,y)}{2mD}) =$$
$$= (4\pi Dh)^{-n/2}\int_{E^n} \exp\left(-\frac{\overline{W}(t-h,x)}{2mD}\right)\exp\left(-\frac{1}{2mD}[\frac{1}{2}m\frac{(y-x)^2}{h} + Q(t,y)h + o''(h)]\right)dx \tag{4.1.20}$$

$$t_o \le t-h < t \le t_1 ,$$

where $\lim_{h\downarrow o} \frac{|o''(h)|}{h} = 0$.

These formulas are to be compared with (3.13) and (3.24). In fact, by Dynkin's theory of α-subprocesses (see [2] and [3]), one can prove that, if (A1), (B3) hold, then

$$p(s,x;s+h,y) = (4\pi Dh)^{-n/2}\exp\left(-\frac{1}{2mD}\left[\frac{1}{2}m\frac{(y-x)^2}{h} + Q(s,x)h + o'(h)\right]\right). \quad (4.1.21)$$

In (4.1.13) the functional

$$\alpha_t^s \triangleq \exp\left(-\int_s^t \frac{Q(\sigma,\tilde{\xi}_{sx}(\sigma))}{2mD}\, d\sigma\right)$$

is an example of contracting, right-continuous, multiplicative functional of the type of those which occur in the theory of α-subprocesses (see [3]). In such α-subprocesses, the trajectories of the original process (here, solution of (4.1.2)) are terminated in a random matter ; here α_t^s can be shown to be the conditional probability that a trajectory starting at (s,x) does not terminate during the time interval [s,t], given that all phenomena connected with the process during the time interval [s,t] are known. From (4.1.14), (4.1.15) it follows that the probability that a trajectory, starting from x at time s, does terminate during the time interval [s,s+h] is equal to $(Q(s,x)/2mD)h$ up to an $o(h)$. Thus the function $Q/2mD$ is called *the termination density*. Therefore, in (4.1.21), $p(s,x;s+h,y)$ is the product of the transition density of the Brownian motion (corresponding to $Q=0$) and of the *non-termination density* $1-(Q/2mD)h$ up to an $o(h)$. This remark gives a simple intuitive meaning to the formulas (4.1.19), (4.1.20).

4.3. Feynman's Formula.

Formula (4.1.13) is a variant of a result of Kac [17] who was trying to understand Feynman [12]. A basic formula in Feynman's theory of *paths integrals* [12] has formal similarities with (4.1.20) (see formula (4-5) on p.77 of [12]). It is, but for a change of the notation

$$\exp\left(-\frac{\hat{W}(t,y)}{2miD}\right) =$$
$$=(4\pi iDh)^{-n/2}\int_{E^n}\exp\left(-\frac{\hat{W}(t-h,x)}{2miD}\right)\exp\left(-\frac{1}{2miD}\left[\frac{1}{2}m\frac{(y-x)^2}{h}+Q(t,y)h+o''(h)\right]\right)dx \quad (4.1.22)$$
$$t_o \le t-h < t \le t_1 ,$$

with

$$\hat{W}(t,y) = V^*(t,y) + iW^*(t,y),$$

where V^* and W^* are $C^{1,2}$ real valued functions on $\overline{S}$; $D = \hbar/2m$ ($\hbar$: the reduced Planck's constant).

The function ψ, given by

$$\psi(t,y) = \exp\left(-\frac{\hat{W}(t,y)}{2miD}\right),$$

is the *wave-function* of Quantum mechanics for the motion of a non-relativistic particle in a field of forces derived from the potential Q .

Except for the underlying idea of semi-group, (4.1.22) is *deeply different* from (4.1.20), in spite of the formal similarity. Feynman's formula is an expression of what is called in Physics the *principle of interferences* , in contrast with what physicists call *dissipative processes*.

5. SCHRÖDINGER SYSTEM AND CONTROLLABILITY RESULTS.

5.1. Schrödinger System.

In this sub-section, we shall continue to assume that there exist pairs of functions (W,v^*), $(\overline{W},\overline{v}^*)$ satisfying the conditions of the Verification Theorems (i.e., $v^*,\overline{v}^*$ are optimal controls). As in Ref. [14], define

$$\rho^*(t,x) \triangleq \overline{\rho}(t,x)\rho(t,x) = \exp\left(-\frac{W^*(t,x)}{mD}\right) \tag{5.1.1}$$

where

$$W^*(t,x) \triangleq \frac{\overline{W}(t,x) + W(t,x)}{2}, \qquad (t,x) \in \overline{S} .$$

Like W^*, ρ^* is in $C^{1,2}(S)$ and continuous on $\overline{S}$; further, it satisfies the boundary conditions

$$\rho^*(t_o,x) = \exp\left(-\frac{\overline{W}_o(x) + W(t_o,x)}{2mD}\right) = \overline{\rho}_o(x)\exp\left(-\frac{W(t_o,x)}{2mD}\right),$$

$$\rho^*(t_1,x) = \exp\left(-\frac{\overline{W}(t_1,x) + W_1(x)}{2mD}\right) = \exp\left(-\frac{\overline{W}(t_1,x)}{2mD}\right)\rho_1(x) .$$

Now, assuming (A1), (A2) hold, using (3.13), (3.24), we have

$$\rho^*(t_o,x) = \overline{\rho}_o(x)\int_{E^n} p(t_o,x;t_1,y)\rho_1(y)dy \tag{5.1.2}$$

$$\rho^*(t_1,y) = \rho_1(y)\int_{E^n} p(t_o,x;t_1,y)\overline{\rho}_o(x)dx \tag{5.1.3}$$

$$x,y \in E^n .$$

Equations (5.1.2), (5.1.3) are a pair of functional equations which form a *Schrödinger system* [8][16], which we now describe in more generality.

Jamison has proved the following [11].

Theorem 5.1.A (Jamison). Suppose M is a σ-compact metric space, that μ_0 and μ_1 are probability measures on its σ-field Σ of Borel sets, and that q is an everywhere continuous, strictly positive function on M×M. Then there is a unique pair μ, π of measures on Σ×Σ for which

(a) μ is a probability measure and π is a σ-finite product measure.

(b) $\mu(B\times M) = \mu_0(B)$, $\mu(M\times B) = \mu_1(B)$, $B \in \Sigma$.

(c) $d\mu/d\pi = q$.

Let $M \triangleq E^n$, $\Sigma \triangleq \mathcal{B}$ and $q(x,y) \triangleq p(t_0,x;t_1,y)$, $(x,y) \in E^n \times E^n$. It follows from (c) that

$$\mu(B_0 \times B_1) = \int_{B_0\times B_1} p(t_0,x;t_1,y)d\pi(x,y), \qquad B_0,B_1 \in \mathcal{B},$$

and since π is a σ-finite product measure, there exist measures ν_0 and ν_1 on $\mathcal{B}$ such that

$$\mu(B_0 \times B_1) = \int_{B_0\times B_1} p(t_0,x;t_1,y)d\nu_0(x)d\nu_1(y), \qquad B_0,B_1 \in \mathcal{B}. \tag{5.1.4}$$

Obviously, the pair (ν_0,ν_1) is not unique : for any pair (k_0,k_1) or real numbers with $k_0k_1 = 1$, the factor measures $k_0\nu_0$, $k_1\nu_1$, will produce the same π . *In case* π *is finite*, we can eliminate this inconvenience by normalizing the ν_0,ν_1, by the condition

$$\int d\nu_0(x) = \int d\nu_1(y) = [\int d\pi(x,y)]^{1/2}.$$

In the general case, by Fubini's theorem, (5.1.4) can also be written

$$\mu(B_0\times B_1) = \int_{B_0} d\nu_0(x) \int_{B_1} p(t_0,x;t_1,y)d\nu_1(y), \qquad B_0,B_1 \in \mathcal{B},$$

so that (b) rewrites

$$\mu_0(B_0) = \int_{B_0} d\nu_0(x) \int_{E^n} p(t_0,x;t_1,y)d\nu_1(y), \qquad B_0 \in \mathcal{B}, \tag{5.1.5}$$

$$\mu_1(B_1) = \int_{B_1} d\nu_1(y) \int_{E^n} p(t_0,x;t_1,y)d\nu_0(x) \qquad B_1 \in \mathcal{B}. \tag{5.1.6}$$

Denote by λ the n-dimensional Lebesgue measure†. Suppose that $\mu_i \ll \lambda$, $i = 0,1$.

---

† The argument can be easily extended to the case where λ is any σ-finite measure.

Let $f \triangleq d\mu_o/d\lambda$, $g \triangleq d\mu_1/d\lambda$. Since $d\mu/d(\nu_o \times \nu_1) = q$, $d(\nu_o \times \nu_1)/d\mu = 1/q$, from which it easily follows that $\nu_i \ll \lambda$, $i = 0,1$. Let $\tilde{\tilde{\rho}} \triangleq d\nu_o/d\lambda$ and $\tilde{\rho}_1 \triangleq d\nu_1/d\lambda$. Then (5.1.5) and (5.1.6) are equivalent to

$$f(x) = \tilde{\tilde{\rho}}_o(x) \int_{E^n} p(t_o,x;t_1,y)\tilde{\rho}_1(y)dy \ , \tag{5.1.7}$$

$$g(y) = \tilde{\rho}_1(y) \int_{E^n} p(t_o,x;t_1,y)\tilde{\tilde{\rho}}_o(x)dx \ , \tag{5.1.8}$$

$$x,y \in E^n \ .$$

Thus, according to (5.1.2), (5.1.3) and (5.1.7), (5.1.8), $\bar{\rho}_o$, ρ_1, are solutions to the Schrödinger system in the case where

$$f(x) = \rho^*(t_o,x) \ , \tag{5.1.9}$$

$$g(x) = \rho^*(t_1,x) \ . \tag{5.1.10}$$

From the definitions of f and g and from the fact that μ_o and μ_1 are probability measures, it follows that $\int f d\lambda = \int g d\lambda = 1$. Then, from (5.1.9), (5.1.10), (5.1.2) and (5.1.3), and Fubini's theorem, we see that conditions (5.1.9), (5.1.10) require that $\bar{W}_o$ and W_1 be such that

(C1) $$\int_{E^n \times E^n} \bar{\rho}_o(x)p(t_o,x;t_1,y)\rho_1(y)dx\,dy = 1 \ .$$

Now consider $\rho^*(t,x)$. By direct computation using (3.7), (3.8), (3.16) and (5.1.1), one can readily verify that the function ρ^* satisfies

$$\frac{\partial \rho^*}{\partial t} = -\sum_{i=1}^{n} \frac{\partial}{\partial x_i}(v_i^*(t,x)\rho^*) + D\Delta\rho^* \ , \tag{5.1.11}$$

$$\rho^*(t_o,\cdot) = \exp\left(-\frac{\bar{W}_o}{2mD}\right)\exp\left(-\frac{W(t_o,\cdot)}{2mD}\right) \tag{5.1.12}$$

$$\rho^*(t_1,\cdot) = \exp\left(-\frac{\bar{W}(t_1,\cdot)}{2mD}\right)\exp\left(-\frac{W_1}{2mD}\right) \tag{5.1.13}$$

where $\bar{W}_o$ and W_1 were given at the outset of Problems 1 and 2 (not necessarily satisfying (C_1)) .

5.2. Controllability Problems for a Fokker-Planck Equation.

The Theorem of Jamison motivates us to consider the following

Controllability Problem 5.2.A. Given non-negative bounded continuous initial data ϕ_o and terminal data ϕ_1, satisfying

$$\int \phi_o(x)dx = \int \phi_1(y)dy = 1 \tag{5.2.1}$$

find $\hat{v} : S \to E^n$ such that the equation

$$\frac{\partial \phi}{\partial t} = - \sum_{i=1}^{n} \frac{\partial}{\partial x_i} (\hat{v}_i(t,x)\phi) + D\Delta\phi \tag{5.2.2}$$

has a solution satisfying the initial condition

$$\phi(t_o,\cdot) = \phi_o$$

and *also* the terminal condition

$$\phi(t_1,\cdot) = \phi_1.$$

In the specific case where (A1), (A2) and (C1) hold, and

$$\phi_o = \rho^*(t_o,\cdot)$$

$$\phi_1 = \rho^*(t_1,\cdot)$$

$\hat{v} = v^*$, the optimal control for Problem 1, is a solution to Problem 5.2.A .

Below, we let μ_o, μ_1 be the probability measures on $\mathcal{B}$ defined by†

$$\mu_o(B_o) = \int_{B_o} \phi_o(x)dx, \qquad \mu_1(B_1) = \int_{B_1} \phi_1(y)dy, \qquad B_o, B_1 \in \mathcal{B} .$$

We let π be the σ-finite product measure associated with μ_o, μ_1 by Theorem 5.1.A with $M = E^n$, $\Sigma = \mathcal{B}$, $q = p(t_o,\cdot;t_1,\cdot)$, and λ be the n-dimensional Lebesgue measure.

By arguments similar to ones of Jamison in (11), which rely on results of Burling (10), one can prove

Proposition 5.2.B. There exists a solution to the Controllability Problem 5.2.A if one of the following conditions holds :

(C2) ϕ_o and ϕ_1 have compact support.

(C3) There is a factorization of π - say $\pi = \nu_o \times \nu_1$ - such that $\nu_i \ll \lambda$, $i = 0,1$, and the function $d\nu_i/d\lambda$, $i = 0,1$, are continuous and bounded.

The proof of Proposition 5.2.B will be given in another publication. It states that, for $f = \phi_o$ and $g = \phi_1$ in (5.1.7), (5.1.8), if one of the conditions : (C2) or (C3) holds, then the functions h and $\bar{h}$ given by

$$h(s,x) \triangleq \int_{E^n} p(s,x;t_1,y)\tilde{\rho}_1(y)dy \quad \text{in} \quad [t_o,t_1) \times E^n , \tag{5.2.3}$$

$$\bar{h}(t,y) \triangleq \int_{E^n} p(t_o,x;t,y)\tilde{\tilde{\rho}}_o(x)dx \quad \text{in} \quad (t_o,t_1] \times E^n , \tag{5.2.4}$$

† Remember that ϕ_o and ϕ_1 are non-negative bounded and continuous.

with $\tilde{\tilde{\rho}}_o$, $\tilde{\rho}_1$ as in (5.1.7), (5.1.8), and $h(t_1,x) = \tilde{\rho}_1(x)$, $\bar{h}(t_o,y) = \tilde{\tilde{\rho}}_o(y)$, are solutions to the Cauchy problems

$$\left.\begin{aligned} &\frac{\partial h}{\partial s} = -D\Delta h + \frac{Q}{2mD}h\,, \quad \text{in} \quad [t_o,t_1)\times E^n\,, \\ &h(t_1,x) = \tilde{\rho}_1(x)\,, \text{ in } E^n\,. \end{aligned}\right\} \tag{5.2.5}$$

$$\left.\begin{aligned} &\frac{\partial \bar{h}}{\partial t} = D\Delta\bar{h} - \frac{Q}{2mD}\bar{h}\,, \quad \text{in} \quad (t_o,t_1]\times E^n\,, \\ &\bar{h}(t_o,y) = \tilde{\tilde{\rho}}_o(y)\,, \text{ in } E^n\,. \end{aligned}\right\} \tag{5.2.6}$$

Further, $h(s,x) > 0$ in $[t_o,t_1)\times E^n$, and $\bar{h}(t,y) > 0$ in $(t_o,t_1]\times E^n$.

Letting

$$\hat{v}_i(t,x) = \frac{2D}{h(t,x)}\,\frac{\partial h(t,x)}{\partial x_i}\,, \qquad i = 1,\ldots n\,,\ (t,x)\in S\,, \tag{5.2.7}$$

$$\phi(t,x) = \bar{h}(t,x)h(t,x)\,, \qquad (t,x)\in \bar{S}\,, \tag{5.2.8}$$

the proof of Proposition 5.2.B shows that $\hat{v}$ given by (5.2.7) is a solution to the Controllability Problem 5.2.A, and that ϕ given by (5.2.8) is a corresponding solution of (5.2.2) satisfying the given end conditions.

5.3. Controllability Problem for a Stochastic Differential Equation.

Consider the following system of stochastic differential equations

$$d\xi = \hat{v}(t,\xi(t))dt + \sqrt{2D}\,dw\,, \qquad t_o \leq t \leq t_1 \tag{5.3.1}$$

with the initial data

$$\xi(t_o) = \xi_o \quad \text{a.s.} \tag{5.3.2}$$

Assume

(D1) The functions $\hat{v}$ and $\frac{\partial \hat{v}}{\partial x_i}$, $(i = 1,\ldots n)$, are bounded in $\bar{S}$ and uniformly Lipschitz continuous in (t,x) in compact subsets of $\bar{S}$.

From the stochastic representation of solutions of a parabolic equation (see [6]) we have the following

Theorem 5.3.A. Under the condition (D1), the transition probability function of the solution of the stochastic differential system (5.3.1) has density, i.e.,

$$P\{\xi_{sx}(t)\in B\} = \int_B \hat{q}(s,x;t,y)dy \qquad (s<t) \tag{5.3.3}$$

for any Borel set B, and $\hat{q}(s,x;t,y)$ is the fundamental solution of equation (5.2.2).

Theorem 5.3.A is a direct consequence of Theorem 5.4, p.149, in Friedman (6), and of the remark at the end of Theorem 2.A .

We have also

Proposition 5.3.B. Let (D1) hold and let ξ_o be independant of $F(w(t), t_o \leq t \leq t_1)$, $E|\xi_o|^2 < \infty$. Suppose the probability distribution - say μ_o - of ξ_o has density ϕ_o, continuous and bounded. Let ϕ be the solution of (5.2.2) with the initial data ϕ_o, and ξ be the solution of (5.3.1) with the initial data ξ_o. Then, for any $t_o \leq t \leq t_1$, the probability distribution of $\xi(t)$ has density, and that density is equal to $\phi(t,\cdot)$.

Proof. From Theorem 3.1, p.109 of (6) we deduce

$$P\{\xi(t) \in B\} = \int_{E^n} P\{\xi_{t_o x}(t) \in B\} d\mu_o(x) , \quad B \in \mathcal{B} , \quad t_o < t \leq t_1 .$$

Then, from (5.3.3)

$$P\{\xi(t) \in B\} \triangleq P_t^*(B) = \int_{E^n} [\int_B \hat{q}(t_o,x;t,y)dy]\phi_o(x)dx$$

for any $B \in \mathcal{B}$, $t_o < t \leq t_1$, where P_t^* denotes the probability distribution of $\xi(t)$. By Fubini's theorem, this is also

$$\int_B [\int_{E^n} \hat{q}(t_o,x;t,y)\phi_o(x)dx] dy = \int_B \phi(t,y)dy,$$

which concludes the proof.

Proposition 5.3.B motivates us to consider the following

Controllability Problem 5.3.C. Let ξ_o be independent of $F(w(t) , t_o \leq t \leq t_1)$, $E|\xi_o|^2 < \infty$, with given probability distribution μ_o on $\mathcal{B}$, having density ϕ_o continuous and bounded. Let μ_1 be another given probability measure on $\mathcal{B}$, having density ϕ_1. Find $\hat{v} : \overline{S} \to E^n$, satisfying (D1) such that the solution ξ to (5.3.1), (5.3.2) satisfy the end condition

$$P\{\xi(t_1) \in B\} = \mu_1(B) \quad \text{for all} \quad B \in \mathcal{B} .$$

As a direct consequence of Proposition 5.3.B, we have

Proposition 5.3.D. If $\hat{v}$ is a function defined on $\overline{S}$, solution to the Controllability Problem 5.2.A with ϕ_o, ϕ_1 as in Problem 5.3.C, and if $\hat{v}$ satisfies (D1), then $\hat{v}$ is a solution to Controllability Problem 5.3.C .

Note : If $\hat{v}$ is a solution to the Controllability Problem 5.2.A satisfying the assumptions of Proposition 5.3.D, our assumption on ϕ_o requires that the given ϕ_1 be continuous and bounded (by Il'in Theorem of Ref. (5)).

In the specific case where (A1), (A2) and (C1) hold, and

$$\phi_o = \rho^*(t_o,\cdot) ,$$

$$\phi_1 = \rho^*(t_1,\cdot) ,$$

and $\hat{v} = v^*$ (the optimal control for Problem 1) satisfies (D1), then $\hat{v} = v^*$ is a solution to Problem 5.3.C.

6. A CONTROLLABILITY RESULT FOR THE SCHRÖDINGER EQUATION.

Fact II of Ref.(20) motivates us to consider the following

Controllability Problem 6.A. Given non-negative bounded continuous initial data ϕ_o and terminal data ϕ_1, satisfying

$$\int \phi_o(x)dx = \int \phi_1(y)dy = 1 \tag{6.1}$$

find $Q^* : S \to E^1$ such that the Schrödinger equation

$$\frac{\partial\psi}{\partial t} = iD\Delta\psi + i\,\frac{Q^*}{2mD}\,\psi \quad , \quad D \triangleq \hbar/2m , \tag{6.2}$$

has a solution satisfying the initial condition

$$\| \psi(t_o,\cdot) \|^2 = \phi_o$$

and *also* the terminal condition

$$\| \psi(t_1,\cdot) \|^2 = \phi_1 .$$

For that problem, we have

Proposition 6.B. If one of the conditions : (C2) or (C3) holds, then there exists a solution to the Controllability Problem 6.A.

Proof. Let Q be arbitrary, continuous on $\overline{S}$ and satisfying (A1), (A2) (for instance let $Q \equiv 0$).

Solve the Schrödinger system

$$\phi_o(x) = \overline{\varphi}_o(x) \int_{E^n} p(t_o,x;t_1,y)\, \varphi_1(y)dy , \tag{6.3}$$

$$\phi_1(y) = \varphi_1(y) \int_{E^n} p(t_o,x;t_1,y)\, \overline{\varphi}_o(x)dx , \tag{6.4}$$

for $\overline{\varphi}_o$, φ_1 (with p associated with the given Q) .

Assume one of the conditions : (C2) or (C3) holds.
Then, recalling equations (5.2.3), (5.2.4) with $\tilde{\rho}_1 = \varphi_1$, $\tilde{\tilde{\rho}}_o = \overline{\varphi}_o$, the functions

$\tilde{\varphi}$ and $\tilde{\tilde{\varphi}}$ given by

$$\tilde{\varphi}(s,x) \triangleq \int_{E^n} p(s,x;t_1,y)\,\varphi_1(y)dy \quad \text{in} \quad [t_o,t_1)\times E^n , \tag{6.5}$$

$$\tilde{\tilde{\varphi}}(t,y) \triangleq \int_{E^n} p(t_o,x;t,y)\,\overline{\varphi}_o(x)dx \quad \text{in} \quad (t_o,t_1]\times E^n , \tag{6.6}$$

with $\tilde{\varphi}(t_1,x) = \varphi_1(x)$, $\tilde{\tilde{\varphi}}(t_o,y) = \overline{\varphi}_o(y)$, are solutions to the Cauchy problems

$$\frac{\partial\tilde{\varphi}}{\partial s} = -D\Delta\tilde{\varphi} + \frac{Q}{2mD}\tilde{\varphi} , \quad \text{in } [t_o,t_1)\times E^n , \tag{6.7}$$

$$\tilde{\varphi}(t_1,x) = \varphi_1(x) , \quad \text{in } E^n . \tag{6.8}$$

$$\frac{\partial\tilde{\tilde{\varphi}}}{\partial t} = D\Delta\tilde{\tilde{\varphi}} - \frac{Q}{2mD}\tilde{\tilde{\varphi}} , \quad \text{in } (t_o,t_1]\times E^n, \tag{6.9}$$

$$\tilde{\tilde{\varphi}}(t_o,y) = \overline{\varphi}_o(y) , \quad \text{in } E^n . \tag{6.10}$$

Further, $\tilde{\varphi}(s,x) > 0$ in $[t_o,t_1)\times E^n$, and $\tilde{\tilde{\varphi}}(t,y) > 0$ in $(t_o,t_1]\times E^n$.

Then, define $\tilde{W}$, $\tilde{\tilde{W}}$ by

$$\tilde{\varphi}(t,x) = \exp\left(-\frac{\tilde{W}}{2mD}\right), \text{ in } [t_o,t_1)\times E^n, \tag{6.11}$$

$$\tilde{\tilde{\varphi}}(t,x) = \exp\left(-\frac{\tilde{\tilde{W}}}{2mD}\right), \text{ in } (t_o,t_1]\times E^n . \tag{6.12}$$

Now define $\tilde{V}^*$ and $\tilde{W}^*$ on S by

$$\tilde{V}^* = (1/2)(\tilde{\tilde{W}} - \tilde{W}), \tag{6.13}$$

$$\tilde{W}^* = (1/2)(\tilde{\tilde{W}} + \tilde{W}). \tag{6.14}$$

Let

$$\tilde{Q} \triangleq Q + 2\left[D\Delta\tilde{W}^* - \frac{1}{2m}(\text{grad }\tilde{W}^*)^2\right] , \tag{6.15}$$

$$\tilde{\psi}(t,x) \triangleq \exp\left(\frac{i}{2mD}\left(\tilde{V}^*(t,x) + i\tilde{W}^*(t,x)\right)\right), \text{ on S} . \tag{6.16}$$

Then, by Fact II of Ref.(20), $\tilde{\psi}$ is a solution of the Schrödinger equation

$$\frac{\partial\tilde{\psi}}{\partial t} = iD\Delta\tilde{\psi} + i\,\frac{\tilde{Q}}{2mD}\tilde{\psi} . \tag{6.17}$$

Since

$$\|\tilde{\psi}(t,x)\|^2 = \tilde{\varphi}(t,x)\,\tilde{\tilde{\varphi}}(t,x) \tag{6.18}$$

and since $\tilde{\varphi}\tilde{\tilde{\varphi}}$ is continuous on $\overline{S}$ and satisfies

$$\tilde{\varphi}(t_o,x)\,\tilde{\tilde{\varphi}}(t_o,x) = \phi_o(x) \tag{6.19}$$

$$\tilde{\varphi}(t_1,y)\,\tilde{\tilde{\varphi}}(t_1,y) = \phi_1(y) , \tag{6.20}$$

the solution of the Schrödinger equation we have found satisfies the given end conditions.

Therefore $Q^* = \tilde{Q}$ is a solution to the Controllability Problem 6.A.

ACKNOWLEDGMENT

I would like to thank Dr Katherine Kime for her contribution to this paper in the course of a joint research. I am grateful to Dr Marie Pauchard Sigal who read and criticized the manuscript.

REFERENCES

1. W. Fleming and R. Rishel, Deterministic and Stochastic Optimal Control, Springer-Verlag, Berlin, 1975.

2. E.B. Dynkin, Théorie des processus markoviens, Dunod, Paris, 1963.

3. E.B. Dynkin, Markov Processes, Vol.I, Springer-Verlag, Berlin, 1965.

4. E.B. Dynkin, Markov Processes, Vol.II, Springer-Verlag, Berlin, 1965.

5. A.M. Il'in, A.S. Kalashnikov, and O.A. Oleinik, Linear Equations of the Second Order of Parabolic Type, *in* Russian Mathematical Surveys, edited by K.A. Hirsch, Vol. XVII, Macmillan and Co. Ltd, London, 1962.

6. A. Friedman, Stochastic Differential Equations and Applications, Vol.I, Academic Press, New York, 1975.

7. S.K. Mitter, On the Analogy Between Mathematical Problems of Non-Linear Filtering and Quantum Physics, *in* Ricerche di Automatica, Vol.X, No.2, December 1979.

8. E. Schrödinger, Une analogie entre la mécanique ondulatoire et quelques problèmes de probabilités en physique classique, Annales de l'Institut Henri Poincaré, 1932.

9. S. Bernstein, Sur les liaisons entre les grandeurs aléatoires, Verh. Int. Math. Zürich, Band I, 1932.

10. A. Beurling, An Automorphism of Product Measures, Annals of Mathematics, Vol.72, No.1, July 1960.

11. B. Jamison, Reciprocal Processes, Z. Wahrsch. Gebiete, Bd 30, p.65, 1974.

12. R.P. Feynman and A.R. Hibbs, Quantum Mechanics and Path Integrals, McGraw-Hill, New York, 1965.

13. E. Nelson, Dynamical Theories of Brownian Motion, Princeton U.P., Princeton, NJ, 1967.

14. A. Blaquière and A. Marzollo, Introduction à la théorie moderne de l'optimisation et à certains de ses aspects fondamentaux en physique, *in* La pensée physique contemporaine, edited by S. Diner, D. Fargue, G. Lochak, Editions Augustin Fresnel, Moulidars, 1982.

15. K. Yasue, Quantum Mechanics and Stochastic Control Theory, J. Math. Phys., *22*(5), May 1981.

16. J.C. Zambrini, Variational Processes and Stochastic Versions of Mechanics, J. Math. Phys. *27* (9), September 1986.

17. M. Kac, Probability and Related Topics in Physical Sciences, Volume I of Lectures in Applied Mathematics, Proceedings of the Summer Seminar, Boulder, Colorado, 1957, Interscience Publishers, Ltd, London, 1959.

18. L. de Broglie, Une tentative d'interprétation causale et nonlinéaire de la mécanique ondulatoire, Gauthier-Villars, Paris, 1956.

19. A. Blaquière and A. Marzollo, An Alternative Approach to Wave Mechanics of a Particle at the Non-Relativistic Approximation, *in* Proc. of the 4th International Seminar on Mathematical Theory of Dynamical Systems and Microphysics, Udine, 1985, Springer-Verlag, Wien, 1987.

20. K. Kime and A. Blaquière, From two Stochastic Optimal Control Problems to the Schrödinger Equation, *in* Proc. of the 3^{d} International Workshop of the Bellman Continuum, this volume, Lecture Notes in Control and Information Sciences, Springer-Verlag, Berlin, (to appear).

21. A.J. Krener, 17th Conference on Stochastic Processes and their Applications, University of Rome, June 27 - July 1, 1988.

APPENDIX I. TRANSITION DENSITY OF AN OPTIMALLY CONTROLLED PROCESS.

(see Paragraph 4.1) .

Under assumption (B 2), $\exp \zeta_s^t(\theta)$ is a martingale[†] so that

$$\tilde{E}[\exp \zeta_s^T(\theta) | F_t] = \exp \zeta_s^t(\theta), \qquad t_o < s < t < T < t_1 , \tag{I.1}$$

relative to the probability space $(\Omega, F, \tilde{P})$. F_t is an increasing family of sub-σ-algebras of J to which $\tilde{w}$ is adapted.

From (a) of Theorem 4.1.A, $\exp \zeta_s^T(\theta)$ is $\tilde{P}$-integrable and from the definition of $\tilde{E}[\ |F_t]$ we have[††]

$$\int_{\xi(t)\in B} \tilde{E} \exp \zeta_s^T(\theta) | F_t] \tilde{P}(d\omega) = \int_{\xi(t)\in B} \exp \zeta_s^T(\theta)\tilde{P}(d\omega), \qquad B \in \mathcal{B} . \tag{I.2}$$

From (I.1) and (I.2) it follows that

$$\int_{\xi(t)\in B} \exp \zeta_s^T(\theta)\tilde{P}(d\omega) = \int_{\xi(t)\in B} \exp \zeta_s^t(\theta)\tilde{P}(d\omega) \tag{I.3}$$

$$t_o < s < t < T < t_1 , \qquad B \in \mathcal{B} .$$

Then, following Fleming and Rishel (Ref.(1),p.143) we let

$$\beta(t,y) \triangleq \tilde{E}[\exp \zeta_s^t(\theta) | \tilde{\xi}(t) = y] , \qquad y \in E^n . \tag{I.4}$$

For any $B \in \mathcal{B}$

$$P\{\xi(t) \in B\} = \int_{\xi(t)\in B} \exp \zeta_s^T(\theta)\tilde{P}(d\omega) = \int_{\xi(t)\in B} \exp \zeta_s^t(\theta)\tilde{P}(d\omega) =$$

$$= \int_B \beta(t,y)\tilde{\pi}_t(dy) ,$$

where $\tilde{\pi}_t$ denotes the distribution of the n-dimensional random vector $\tilde{\xi}(t)$, defined by

$$\tilde{\pi}_t(B) = \tilde{P}\{\tilde{\xi}(t) \in B\} , \qquad B \in \mathcal{B} .$$

The left side is $\pi_t(B)$. Therefore π_t is absolutely continuous with respect to $\tilde{\pi}_t$, and

$$\pi_t(dy) = \beta(t,y)\tilde{\pi}_t(dy) .$$

† See for instance : A. Bensoussan, Stochastic Control by Functional Analysis Methods, North-Holland Pub. Co., Amsterdam, 1982.

†† Remember that ξ and $\tilde{\xi}$ have the same sample functions, i.e., $\xi(\cdot,\omega)=\tilde{\xi}(\cdot,\omega)$.

In particular, consider initial data $\tilde{\xi}(s) = \xi(s) = x$. Then $\tilde{\pi}_t$ has the density $k(s,x;t,\cdot)$, and accordingly π_t has the density $q^*(s,x;t,\cdot)$, with

$$q^*(s,x;t,y) = \beta(t,y)k(s,x;t,y) \ . \tag{I.5}$$

Therefore, from (I.4),(I.5) and (4.1.7) we conclude that the process ξ governed by

$$d\xi = v^*(t,\xi(t))dt + \sqrt{2D}\, dw \quad \text{relative to } (\Omega, F, P)$$

has the transition probability densité q^* :

$$q^*(s,x;t,y) =$$

$$= \frac{\exp(-\frac{W(t,y)}{2mD})}{\exp(-\frac{W(s,x)}{2mD})} \tilde{E}\left[\exp\left(-\int_s^t \frac{Q(\sigma,\tilde{\xi}_{sx}(\sigma))}{2mD}\, d\sigma\right)\middle|\tilde{\xi}_{sx}(t) = y\right] k(s,x;t,y). \tag{I.6}$$

Note, in paragraph 4, one can relax condition (B 2) and make use of the following arguments which follow closely ones of Mitter in Ref. [7] :

Relation (4.1.13) holds under assumptions weaker than the ones of paragraph 4.2. Indeed, since $\rho = \exp(-W/2mD)$ is a solution to the Cauchy problem (3.6), (3.7), with ρ_1 continuous and bounded, if Q continuous on $\overline{S}$ satisfies (A1) and

(A'2) Q is uniformly Hölder continuous in (s,x) in compact subsets of $\overline{S}$, then according to Theorem 5.3, p.148, of Ref.[6], (4.1.13) holds.

It follows that

$$\int_\Omega \exp \zeta_s^T(\theta_{sx})\tilde{P}(d\omega) = 1, \qquad t_o < s < t < T < t_1 \ , \tag{I.7}$$

with

$$\exp \zeta_s^T(\theta_{sx}) \triangleq \frac{\exp\left(-\frac{W(T,\tilde{\xi}_{sx}(T))}{2mD}\right)}{\exp(-\frac{W(s,x)}{2mD})} \exp\left(-\int_s^T \frac{Q(\sigma,\tilde{\xi}_{sx}(\sigma))}{2mD}\, d\sigma\right). \tag{I.8}$$

and, as in paragraph 4.1, assuming (B1),

$$\zeta_s^T(\theta_{sx}) = \int_s^T \theta_{sx}(\sigma)d\tilde{w}(\sigma) - \frac{1}{2}\int_s^T |\theta_{sx}(\sigma)|^2\, d\sigma, \tag{I.9}$$

$$\theta_{sx}(t) = \frac{1}{\sqrt{2D}} v^*(t,\tilde{\xi}_{sx}(t)) = -\frac{1}{m\sqrt{2D}} \text{grad } W(t,\tilde{\xi}_{sx}(t)) \ . \tag{I.10}$$

Here, we suppose that the standard F_t n-dimensional Brownian motion $\tilde{w}$ relative to $(\Omega, F, \tilde{P})$ is defined on [s,T] , with $\tilde{w}(s) = 0$.

For each (s,x), let P_{sx} be absolutely continuous with respect to $\tilde{P}$, with

$$P_{sx}(d\omega) = \exp \zeta_s^T(\theta_{sx})\widetilde{P}(d\omega), \qquad t_o < s < t < T < t_1 .$$

In view of (I.7), for each (s,x), P_{sx} is a probability measure on (Ω, F) .

Therefore, by the Girsanov theorem (Ref.(6)), the process w_{sx} given by

$$w_{sx}(t) = \widetilde{w}(t) - \int_s^t \theta_{sx}(\sigma)d\sigma$$

is a standard n-dimensional Brownian motion (with $w_{sx}(s) = 0$) with respect to P_{sx} . By arguments similar to the ones of paragraph 4.1, it follows that

$$\widetilde{\xi}_{sx}(t) = x + \int_s^t v^*(\sigma,\widetilde{\xi}_{sx}(\sigma))d\sigma + \sqrt{2D}\, w_{sx}(t) .$$

In other words, the stochastic process $\widetilde{\xi}_{sx}$, $s < t < T$, on the probability space (Ω, F, P_{sx}) is a solution of

$$d\xi = v^*(t,\xi(t))dt + \sqrt{2D}\, dw_{sx}$$

$$\xi(s) = x .$$

By standard theorems of stochastic calculus, one can prove that, under (A1), (A'2), (B1), $\exp \zeta_s^t(\theta_{sx})$, $s < t < T$, is a martingale. Based on this fact, formula (I.6) can be obtained by arguments similar to the ones before.

APPENDIX II. NONUNIQUENESS OF THE SOLUTIONS TO PROBLEM 5.2.A.

Proposition 5.2.B gives sufficiency conditions for a solution to Controllability Problem 5.2.A to exist. Clearly, the solution we have constructed in paragraph 5.2 under such conditions is non unique since it depends on the function p which, itself, is (uniquely) determined by the choice of the potential function Q satisfying (A1), (A2). In other words, there exists a family of such solutions indexed by Q .

Two questions arise :

1. Does there exist solutions to Problem 5.2.A not belonging to that family ? and
2. If it is the case, why have we priviledged this family ?

An answer to point 1 is readily provided by a remark of Prof. H. Sussmann, reported by Dr K. Kime.

Assume ϕ_o, $\phi_1 \in C_o^2(E^n)$, ϕ_o, ϕ_1 non negative with compact supports A_o, A_1. Let

$$\phi(t,x) = \frac{\phi_o(x)(t_1-t) + \phi_1(x)(t-t_o)}{t_1-t_o}, \qquad (t,x) \in \overline{S} .$$

Now, suppose $E^n = E^3$ and find $\hat{v}_i$, i = 1,2,3, such that

$$\frac{\partial\phi}{\partial t} = D\Delta\phi - \sum_{i=1}^{3} \frac{\partial}{\partial x_i} (\hat{v}_i(t,x)\phi) .$$

One solution is to look for $\hat{v}$ of the form $\hat{v} = (\hat{v}_1,0,0)$; then

$$\frac{\partial}{\partial x_i} (\hat{v}_1(t,x)\phi) = D\Delta\phi - \frac{\partial\phi}{\partial t} .$$

The right-hand side is a function continuous with compact support, by the definition of ϕ.

Then

$$\hat{v}_1(t,x)\phi(t,x) - \hat{v}_1(t,a)\phi(t,a) = \int_a^x [D\Delta\phi(y,t) - \frac{\partial\phi}{\partial t}(y,t)]dy .$$

Take a outside $A_o \cup A_1$, then $\phi(t,a) = 0$.

Provided that $\phi(t,x) \neq 0$, we have

$$\hat{v}_1(t,x) = \frac{1}{\phi(t,x)} \int_a^x [D\Delta\phi(y,t) - \frac{\partial\phi}{\partial t}(y,t)]dy .$$

The solution thus obtained needs not belong to the above family.

Note that if ϕ_o, ϕ_1 are not C^2, the linear interpolation will not yield a $C^{1,2}$ function ; if ϕ_o, ϕ_1 are merely continuous, the argument above can be extended by constructing some $C^{1,2}$ function ϕ such that $\lim_{t\to t_o} \phi(\cdot,t) = \phi_o$ and $\lim_{t\to t_1} \phi(\cdot,t) = \phi_1$.

We are indebted to Prof. H. Sussmann for his comment.

Concerning point 2, our interest in the family constructed in paragraph 5.2 is motivated by a connection between Controllability Problems 5.2.A and 6.A, which is introduced by Fact I and Fact II of Ref. (20). Having in view the study of Problem 6.A, we have been led by Fact I and Fact II to restricting the class of functions ϕ for Problem 5.2.A to those functions which are the product of two functions $\varphi, \overline{\varphi} \in C^{1,2}(S)$, solutions of

$$\frac{\partial\varphi}{\partial t} = - D\Delta\varphi + \frac{Q}{2mD}\varphi ,$$

$$\frac{\partial\overline{\varphi}}{\partial t} = D\Delta\overline{\varphi} - \frac{Q}{2mD}\overline{\varphi}$$

for some function Q .

NONDEMOLITION MEASUREMENTS, NONLINEAR FILTERING AND DYNAMIC PROGRAMMING OF QUANTUM STOCHASTIC PROCESSES

Viacheslav P. Belavkin
Applied Mathematics Department, M.I.E.M.,
B. Vusovski 3/12 Moscow 109028 USSR

The class of continuous nondemolition measurements v_t in quantum stochastic systems is characterized in terms of Hudson-Parthasarathy stochastic calculus. Two types of such measurements of a quantum stochastic process are derived: a Poissonian counting measurement and a Brownian indirect observation. The corresponding nonlinear filtering equations are derived in semi-martingal and density-matrix form, and a posterior Schrödinger equation is found. A quantum continuous Bellman equation is derived for the solution of the problem of optimal control of a quantum stochastic process with nondemolition measurements. The solution of this equation $u^o(t,u^t,\rho)$ together with the solution of the corresponding nonlinear filtering problem $\rho = \hat{\rho}_t(w^t, v(t))$, $w^t = (u^t, v^t)$ defines the optimal control strategy $d^o(t, w^t, v(t)) = u^o(t, u^t, \hat{\rho}_t(w^t, v(t)))$.

Quantum nondemolition principle.

The problem of description of continuous observation in quantum dynamic systems can be effectively solved in the framework of quantum stochastic calculus of nondemolition processes developed for general linear boson models in [1 - 3]. The idea of nondemolition quantum measurements which is intensively discussed in physical literature in connection with the problem of gravitation waves distinction [4,5], can be usually reduced to the condition of consistent measurability for a given family of physical variables x_t related to different times $t \in \mathbb{R}$. The mathematical definition of such self-nondemolishing quantum observation given in [6], is equivalent to the condition of pairwise commutativeness for corresponding self-adjoint operators X_t which represent these variables in a certain, generally speaking, extended Hilbert space $\mathcal{H}$. In such a form we can represent any real random process $x_t(\omega)$ defined on the probability space $(\Omega, \mathcal{F}, P)$ (it is sufficient to take the space $L^2(\Omega, \mathcal{F}, P)$ as $\mathcal{H}$, and the operators of multiplication on $x_t(\omega)$ as X_t), and the class of nondemolition measurements is exhausted by operator representations of classical random processes in this described narrow sense.

In order to explain the non-commutative generalization of the nondemolition measurements principle defined in [7] by means of a general quantum random process with respect to a non-increasing family $\mathcal{A}_t$ of operator subalgebras $\mathcal{A}_t \subseteq \mathcal{A}_s$, $t \geqslant s$, of the algebra $\mathcal{B}(\mathcal{H})$ of all the bounded operators, we consider the indexified pair X_t, Y_t of Hermitian operators in $\mathcal{H}$.

The process $Y = (Y_t)$ is called nondemolishing with respect to the process $X = (X_t)$, if Y_t commutates with X_t and with all future X_s, $s > t$, for each $t \in \mathbb{R}$. By choosing as $\mathcal{A}_t$ the operator algebra generated by the family $\{X_s, s \geqslant t\}$, we obtain the non-increasing family $\{\mathcal{A}_t\}$, with respect to which the nondemolition condition can be formulated in the form $Y_t \in \mathcal{A}_t'$ for all t, where $\mathcal{A}_t' = \{X_s, s \geqslant t\}'$ is the commutant of the algebra $\mathcal{A}_t$ (every operator Y_t, commutative with $\{X_s, s \geqslant t\}$, is also commutative with $\mathcal{A}_t$). Such a nondemolition process Y called in [7] a process with respect to $\mathcal{A} = (\mathcal{A}_t)$, can be described by the non-commutative family $\{Y_t\}$, if only $Y_t \notin \mathcal{A}_t$ even for a single t. It is interpreted as a process of subsequent, in general case self-demolishing, indirect mesurements with respect to a non-demolition quantum dynamic system, the present and future of which are described by the algebras $\mathcal{A}_t$. We note that the process Y non-demolishing with respect to X can be non-commutative with the past X_s, $s < t$, even in case of direct measurements $Y_t \in \mathcal{A}_t$ for all t, when it is commutative, if only $\mathcal{A}_t$ is not generated by the family $\{Y_s, s \geqslant t\}$ for all t, as it is in case of self-nondemolition process $Y_t = X_t$ for all t.

Such a weakened notion of nondemolition observation which does not demand due to the causality principle the measurability (commutativeness) of the present process Y_t consistent with the unobservable past X_s, $s < t$, permitted to formulate and to solve in the linear case the simplest problems of the filtering theory and control in open quantum systems not only in discrete, but also in continuous time [3].

The sufficiently general models of continuous quantum observation obtained in [9] as the limit $\Delta t \to 0$ when discrete subsequent indirect measurements were considered and which turned out to be nondemolishing in the weakened sense mentioned above, as well the possibility of their description [10] within the frames of quantum stochastic calculus [11, 12] indicate that it is possible to construct some general stochastic calculus of nondemolition processes with respect to a given quantum random process X. Below we give the scheme of such calculus with respect to a quantum Markovian process represented in the Fock space, then we formulate a problem of nonlinear filtering in continuous time as a problem of finding a posterior state on the algebra $\mathcal{A}$ generated by unobservable operators X_t for fixed t, and derive the stochastic differential equation for a posterior density matrix. In the commutative case this equation coincides with the corresponding equation of nonlinear filtering for a Markovian process which is described a prior by the Kolmogorov-Feller equation, in the pure non-commutative case $\mathcal{A} = \mathcal{B}(\mathcal{K})$ it is reduced to a nonlinear equation for the probability amplitude which generalizes the Schrödinger equation. This equation describes the a postrior density matrix of a Gaussian boson system with a linear coherent channel under the Gaussian indirect observation which is described by the quantum analogue of the Kalman-Busy filter obtained for a quantum oscillator in [5, 6]. We note that quantum a posterior dynamics obtained in such a way which gives the solution of optimal nondemolition filtration problem for quantum random processes according to the quadratic criterion coincides for conservative (Hamiltonian) systems with a prior dynamics described by the Schrödinger equation. The problem of nondemolition filtration in quantum Hamiltonian systems in the sense of self-nondemolition of the observable process without the quantum dynamics reduction was considered in [13, 14].

1. Nondemolition observation and a posterior dynamics of a quantum particle.

Before we develop regorously the quantum stochastic calculus of general nondemolition processes in the Fock space, we consider a special case of physical interest: namely, a quantum m-dimensional particle of mass μ with the potential $\varphi(x)$ in a Boson reservoir modelling the measurement apparatus. The position of the particle $x = (x^i)$ is observed indirectly together with the white noise by measuring a self-nondemolition vector process, which satisfies the equations $\dot{Q}_i = \sqrt{2\lambda}X^i + 2\mathrm{Re}\,\dot{A}_i^+$, or in terms of Ito differentials:

$$dQ_i(t) = \sqrt{2\lambda}\,X^i(t)dt + 2\mathrm{Re}\,dA_i^+(t), \qquad Q_i(0) = 0, \tag{1.1}$$

where $X(t)$ is the vector of coordinate operators in the Heisenberg picture, $2\mathrm{Re}\,A_i^+(t) = A_i^+(t) + A_i^+(t)^*$ is the operator representation of a standard Winer vector process in the Fock space with the canonical annihilation $\dot{A}_-^k(t)$ and creation $\dot{A}_i^+(t)$ operator-functions, which are generalized derivatives of the processes $A_-^{i*} = A_i^+$ with $[A_-^k(t), A_i^+(s)] = \delta_i^k \min(t,s)$. The apparatus effects on a quantum particle by means of the perturbation forces $f_i = \hbar\sqrt{2\lambda}\,\mathrm{Im}\dot{A}_i^+$, which are white noises of the intensity $\lambda\hbar^2/2$, proportional to the measurement accuracy λ, so that the momentum operators $P_i(t) = \mu\dot{X}^i(t)$ in the Heisenberg picture satisfy the quantum Newton-Langevian equations $\dot{P}_i + F_i = f_i$, $F = \nabla\varphi$, or in the form of quantum stochastic differential equations (QSDE):

$$dP_i(t) + F_i(X(t))dt = \hbar\sqrt{2\lambda}\,\mathrm{Im}\,dA_i^+, \tag{1.2}$$

with initial $P_i(0) = \hbar\nabla_i/j$, $j = \sqrt{-1}$, $X^i(0) = x^i$, $\mathrm{Im}A_i^+ = (A_i^+ - A_-^i)/2j$, $\varphi_i' = \nabla_i\varphi$. As it follows from the next section, the equations (1.2) are uniquely possible QSDE for a quantum particle with the Hamiltonian $P^2/2\mu + \varphi(X)$ to be nondemolished by the measurement of the commutative stochastic process Q defined by (1.1) in the sense of

$$[P_i(s), Q_k(t)] = 0, \quad [X_i(s), Q_k(t)] = 0, \quad \forall i,k; \qquad s \geq t.$$

It means that the observable in such a way quantum particle is an open quantum system and its a prior state in the Schrödinger picture is a mixed state for $\lambda > 0$ even if the initial state is purely described, i.e. described for the particle by the wave function $\psi(x)$ and for the reservoir by the vacuum Fock vector.

By taking into account the measurement data, one can obtain a posterior quantum state of the particle which as we shall prove it is a pure state for a given initial ψ and a vacuum Fock state. This a posterior state is described by a posterior stochastic wave function $\hat{\psi}(t,x)$, satisfying the new nonlinear stochastic (posterior) Schrödinger equation :

$$d\hat{\Psi}(x)+\left(\frac{\lambda}{4}\tilde{x}^2+\frac{i}{\hbar}\varphi(x)-\frac{j\hbar}{2\mu}\Delta\right)\hat{\Psi}(x)dt=\sqrt{\frac{\lambda}{2}}\hat{\Psi}(x)\tilde{x}\,d\tilde{Q}\ , \tag{1.3}$$

where $\tilde{x}(t) = x - \hat{q}(t)$, $\hat{q}(t) = \int x\,|\hat{\Psi}(t,x)|^2\,dx$ is a posterior (stochastic) mean value of the position at $t>0$, and $\tilde{Q}$ is an innovating process defined as the vector Winer process by the Ito equation:

$$d\tilde{Q}_i(t) = dQ_i(t)-\sqrt{2\lambda}\,\hat{q}^i(t)dt, \qquad \tilde{Q}_i(0) = 0\ . \tag{1.4}$$

By using the table $dQ_i dQ_k = \delta_{ik}dt$ in the Ito formula

$$df(\hat{\Psi}) = f'(\hat{\Psi})d\hat{\Psi} + \frac{1}{2}f''(\hat{\Psi})(d\hat{\Psi})^2$$

for the function $f(\hat{\Psi}) = |\hat{\Psi}|^2$, one obtains the stochastic (posterior) equation of continuity for the stochastic density function $\hat{\rho}(t,x) = |\hat{\Psi}(t,x)|^2$:

$$d\hat{\rho}(x)+\nabla(V(x)\hat{\rho}(x))dt = \sqrt{2\lambda}\,\hat{\rho}(x)\tilde{x}d\tilde{Q}, \tag{1.5}$$

where $V^i(x) = \nabla_i S(x)/\mu$ are the velosity functions defined by the action $S(t,x) = \hbar\,\mathrm{Arg}\,\hat{\Psi}(t,x)$ satisfying the Hamilton-Jacobi equation:

$$\frac{\partial}{\partial t}S + (\nabla S)^2/2\mu + \varphi = \hbar^2\Delta\sqrt{\hat{\rho}}\ /2\mu\sqrt{\hat{\rho}}\ . \tag{1.6}$$

The real equation (1.5) together with (1.6) is equivalent to the complex equation (1.3) for $\hat{\Psi}(x) = \sqrt{\hat{\rho}(x)}\exp\{jS(x)/\hbar\}$, and its nonlinearity due to the dependence of $\hat{q}$ on $\hat{\rho}$ only is connected with normalization preserving property $\int\hat{\rho}(x)dx = \hat{1}$. By multiplying $\hat{\rho}(x,t)$ on the positive process $\hat{r}(t)$ defined by the equation:

$$d\hat{r}(t) = \sqrt{2\lambda}\,\hat{r}(t)\hat{q}(t)dQ(t) \tag{1.7}$$

one obtains the linear stochastic continuity equation for non-normalized posterior density $R(t,x) = \hat{r}(t)\hat{\rho}(t,x)$:

$$dR(x) + \nabla(V(x)R(x))dt = \sqrt{2\lambda}\,R(x)xdQ, \tag{1.8}$$

corresponding to the following linear stochastic Schrödinger equation for the normalized by $\sqrt{\hat{r}}$ posterior wave function $\Psi(x)=R(x)^{1/2}\cdot\exp jS(x)/\hbar$:

$$d\Psi(x)+(\frac{\lambda}{4}x^2+\frac{j}{\hbar}\varphi(x)-\Delta\frac{j\hbar}{2\mu})\Psi(x)dt=\sqrt{\frac{\lambda}{2}}\Psi(x)x\,dQ \qquad (1.9)$$

By taking into account that a posterior quantum pure state corresponds to a nonnormalized stochastic wave function $\hat{c}(t)\Psi(t,x)$ up to an arbitrary complex stochastic multiplicator $\hat{c}$, one should identify it with the complex velosity vector function:

$$W(t,x)=\frac{\hbar}{\mu}\nabla\ln\hat{\Psi}(t,x)=\frac{\hbar}{\mu}\nabla\ln\Psi(t,x)=U(t,x)+jV(t,x)$$

satisfying the following system of quasi-linear stochastic equations:

$$dW(x)+(\frac{\hbar}{\mu}\lambda x+\frac{j}{\mu}F(x)-\frac{j}{2}(\nabla W^2(x)+\frac{\hbar}{\mu}\Delta W(x)))dt=\sqrt{\frac{\lambda}{2}}\frac{\hbar}{\mu}dQ \qquad (1.10)$$

which can be obtained from (1.3) or (1.9) by using the Ito formula for the logarithm $f = \ln$.

Let us find the solution of the equation (1.10) in case of linear potential force $F(x) = \mu g - \hbar\varkappa x$ and the Gaussian coherent initial wave function:

$$\Psi(x)=(2\pi\sigma^2)^{-m/4}\exp\left\{-\frac{1}{4\sigma^2}(x-q)^2+jpx/\hbar\right\},$$

defined by the mean values q, p and the dispersion σ^2 of the wave packet in the coordinate representation. We find the solution of (1.10) with the initial condition:

$$W(0,x)=\hat{1}\frac{\hbar}{\mu}\nabla\ln\Psi(x)=\hbar(q-x)\,\hat{1}/2\mu\sigma^2+jp\,\hat{1}/\mu \qquad (1.11)$$

in the linear form $W(t,x)=\hat{w}(t)+\omega(t)x\hat{1}$. By inserting $\nabla W^2 = 2\omega W$, $\Delta W = 0$ into (1.10), one obtains the following equations for the coefficients $\hat{w}(t)$ and $\omega(t)$:

$$d\hat{w}+j(g-\omega\hat{w})dt=\sqrt{\frac{\lambda}{2}}\frac{\hbar}{\mu}dQ, \quad \hat{w}(0)=(\frac{\hbar q}{2\mu\sigma^2}+\frac{jp}{\mu})\hat{1}, \qquad (1.12)$$

$$d\omega/dt+\hbar\lambda/\mu = j(\hbar\varkappa/\mu+\omega^2), \quad \omega(0)=-\hbar/2\mu\sigma^2 \qquad (1.13)$$

The first equation of this system is a complex linear filtration equation, while the second is a complex Rickatti equation having the unique solution:

$$\omega(t) = j\alpha \frac{\omega(0) + j\alpha \operatorname{th} \alpha t}{j\alpha + \omega(0) \operatorname{th} \alpha t}, \quad \alpha = \sqrt{\frac{\hbar}{\mu}} (\varkappa + j\lambda)^{1/2}, \tag{1.14}$$

satisfying the normalizability condition $\operatorname{Re}\omega < 0$ for the corresponding wave function:

$$\Psi(t,x) = \exp\{\mu(\hat{w}(t)x + \omega(t)\, x^2/2)/\hbar\} .$$

Hence the posterior wave function $\hat{\varphi}(x) = \Psi(x)/\|\Psi\|$ has the form of the Gaussian wave packet:

$$\hat{\varphi}(t,x) = c(t)\exp\Big\{\frac{1}{\hbar}(\mu\,\omega(t)(x-\hat{q}(t))^2/2 + j\hat{p}(t)x)\Big\}, \tag{1.15}$$

where $\hat{c}(t) = (2\pi\sigma^2(t))^{-m/4}$ up to the unessential stochastic phase multiplier, $\sigma^2(t) = -\hbar/2\mu \operatorname{Re}\omega(t)$ is a posterior coordinate dispersion $\sigma^2 = \widehat{q^2} - \hat{q}^2$ giving the minimal square error of the position estimation, $\hat{q} = -\operatorname{Re}\hat{w}/\operatorname{Re}\omega$ is a posterior mean coordinate $\hat{q}$, and $\hat{p} = \mu\operatorname{Im}(\hat{w} + \omega\hat{q})$ is a posterior mean momentum

$$\hat{p} = \frac{\hbar}{j}\int \hat{\varphi}(x)^* \nabla \hat{\varphi}(x)dx = \mu \int V(x)\,|\varphi(x)|^2 dx.$$

Note that $\hat{q}$ satisfies the equation $\mu U(\hat{q}) = \hbar\nabla \ln\sqrt{\hat{\rho}}\,(\hat{q}) = 0$, which defines the maximum of a posterior density $\hat{\rho}$ (and R), and $\hat{p}$ coincides due to linearity of V with the value $\mu V(\hat{q}) = \nabla S(\hat{q})$ of the classical momentum function $\nabla S(x)$ at $x=\hat{q}$, giving the optimal velocity estimation $V(q) = \hat{p}(t)/\mu$ of the observable quantum particle.

By taking into account that $\hat{w} = j\hat{p}/\mu - \omega\hat{q}$, $jd\hat{p} - \mu\omega d\hat{q} = \mu(d\hat{w} + \hat{q}d\omega) = \sqrt{\frac{\lambda}{2}}\,\hbar d\tilde{Q} - (\omega\hat{p} + j(\mu g - \hbar\varkappa\hat{q}))dt$, one obtains the following a posterior system of Hamilton-Langevian equations:

$$\hat{p}dt - \mu d\hat{q} = \hbar\sqrt{\lambda/2}\, d\tilde{Q}/\operatorname{Re}\omega , \tag{1.16}$$

$$d\hat{p} + \mu g dt = \hbar(\varkappa q dt - \sqrt{\lambda/2}\,\operatorname{Im}\omega d\tilde{Q}/\operatorname{Re}\omega),$$

which together with (1.14) define a posterior dynamics of quantum observable particle in the Gaussian coherent initial state, defined by $\hat{p}(0) = p$ and $\hat{q}(0) = q$. We note that a posterior momentum dispersion $\tau^2 = \widehat{p^2} - \hat{p}^2$,

$$\widehat{p^2} = \hbar \int |\nabla \Psi(x)|^2 dx = \mu \int \; | W(x)|^2 \; |\Psi(x)|^2 \; dx \; ,$$

gives the minimal square error $\tau^2(t) = - \hbar\mu |\omega(t)|^2 / 2\mathrm{Re}\,\omega(t)$ of the momentum estimation, satisfying the Heisenberg inequality $\delta^2(t) \; \tau^2(t) \geq \hbar^2/4$. If $\lambda \neq 0$, these dispersions have the finite limits:

$$\delta^2_\infty = \hbar/2\mu \; \mathrm{Im}\alpha = (\hbar/2\mu(\sqrt{\varkappa^2 + \lambda^2} - \varkappa))^{1/2} \tag{1.17}$$

$$\tau^2_\infty = \hbar |\alpha|^2 \mu / 2\mathrm{Im}\,\alpha = \hbar(\hbar\mu(\varkappa^2 + \lambda^2)/2(\sqrt{\varkappa^2 + \lambda^2} - \varkappa))^{1/2},$$

corresponding to the stationary solution $\omega = j\alpha$.

The obtained a posterior quantum dynamics (1.16) helps to reduce the quantum optimal control problem of this particle by gravitational strength $g = u(t)$ to the solution of corresponding classical stochastic problem, if one considers only Gaussian coherent initial states and mean square cost functionals. Due to the linearity of the system (1.16) the optimal control strategy $u^o(t)$ is a linear function $u^o(t) = -L(t)\hat{q}(t)$ on posterior position $\hat{q}(t)$, where the matrix $L(t)$ is the control matrix for the classical deterministic controlled system:

$$p - \mu\dot{q} = 0, \quad \dot{p} + \mu u = \hbar\varkappa q,$$

minimizing the corresponding quadratic cost functional. Such a solution of the feedback control problem for the indirectly observable quantum oscillator was found in [6].

2. QND stochastic calculus in Fock space.

Now we shall develop the general quantum nondemolition (QND) stochastic calculus in the Fock space $\mathcal{F} = \Gamma(\mathcal{E})$ over the Hilbert space $\mathcal{E} = \mathcal{L}^2(\mathbb{R}_+ \to \mathbb{C}^m)$ of square-integrable complex vector-functions $t \mapsto \varphi(t) = (\varphi^k_t), k=1,\ldots,m$ on $t \geq 0$. As in the previous section, we shall use the tensor notations in which a vector $\varphi \in \mathcal{F}$ is represented by all the tensors $\varphi^{k_1,\ldots,k_n}_{t_1,\ldots,t_n} = \bar{\varphi}^{t_1,\ldots,t_n}_{k_1,\ldots,k_n}$, $n \in \mathcal{N}$ with

$$\langle \mathcal{X} | \varphi \rangle = \sum_{n=0}^{\infty} \int \dots \int_{t_1 < \dots < t_n} \mathcal{X}_{k_1,\dots,k_n}^{t_1,\dots,t_n} \varphi_{t_1,\dots,t_n}^{k_1,\dots,k_n} dt_1 \dots dt_n$$

(the sum is taken also over identical co- and contra-variant indexes $k_i = 1,\dots,m$).

One should consider φ as a tensor-valued function $\tau \mapsto \varphi(\tau)$ on $\tau = (t_1 < \dots < t_n)$ with $\varphi(\tau) \in \mathbb{C}^{m \times n}$. Let us denote $\delta_\emptyset$ the vacuum function $\delta_\emptyset(\tau) = 0$, if $\tau \neq \emptyset$, $\delta_\emptyset(\emptyset) = 1$, $\mathcal{F}_t = \Gamma(\mathcal{E}_t)$ and $\mathcal{F}_r^t = \Gamma(\mathcal{E}_r^t)$ the Fock spaces over orthogonal subspaces $\mathcal{E}_t = \{\varphi(s) = 0,\ s \leq t\}$, $\mathcal{E}_r^t = \{\varphi(s) = 0,\ s \notin]r,t]\}$, and $\mathcal{F}^t = \mathcal{F}_0^t$ over $\mathcal{E}^t = \mathcal{E}_0^t$, so that $\mathcal{F} = \mathcal{F}^r \otimes \mathcal{F}_r^t \otimes \mathcal{F}_t$ in accordance with $\mathcal{E} = \mathcal{E}^r \oplus \mathcal{E}_r^t \oplus \mathcal{E}_t$ for all $r < t$.

Let us define the following basic quantum martingals with respect to the filtration $(\mathcal{F}^t)$ and the state-vector $\delta_\emptyset$: the creation processes $\hat{A}_i^+(t)$, the annihilation processes $\hat{A}_-^k(t)$ and the preservation processes $\hat{A}_i^k(t)$, where $i,k = 1,\dots,m$, described on the product functions $\varphi(\tau) = \bigotimes_{t \in \tau} \varphi(t)$ by

$$\hat{A}_i^+(t)\varphi(\tau) = \sum_{s \in \tau} \varphi(\tau^s) \otimes \mathcal{X}_i^t(s) \otimes \varphi(\tau_s)$$

$$\hat{A}_-^k(t)\varphi(\tau) = \int_0^t \varphi^k(s) ds\, \varphi(\tau) \tag{2.1}$$

$$\hat{A}_i^k(t)\varphi(\tau) = \sum_{s \in \tau} \varphi(\tau^s) \otimes \varphi_s^k\, \mathcal{X}_i^t(s) \otimes \varphi(\tau_s),$$

where $\tau^s = \{t_k \in \tau : t_k < s\}$, $\tau_s = \{t_k \in \tau : t_k > s\}$, $\mathcal{X}_i^t(s) = 0$, $s > t$, $\mathcal{X}_i^t = \delta_i$, $s \leq t$ with $\delta_i = (\delta_i^k)$.

By denoting $\hat{A}^+(f_+,t) = f_+^i \hat{A}_i^+(t)$, $\hat{A}_-(f^-,t) = f_k^- \hat{A}_-^k(t)$, and $\hat{\Lambda}(e,t) = e_k^i \hat{A}_i^k(t)$, where $f_+ = (f_+^i) \in \mathbb{C}^m$ are m-columns, $f^- = (f_k^-) \in \mathbb{C}^m$ are m-rows, and $e = (e_k^i)$ are complex mxm-matrices, one obtains the canonical commutation relations [11] for the processes $\hat{A}^+$, $\hat{A}_-$ and $\hat{\Lambda}$, which we write in the following compact form

$$[\hat{A}(f,t), \hat{A}(g,t)] = \hat{A}([f,g], t) \tag{2.2}$$

of the operator representation in Fock space

$$\hat{A}(f,t) = \hat{A}_-^+(f_+^-,t)+\hat{A}^+(f_+,t) + \hat{A}_-(f^-,t) + \hat{\Lambda}(e,t) \tag{2.3}$$

with $\hat{A}_-^+(f_+^-,t) = f_+^- t\hat{1}$ for the Lie $\star$-algebra of block-matrices

$$f = \begin{pmatrix} 0 & f^- & f_+^- \\ 0 & e & f_+ \\ 0 & 0 & 0 \end{pmatrix}, \qquad f^\star = \begin{pmatrix} 0&0&1\\0&1&0\\1&0&0 \end{pmatrix} f^* \begin{pmatrix} 0&0&1\\0&1&0\\1&0&0 \end{pmatrix} = \begin{pmatrix} 0 & f_+^* & f_+^{-*} \\ 0 & e^* & f^{-*} \\ 0 & 0 & 0. \end{pmatrix} \tag{2.4}$$

Here f_+^- is a complex multiplier of the identity operator $\hat{1}$ in the Fock space, $f_+^{-\,*} = \bar{f}_+^-$, $f^{\star -} = f_+{}^*$ is the m-row with $f^{\star -}_k = \bar{f}_+^k$, $f^\star{}_+ = f^{-*}$ is the m-column with $f^{\star i}_+ = \bar{f}_k^-$, and $e^\star = e^*$ is mxm-matrix with $e^{\star i}_k = e^{k*}_i$, so f is an operator and $f^\star$ is a conjugated one in an indefinite space $\mathbb{C}^{1+m+1}$ of $\chi = (\chi_-, \chi_i, \chi_+)$ with the pseudo-scalar product $(\chi \mid \varphi) = \chi_- \cdot \varphi^- + \chi_j \cdot \varphi^j + \chi_+ \cdot \varphi^+$, where $\varphi^- = \bar{\varphi}_+$, $\varphi^k = \bar{\varphi}_k$, $\varphi^+ = \bar{\varphi}_-$. Due to the so defined conjugation one obtains the $\star$-property $\hat{A}(f,t)^* = \hat{A}(f^\star,t)$ for the Lie representation $\hat{A}(f,t) = f^\lambda_\nu \hat{A}^\nu_\lambda(t)$, where $\lambda,\nu \in \{-,1,\dots,m,+\}$ are the summation indexes of the 1+m+1-dimensional indefinite space.

The multiplication table for Ito differentials $d\hat{A}^\nu_\lambda(t) = \hat{A}^\nu_\lambda(t+dt) - \hat{A}^\nu_\lambda(t)$ of the processes $\hat{A}_-^+(t) = t\hat{1}$, $\hat{A}_i^+(t)$, $\hat{A}_-^k(t)$ and $\hat{\Lambda}(t)$ in Hudson-Parthasarathy quantum stochastic calculus can be written simply in terms of $\hat{A}(f,dt) = f^\lambda_\nu d\hat{A}^\nu_\lambda(t)$ as:

$$\hat{A}(f,dt)\hat{A}(g,dt) = \hat{A}(fg,dt) \,. \tag{2.5}$$

Now let us consider an initial Hilbert space $\mathcal{H}^0 = \mathcal{K}$, denoting $\mathcal{H}^t = \mathcal{K} \otimes \mathcal{F}^t$ and $\hat{F}^t : \mathcal{H}^t \to \mathcal{H}^t$ the corresponding components of an adapted process $\hat{F}(t) = \hat{F}^t \otimes \hat{1}_t$, and let us regard the operators $A(t) = A^t \otimes \hat{1}_t$ acting in $\mathcal{H} = \mathcal{K} \otimes \mathcal{F}$ by multiplying each $\hat{A}^\lambda_\nu(t)$ on the identity operator I in $\mathcal{K}$, so that the quantum stochastic integral $\int_0^t A(\hat{F}(s),ds)$ is defined as in [11] by the sum:

$$\int_0^t \hat{F}^\lambda_\nu dA^\nu_\lambda = \int_0^t (\hat{F}_+^- ds + \hat{F}^i_+ dA^+_i + \hat{F}_k{}^- dA^k_- + \hat{E}^i_k d\Lambda^k_i), \tag{2.6}$$

$\Lambda^k_i \equiv A^k_i$ for adopted weakly measurable locally square-integrable functions $t \mapsto \hat{F}^{\lambda}_{\nu}(t)$ with $\hat{F}^{\lambda}_{\nu}(t)$ acting in $\mathcal{H}$ for $\lambda \in \{-, i\}$, $\nu \in \{k, +\}$, $i, k = 1, \dots, m$, and $\hat{F}^{\lambda}_{\nu}(t) = 0$ for other indexes λ, ν. The Ito formula [11] defines the product $\hat{X}(t)\hat{Y}(t)$ of integrals $\hat{X}(t) = \int_0^t \hat{D}^{\lambda}_{\nu} dA^{\nu}_{\lambda}$ and $\hat{Y}(t) = \int_0^t \hat{F}^{\lambda}_{\nu} dA^{\nu}_{\lambda}$ in the form of an integral $\int_0^t d(\hat{X}\hat{Y})$ over:

$$d(\hat{X}\hat{Y}) = (\hat{D}^{\lambda}_{\nu}\hat{Y} + \hat{X}\hat{F}^{\lambda}_{\nu} + \hat{D}^{\lambda}_{\mu}\hat{F}^{\mu}_{\nu})dA^{\nu}_{\lambda} = d\hat{X}\hat{Y} + \hat{X}d\hat{Y} + d\hat{X}d\hat{Y} \, . \qquad (2.7)$$

The unitary quantum stochastic evolution $U(t): \psi \otimes \varphi \mapsto U(t)(\psi \otimes \varphi)$ of the combined system consisting of pure states $\psi \in \mathcal{K}$ and a Boson reservoir is described by the quantum stochastic differential equation (QSDE):

$$dU(t) = \hat{L}^{\lambda}_{\nu}(t)U(t)dA^{\nu}_{\lambda}(t) \equiv A(\hat{L}(t)U(t), dt) \qquad (2.8)$$

with $U(0) = I \otimes \hat{1} \equiv \hat{I}$ and adopted processes $\hat{L}^{\lambda}_{\nu}(t) = \hat{L}^{t\lambda}_{\nu} \otimes \hat{1}$, ($\hat{L}^{\lambda}_{\nu} = 0$, if $\lambda = +$ or $\nu = -$) satisfying the following conditions.

<u>THEOREM</u> 1. If $U(t)$ is a unitary process in the Hilbert space $\mathcal{H}$: $U(t)^{*}U(t) = \hat{I} = U(t)U(t)^{*}$, then $\hat{Z}^{t} = \hat{L}^{t} + \hat{I} \otimes \delta$ forms a pseudounitary block-operator $\hat{Z} = (\hat{Z}^{\lambda}_{\nu})$ in the indefinite space $\mathcal{H} \otimes \mathbb{C}^{1+m+1}$ for each t : $\hat{Z}^{\star}\hat{Z} = \hat{I} \otimes \delta = \hat{Z}\hat{Z}^{\star}$, where $\delta = (\delta^{\lambda}_{\nu})$, or in terms of $\hat{L}$:

$$\hat{L}^{\star\lambda}_{\nu} + \hat{L}^{\lambda}_{\nu} + \hat{L}^{\star\lambda}_{\mu}\hat{L}^{\mu}_{\nu} = 0 = \hat{L}^{\lambda}_{\nu} + \hat{L}^{\star\lambda}_{\nu} + \hat{L}^{\lambda}_{\mu}\hat{L}^{\star\mu}_{\nu} \qquad (2.9)$$

for all λ, ν and t. The conditions (2.9) are not only necessary but also sufficient for the existence of the unitary solution of (2.6) at least in case of local boundedness of the operator-valued functions $t \mapsto \hat{L}^{t}$.

<u>PROOF</u>. (2.9) follows from the Ito formula of (2.8) as $U(t) = U^{t} \otimes \hat{1}_{t}$, $U^{t} = J^{\star}_{t} : \overleftarrow{\exp} \int_0^t \hat{L}^{s} ds : J_{t}$ where $J_{t}(\tau_{-}, \tau, \tau_{+}) = \delta_{\emptyset}(\tau_{-})\psi(\tau)$ is the isometry: $\mathcal{H}^{t} \to \mathcal{K} \otimes \Gamma(\mathcal{G}^{t})$, $\mathcal{G}^{t}$ is the indefinite space of square-integrable 1+m+1-dimensional functions $s \mapsto (\varphi^{\nu}_{s})$, $\nu = -, 1, \dots, m, +$, with $\varphi^{\nu}_{s} = 0$ for $s > t$, $\Gamma(\mathcal{G}^{t})$ is the space of tensor-valued functions $\varphi(\tau_{-}, \tau, \tau_{+}) = (\varphi^{k_1 \dots k_n}_{t_1 \dots t_n})(\tau_{-}, \tau_{+}) \equiv \varphi(\tau_{-}, \tau_{+})(\tau)$ with the pseudo-Eucleadian metric defined by the integral:

$$\alpha|\varphi) = \iint \langle \mathcal{X}(\tau_{+}, \tau_{-}) \mid \varphi(\tau_{-}, \tau_{+}) \rangle d\tau_{-} d\tau_{+}$$

over the finite subsets $\tau_{-}, \tau_{+} \subset [0, t]$, $d\tau = dt_1 \dots dt_n$ for $\tau = (t_1 < \dots \dots < t_n)$, $J^{\star}_{t}\varphi(\tau) = \int \varphi(\tau_{-}, \tau, \emptyset) d\tau_{-}$, and $:\overleftarrow{\exp} \int_0^t \hat{L}(s) ds:$ is the natural representation on $\mathcal{K}\otimes\Gamma(\mathcal{G}^t)$ of recursively defined chronological products

$$\hat{Z}\,{}^{\lambda_1\dots\lambda_n\lambda}_{\nu_1\dots\nu_n\nu}(s_1,\dots,s_n,s) = \hat{Z}^s{}^{\lambda}_{\nu}\,\hat{Z}\,{}^{\lambda_1\dots\lambda_n}_{\nu_1\dots\nu_n}(s_1\dots s_n).$$

Now let us consider the output canonical processes $B^{\lambda}_{\nu}(t)=U(t)^*\cdot A^{\lambda}_{\nu}(t)U(t)$ and define the integral $\int_0^t B(F(s),ds)$ as the integral (2.6) with $\hat{F}(t) = U(t)F(t)U(t)^*$ in the Heisenberg picture:

$$\int_0^t B(F(s),ds) = U(t)^*\int_0^t A(\hat{F}(s),ds)\,U(t), \text{ if } [\hat{F}^{*}_{\mu}(s),\hat{L}^{\lambda}_{\nu}(t)] = 0,\ \forall s\leq t. \quad (2.10)$$

<u>THEOREM</u> 2. The integral (2.10) with adopted F(s) can be represented as the integral (2.6) in the form $\int_0^t A(Z(s)^*F(s)Z(s),ds)$,

where $Z^{\lambda}_{\nu}(t) = L^{\lambda}_{\nu}(t) + \hat{I}\,\delta^{\lambda}_{\nu}$, $L(t)= U(t)^*\hat{L}(t)U(t)$, so

$$Z^*FZ = L^*FL + L^*F + FL + F\,. \quad (2.11)$$

<u>PROOF</u>. It is a direct consequence of the Ito formula (2.7).

Let us denote $X(t) = X(0) + \int_0^t A(D(s),ds)$ a quantum process, defined by the initial operator $X(0) = X\otimes\hat{1}$ with $X\in\mathcal{B}(\mathcal{K})$ and adapted $D^{\lambda}_{\nu}(t)$, and $Y(t) = Y(0) + \int_0^t B(F(s),ds)$ an output self-adjoint process, defined by the differential:

$$dY(t) = A(Z(t)^*F(t)Z(t),dt) \quad (2.12)$$

with $F(t)^* = F(t)$, $Y(0)^* = Y(0)$, where $Z^{\lambda}_{\nu}(t)$ are assumed to be nondemolished by $Y(t)$: $[Y(s),Z^{\lambda}_{\nu}(t)] = 0$ for all $s\leq t, \lambda, \nu$, as it is in the case $L^{t}{}^{\lambda}_{\nu} = \hat{1}^t\otimes L^{\lambda}_{\nu}$, $\hat{F}^{s}{}^{\varkappa}_{\mu} = \hat{f}^{s}{}^{\varkappa}_{\mu}\otimes I$, $Y(0) = c\,\hat{1}$.

<u>THEOREM 3</u>. The output process (2.12) is nondemolished with respect to the process X(t) iff $[X(0),Y(0)] = 0$, and

$$D(t) = Z(t)^*G(t)Z(t) - X(t)\otimes\delta\,, \quad (2.13)$$

the block operators $G(t) = (G^{\lambda}_{\nu}(t))$ commute with $F(t) = (F^{\lambda}_{\nu}(t))$ and all $Y(s)\otimes\delta = (Y(s)\,\delta^{\lambda}_{\nu})$, $s\leq t$.

<u>PROOF</u>. It is an application of the Ito formula (2.7) to $d[X(t),Y(t)] = 0$ and $[dX(t),Y(t)] = 0$.

Now we consider a family $\{Y_i(t),\ i=1,\dots,m\}$ of output self-adjoint and nondemolition processes, defined by (2.12) with $F_i(t) = F_i(t)^*$, satisfying the following conditions.

THEOREM 4. A family Y_i of the processes (2.12) is self-nondemolition: $[Y_i(s),Y_k(t)] = 0$ for all i,k and s,t, iff the block operators $F_i(t)$ commute $[F_i(t),F_k(t)] = 0$ for all i,k,t and with $Y_j(s)\delta$ for all $s \leq t$, $j=1,\dots,m$.

PROOF. It is a consequence of the Proof of Theorem 3.

Note, that if the process $F_{i\,\nu}^{\;\lambda}(t)$ commute with $Z_\mu^*(t)$ as it takes place in case of the operators $F_{i\,\nu}^{\;\lambda}$ affiliated with the algebra $\mathcal{B}^t = \{Y_i(s)\,|\,s \leq t,\ i=1,\dots,m\}''$, then the output integrals $\int_0^t B(F(s),ds)$ can be represented as in (2.6) by

$$\int_0^t F_\nu^\lambda\, dB_\lambda^\nu = \int_0^t F_+^- ds + F_+^i dB_i^+ + F_k^- dB_-^k + E_k^i dN_i^k , \qquad (2.13)'$$

$E_k^i \equiv F_k^i$, where $B_i^+(t)$ are creation, $B_-^k(t)$ are annihilation and $N_i^k(t) = B_i^k(t)$ are preservation output processes, defined as

$$dB_i^+(t)^* = Z_+^i(t)dt + V_k^i(t)dA_-^k(t) = dB_-^i(t)\ ,\quad V_k^i \equiv Z_k^i$$

$$dN_i^k = Z_+^{i*}Z_+^k dt + Z_l^{i*}Z_+^k dA_l^+ + Z_+^{i*}Z_j^k dA_-^j + Z_l^{i*}Z_j^k d\Lambda_l^j\ . \qquad (2.14)'$$

3. Stochastic calculus of quantum open systems.

Let $\mathcal{A} = \mathcal{B}(\mathcal{K})$ be the algebra of bounded operators describing an open dynamic system by a family $\iota = (\iota_t)$ of normal representations $\iota_t : \mathcal{A} \to \mathcal{B}(\mathcal{K} \otimes \mathcal{F})$. We assume that the family ι is generated by linear w^*-continuous maps α_k^i, $\beta_+^i = \beta^i = \beta_i^{-*}$, $\gamma : \mathcal{A} \to \mathcal{B}(\mathcal{K} \otimes \mathcal{F})$ defining a qua tum Ito equation [11,12]:

$$d\,\iota(X) = \alpha_k^i(X)d\Lambda_i^k + \beta_+^i(X)d\,A_i^+ + \beta_k^-(X)dA_-^k + \gamma(X)dt, \qquad (3.1)$$

with respect to the standard gauge $\Lambda_i^k(t)$, Bose annihilation $A_-^k(t)$ and Bose creation $A_i^+(t)$ processes in $\mathcal{K} \otimes \mathcal{F}$, the increments of which satisfy the multiplication table

$$d\Lambda_i^k d\Lambda_j^l = \delta_j^k d\Lambda_i^l,\quad dA_-^k d\Lambda_i^l = \delta_i^k dA_-^l,\quad d\Lambda_i^k dA_j^+ = \delta_j^k dA_i^+,\quad dA_-^k dA_i^+ = \delta_i^k \hat{I} dt$$

with zero products, where $\hat{I} = I \otimes \hat{1}$ is the identity operator in $\mathcal{H}$.

THEOREM 5. Equation (3.1) with the initial condition $\iota_0(X) = X \otimes \hat{1}$ for all $X = X^*$ generates an open adapted dynamic system $(\mathcal{A}, \iota)$, $\iota_t(X) = \iota^t(X) \otimes \hat{1}_t$, iff the adapted weakly measurable and locally square-integrable maps $t \mapsto \alpha_{t\,k}^{\,i}(X)$, $\beta_{t\,+}^{\,i}(X) = \beta_{t\,\bar{i}}(X)$, $\gamma_t(X)$ depend on X in the following way:

$$\alpha_t(X) = \varkappa_t(X) - \hat{\delta}\otimes\iota_t(X), \qquad \beta_t(X) = \varkappa_t(X)\hat{Z}_t - \hat{Z}_t\,\iota_t(X)$$
$$\gamma_t(X) = \hat{Z}_t^*\,\varkappa_t(X)\hat{Z}_t - \iota_t(X)(\tfrac{1}{2}\hat{Z}_t^*\hat{Z}_t + i\hat{H}_t) - (\tfrac{1}{2}\hat{Z}_t^*\hat{Z}_t - i\hat{H}_t)\iota_t(X) \tag{3.2}$$

Here $\varkappa_t(X) = \hat{V}_t^*\hat{\delta}\otimes\iota_t(X)\hat{V}_t$, $\hat{V}_t^* = \hat{V}_t^{-1}$, $\hat{Z}_t$ and $\hat{H}_t = \hat{H}_t^*$ are defined uniquely by some adapted operator-valued functions $V_t^* = V_t^{-1}$, Z_t and $H_t = H_t^*$ up to operators $\hat{v}_t$, $\hat{z}_t$ and $\hat{h}_t$ from the commutant $\mathcal{A}'_t$ of $\iota_t(\mathcal{A})$:

$$\hat{V}_t = \hat{v}_t V_t\,, \quad \hat{Z}_t = V_t^*(Z_t + \hat{z}_t)\,, \quad \hat{H}_t = H_t + \operatorname{Im}\hat{z}_t^* Z_t + \hat{h}_t\,. \tag{3.3}$$

The proof of the theorem is based on the quantum Ito formula. Note, that in the Markovian stationary case $\alpha_{k\,t}^{\,i}$, $\beta_t^{\,i}$ and γ_t are the maps $\mathcal{A} \to \iota_t(\mathcal{A})$, and V_t, Z_t, H_t can be defined as images of some operators $V^* = V^{-1}$, Z and $H = H^*$ from $\mathcal{A}$:

$$V_{tk}^{\,i} = \iota_t(V_k^i), \quad Z_t^{\,i} = \iota_t(Z^i), \quad H_t = \iota_t(H)\,. \tag{3.4}$$

Let us regard a quantum process $Y = (Y_t)$ in $\mathcal{H}$, which is adapted in the sense that $Y_t = Y_t^*$ is affiliated with $\mathcal{B}(\mathcal{H}^t)\otimes\hat{1}_t$, and nondemolishing (QND) with respect to $(\mathcal{A}, \iota)$: $X_s Y_t = Y_t X_s$ for all $t \leq s$ and $X_s \in \iota_s(\mathcal{A})$. We assume that the process Y is generated by the quantum Ito equation:

$$dY_t = E_t d\Lambda_t + 2\operatorname{Re}F_t dA_t^+ + G_t dt \equiv E_{tk}^{\,i}\,d\Lambda_i^k(t) + F_t^{\,i} dA_i^+(t) + F_{t\,k}^* dA_-^k(t) + G_t dt \tag{3.5}$$

with the corresponding (adapted, local square-integrable) operator-valued functions $t \mapsto E_t$, F_t, G_t. The following theorem is a consequence of the QND principle and the Ito formula.

THEOREM 6. The adapted process Y_t, satisfying equation (3.4) is QND with respect to the system (3.1), iff

1) it is QND with respect to the processes $\hat{V}_t$, $\hat{Z}_t$ and $\hat{H}_t$ up to $\hat{v}_t$, $\hat{z}_t$ and $\hat{h}_t \in \mathcal{A}'_t$, i.e. if

$$[V_s, Y_t] = 0, \; [Z_s, Y_t] = 0, \quad [H_s, Y_t] = 0, \quad s \geq t \tag{3.6}$$

for the operators V_t, Z_t and H_t, defining (3.2) by (3.3), $Y_o = y\hat{I}$ for $y \in \mathbb{R}$, and

2) $V_t E_t V_t^* = \hat{a}_t, \quad V_t F_t = \hat{a}_t Z_t + \hat{b}_t, \quad G_t = Z_t^* \hat{a} Z_t + 2\mathrm{Re}\hat{b}_t^* Z_t + \hat{c}_t, \quad t \in \mathbb{R}_+,$ (3.7)

where $\hat{a}_t = \hat{a}_t^*$, $\hat{b}_t$ and $\hat{c}_t = \hat{c}_t^*$ are adapted and commuting with $X_t \in \mathcal{A}_t$ for all t :

$$[a_t, X_t] = 0, \quad [b_t, X_t] = 0, \quad [c_t, X_t] = 0 . \tag{3.8}$$

Condition (3.6) for the Markovian case (3.4) can be omitted. The discovered structure (3.6) of QND processes (3.5) allows us to consider the following three basic types of QND observations:
a QND counting observation $Y_i(t) = N_i(t)$, where

$$dN_i = V_k^{i*} V_l^i d\Lambda_k^l + 2\mathrm{Re} V_k^{i*} Z^i dA_k^+ + Z^{i*} Z^i dt, \qquad N_i(0) = 0 \tag{3.9}$$

($a = \delta_i \hat{I}$, $b = 0$, $c = 0$), a QND diffusion observation $Y_i(t) = Q_i(t), Q_i(t) = 2\mathrm{Re}B_i^+(t)$ where $B_i^+(t) = B_-^i(t)^* \equiv B_i(t)$,

$$dB_-^i = V_k^i dA_-^k + Z^i dt, \quad dB_i^+ = V_i^{*k} dA_k^+ + Z_i dt, \quad B_-^i(0) = 0 = B_i^+(0) \tag{3.10}$$

($\hat{a} = 0$, $\hat{b} = \delta_i \hat{I}$, $\hat{c} = 0$), and a QND time observation $Y_t = t\hat{I}$ ($\hat{a} = 0$, $\hat{b} = 0$, $\hat{c} = \hat{I}$). As it follows from the next theorem, the basic QND processes $N_i(t)$, $Q_i(t)$, $t\hat{I}$ are commutative, but not mutually commutative with the multiplication table for their increments reads

$$dN_i dN_i = dN_i, \quad dQ_i dN_i = dB_i, \quad dN_i dQ_i = dB_i^*, \quad dQ_i dQ_i = \hat{I} dt$$

and zero other products.

<u>THEOREM</u> 7. A family $Y = (Y_1, \ldots, Y_m)$ of QND processes $Y_i(t)$ satisfying Ito equations (3.4) is mutually commutative $[Y_i(t), Y_k(s)] = 0$, iff the operators $\hat{a}_i$, $\hat{b}_i$, $\hat{c}_i$ defining by (3.7) the coefficients E_i, F_i, G_i satisfy the QND conditions:

$$[\hat{a}_i(s), Y_k(t)] = 0, \; [\hat{b}_i(s), Y_k(t)] = 0, \; [\hat{c}_i(s), Y_k(t)] = 0, \; s > t \tag{3.11}$$

$$\hat{a}_i \hat{a}_k = \hat{a}_k \hat{a}_i, \quad \hat{a}_i \hat{b}_k = \hat{a}_k \hat{b}_i, \quad \hat{b}_i \hat{b}_k = \hat{b}_k \hat{b}_i, \qquad t \in \mathbb{R}_+ \tag{3.12}$$

Moreover, the adapted nondemolished process Z_t in (3.7) defining (3.2) by (3.3) can be chosen in such a way that $\hat{a}_i(t)\hat{b}_i(t)=0$ for all i and t. If $\hat{a}_i(t)$, $\hat{b}_i(t)$ and $\hat{c}_i(t)$ commute with Z_t for all t, as it can be done in Markovian case (3.4), then the processes $Y_i(t)$ satisfy the equivalent Ito equation:

$$dY_i(t) = \hat{a}_i(t)dN_t + 2\mathrm{Re}\hat{b}_i(t)dB_t^+ + \hat{c}_i(t)dt, \tag{3.13}$$

$Y_i(0) = y_i I$ with respect to the canonical QND processes (3.9), (3.10).

The family $Y(t) = (Y_1(t),\dots,Y_n(t))$ of mutually commutative QND processes (3.13) is called QND filter, if the operators $\hat{a}_i(t)$, $\hat{b}_i(t)$ and $\hat{c}_i(t)$ are affiliated with the Abelian algebra $\mathcal{B}_t = \{Y(s)\}''_{s\leq t}$ for each t. In that case of $\hat{a}_i(t)$, $\hat{b}_i(t)$ and $\hat{c}_i(t)$ defined by the corresponding functions $a_i(v^t)$, $b_i(v^t)$ and $c_i(v^t)$ on the trajectory space $\mathcal{V}^t$ of the observed values $v(s)$ of $Y(s)$, $s\leq t$, we have:

$$a_i(t) = \int^{\oplus} a_i(v^t)I_{v^t}, \quad b_i(t) = \int^{\oplus} b_i(v^t)I_{v^t}, \quad c_i(t) = \int^{\oplus} c_i(v^t)I_{v^t},$$

where $I = \int^{\oplus} I_{v^t}$ is the orthogonal identity resolution for $Y^t = \{Y(s) \mid s\leq t\}$. The condition $e_i^k(t)f_+^i(t) = 0$ meaning either $e_i^k(v^t)=0$ or $f_+^i(v^t)=0$ decompose a QND filter on a jumping and continuous orthogonal parts:

$$dY_i(v^t) = e_i(v^t)dN_t(v^t) + f_i(v^t)dt, \qquad v^t \in \mathcal{V}_{\parallel}^t$$

$$dY_i(v^t) = 2\mathrm{Re}b_i(v^t)dB_t^+(v^t) + c_i(v^t)dt, \qquad v^t \in \mathcal{V}_{\perp}^t$$

where $\mathcal{V}^t = \mathcal{V}_{\parallel}^t \cup \mathcal{V}_{\perp}^t$ is a measurable disjunction for each i and t and the increments $dN_t(v^t)$, $dB_t(v^t)$, $dB_t(v^t)^*$ are defined in (3.9), (3.10) by the corresponding components of the decompositions:

$$V_t = \int^{\oplus} V(v^t), \quad Z_t = \int^{\oplus} Z(v^t), \quad H_t = \int^{\oplus} H(v^t).$$

Note that as it follows from (3.12), $b_i(v^t) = 0$ for all i and fixed v^t provided that $a_i(v^t) \neq 0$ for such v^t and some i, and $a_i(v^t) = 0$ for all i and fixed v^t, provided that $b_i(v^t) \neq 0$ for such v^t and some i so that the subsets $\mathcal{V}_{\parallel}$ and $\mathcal{V}_{\perp}$ do not depend on t.

4. QND filtering and optimal control in Fock space.

Let us consider an initial normal state φ on $\mathcal{A}=\mathcal{B}(\mathcal{K})$ and vacuum state ω on $\mathcal{B}(\mathcal{F})$, defined by normalized vectors $\psi\in\mathcal{K}$ and $\delta_\emptyset\in\mathcal{F}$, denote by ω_r^s, ω^t the corresponding states on $\mathcal{N}_r^s=\mathcal{B}(\mathcal{F}_r^s)$, $\mathcal{N}_t=\mathcal{B}(\mathcal{F}_t)$ and $\omega^t=\omega_0^t$ on $\mathcal{N}^t=\mathcal{N}_0^t$. The open dynamic system $(\mathcal{A},\iota)$ defined by a family $\iota=(\iota_t)$ of representations $\iota_t:\mathcal{A}\to\mathcal{B}(\mathcal{K}\otimes\mathcal{F})$ is a quantum stochastic system with respect to the filtration $\{\mathcal{A}\otimes\mathcal{N}^t\}$, and the nondemolition process, defined by (3.4)-(3.8) is a regular semimartingal with respect to $\{\mathcal{A}\otimes\mathcal{N}^t\}$. We denote by $\langle X\rangle_t$ the conditional expectations on the von Neumann decomposable algebras $\mathcal{A}_t\otimes\mathcal{B}^t$ with respect to their central subalgebras $\mathcal{B}^t$, which are defined by projectors E^t on subspaces $\mathcal{K}^t\subseteq\mathcal{K}\otimes\mathcal{F}^t{}_t$, generated by the action of $\mathcal{B}^t$ on $\psi\otimes\delta_\emptyset$: $\varepsilon^t(X)=E^tX\xi$, for all $\xi\in\mathcal{K}^t$, and $\hat{\varphi}_t=\varepsilon^t\circ\iota_t$ the posterior state on $\mathcal{A}$, identified with the density matrix due to its normality.

THEOREM 8. Let $(\mathcal{A},\iota,N)$ be a quantum stochastic process (3.1) with counting nondemolition measurement (3.9), and $\langle Z_t^*Z_t\rangle_t\in\mathcal{B}^t$ is invertable for all t. Then the posterior mean value $\hat{q}_t=\langle\iota_t(X)\rangle_t$ satisfies the following stochastic filtration equation:

$$d\langle\iota_t(X)\rangle_t=\langle\delta_t(X)\rangle_t d\tilde{N}_t+\langle\gamma_t(X)\rangle_t dt, \tag{4.1}$$

where $d\tilde{N}_t=dN_t-\langle Z_t^*Z_t\rangle_t dt$,

$$\delta_t(X)=Z_t^*\,\iota_t(X)Z_t/\langle Z_t^*Z_t\rangle_t-\iota_t(X)\,. \tag{4.2}$$

In the Markovian case (3.4) the posterior density matrix $\hat{\varphi}_t$ satisfies the recursive stochastic equation in Ito sense:

$$d\hat{\varphi}_t+(K\hat{\varphi}_t+\hat{\varphi}_tK^*-Z\hat{\varphi}_tZ^*)dt=(Z\hat{\varphi}_tZ^*/\widehat{Z^*Z}-\hat{\varphi}_t)d\tilde{N}_t, \tag{4.3}$$

where $K=iH+Z^*Z/2$, $\widehat{Z^*Z}=\hat{\varphi}_t(Z^*Z)$, which has the solution $\hat{\varphi}_t=\hat{\psi}_t\otimes\hat{\psi}_t^*$ for $\varphi_0=\psi\otimes\psi^*$, if $\hat{\psi}_t$ satisfies the nonlinear Ito equation:

$$d\hat{\psi}_t+(i\tilde{H}_t+\tilde{Z}_t^*\tilde{Z}_t/2)\hat{\psi}_t dt=\tilde{Z}_t\hat{\psi}_t d\tilde{N}_t/\|Z\hat{\psi}_t\|,\qquad\hat{\psi}_0=\psi, \tag{4.4}$$

where $\tilde{H}_t=H-(\hat{\psi}_t|Z^*Z\hat{\psi}_t)^{1/2}\mathrm{Im}Z$, $\tilde{Z}_t=Z-(\hat{\psi}_t|Z^*Z\hat{\psi}_t)^{1/2}$.

THEOREM 9. Let $(\mathcal{A},\iota,Q)$ be a quantum stochastic process (3.1) with diffusion nondemolition measurement $Q_t=2\mathrm{Re}B_t$, defined by (3.10).

Then the posterior mean value $\hat{q}_t = \langle \iota_t(X)\rangle_t$ satisfies the following stochastic filtration equation:

$$d\,\langle \iota_t(X)\rangle_t = \langle \Theta_t(X)\rangle_t d\tilde{Q}_t + \langle \Upsilon_t(X)\rangle_t dt, \qquad (4.5)$$

where $d\tilde{Q}_t = dQ_t - \langle 2\mathrm{Re}Z_t\rangle_t dt$,

$$\Theta_t(X) = 2\mathrm{Re}\,\iota_t(X)(Z_t - \langle Z_t\rangle_t). \qquad (4.6)$$

In the Markovian case (3.4) the posterior density matrix satisfies the recursive stochastic equation in Ito sense:

$$d\hat{\phi}_t + (K\hat{\phi}_t + \hat{\phi}_t K^* - Z\hat{\phi}_t Z^*)dt = 2\mathrm{Re}(Z - \hat{z}_t)\hat{\phi}_t d\tilde{Q}_t \;, \qquad (4.7)$$

where $K = iH + Z^*Z/2$, $\hat{z}_t = \hat{\phi}_t(Z)$, which has a solution $\hat{\phi}_t = \hat{\psi}_t \otimes \hat{\psi}_t^*$ for $\hat{\phi}_o = \psi\otimes\psi^*$, if $\hat{\psi}_t$ satisfies the nonlinear Ito equation:

$$d\hat{\psi}_t + (i\tilde{H}_t + \tilde{Z}_t^*\tilde{Z}_t/2)\hat{\psi}_t\,dt = \tilde{Z}_t\hat{\psi}_t d\tilde{Q}_t, \qquad \hat{\psi}_o = \psi \;,$$

where $\tilde{H}_t = H - (\hat{\psi}_t|\mathrm{Re}Z_t\,\hat{\psi}_t)\mathrm{Im}Z$, $\tilde{Z}_t = Z - (\hat{\psi}_t|\mathrm{Re}Z_t\,\hat{\psi}_t)$.

The linear continuous filtration for the Gaussian ϕ_o and canonical Z was considered in [1-3], and the general formulation of quantum nonlinear filtration for a quantum Markovian partially observable controlled objects in operational approach was given in [16].

Let us consider a quantum controlled process over the algebra $\mathcal{A} = \mathcal{B}(\mathcal{K})$ described by the family of normal representations $\iota_t : \mathcal{A} \to \mathcal{A}\otimes\mathcal{N}^t\otimes\mathcal{C}^t$, where $\mathcal{C}_t$ is a C^*-algebra of continuous functions $\mathcal{U}^t \to \mathbb{C}$, $\mathcal{N}^t = \mathcal{B}(\mathcal{F}^t)$ for the Fock space $\mathcal{F}^t$. Let $\mathcal{U}_r^s \subseteq \times_{r\leq t<s}\mathcal{U}(t)$ be a Hausdorf space of controlling processes $u_r^s = \{u(t)|\,r\leq t<s\}$ such that $\mathcal{U}^t \times \mathcal{U}_t^s = \mathcal{U}^s$ for all $t,s>0$, where $u(t)\in\mathcal{U}(t)$, $\mathcal{U}^t = \mathcal{U}_o^t \ni u^t$ and $\mathcal{U}_t = \mathcal{U}_t^\infty \ni u_t$, $\mathcal{U} = \mathcal{U}_o$. We consider a quantum controlled process $X_t(u^t) = \iota_t(u^t, X)$ over the algebra $\mathcal{A}\ni X$, $u^t\in\mathcal{U}^t$, described by the Hudson-Parthasarathy dynamic equation:

$$dX_t = \hat{A}_t d\Lambda_t + 2\mathrm{Re}\hat{B}_t dA_t^* + \hat{C}_t(u(t))dt, \qquad X_o = X\otimes\hat{1}\,, \qquad (4.8)$$

where $\hat{A}_t(u^t) = \alpha_t(X)(u^t)$, $B_t(u^t) = \beta_t(X)(u^t)$, $C_t(u^t, u(t)) = \Upsilon_t(X, u(t))(u^t)$ are defined in a standard way (3.2), (3.3) by the operator-valued continuous adapted functions $V_t(u^t) = \iota_t(u^t, V)$, $Z_t(u^t) = \iota_t(u^t, Z)$, $H_t(u^t) = \iota_t(u^t, H)$ with unitary $V\in\mathcal{A}$, $Z\in\mathcal{A}$ and self-adjoint $H(u(t))\in\mathcal{A}$. We shall assume that the control process

$\dot{u}(t)$ is defined by strategies $u_t = d_t(w^t, v_t) = \{\mathcal{d}_t(s, w^t, v_t^s, v(s)) | s \geq t\}$, where $w^t = (u^t, v^t)$, $v^t = v_o^t$, $v_r^s = \{v(t) | r \leq t < s\}$ are the results of non-demolition measurements $v(t)$ on the interval $[r, s[$, $v_t = v_t^\infty$, described by a commutative process $Y(t)$, satisfying the equations either $dY(t) = \hat{a}(t)dN_t + \hat{c}(t)dt$, or $dY(t) = 2\mathrm{Re}\, \hat{b}(t)dB_t + \hat{c}(t)dt$ with invertable $\hat{a}(t, u^t)$, $\hat{b}(t, u^t)$, $\hat{c}(t, u^t) \in \mathcal{B}^t = \{Y(s) | s \leq t\}''$ defined by the corresponding real-valued functions $a(w^t, v(t))$, $b(w^t, v(t))$, $c(w^t, v(t))$.

Let us consider the optimal control problem with the operator-valued risk $u \in \mathcal{U} \longmapsto R_t(u) \in \mathcal{A}_t(u) = \{ \iota_s(\mathcal{A}, u^s) \backslash s \geq t\}''$ satisfaing the equation

$$R_{t_o}(u) = \int_{t_2}^{t_1} S_t(u^t, u(t))dt + R_{t_1}(u), \tag{4.9}$$

where $S_t(u^t, u(t)) = \iota_t(S(u(t)), u^t)$ for a continuous $\mathcal{A}$-valued self-adjoint function $S(u(t)) = S(u(t))^*$. The optimal control strategy $\mathcal{d}_t^o$ of the extremal problem

$$\langle \phi \otimes \omega, R_t(u^t, \mathcal{d}_t(w^t, v_t)) \rangle = \inf, \tag{4.10}$$

where ϕ is an initial normal state on $\mathcal{A}$, and ω is the vacuum state on $\mathcal{N} = \mathcal{B}(\mathcal{F})$. This solution can be found by the dynamic programming method as a solution of the following Bellman continuous inverse-time equation.

<u>THEOREM</u> 10. Let $R(t, w^t, \mathcal{d}_t)$ be the averaged $\mathcal{A}$-valued risk uniquely defined for the strategy $\mathcal{d}_t$ by

$$\langle \phi \otimes \omega, \iota_t(u^t, R(t, w^t, \mathcal{d}_t)) \rangle = \langle \phi \otimes \omega, R_t(u^t, \mathcal{d}_t(w^t, v_t)) \rangle$$

due to the Markov condition for $X_t(u^t)$ with respect to $\omega = \omega^t \otimes \omega_t$, and

$$\hat{r}_t(w^t, \mathcal{d}_t) = \varepsilon^t[R_t(u^t, \mathcal{d}_t(w^t, v_t))] = \langle \hat{\phi}_t(w^t), R(t, w^t, \mathcal{d}_t) \rangle$$

be the posterior risk, corresponding to the strategy $\mathcal{d}_t$, where ε^t is the conditional expectation on $\mathcal{A}_t \vee \mathcal{B}^t$ with respect to the commutative algebra $\mathcal{B}^t$ and

$$\hat{\phi}_t(w^t, v(t)) = \varepsilon^t \circ \iota_t(u^t)(v^t)$$

be controlled a posterior state on $\mathcal{A}$ for a $w^t = (u^t, v^t)$. Then

$$\inf_t \langle \hat{\phi}_t(w^t, v(t)), R(t, w^t, d_t) \rangle = s(t, \hat{\phi}_t(w^t, v(t))),$$

where the functional $s(t,\phi)$ satisfies the following Bellman equation:

$$-\partial_t s(\phi) = \inf_{u(t)} \langle \phi , S(u(t)) + \Lambda(u(t),\delta)s(\phi)\rangle + \langle \phi, Z^+Z\rangle \Delta s(\phi) \quad (4.11)$$

in case of counting observation and

$$-\partial_t s(\phi) = \inf_{u(t)} \langle \phi , S(u(t)) + (u(t),\delta)s(\phi)\rangle + \tfrac{1}{2}\langle \phi, \Theta(\delta)\rangle^2 s(\phi) \quad (4.12)$$

in case of diffusion observation.

Here $\partial_t = \partial/\partial t$, $\delta = \delta/\delta\phi$, $\Delta s(\phi) = s(Z\phi Z^*/\langle \phi, Z^*Z\rangle) - s(\phi)$,

$$\Lambda(u(t), \delta)s(\phi) = \delta s(\phi) - 2\mathrm{Re}K(u(t))^* \delta s(\phi)$$

$$\Gamma(u(t), \delta)s(\phi) = Z^*\delta s(\phi) Z - 2\mathrm{Re}K(u(t))^* \delta s(\phi)$$

$$\Theta(\delta)s(\phi) = 2\mathrm{Re}(Z - \langle \phi, Z\rangle)^* \delta s(\phi)$$

and $K(u(t)), K(u(t))^*$ are defined by $\pm iK(u(t)) + Z^*Z/2$ and $\phi_t(w_t)$ is a posterior state on $\mathcal{A}$ for controlled and observed data $w^t = (u^t, v^t)$ satisfying the corresponding nonlinear filtering equations: either (4.3) or (4.7) written respectively in the form:

$$d\hat{\phi}_t + \hat{\phi}_t \Lambda(u(t))dt = \hat{\phi}_t \Delta\, dN_t \; ; \quad d\hat{\phi}_t + \hat{\phi}_t \Gamma(u(t))dt = \hat{\phi}_t \Theta\, dQ_t$$

where $\hat{\phi}_0 = \phi$.

The linear programming for Gaussian ϕ, canonical Z and quadratic $S(u(t))$ was considered in [3]. The general formulation of quantum dynamic programming for the partially observable controlled quantum objects in operational approach was given in [16].

References

1. V.P.Belavkin, Optimal quantum filtration of Markovian signals. Problems of Control and Information Theory, v.7(5),p.345-360 (1978).
2. V.P.Belavkin, Quantum filtering of Markovian signals with quantum white noises. Radiotechnika i Elektronika, v.25 (7), p.1445-1453 (1980) (in Russian).
3. V.P.Belavkin, Nondemolition measurement and control in quantum dynamic systems. In: Information complexity and control in quantum physics, ed. A.Blaquiere, S.Diner, G.Lochak, Springer-Verlag, Wien - New York, p.311-336 (1987).

4. V.P. Braginsky, The prospects for high sensitivity gravitational antennae. In: Gravitational Radiation and Gravitational Collapse, ed. by C.De Witt-Morette, p.28, Reidel, Dordrecht, 1974.
5. C.M.Caves, Quantum nondemolition measurements. In: Quantum Optics, Experimental Gravitation and Measurement Theory, ed. by P.Meystre and M.D.Sculy, Plenum Press, New York, 1982.
6. A.Barchielli, Continuous observation in quantum mechanics: An application to gravitational-wave detectors. Phys. Rev. D, v.32 (2), p.347-367 (1984).
7. A.S.Holevo, Quantum estimation. In: Advances in Statistical Signal Processing, v.1, p.157-202 (1987).
8. V.P.Belavkin, Reconstruction theory for a quantum stochastic process. Theor. Math. Phys., v.62 (3), p.275-289 (1985).
9. A.Barchielli, L.Lanzand and G.M.Prosperi, Statistics of continuous trajectories in quantum mechanics: Operator-valued Stochastic Processes. Foundation of Physics, 13 (18), p.779-812 (1983).
10. A.Barchielli and G.Lupieri, Quantum stochastic calculus, operator-valued stochastic processes and continual measurements in quantum mechanics, J. Math. Phys. (1985).
11. R.S.Hudson and K.R.Parthasarathy, Quantum Ito's formula and stochastic evolution, Commun. Math. Phys., 93, 301 (1984).
12. H.Maassen, Quantum Markov processes on Fock space described by integral kernels.In:Quantum Probability and Applications II, ed. by L.Accardi and W.von Waldenfelds, Springer LNM 1136 (1985).
13. J.W.Clark, C.K.Ong, T.J.Tarn and G.M.Huang, Quantum nondemolition filters. Math. Systems Theory, 18, p.33-55 (1985).
14. J.W.Clark,and T.J.Tarn, Quantum nondemolition filtering. In: Information and Control in Quantum Physics, ed. by A.Blaquiere, S.Diner, G.Lochak, Springer-Verlag, Wien-New York, 332-346 (1987).
15. H.Umegaki, Conditional expectation in operator algebras, I.Tohoku Math. J., (6), p.177-181 (1954); II. (3), p.86-100, (1956).
16. V.P.Belavkin, Theory of the control of observable quantum systems. Automatica and Remove Control, 44(2), p.178-188 (1983).

GLOBAL METHODS TO IMPROVE CONTROL AND OPTIMAL CONTROL OF RESONANCE INTERACTION OF LIGHT AND MATTER.

V.F. Krotov
Institute of Control Sciences
Profsouznaya 65, Moscow, USSR

This paper presents computational methods for *optimization of control*, connected with development and application of an approach in which sufficient conditions for global minima of functionals in variational calculus and optimal control theory are used. The first results were given at the beginning of the sixties [1-5], see also [6,7]. The main element which is looked for in this approach is a so called solving function depending on the state and the argument (time) of the process under consideration. Having properly chosen this function an optimal solution is found through maximization of some scalar function of state, control, and time with respect to first two variables.

Making use of these methods the problem of optimal control for resonance interaction of radiation with a quantum system is being investigated. An iteration method for solving this problem, applicable to a large dimensional system, is proposed. The author presents the numerical solutions of the following problems :
1) the obtaining of the maximal inverse population of a three level system excited by three fields with relaxation (dimension of a phase vector is equal to $n = 9$) ;
2) the obtaining of the maximal population of the first oscillating zone of a molecule of a spherical top type, excited by one external field (dimension of a phase vector equal to $n = 15202$).

1. PROBLEM STATEMENT.

Let us consider a triple of variables $t \in T$, $x \in X$, $u \in U$ and pair of functions $v = (x(t),u(t))$, $v : T \to X \times U$. The latter is called an admissible controlled process if the following conditions are satisfied :

$$(t,x(t),u(t)) \in V, \qquad \forall\, t \in T \tag{1.1}$$

where V is a given subset of the direct product $T \times X \times U$. A process equation or equation of motion is satisfied. Two kinds of those equations are considered :
a) The multistage (discrete argument) process. Here the set T is a real sequence $\{t_0, t_0+1, t_0+2,\ldots,t_1\}$ and the process equation has the form

$$z(t) = x(t+1) - f[t,x(t),u(t)] = 0 \qquad (1.2)$$

$$t = t_o,\ t_o+1,\dots,t_1-1,t_1,\ x(0) = x_o$$

where $f(t,x,u)$ is a given function, $f : T\times X\times U \to X$ and x_o is a given element from the set X .

b) The continuous argument process. Here T is an interval $[t_o,t_1]$ of the real line, X, U are real vector spaces R^n and R^r, respectively, and the process equation has the form :

$$z(t) = \dot{x}(t) - f[t,x(t),u(t)] = 0 \qquad (1.3)$$

$$t \in (t_o,t_1),\quad x(t_o) = x_o$$

where $f(t,x,u)$ is a given vector function.

In the latter case some additional requirements must be imposed on v to satisfy the equation (1.3). Usually this is piecewise continuity of the function u(t) and continuity and piecewise differentiability of the function x(t), of measurability of u(t) and absolute continuity of x(t).

The set of all admissible processes is denoted by D, t is called the process argument, x the state, and u the control. The function x(t) and u(t) are called the trajectory and the program control, respectively. To unify the notation we use the sign S which will stand either for the sum $\sum_{t=t_o}^{t_1-1}$ or for the integral $\int_T dt$.

Moreover, we use the notation $[t_o,t_1]$, (t_o,t_1) etc. not only for continuous time but also for discrete sequences with included or excluded initial and end points. We also admit the following notation : the superscript denotes the intersection of a given set with the set of constant values of a given variable, and the subscript denotes the projection of the set on the subset of this variable. For example V_x^t is the projection of the intersection of V and the set t = constant on X . D_x is the set of the admissible process trajectories. The same letters used in multiplication will denote the summation over the repeated argument.

On the set D we define the functional :

$$J(v) = Sf^o(t,x(t),u(t)) + F(x(t_1)) \qquad (1.4)$$

where $f^o(t,x,u)$ and $F(x)$ are given real functions which are continuous when the continuous case is considered.

We introduce also a set E of processes v with the following properties :

1. $D \subseteq E$;
2. There exists a sufficiently simple algorithm for construction of processes $v \in E$;
3. A functional $\rho(v)$ is defined which satisfies the conditions :

$$\rho(v) = 0 \quad \text{for} \quad v \in \mathcal{D}$$

$$\rho(v) > 0 \quad \text{for} \quad v \in E \setminus \mathcal{D} ,$$

and is called the distance between the element $v \in E$ and the set D ;
4. A functional $J'(v)$ satisfying $J'(v) = J(v)$ for $v \in D$ is defined.

In this paper we define the set E as follows. The pair of functions $v = (x(t),u(t)) \in E$ satisfies restrictions (1.1) but not necessarily (1.2) or (1.3). Although it is enough in the discrete time, one has to say more about the properties of the functions x(t) and u(t) in the continuous time case. Namely, we assume that both functions are piecewise continuous. The distance ρ is defined as :

$$\rho(v) = S|z(t)| + \sum_{t \in \beta, i} |x^i(t)| \Big|_{t-0}^{t+0} \tag{1.5}$$

where $z(t)$ is as in equations (1.2) or (1.3), and β is the set of arguments in which the function $x(t)$ jumps. The second element in the sum (1.5) is taken into account only in the continuous case.

Two subproblems may be associated with the above introduced problem.

(I) Construction of an admissible process. This consists in finding a sequence $\{v_s\} \subset E$ which converges to D :

$$\rho(v_s) \to 0 \tag{1.6}$$

This is one of the basic problems in control theory : find a control strategy and trajectory resulting from it which satisfy the given constraints. In mathematical language, it is the problem of finding the solution of the open system of differential (difference) equations which satisfies the constraints.

It is also often required that during the convergence some criterion is minimized.

(P) Construction of an optimal admissible process. Besides the conditions of subproblem (I) it is required that the sequence $\{v_s\} \subset E$ satisfies :

$$J(v_s) \to d = \inf_D J(v) \tag{1.7}$$

In particular, when a minimum of the functional $J(v)$ exists on D, then it is required to find a solution $\bar{v} \in D$ such that :

$$J(\bar{v}) = \min_{v \in D} J(v) \tag{1.8}$$

Also approximate variants of those subproblems will be considered. Let us introduce a set $D_\varepsilon(\rho)$ of processes which satisfy the following conditions :

$$v \in E, \ \rho(v) < \varepsilon \quad , \quad \varepsilon > 0 \tag{1.9}$$

We call $D_\varepsilon(\rho)$ an ε-extension of the set D in the metric ρ .

The approximate variant of the subproblem (I) is : find an ε-extended solution $\bar{v} \in D_\varepsilon(\rho)$. In the approximate variant of the subproblem (P) it is required that the solution is also η-optimal on $D_\varepsilon(\rho)$:

$$J(v) - d_\varepsilon(\rho) < \eta \ , \qquad \eta > 0, \qquad d_\varepsilon(\rho) = \inf_{D_\varepsilon(\rho)} J(v) \tag{1.10}$$

Also two other variants can be independently considered : $\varepsilon = 0$, i.e., η - optimality on strict D, and $\eta = 0$, i.e., strong optimality on ε - extension of D . In connection with the above definitions arises a question of the problem correctness, that is whether $\bar{v} \in D_\varepsilon(\rho) \to \bar{v} \in D$ when $\varepsilon \to 0$.

We call a solution to the subproblem (P) an optimal program control in agreement with one of its parts, i.e., program control u(t).

In control theory a solution in feedback form also plays an important role. Let us assume that there is given a function u(t,x), $u: T \times X \to U$. Let us further assume that there exists a solution x(t) of the system (1.2) or (1.3) with u(t,x) inserted for u(t) to form a program control $u(t) = u(t,x(t))$ such that $v = (x(t),u(t)) \in D$. We say that the process v is associated with the control (or feedback policy) function u(t,x).

Let B be a set of initial conditions (t_o,x_o) and let there exist a family of optimization problems with an initial condition $(t_o,x_o) \in B$ and the rest of the problem conditions fixed. We shall include the dependence on the initial conditions in notation such as $D(t_o,x_o)$ $d(t_o,x_o)$, $v(t_o,x_o)$, etc. For every (t_o,x_o), let there exist the unique process $v(t_o,x_o) \in D(t_o,x_o)$ associated with u(t,x). Then we call the latter a control synthesizing function or simply a control synthesis.

Let a synthesis $\bar{u}(t,x)$ satisfy the following conditions :

$$J(\bar{v}(t_o,x_o)) - d(t_o,x_o) \leq \varepsilon \ , \quad \varepsilon > 0, \quad \forall t_o,x_o \in B \ .$$

The function $\bar{u}(t,x)$ will be called an ε-optimal control synthesis and for $\varepsilon = 0$ an optimal synthesis. That means that a construction of an optimal synthesizing function is equivalent to the solution of a family of optimal program control problems with an initial condition (t_o,x_o) .

2. BOUNDING AND SOLVING FUNCTIONS, SUFFICIENT CONDITIONS OF OPTIMALITY.

Let us introduce a class Π of real functions $\varphi(t,x)$: (continuously differentiable in the continuous case) such that the following exist :

$$R(t,x,u) = \begin{cases} \partial\varphi/\partial x^i \cdot f^i(t,x,u) - f^o(t,x,u) + \partial\varphi/\partial t & \text{(continuous)} \\ \varphi[t+1,f(t,x,u)] - \varphi(t,x) - f^o(t,x,u) & \text{(discrete)} \end{cases} \quad (2.1)$$

$$G(x) = \varphi(t_1,x) + F(x) \quad (2.2)$$

$$\mu(t) = \sup_{(x,u)\in V^t} R(t,x,u), \quad m = \inf_{x \in V_x^{t=t_1}} G(x) \quad (2.3)$$

$$L(\varphi,v) = G(x(t_1)) - S \cdot R(t,x(t),u(t)) - \varphi(t_o,x_o) \quad (2.4)$$

$$\ell(\varphi) = m - S \cdot \mu(t) - \varphi(t_o,x_o) \quad (2.5)$$

$$\tilde{V}^t = \operatorname{Arg} \max_{(x,u)\in V^t} R(t,x,u) \quad (2.6)$$

$$\tilde{E} : v \in E, \ (x(t),u(t)) \in \tilde{V}^t, \ t \in [t_o,t_1), \ x(t) \in \operatorname{Arg} \min_{V_x^{t=t_1}} G(x) \quad (2.7)$$

$$\tilde{u}(t,x) \in \tilde{U}(t,x) = \operatorname{Arg} \max_{u \in V^{tx}} R(t,x,u) \quad (2.8)$$

$$P(t,x) = \sup_{u \in V_x^t} R(t,x,u) \quad (2.9)$$

The functions $\varphi(t,x) \in \Pi$ will be called bounding functions.

The above introduced values and variables have some properties which are useful when analising the problems considered in the paper. They will be reviewed below.

1/ $L(\varphi,v) = J(v)$, $\forall \varphi \in \Pi$, $v \in D$. This equation defines a family of functionals $J(v)$ on D . For a given (nonoptimal) process v_o it is possible to choose $L(\varphi,v_o)$ in such a way that it is obvious how to improve the process v_o, i.e., how to choose a v such that $L(\varphi_o,v) = J(v) < J(v_o) = L(\varphi_o,v_o)$ and also how to guess a rule of choosing it. These bounding functions will be called improving functions. It is possible to construct methods for approximate solution of the problem (P) in D by sequential improvements. Besides, $L(\varphi,v)$ is used to define $J(v)$ on E (outside D) as $J'(v) = L(\varphi,v)$. This is used in the proof of sufficient optimality conditions and in some algorithms.

2/ $J(v) \geq \ell(\varphi)$, $\forall v \in D$, $\varphi \in \Pi$, i.e., to any bounding function φ corresponds a lower bound for the functional J on D . This inequality can be used to obtain sufficient optimality conditions and directly to obtain global bounds for the criterion. From these bounds it is possible to find the best :

$$\bar{\ell} = \sup I(\varphi), \quad \varphi \in \Pi . \tag{2.10}$$

3/ A sufficient condition of optimality. Assume that there are given a function $\bar{\varphi} \in \Pi$ and a process $\bar{v} = \{\bar{x}(t), u(t)\} \in D$, such that

$$R(t,\bar{x}(t),\bar{u}(t)) = \max_{x,u \in V^t} R(t,x,u) = \mu(t), \; t \in [t_o,t_1) \tag{2.11}$$

$$G(x(t)) = \min_{x \in V_x^t} G(x) = m, \quad t = t_1$$

Then

$$J(\bar{v}) = \min_D J(v) = \ell(\bar{\varphi}) = \max_\Pi \ell(\varphi) \tag{2.12}$$

More generally : assume that there exist sequences $\{\varphi_s\} \subset \Pi$ and $\{v_s\} \subset D$ such that :

$$S[R_s(t, x_s(t),u_s(t)) - \mu_s(t)] \to 0 \tag{2.13}$$

$$G(x_s(t_1)) - m_s \to 0 \tag{2.14}$$

Then

$$J(v_s) \to \inf_D J(v) = \bar{\ell} = \lim \ell(\varphi_s) \tag{2.15}$$

These conditions of optimality are the basis of the following approach to the solution of variational problems (the principle of optimality, [2] : for different bounding functions find solutions $\tilde{x}(t)$, $\tilde{u}(t)$ of the family of extremal problems (2.11) with the parameter t and then take $\varphi = \bar{\varphi}$ such that the process $\tilde{v} = (\tilde{x}(t),\tilde{u}(t))$ satisfies the equations (1.2) or (1.3) and the earlier mentioned properties of the functions x(t), u(t) in the continuous time case. (the satisfaction of the condition (1.1) is looked for after construction of $\tilde{v}$). The function $\bar{\varphi}$ is generally essentially nonunique and when specified for different subclasses of Π it leads to different methods of solution. The function $\bar{\varphi}(t,x) \in \Pi$ is called a solving function and the set of all these functions is called $\bar{\Pi}$. Finding a pair $\bar{v} \in D$, $\bar{\varphi} \in \Pi$ means that the pair of the dual problems (2.12) has been solved.

4/ According to (2.7) for any function $\varphi \in \Pi$ there exists a process $\tilde{v} = (\tilde{x}(t),\tilde{u}(t)) \in \tilde{E}$ (perhaps nonunique) whose distance from the set D is $\varepsilon = \rho(\tilde{v})$. It is a solution of the ε-extended problems (I) and (P) where :

$$\min_{v \in D_\varepsilon} J'(v) = \ell(\varphi), \quad J'(v) = L(\varphi, v) \tag{2.16}$$

If a sequence $\{\varphi_s\} \subset \Pi$ is a solution of the dual problem (2.10), then for some sufficiently weak conditions there hold $\rho(\tilde{v}_s) \to 0$, $I(\varphi_s) \to d$, i.e., the sequence $\{\tilde{v}_s\}$ is a solution of the problems (I) and (P). From that numerical algorithms for computing admissible and optimal processes can be built. They will be considered in the sequel.

5/ Let a function $\tilde{u}(t,x)$ implied by $\varphi \in \Pi$ through (2.8) be a synthesizing function on the set B of initial conditions (t_o,x_o). For example this is always true in the discrete case if there are no constraints on states: $V_x^t = X$, $t = t_o+1,\dots,t_1$. Then for all (t_o,x_o) the following is true, see (4) :

$$J(\tilde{v}(t_o,x_o)) - d(t_o,x_o) \leq \Delta(\varphi) = \tag{2.17}$$

$$= S\,[\sup_{V_x^t} P(t,x) - \inf_{V_x^t} P(t,x)\,] + \sup_{V_x^{t=T}} G(x) - \inf_{V_x^{t=T}} G(x)$$

i.e., the synthesis $\tilde{u}(t,x)$ is ε-optimal, $\varepsilon = \Delta(\varphi)$. Minimizing the functional $\Delta(\varphi)$ it is possible to have it sufficiently small.

A group of numerical algorithms for an approximate optimal solution is based on this idea. Let there exist a function $\varphi(t,x)$ which satisfies the conditions :

$$P(t,x) = C(t),\ \forall t\ ;\ \varphi(t_1,x) = -F(x) + C_1 \tag{2.18}$$

where $C(t)$ is a function and C_1 a constant. Then, according to (2.17), the synthesis $\tilde{u}(t,x)$ is optimal. If we take $C(t) \equiv 0$ and $C_1 = 0$, then the function $\varphi(t,x)$ which satisfies (2.18) is the dynamic programming return for an optimal value function with a negative sign. The equation (2.18) is then the Hamilton-Jacobi or the dynamic programming equation in respective cases.

6/ The transformation $\varphi' = \varphi + C(t)$ where $C(t)$ is a differentiable function does not change the values of functionals $I(\varphi)$ and $L(\varphi,v)$ (when v is fixed) nor the sets $\tilde{E}$, $\tilde{U}(t,x), \tilde{V}^t$. From this it is seen that the bounding function φ can be defined in such a way that $\mu(t) \equiv 0$, $m = 0$. Then, the function φ is called normalized.

7/ Along the admissible trajectory $x(t) \in D_x$ a normalized function $\varphi(t,x)$ is nonincreasing and thus :

$$\varphi(t,x(t)) \leq \varphi(t_o,x_o), \quad \forall t, \quad \forall x(t) \in D_x .$$

8/ If the function $\varphi \in \overline{\Pi}$ is normalized, then all optimal trajectories $\overline{x}(t)$ are situated on the surface $\varphi(t,x) = \text{constant} = \varphi(t_o,x_o)$. In particular it means that the optimal trajectories can not cross this surface, not only in the upper but also in the lower direction.

The mathematical facts mentioned above are elementary but they imply nontrivial corollaries. Sufficient conditions of optimality include the basic equalities and inequalities of variational calculus and optimal control theory like the maximum principle equations, the Jacobi conditions, and the Hamilton-Jacobi-Bellman equations. This means that they are quite close to necessary conditions. And in fact, after some natural additional assumptions they become necessary. This observation made it possible to find new classes of solutions for the variational calculus problems and also methods of finding them. These facts were the basis for new ideas of constructing numerical algorithms for computing optimal or simply admissible processes. The mathematical methods which use the bounding functions and related constructions were found efficient not only for problems formulated in this paper but also in many other problems in analysis and synthesis of dynamic system control. It seems that they are as much adequate for solving global problems as the methods which use adjoint equations for local problems.

The equations (2.1) - (2.9), the presented mathematical facts, and resulting new possibilities and nontraditional directions in solving variational calculus and optimal control problems were developed in the papers surveyed here, (1-5). But also earlier papers containing some elements of this theory should be mentioned. The Hamiltonian-Jacobi method in variational calculus and analytical mechanics can be regarded as a first application of solving functions. We can also consider that they are used in Bellman dynamic programming (8) which is a generalization of the Hamiltonian-Jacobi method to the modern problems of control and, in particular, to the problems of optimal control of multistage processes. However, these are solving functions of special types, defined by equation (2.18). They do not cover all possible applications of this theory. Functions of the solving type were used by Caratheodory (8, p.335)for examining local conditions of variational calculus. To those results we can also add the second Lyapunov method for analysis of stability of motion. In it (7) bounding functions were defined and extensively used.

3. RELATIONS TO OTHER OPTIMALITY CONDITIONS.

The relation of the described optimality conditions to Pontryagin's maximum principle (10) is obvious from the following necessary extremum conditions (2.11), see (1,6,7) :

$$R_x(t,x(t),u(t)) \equiv \dot{\Psi} + H_x(t,\Psi(t),\bar{x}(t),\bar{u}(t)) = 0 \tag{3.1}$$

$$\bar{u}(t) \in \text{Arg}\max_u R(t,\bar{x}(t),u) = \text{Arg}\max_u H(t,\Psi(t),\bar{x}(t),u) \tag{3.2}$$

$$\begin{gathered} H(t,\Psi,x,u) = \Psi_i\, f^i(t,x,u) - f^o \\ \Psi(t) = \varphi_x(t,\bar{x}(t)) \end{gathered} \tag{3.3}$$

In the discrete variant the analogous conditions have the form, see (5) :

$$R_x(t,\bar{x}(t),\bar{u}(t)) \equiv H_x\,[t\,,\Psi(t+1),\bar{x}(t),\bar{u}(t)] - \Psi(t) = 0 \qquad t \in [0,T-1] \tag{3.4}$$

$$R_u(t,\bar{x}(t),\bar{u}(t)) \equiv H_u\,[t,\Psi(t+1),\bar{x}(t),\bar{u}(t)] = 0 \tag{3.5}$$

i.e., the maximum principle equations coincide with the maximum conditions given earlier for the function $R(t,x,u)$. Together with the process equations (1.2),(1.3) they form a closed system of equations where the solving function $\varphi(t,x)$ is represented only by its gradient on an optimal trajectory. The analogous coincidence of equations is true in appropriate extensions of the maximum principle (Dubavitzky-Milyutin conditions) and for state constraints, see Khrustalev (7, p.120-136).

Equations (2.11) extend this necessary optimality conditions to the global sufficient conditions which depend on functions $\varphi(t,x)$ such that $\varphi_x(t,\bar{x}(t)) = \Psi(x)$. Simple conditions of this type can be obtained taking a linear solving function $\varphi(t,x) = \Psi(t)x$. They were considered in (11). In (3,7) differential equalities for the matrix $\sigma(t) = \| \varphi_{x^i x^j}(t,\bar{x}(t)) \|$ were given. Their satisfaction guarantees the strong or weak relative minimum of the functional. The (necessary and sufficient) Jacobi conditions of variational calculus are equivalent to the existence of the matrix $\sigma(t)$ in the appropriate cases. Development of these kinds of conditions for a local optimum is given in the papers by Rozenberg (12) and Zeidan (13).

The Bellman dynamic programming equations (8) and the Hamilton-Jacobi partial differential equations of variational calculus coincide with the equation (2.18) which defines a solving function of a special type. Extensive analysis of the relations between the return functions and the solving functions has been done by Girsanov (14).

4. COMPUTER ALGORITHMS BASED ON THE TECHNIQUE OF GLOBAL BOUNDS.

Computer algorithms based on the technique of global bounds can be divided into 3 groups :

(i) Methods of successive improvements of the sweeping type similar to the traditional way presented, for example, by Kelley (22), Eneev (23), Krylov and Chernousko (24), Bryson and Ho (25). However, the choice of an improving function allows one to optimize not in a local (gradient) direction but in a global one.

(ii) Dual methods which are connected with a construction of sequences of solving functions $\{\varphi_s\} \subset \Pi$ maximizing the functional $\ell(\varphi)$ given by (2.5). This way we get the increasing sequence of lower bounds for the functional J on the set D which converges under appropriate conditions to $\inf_D J(v)$. Yet a solution to the problem is not this sequence but the sequence $\{v_s\} \subset E$ which satisfies (1.6) and (1.7). The role of this sequence is played by $\{\tilde{v}_s\} = \{\tilde{x}_s(t), \tilde{u}_s(t)\} \subset \tilde{E}_s$ which is related to $\{\varphi_s(t,x)\}$ through (2.7). This way we get an approximation to $\inf J(v)$, $v \in D$, by an "outside" approximation of an admissible process. Thus we solve not only the problem (P) but also the problem (I) .

(iii) Methods where the ε-optimal feedback control $u(t,x)$ is constructed using the bound (2.17). This leads to minimization of the functional $\Delta(\varphi)$ until it is not greater than a given ε.

4.1. The methods of successive improvements of control.

We start from a description of the methods mentioned of local improvement of control in terms of the improving function. Let us assume that we know an admissible process $v_o = (x_o(t), u_o(t)) \in D$. We want to improve it, i.e. to find a $v = (x(t), u(t)) \in D$ such that $J(v) < J(v_o)$. We replace optimization of the functional $J(v)$ by optimization of $L(v, \varphi)$ given by (2.4) with a suitably chosen function φ. We shall look for v which is sufficiently close to v_o in such a way that the sign of $\Delta J = J(v) - J(v_o)$ is the same as its main linear part :

$$\delta J = \delta L = G_x(x_o(t_1))\delta x(t_1) - S(R_x \delta x(t) + R_u \delta u(t)) \tag{4.1}$$

$$\delta x = x - x_o, \qquad \delta u = u - u_o$$

It is tacitly assumed above that the functions $R(t,x,u)$ and G are differentiable. The formula for δL is given to within the function $\varphi(t,x)$. We require that it complies with the equalities :

$$R_x(t, x_o(t), u_o(t)) = 0 \tag{4.2}$$

$$G_x(x(t_1)) = \Psi(t_1) + F_x(x(t_1)) = 0 \tag{4.3}$$

These equations contain only the gradient of the function $\varphi(t,x)$ on the points of the trajectory $x_o(t)$. The value of $\Psi(t) = \varphi_x(t,x_o(t))$ and the value of (3.1) and (3.4) are determined after replacing $\bar{x}(t),\bar{u}(t)$ by $x_o(t),u_o(t)$. This means that the equations (4.2) and (4.3) are satisfied by the functions of the form $\varphi(t,x) = \Psi_i(t)x^i$, where the vector $\Psi(t) = \{\Psi_i(t)\}$ is determined by (4.2) and (4.3). This function we call local improving for control. Then :

$$\delta J(v_o) = \delta L(v_o, \varphi) = S\, R_u(t,x_o(t),u_o(t))\delta u(t) \tag{4.4}$$

where $R_u(t,x_o(t),u_o(t))$ equals $H_u(t,\Psi(t),x_o(t),u_o(t))$ or $H_u(t,\Psi(t+1),x_o(t),u_o(t))$ for continuous and discrete variants of the problem, respectively.

Let there be given a function $\delta u(t)$ and an infinitely small parameter ε such that :

1. The right-hand side of (4.4) is positive,
2. $u(t,\varepsilon) = u_o + \varepsilon\delta u \in V_u^t$, $t \in T$,
3. $x(t,\varepsilon) \in V_x^t$ where $x(t,\varepsilon)$ is the trajectory determined by the program control $u(t,\varepsilon)$, the equation of motion, and the initial conditions,
4. $v(\varepsilon) = (x(t,\varepsilon),u(t,\varepsilon)) \in D$.

Then there exists $\varepsilon > 0$ such that

$$J(v) < J(v_o), \quad v = v(\varepsilon) \tag{4.5}$$

Without state constraints, i.e. for $V_x^t = X$, $t \in (0,T]$, the improvement of the given program control $u_o(t)$ reduces to the following steps :

(i) Find the trajectory $x_o(t)$ by solving the Cauchy problem (1.2) or (1.3) with $u = u_o(t)$, $x(0) = x_o$. The program control $u_o(t)$ should satisfy $v_o = (x_o(t),u_o(0)) \in D$.

(ii) Find $\Psi(t)$ and $R_u(t,x_o(t),u_o(t))$ by solving the linear Cauchy problem (4.2) with the initial condition (4.3) which determines a local improving function $\varphi = \Psi(t)x$.

(iii) Set a variation of the program control $\delta u(t)$ which makes the right-hand side of (4.4) positive.

(iv) For different $\varepsilon > 0$ solve the problem (I) with $u = u_o + \varepsilon\delta u$. The value of ε should be taken in such a way that (6.5) holds.

The basic part of this algorithm is the "sweeping" solution of the pair of Cauchy problems : the equation of motion from t_o to t_i and the adjoint equation from t_i to t_o . The consecutive repetition of these operations allows one to find the improving sequence $\{v_s\} \subset D$.

The expression (4.4) gives the gradient of the functional in the space of control functions u(t). The method presented can then be considered as an application of the gradient techniques to the above class of problems. A weak point of it is the local character of improvement which is guaranteed only for small variations of the control u(t). This is not only troublesome because the convergence is slow but also because the small variations can be unrealizable, for example when the set V_u^t is finite. This deficiency can be avoided when the globally improving functions are used.

It was shown in (26) that the function $\varphi(t,x)$ is globally improving for a given process $v_o = (x_o(t),u_o(t)) \in D$ if it satisfies the following conditions :

$$R(t,x_o(t),u_o(t)) = \min_x R(t,x,u_o(t)), \quad t \in T$$
$$G(x_o(t)) = \max_x G(x), \quad t = t_1 \tag{4.6}$$

A process v = (x(t),u(t)) which is determined by the control $\tilde{u}(t,x) = \arg\max_u R(t,x,u)$ satisfies the inequality $J(v) < J(v_o)$ if the process v_o is not an optimal one. For continuous processes it also holds that $\tilde{u}(t,x) = \arg\max_u H[t, \varphi_x(t,x),x,u]$. That is, when the local improvement was previously realized by a small variation of control in order to increase the function $R(t,x_o(t),u)$, the new control is chosen as a global maximum of R with respect to u . The condition (4.6) which is satisfied by an improving function can be slightly weakened :

$$R(t,x_o(t),u_o(t)) \leq R(t,x(t),u_o(t)), \quad t \in \overline{T}$$
$$G(x_o(t_1)) \geq G(x(t_1)) \tag{4.7}$$

where x(t) is the trajectory determined by $\tilde{u}(t,x)$.

To satisfy the equalities (4.6) it is enough to consider improving functions in the form :

$$\varphi(t,x) = \Psi_i(t)x^i + \sigma_{ij}(t)(x^i-x_o^i(t))(x^j-x_o^j(t))$$

where the coefficients $\Psi(t) = \{\Psi_i\}$, $\sigma_{ij}(t)$, $i,j = \overline{1,n}$, have to be found. It is easy to see that the equations for $\Psi(t)$ implied by (4.6) are the same as (4.2) and (4.3). Determination of the matrix $\sigma(t)$ is not unique. One possibility is to consider the equations :

$$R_{x^ix^j}(t,x_o(t),u_o(t)) = \delta_{ij}\eta \, , G_{x^ix^j}(x_o(t)) = - \delta_{ij}\alpha \, , \quad i,j = \overline{1,n} \tag{4.8}$$

Here δ_{ij} is the Kronecker delta : $\delta_{ij} = 0$ for $i \neq j$, $\delta_{ii} = 1$; η and α are positive constants. The equations (4.8) form the system of $(n+1)n/2$ linear differential (or difference) equations with unknowns $\sigma_{ij}(t) = \sigma_{ji}(t)$ and the given boundary condition at $t = t_1$. These equations together with (4.2), (4.3) and arbitrary positive η, α determine the coefficients of the function $\varphi(t,x)$ such that $x = x_0(t)$ is a relative minimum of $R(t,x,u_0(t))$ and maximum of $G(x)$. Appropriately choosing η we can satisfy inequalities (4.7) and therefore (4.5). This way we obtain the following algorithm for improving a solution :

(i) Set $\eta > 0$, $\alpha > 0$, and find $\Psi(t)$, $\sigma(t)$, $\varphi(t,x), \tilde{u}(t,x)$ by solving the linear Cauchy problem (4.2), (4.3), (4.8) from t_i to t_0 ,

(ii) Find the process $v = (x(t),u(t)) = \tilde{u}[t,x(t)]) \in D$ by solving the Cauchy problem for the equation of motion with $u = \tilde{u}(t,x), x(t_0) = x_0$, from t_0 to t_1 and verify the inequality $J(v) < J(v_0)$.

If it is not satisfied, then choose another α, η, and repeat the calculations. This procedure improves any process which does not satisfy the maximum principle equations or its discrete equivalent.

Consecutively repeating the above algorithm we find an improving sequence $\{v_s\} \subset D$. However, in general it does not converge to $\inf J(v)$, $v \in D$.

Example 1 (27).

The problem is :

$$J = - x^2(2) \to \min$$

s.t.

$$x^1(t+1) = x^1(t) + 2u(t)$$

$$x^2(t+1) = - (x^1(t))^2 + x^2(t) + u^2(t), \quad t = 0,1$$

$$x^1(0) = 3, \quad x^2(0) = 3, \quad |u(t)| \leq 5 .$$

The optimal solution is $\bar{u}(0) = - 2$, $\bar{u}(1) = \mp 5$, $J = - 19$. For this problem the Pontryagin maximum principle does not hold. The Hamiltonian $H(t,u) = \Psi_i(t) f^i(t,x(t),u)$ has for $t = 0$ at $u(0) = - 2$ not the maximum but the minimum. We take $\sigma_{12}(t) = \sigma_{22}(t) = 0$, $\forall t$, $\sigma_{11} = \sigma$ and thus :

$$\varphi(t,x) = \Psi_1(t)x^1 + \Psi_2(t)x^2 + \sigma(t)(x^1 - x_0^1(t))^2/2 .$$

The functions R and G take the form :

$$R(t,x,u) = \Psi_1(t+1)(x+2u) + \Psi_2(t+1)(-x^2+x+u^2) + $$
$$+ 0\cdot5\cdot\sigma(t+1)[x^1+2u-x_o^1(t+1)]^2 - \Psi_i(t)\cdot x_i -$$
$$- 0\cdot5\cdot\sigma(t)[x^1-x_o^1(t)]^2 ,$$
$$G(x) = - x^2 + \Psi_i(2)\cdot x^i + 0\cdot5\cdot\sigma(2)[x^1-x_o^1(2)]^2 .$$

The adjoint equation and the equation for $\sigma(t)$ are as follows :

$$\Psi_1(t) = \Psi_1(t+1) - 2x_o^1(t)\cdot\Psi_2(t+1), \ \Psi(2) = 0 ;$$
$$\Psi_2(t) = \Psi_2(t+1), \ \Psi_2(2) = 1,$$
$$\sigma(t) = - 2\cdot\Psi_2(t+1) + \sigma(t+1) - \eta, \ \sigma(2) = \alpha$$

The results : Iter. No 1 : $u(0) = u(1) = 0$; $J = - 18$. Iter. No 2 - optimum .

We choose a class of nonlinear optimal control problems for which the global improving function satisfying (4.6) has the form $\varphi(t,x) = \Psi_i(t)x^i$. In this case the algorithm presented above is substantially simplified because there is no need to adjust the coefficients η, α nor to solve the system of equations (4.8). The problem functions have now the form :

$$f(t,x,u) = A(t,u)x + B(t,u) \ ; \ F(x) = \lambda x \ ; \ u \in V_u^t$$
$$f^o(t,x,u) = a^o(t,u)x + b^o(t,u) \tag{4.9}$$

An interesting subclass of these problems is connected with the control of quantum systems by the means of a laser radiation. It was investigated and algorithmized using the method described above in (28). In a simulated experiment a good convergence and effectiveness of the method was obtained for very big dimensions of the state vector which reached some ten thousands.

We mention also a class of so called knapsack multivariate problems where the above method seems to be effective :

$$J(v) = \sum_{t=1}^{N} C_t u_t \rightarrow \min \ ; \ u_t \in [0,\beta_t] \ ; \tag{4.10}$$

$$\sum_{t=1}^{N} a_t^i u_t < b^i \ ; \ i = \overline{1,n} \ ; \tag{4.11}$$

and u_t is integer. Application of the above method to these problems was considered by the present author together with Feldman. Introducing a sequence $\{x_t\} \subset R^n$, $t = 0,1,\ldots,N$:

$$x^i_{t+1} = x^i_t + a^i_t u_t \ ; \ x(1) = 0 \ ; \ x^i(N+1) \leq b^i \ , \quad i = \overline{1,n} \qquad (4.12)$$

we can transform the problem (4.10)-(4.11) to the multistage optimization problem where :

$t_o = 0, \quad t_1 = N+1, \quad x = R^n, \quad$ U-set of integers,

$f(t,x,u) = x + a(t)u, \quad f^o = C(t)u, \quad V^t_x = R^n$ for $t < t_1$,

$V^t_x = \{x : x^i \leq b^i\}$ for $t = t_1, \quad V^{tx} = [0,\beta_t]$.

We have :

$$\varphi(t,x) = \Psi_i(t)x^i + \frac{1}{2}\sigma_{ii}(t)(x^i - x^i_o(t))^2 \ ; \ \sigma_{ij} = 0, \quad i \neq j \ .$$

$$R(t,x,u) = \Psi_i(t+1)(x^i + a^i(t)u) + \frac{1}{2}\sigma_{ii}(t+1)(x^i + a^i(t)u - x^i_o(t+1))^2 -$$

$$- \Psi_i(t)x^i - \frac{1}{2}\sigma_{ii}(t)(x^i - x^i_o(t))^2 - C(t)\cdot u \ ;$$

$$G(x) = \Psi_i(t_1)x^i + \frac{1}{2}\sigma_{ii}(t_1)(x^i - x^i_o(t_1))^2 \ .$$

Taking $\eta = 0$ and solving the equations (4.2), (4.3) and (4.8) we get :

$\Psi_i(t) = \text{constant} = \Psi_i = 0$ if $x^i_o(t_1) < b^i$ and

$\Psi_i(t) >$ if $x^i_o(t_1) < b^i$, $i = \overline{1,n}$;

$\sigma_{ii}(t) = \text{constant} = -\alpha_i , \quad \alpha_i > 0$

$R(t,x,u) = -A(t)u^2/2 + B(t,x)u + C(t,x)$

$A(t) = \alpha_i(a^i(t))^2 > 0$

$B(t,x) = -\alpha_i a^i(t)(x^i - x^i_o(t+1)) + \Psi_i a^i(t) - C(t)$

$C(t,x) = \alpha_i[x^i_o(t+1) - x^i_o(t)] + \alpha_i[x^{i2}_o(t+1) - x^{i2}_o(t)]$.

The expression for $R(t,x,u)$ satisfies (4.6). The control $\tilde{u}(t,x)$ is taken as an integer from the interval $[0,\beta_t]$ which is closest to the value $u^*(t,x) = B(t,x)/A(t)$. The values α_i are chosen in such a way that the improved trajectory satisfies the inequalities (4.11) .

Example 2 (29) .

The problem is :

$$J = -[6u_1 + 4u_2 + u_3] \rightarrow \min$$

$$\text{s.t.} \quad \begin{aligned} u_1 + 2u_2 + 3u_3 &\leq 5 \\ 2u_1 + u_2 + u_3 &\leq 4 \\ u_t &= \{0,1\} \end{aligned}$$

and the optimal solution (see Table) :

$$\bar{u}_1 = \bar{u}_2 = 1 \ ; \quad \bar{u}_3 = 0, \qquad J = -10 .$$

Iter Number	Control			Vector α		Functional
	u_1	u_2	u_3	α_1	α_2	
0	0	0	0	1	1	0
1	1	0	0	1	1	- 6
2	1	1	0			- 10

Example 3 (30).

The problem is :

$$J = -[3u_1 + 3u_2 + 13u_3] \to \min$$

s.t.
$$-3u_1 + 6u_2 + 7u_3 \leq 8$$
$$6u_1 - 3u_2 + 7u_3 \leq 8$$
$$0 \leq u_t \leq 5, \quad u_t - \text{integer},$$

and the optimal solution (see Table) :

$$u_1 = u_2 = 0 \ ; \ u_3 = 1, \ J = -13$$

Iter Number	Control			Vector α		Functional
	u_1	u_2	u_3	α_1	α_2	
0	0	0	0	0.3	0.3	0
1	0	0	1			- 13

In this the version with the global improving function quadratic in x (26) is discussed. There exist other versions of this method which are presented in (31).

4.2. The methods of successive improvement of the bounding function.

The method is presented according to (32, 33). Let there exist a function $\varphi_o(t,x) \in \Pi$. We give the operation of improving it, i.e. finding a function $\varphi \in \Pi$ such that $\ell(\varphi) > \ell(\varphi_o)$. We assume that it has the form : $\varphi = \varphi_o + \lambda\gamma$, where $\lambda, \gamma(t,x)$ are a coefficient and a function which should be determined. We introduce a functional :

$$\delta(v) = \int r(t,x(t),u(t)) + \gamma(t,x(t)) \Big|_{t=0}^{t_1} \tag{4.13}$$

$$r(t,x,u) = \gamma_{x_i} \cdot f^i$$

We denote by $R(t,x,u,\lambda)$, $\tilde{E}(\lambda)$, etc. the appropriate constructions associated with $\varphi = \varphi_o + \lambda\gamma$, and also $R_o(t,x,u) = R(t,x,u,0)$ etc. Taking into account (4.13) and (2.5) the increment $\Delta\ell = \ell(\lambda) - \ell_o$ can be written in the form :

$$\Delta\ell = \lambda\delta(v) + [L_o(v) - \ell_o] , \quad v \in \tilde{E}(\lambda) \tag{4.14}$$

From this it follows that $\ell(\varphi) > \ell(\varphi_o)$ if at least for one $v \in \tilde{E}(\lambda)$:

$$\lambda\delta(v) > 0 \tag{4.15}$$

Fitting λ and δ which satisfy the above inequality will be called an elementary bound improving operation. In the sequel for simplicity we consider only the case when the set $\tilde{E}$ contains only one element $\tilde{v}(\lambda) = (\tilde{x}(t,\lambda),\tilde{u}(t,\lambda))$, i.e. the function $R(t,x,u,\lambda)$ has only one maximum. We also denote $\delta(\lambda) = \delta(\tilde{v}(\lambda))$. Under sufficiently general conditions the function $\delta(\lambda)$ is lower semicontinuous at $\lambda = 0$. Thus if we define the function γ to satisfy :

$$\delta(0) = S\ r(t,\tilde{x}_o(t),\tilde{u}_o(t)) + \gamma(t,\tilde{x}_o(t)) \Big|_0^{t_1} > 0 \tag{4.16}$$

then for a sufficiently small $\lambda > 0$ the inequality (4.15) holds and therefore $\ell(\varphi) > \ell(\varphi_o)$. We see then that the elementary operation can be done in two steps. In the first $\gamma(t,x)$ is chosen according to (4.16) and in the second a $\lambda > 0$ is taken.

It is easier to interpret the idea of elementary operation when the improving component is taken in the form $\gamma = \nu_i(t)x^i$ and the functional $\delta(v)$ in the form :

$$\delta(v) = \int_{t_o}^{t_1} \nu(t)\tilde{z}(t)dt + \sum_{t\in\beta} \nu(t)\tilde{x}(t) \Big|_{t-1}^{t+1} ; \quad \sum_{t=0}^{T-1} \nu(t+1)\tilde{z}(t) \tag{4.17}$$

where $\tilde{z}(t)$ is related to the process $\tilde{v}$ through (1.2) or (1.3), β is the set of points of discontinuity of the function $\tilde{x}(t)$. The first and the second expressions correspond to the continuous and the discrete processes, respectively.

It follows from (4.17) that if there exists a value $t = \tau$ such that $\tilde{z}(\tau) \neq 0$ or in the continuous time $\tilde{x}(\tau+0) - \tilde{x}(\tau-0) \neq 0$, then the improvement of the function $\varphi_o(t,x)$ can be achieved by adding a linear term $\gamma(t,x) = \nu_i(t)x^i$ where the function $\nu(t)$ is taken to keep the right hand side of (4.17) positive for $\tilde{v} = \tilde{v}_o(\tilde{x}_o(t),\tilde{u}_o(t))$.

The use of $\gamma(t,x)$ in more complicated cases is necessary only when the maximum of the function $R_o(t,x,u)$ is not unique.

A weak point in this method is the necessity to maximize the function $R(t,x,u)$ for every t in order to form the process $\tilde{v} = (\tilde{x}(t),\tilde{u}(t))$ or more generally - the set E . Therefore the method can be applied only to the problems where this maximization can be performed analytically or there exist efficient numerical procedures for doing it.

Repeating consecutively the elementary improving operations we get the sequence $\{\varphi_s\}$ for which the value $\ell(\varphi)$ increases.

There exist theorems in which it is shown that under some stronger conditions for γ and λ the above sequences ensure a solution to the problem (P) for a wide class of systems. Namely, the sequence $\{\tilde{v}_s\} = \{\tilde{x}_s(t),\tilde{u}_s(t)\}$ corresponding to $\{\varphi_s\}$ by (2.7) is a generalized solution to the problem (I) in the sense of (1.5) and (1.6), and to the problem (P) in the sense $\ell(\varphi_s) \to \lim_{\varepsilon\to 0} d_\varepsilon(\rho)$ where $d_\varepsilon(\rho)$ is given by (1.10).

Exemple 4.

Find a solution to the system :

$$\dot{x} = u, \quad x(0) = x_o > 0,$$
$$x(1) = 0$$

which minimizes the functional

$$J = \int_0^1 (u^2 + x^2)dt$$

Here x and u are scalar functions. We look for a solution in the form of a sequence:

$$\varphi_s = \Psi(t)x, \quad \gamma_s = \nu_s x$$

We have :

$$R_s(t,x,u) = \Psi_s u - u^2 - x^2 + \dot{\Psi}_s x$$
$$\tilde{x}_s = (1/2)\dot{\Psi}_s$$
$$u_s = (1/2)\Psi_s$$

$$\mu_s(t) = R(t,\tilde{x}_s,\tilde{u}_s) = (\Psi_s^2 + \dot{\Psi}_s^2)/4$$

$$\ell_s = -\Psi_s(0)x_o - \int_0^1 (\Psi^2 + \dot{\Psi}^2)dt$$

$$\tilde{z} = \dot{\tilde{x}}_s - \tilde{u}_s$$

$$\Delta_s = \Delta_s^1 + \Delta_s^2$$

$$\Delta_s^1 = \int_0^1 |\tilde{z}_s| dt$$

$$\Delta_s^2 = |\tilde{x}(1)| + |\tilde{x}_s(0) - x_o|$$

$$\delta_s(x,u) = -\nu_s(1)\tilde{x}_s(1) - \nu_s(0)(x_o - \tilde{x}(0)) - \int_0^1 \nu_s(t)\tilde{z}_s dt$$

$$\tilde{\delta}_s = \delta_s(\tilde{x}_s,\tilde{u}_s)$$

$$\Delta\tilde{x}_s = \dot{\nu}_s/2$$

$$\Delta\tilde{u}_s = \nu_s/2$$

$$R_s(t,x,u,\lambda) = R_s(t,x,u) + \lambda r_s(t,x,u)$$

$$r_s(t,x,u) = \nu_s u + \dot{\nu}_s x$$

$$\tilde{x}_s(\lambda) = \tilde{x}_s + \lambda\Delta\tilde{x}_s$$

$$\tilde{u}_s(\lambda) = \tilde{u}_s + \lambda\tilde{u}_s \ .$$

The value λ_s is taken to satisfy the condition :

$$\tilde{\delta}_s(\lambda) = \delta_s(\tilde{x}_s + \lambda\Delta\tilde{x}_s, \tilde{u}_s + \lambda\Delta\tilde{u}_s) = 0$$

which is in this case an elementary improving operation (49). We have :

$$\tilde{\delta}_s(\lambda) = \delta_s(\tilde{x}_s,\tilde{u}_s) + \lambda\delta_s(\Delta\tilde{x}_s,\Delta\tilde{u}_s)$$

$$\lambda_s = -\delta_s(\tilde{x}_s,\tilde{u}_s)/\delta_s(\Delta\tilde{x}_s,\Delta\tilde{u}_s)$$

$$\Psi_{s+1} = \Psi_s + \lambda_s\nu_s \ .$$

The function R(t,x,u) for any Ψ has the unique maximum at $\tilde{x}(t)$, $\tilde{u}(t)$ and thus the elementary operation is solvable in the class of linear functions $\gamma(t,x) = \nu(t)x$ for any Ψ which does not ensure the strict optimum. We provide the specific iterations starting from $\Psi_o = 0$.

Iteration 1.

We have :

$$\tilde{x}_o(t) = \tilde{u}_o(t) = 0$$

$$\Delta_o^1 = 0, \quad \Delta_o^2 = x_o - \tilde{x}_o(0) = x_o, \quad \tilde{z}(t) \equiv 0, \quad \ell_o = 0$$

$$\tilde{\delta} = \delta_o(\tilde{x}_o,\tilde{u}_o) = - \nu_o(0)x_o + \int_o^1 \nu_o(t)\tilde{z}_o(t)dt = - \nu_o(0)x_o .$$

The condition (4.16) is satisfied for $\nu_o(0) = - 1$. For other values of t the function $\nu_o(t)$ can be defined arbitrarily. We define it in a simple way : $\nu_o(t) = - 1$. We have $\tilde{\delta}_o = x_o$, $\Delta\tilde{x}_o = 0$, $\Delta\tilde{u}_o = - 1/2$, $\delta_o(\Delta\tilde{x}_o,\Delta\tilde{u}_o) = - 1/2$, $\lambda_o = 2x_o$. Hence $\Psi_1(t) = 0+\lambda_o\nu_o = - 2x_o$. Moreover $\tilde{x}_1(t) \equiv 0$, $\tilde{u}_1(t) = -y_o$, $\Delta_1^1 = x_o$, $\ell_1 = x_o^2 > \ell_o = 0$.

This way in the first iteration the value of l increased but the pair x,u did not move closer to D, neither in the boundary conditions, i.e. in the norm Δ^2, nor in the integral norm Δ^1.

Iteration 2.

We have :

$$\tilde{\delta}_1 = \delta_1(\tilde{x}_1,\tilde{u}_1) = - \nu_1(0) + x_o \int_o^1 \nu_1(t)dt .$$

According to (4.16) and requirements of normalization [33] (first way) :

$$\nu_1(0) = - 1, \quad \nu_1(t) \equiv 1 \quad \text{for} \quad t \in (0,1) .$$

This function is discontinuous and does not comply with the conditions of the elementary operation. Therefore we take as $\nu_1(t)$ a continuous function from the approximating sequence $\{1-2(t-1)^k, k = 2, 4, 6,... \}$. We choose the function which is simplest for computing, i.e. $\nu_1(t) = 1-2(t-1)^2$.

We have : $\tilde{\delta}_1 = 4/3\ x_o$

$$\Delta x_1 = 2(1-t) \qquad \Delta u_1 = 1/2 - (t-1)^2$$

$$\delta_1(\Delta x_1,\Delta u_1) = - 29 \qquad \lambda_1 = \tilde{\delta}_1/\delta_1(\Delta x_1,\Delta u_1) = 40/87\ x_o$$

$$\tilde{x}_2(t) = \lambda_1\Delta u_1 = 80/87\ x_o(1-t), \qquad \tilde{u}_2(t) = \tilde{u}_1+\lambda_1\Delta u = -1/87\ x_o[67+40(t-1)^2]$$

The estimate of the distance from D is :

$$\Delta_1^1 = 10/87\ x_o \approx 1/9\ x_o , \quad \Delta_2^2 = x_o - \tilde{x}_2(0) = 7/87\ x_o \approx 7/90\ x_o .$$

The lower bound is $\ell_2 \approx x_o^2$. Therefore in the second iteration the pair $\tilde{x}$, $\tilde{u}$ was moved substantially closer to D, approximately 10 times in each criterion.

The above method was applied for developing algorithms for solving integral assignment, scheduling, traveling salesman problems (34), different optimization problems of space maneuvers (35) and distributed parameters systems (36) .

4.3. Methods of ε-optimal control synthesis.

We want to find an ε-optimal control synthesizing function $\tilde{u}(t,x)$. We consider the case when there are no state constraints including in it also boundary constraints, i.e. $V_x^t = X, \forall t, x \in T \times X$. Other problems can be solved by this method using penalty functions. We showed above that this problem can be solved using the bounding expression (2.17) and minimizing the functional $\Delta(\varphi)$ until it has the value $\Delta(\tilde{\varphi}) = \varepsilon$. Then the synthesizing function $\tilde{u}(t,x) = \arg \max_{u \in V^{tx}} R(t,x,u)$ is ε -optimal. The problem of finding an optimal control synthesis is therefore reduced to the minimization of the functional $\Delta(\varphi)$. The lower bound for the latter is zero. This bound is attained when in the class Π or its above mentioned refinements there exists a solution of the dynamic programming equation (2.18) or a sequence which approximates this solution in the sense of $\Delta(\varphi)$.

There exist numerical algorithms which use this approach. One of these algorithms (4, 6, 37, 38) is the following. The desired function $\varphi(t,x)$ is taken as an interpolating polynomial in the space $X = R^n$. Its parameters depend on t and are determined from the equation :

$$P(t,x_\ell(t)) = 0 \ ; \quad G(x_\ell(t_1)) = 0 \tag{4.18}$$

where $\{x_\ell\}$ is a given set of interpolation knots, $P(t,x)$ and $G(t,x)$ are given by (2.9) and (2.2). The equations (4.18) form a system of normal differential (difference) equations in the parameters of the function $\varphi(t,x)$ with the given boundary conditions for $t = t_1$. Solving this system we get the function $\varphi(t,x)$, the corresponding control synthesis $\tilde{u}(t,x)$, and the bound $\Delta(\varphi)$. If the latter is too big, then the computations are repeated with a better set. This is reiterated until we get $\Delta(\varphi) < \varepsilon$.

A second algorithm which was used in some interesting applied problems (7, ss. 349-357) consists in solving the problem $\Delta(\varphi) \to \min$ by the Ritz method. Then a class of functions $\varphi(t,x) = \xi(t,x,a)$ depending on a parameter a is taken. The functional $\Delta(a) = \Delta(\xi(t,x,a))$ is computed and the minimal value of $\Delta(\varphi)$ is found using the mathematical programming method in this class.

The possibilities of using the above methods are limited because of the operations $\sup_x P(t,x)$ and $\inf_x P(t,x)$ which are in (2.17). For many specific problems (39, 40, 41) these operations can be performed analytically. In these cases it is much easier to realize and justify the algorithms for solving the problem $\Delta(\varphi) \to \min$ in the class of the bounding functions which are quadratic in x and moreover to get exact solutions in the form of minimizing sequences of control synthesizing functions.

5. OPTIMAL CONTROL FOR RESONANCE INTERACTION OF LIGHT AND MATTER.

5.1. Introduction

At present there exists a vast domain of new techniques based on laser application : isotope separation, photochemistry, pure substance production, detecting of single atoms and molecules and others. This makes the problem of the most effective utilization of lasers in such processes extremely important and causes the formation of a new type of problem of optimal control theory, namely, control of quantum systems by making use of macroscopic electromagnetic fields. Consideration of such problem statements is presented in (42). The authors of paper (43), using the sufficient conditions of optimality, the method of multiple maxima and the Pontryagin maximum principle, have analytically found the optimal regimes for some small dimension (dimensionality 2-3) models of photoprocesses in gases. Optimal excitation of the three level system set up by two resonant fields had been studied in (56) in the class of constant controls (amplitudes and detuning of resonant fields, relaxation constants). In this paper the authors describe the solution method for a given class of nonlinear optimizing problems. This method is practically applicable to systems of large dimensions, i.e. for a great number of energy levels.

Papers (44, 45, 46) deal with the analysis of some different systems, namely, optimal laser heating of macroscopic volumes of a substance (i.e. a nonquantum system and nonresonance interaction).

5.2. Statement of the problem.

A quantum system (atom, molecule) is considered to be interacting with a macroscopic electromagnetic vector field E of the following form

$$E = \sum_{j=1}^{p} u^j(t)\exp(i\omega_j t) , \qquad (5.1)$$

where $u^j(t)$ vary insignificantly in times $\sim \omega_j^{-1}$. Interaction is assumed to be electrodipole. The state of the system is described by a matrix of density $\hat{x}$ representing the interaction, which satisfied the kinetic equation

$$\frac{d\hat{x}}{dt} + \hat{\Gamma}\hat{X} = -\frac{i}{\hbar}[\hat{V},\hat{X}], \tag{5.2}$$

where $\hat{V} = -\hat{d}E$ is an operator of the quantum system interaction with a radiation field, $\hat{d}$ is an operator of a dipole moment in interaction representation, $\hat{\Gamma}$ is a linear operator of the quantum system interaction with a dissipating system and $[\hat{V},\hat{X}] = \hat{v}\hat{x} - \hat{x}\hat{v}$ is a commutator of the two operators (47). Assume that one or several resonance conditions are being fulfilled

$$\sum_j n_k^j \omega_j = \omega_{mn} + \nu^k . \tag{5.3}$$

where n_k^j are positive integers and ω_{mn} are naturel frequencies of the system. Then in resonance approximation and under the condition that $|\nu^k| \ll \omega_{mn}$ and $|V_{mn}| \ll \hbar\omega_{mn}$ (V_{mn} is a matrix element of transition from the state m to n, truncated equations for the density matrix have the form of 48 ; summation is assumed with respect to repeated indices) :

$$\dot{x}^i = A_j^i(t, u)x^j \quad ; \; x^i(0) = x_o^i \; ; \; i = \overline{1,q}$$
$$A_j^i = a_j^i + B_{j\alpha_1\alpha_2}^{ix_1x_2} \cdot u_{x_1}^{\alpha_1} \cdot u_{x_2}^{\alpha_2} + A_{j\alpha_1}^{i(1)x_1} \cdot u_{x_1}^{\alpha_1} +$$
$$+ A_{j\alpha_1\alpha_2}^{i(2)x_1x_2} \cdot u_{x_1}^{\alpha_1} \cdot u_{x_2}^{\alpha_2} \; \dots \tag{5.4}$$

where x^i, $(i \div 1,q)$ are real and imaginary parts of the density matrix components, related to resonant levels as well as to nonresonant ones strongly connected by relaxation. $x_1,\dots, x_s$ are vector indices, varying from 1 to 3 and $\alpha_1,\dots, \alpha_s$ varying from 1 to p . The first term of the right part in equation (4) describes relaxation effects, the second one describes a dynamical Stark shift of the equations, the third - contribution of one photon processes, the fourth - contribution of two photon processes and so on. Field amplitudes are assumed as limited ones $0 < U_\alpha^x < M_\alpha^x$ where M_α^x are constants. It is quite natural by virtue of resonance approximation that $|V_{mn}| \ll \hbar\omega_{mn}$. We are given a period of time (0,T) and a state of the system at the initial moment of time. M_α^x are considered sufficiently small for preserving the resonance conditions.

Thus, the interaction of light with matter is described by the components of the density matrix $X_i(t)$, i.e. the state of the process meeting equations (5.4) with initial conditions X_i^o , where amplitudes of the quasimonochromatical field $u_j(t)$ are limited and play the role of controls. The elements of the set D of admissible processes are described by the pairs $v = (x(t), U(t))$. We define the functional

$$I(v) = \int_0^T x_i f_i(t,U)dt + x_i(T)\ell_i \tag{5.5}$$

This functional, in particular cases includes polarization of the system, population of the levels in the moment of time T, their mean values for a period of the time T and also any linear combinations, for example, an inverse population of the levels m and n, $X_m(T) - X_n(T)$. Since level population is an important stage in the development of chemical reactions, controlled by laser radiation (49)as well as a laser spark and multiphoton ionization (50, 51), the choice of functional type I(v) includes a significant part of cases interesting from the physical view point.

We are to find the sequence v_s on which the functional (5.5) tends to its least value :

$$\lim_{s\to\infty} I(v_s) = \inf_{v\in D} I(v) \tag{5.6}$$

Note some essential features of the formulated problem :

1/ no limits for phase coordinates in the time interval [0,T],

2/ independence of control limits of time and phase coordinates,

3/ linearity of process equations and functional depending on the phase coordinates with the fixed program of control U(t).

In that case if the modes, utilizing most efficiently the energy applied to a process are to be found, then the functional is presented in the form : $I_1(v) = - I(v)/(\int_0^T u_{\alpha_1}^{x_1} u_{\alpha_1}^{x_1} dt)$. In this case the optimization problem looses its property (3), i.e. it becomes nonlinear in the phase coordinates.

It is also possible that there will be limits imposed on power consumption by controls, then functional (5.5) preserves its form but a new variable $\dot{x}_{q+1} = u_{\alpha_1}^{x_1} u_{\alpha_1}^{x_1}$ is introduced and it is restricted by $x_{q+1}(T) \leq EM$ where EM is a constant. In this statement the optimization problem looses its property 1 .

5.3. Description of the method.

Let us analyse only the case when power consumption in controlling is inessential, i.e. the functional has the form of (5.5) and process equations can be written (sub/superscripts omitted) in the following form : $\dot{x} = A(u,t)x + B(u,t)$, i.e. this problem has the form (4.9) and its solution requires the application of the control successive improvement method. As it has been previously shown, the choice of the solvable function can be limited for problems of this kind : $\varphi(t,x) = \Psi_i(t)\cdot x^i$. The corresponding constructions are :

$$R(t,x,u) = [\dot{\Psi}_j + \Psi_i A^i_j(t,u) - f_j(t,u)]x^j \; ; \tag{5.7}$$

$$\tilde{u}(t,x) = \arg \max_{0 \leq u \leq M} R(t,x,u) \; ; \tag{5.8}$$

$$G(x) = [\ell_i(T) + \Psi_i(T)]\cdot x^i \tag{5.9}$$

Under the assigned program $u_o(t)$ conditions (5.6) are equivalent to the equalities :

$$\dot{\Psi}_i + A^j_i(t,u_o(t))\Psi_j - f_i(t,u_o(t)) = 0 \; ; \; \Psi_i(T) = -\ell_i(T), \quad j,i = \overline{1,q} \tag{5.10}$$

The improvement algorithm has the form :

1/ Assign $\varphi = \varphi_o(t,x) = \Psi_{io}(t)\cdot x^i$ and from (5.8) find $\tilde{u}_o(t,x)$;

2/ Solving the Cauchy problem for the system (5.4) of the closed control $u = u_o(t,x)$ we find a trajectory $x_o(t)$ and the control program $u_o(t) = \tilde{u}(t,x_o(t))$. Instead of item 1, the initial control program $u_o(t)$ can be directly assigned, and by virtue of item 2 with $u = u_o(t)$;

3/ Iterating (5.10) find the vector-function $\Psi(t)$;

4/ Repeating 1/ and 2/ with $\varphi(t,x) = \Psi_i(t)\cdot x^i$ find the process $v = (x(t),u(t))$, satisfying the inequality $I(v) < I(v_o)$; $u_o = (x_o(t),u_o(t))$ if the process v_o is not the solution of the maximum principle.

Repeating this procedure we obtain the improvement sequence $\{v_k\}$ $I(v_{k+1}) < I(v_k)$. The procedure of improvement leads to the solution equations of the maximum principle. The practical signal for completing the iterations is the repetition of the values $I(v_{k+1}) = I(v_k)$.

5.4. Computation features of the method.

1) It was proved in (52) that $I^{(k)} - I^{(k+1)} = \Delta_1 + \Delta_2 + \Delta_3$. In this problem $\Delta_1 \equiv 0$, $\Delta_2 > 0$ by construction, whereas Δ_3 is equal to

$$\Delta_3 = \int_0^T (x^j_{(k)} - x^j_{(k+1)})(\dot{\Psi}_j + A^i_j \Psi_i - f^o)dt, \tag{5.11}$$

A^i_j from (5.4). In other words, Δ_3 is equal to the sum of integrals of phase coordinate differences in (K) and (k+1) interations multiplied by an error of integrating the equations (5.10) on the (k^{th})-iteration. If integration (5.10) was accurate then $\Delta_3 \equiv 0$ while insufficient accuracy of integration Δ_3 may produce a negative contribution (especially in the neighbourhood of the extremal when Δ_2 is decreasing). Therefore if $I^{(k+1)} > I^{(k)}$ then it is desirable to go back to the (k^{th})-iteration and to solve (5.10) with higher accuracy.

2) Step 2/ of the algorithm in the event when only one-photon transitions, is realized in such a way, if $A^{(1)x}_{ij\alpha} \cdot x^{(k)}_j \cdot \Psi^{(k-1)}_i > 0$ then $u^x_\alpha = M^x_\alpha$, and if consideration of the Stark shift level or contribution of the two-photon processes transforms this step of the method into a quadratic programming problem while consideration of the S-photon programming program.

3) Note that this method also guarantees the finding of local minimals only (Pontryagin's extremals), and therefore multiple iteration with different initial controls $u^o(t)$ is desired to be followed by the selection of the best solution from the obtained ones. Within the framework of this method, the problem of an absolute minimum remains unsolved. Its solution requires either utilization of sufficient optimality conditions with additional defining of $\varphi(t,x)$ (12), or the analysis of the group properties of the problem (54). It is possible to use method (56) containing some stronger operation of improvement than in (52).

Example 1.

We shall consider a three level system being excited by three resonance fields. The external field $E = \sum_{j=1}^{3} u^j(t)\exp(i\omega^j t)$ is assumed to be linearly polarized with amplitudes of the quasimonochromatic field $u^j(t)$ being limited by M_j and detunings $\nu_i = \omega_{ij} - \omega_k < \omega_{ij}$ being constant. The maximum population of level (1) at time T is to be found. At the initial time population of level (3) is equal to 1 and levels 1 and 2 are not populated ; $T = 3$, $(Vid/\hbar)max = 1$, relaxation constant $\chi_i = 0.3$. The initial value of the functional (with all the three controls being maximal) is equal to 0.293 , then at the first iteration it is 0.475 , at the second 0.482 , the third 0.483 , the fourth 0.487 , the fifth 0.492 , the sixth 0.492 . Optimization in the class of constant controls provides for $I = 0.303$. The dynamics of functional changes in the iterations of this problem are the following : zero : (-0.417), first : (of 0.163), second : (+ 0.254), third : (+ 0.259) . Note that on the graph for the field U^3 the part of a so-called "pseudosliding" mode is in the beginning and it is characterized by frequent switchings of controls from maximum to minimum. It is not connected with the existence of a real sliding mode, since it is the particularity of application of this computational method.

The experience of this method of application reveals that the greatest improvement is achieved as a rule in the first iteration and the iteration number does not exced 10.

As is known (55) the population of the levels in a quantum system in a resonance constant electromagnetic field is exposed to oscillations with the Rabi frequences (for example, for one photon transitions in a two-level system $\omega_r^2 = \nu^2 + (dE/\hbar)^2$. In fact we consider the system with the Rabi variable frequencies and control then by changing the external field amplitudes, trying to compensate the influence of relaxation effects.

Note as well that since we used the resonance approximation, the obtained controls and phase coordinates should be slightly varying functions in times $\sim \omega_j^{-1}$, however the periods with instantenuous on/off switchings of the fields are admissible (55).

Example 2.

As the second example we shall consider the system being excited by one external field of a molecule of a spherical top type. Its oscillating rotating spectrum has a zone structure : each oscillating component has its corresponding rotating structure, whose component, in its turn, is (2J+1) times degenerated by a magnetic quantum number m (J is a rotary quantum number). The field is to be considered linearly polarized and resonant of the transition $v = 0$, $v = 1$ (v is an oscillating quantum number). The selection rules permit transitions with an increment in $\Delta m = 0$, $\Delta v = \pm 1$, $\Delta J = \pm 1$. At the initial time the population levels are defined by the Boltzmann distribution at a temperature of 100 °K, $B = 1^-$ (B is a rotating constant), $E = 10^3$ (E is an energy difference of the first and zero oscillating levels). The control which maximizes the population of oscillating zones $v = 1$ at time $T = 10$ is to be found. (time is measured in the units equal to the inverted Rabi frequency of the system $(\hbar/dE_M)$, d is a dipole moment of a transition $v = 0 \rightarrow v = 1$, E_M is a maximal value of the external field, with $M = 1$). Relaxation is neglected and, as a rule it can be always achieved in gases at a sufficiently low pressure. Dimension of the process equations system is to be determined by a maximal rotating number J_M and it is equal to $G \cdot J_m^2 + 4 \cdot J_M + 2$. In our computations $n = 50$ and the system dimension $n = 15202$. In the absence of the external field the functional value is equal to 0.363. The control in the zero interation is assumed to be $U(t) = 1$ and here the functional equals 0.868. The iteration process converges in the first iteration and provides for a functional value which is equal to 1. Note that in four periods the control has a "pseudosliding" form (see example 1) and the duration of these periods is sufficiently great. As has been previously indicated, these periods have corresponding parts of special trajectories of the system.

It should be also noted that the solution of this problem for systems of such dimensions involves meeting high requirements of a configuration of an applied computer (for this problem solution 1000 integration steps require 800 Kb of operational storage, 100 Mb of external storage, 3 hr 15 min of computation time of EC-1045).

The above analysis of this problem makes it possible to draw the following conclusions :

1) Programmed control of the resonance light radiation effecting the quantum systems allows the essential increase of this effect efficiently.

2) The phase dimension of the corresponding problems of optimal control quickly increases with the growth of the number of system quantum states. The systems with several dozens of such states present an applied interest. The problems of optimal control correspond to such systems with dimensions of several thousands.

3) The described method of solving the problems of optimal control provides for a regular computation algorithm allowing the solution of such problems.

REFERENCES.

1. Krotov V.F., Methods of solution of variational problems on the basis of sufficient conditions for absolute minimum. I. Avtomat. i Telemeh., 23 (1962) 12, 1571-1583.
2. Krotov V.F., Methods of solving variational problems. II. Sliding regimes. Avtomat. i Telemeh., 24 (1963) 5, 581-598.
3. Krotov V.F., Methods for variational problem solution based on absolute minimum sufficient conditions. III. Avtomat. i Telemeh., 25 (1964) 7, 1037-1046.
4. Krotov V.F., Approximate synthesis of optimal control. Avtomat. i Telemeh., 25 (1964) 11, 1521-1527.
5. Krotov V.F., Sufficient optimality conditions for discrete control systems. Dokl. AN SSSR, 172 (1967) 1, 18-21.
6. Krotov V.F., Bukreev V.Z., Gurman V.I., New Variational Methods in Flight Dynamics. Mashinostroyenie, Moskva, 1969 (English transl. : NASA, Transl. TTF-657, 1971).

7. Krotov V.F., Gurman V.I., Methods and Problems of Optimal Control. Nauka, Moskva, 1973 (in Russian).
8. Bellman R., Dynamic Programming. Princeton University Press, New Jersey, 1957.
9. Ioffe A.D., Tihomirov V.M., Theory of Optimal Problems. Nauka, Moskva, 1974 (English transl.: North Holland, Amsterdam, 1973).
10. Pontryagin L.S., Boltianski V.G., Gamkrelidze P.V., Mishchenko E.F., Mathematical Theory of Optimal Processes. Fizmagtiz, Moskva, 1961 (English transl. : Interscience Publishers Inc. New York, 1962).
11. Mereau P.M., Powers W.F., A direct sufficient condition for free final time optimal control problems. SIAM J. Control Optim., 14 (1976) 4, 613-622.
12. Rozenberg G.S., On the necessity and the sufficient conditions for minimum in variational problems. Differencialnye Uravnenija, (1968) 2 (transl. to English as Differential Equations).
13. Ziedan V., 1st and 2nd order sufficient conditions for optimal control and the calculus of variations. Appl. Math. Optim., 11 (1984) 3, 209-226.
14. Girsanov I.V., On a relation between Krotov and Bellman functions in the dynamic programming method. Vestnik Moskov. Univ., Ser. I, Mat. Meh., (1968), 56-59 (in Russian).
15. Khrustalev M.M., On sufficient conditions of global maximum. Dok. AN SSSR, Ser. Mat., 174 (1967) 5 (in Russian).
16. Ioffe A.D., Convex functions occuring in variational problems and the absolute minimum problem. Mat. Sb., 88 (1972), 194-210.
17. Vinter R.B., Weakest conditions for existence of Lipchitz continuous Krotov functions in optimal control theory. SIAM J. Control Optim., 21 (1983) 2, 215-234.
18. Vinter R.B., New global optimality conditions in optimal control theory. SIAM J. Control Optim., 21 (1983) 2, 235-245.
19. Vinter R.B., Lewis R.M., A verification theorem which provides a necessary and sufficient condition for optimality. IEEE Trans. Autom. Control, AC-25 (1980) 1, 84-89.
20. Klotzler R., Starke Qualitat in der Steuerungstheorie. Math. Nachr., 95 (1980), 253-263.
21. Khrustalev M.M., On sufficient optimality conditions in the problem with phase coordinates constraints. Avtomat. i Telemeh., (1967) 4, 1829 (in Russian).
22. Kelley H.J., Gradient theory of optimal flight path. ARS J., 30 (1960) 10.

23. Eneev T.M., On application of gradient method in optimal control theory problems. Kosmicheskiye Issledovania, 4 (1966) 5 (in Russian).

24. Krylov I.A., Chernousko F.L., On a method of consecutive approximations for solving optimal control problems. Z. Vycisl. Mat. i Fiz., 2 (1962) 6, 1132-1139 (in Russian).

25. Bryson A.E., Ho Y. Ch., Applied Optimal Control. Hemisphere Publ. Corp., Washington, 1975.

26. Krotov V.F., Feldman I.N., An iterative method for solving problems of optimal control. Izv. AN SSSR, Technic. Kibern. (1983) 2, 162-167 (in Russian).

27. Propoi A.I., Elements of the Theory of Optimal Discrete Systems. Nauka, Moskva, 1981 (in Russian).

28. Kazakov V.A., Krotov V.F., Optimal control of resonance interaction of the light and a substance. Avtomat. i Telemeh., (1987) 4, 9-15.

29. Emelicev V.A., Komlik V.I., A Method for Constructing a Sequence of Plans for Solving Discrete Optimal Problems. Nauka, Moskva, 1981 (in Russian).

30. Wagner G., Principles of Operations Research with Application to Managerial Decisions. Prentice Hall, Englewood Cliffs, 1969.

31. Gurman V.I., Baturin V.A., Danilina E.V., New Methods of Improving Controlled Process. Nauka, Novosibirsk, 1987 (in Russian).

32. Krotov V.F., Computing algorithms for solution and optimization of a controlled system of equations. I. Izv. AN SSSR, Technic. Kibern., (1975) 5, 3-15 (in Russian).

33. Krotov V.F., Computing algorithms for solution and optimization of a controlled system of equations. II. Izv. AN SSSR, Technic. Kibern., (1975) 6, 3-13 (in Russian).

34. Krotov V.F., Sergeev S.I., Computing algorithms for solution of some problems in linear and linear-integer programming. Avtomat. i Telemeh., Part I (1980) 12, 86-96, Part II (1981) 1, 86-96, Part III (1981) 3, 83-94, Part IV (1981) 4, 103-112.

35. Egorov V.A., Gusev L.I., The Dynamics of the Flight between the Earth and the Moon. Nauka, Moskva, 1980 (in Russian).

36. Gukasian M.H., On computing algorithms for optimization in linear controlled systems with distributed parameters. *in* : Investigations in Mechanics of a Solid Deforming Body. Erewan, 1981, 93-97.

37. Bukreev V.Z., Concerning certain method of approximate synthesis of optimal control. Avtomat. i Telemeh., (1968) 11, 5-13.

38. Bukreev V.Z., Rozenblat G.M., Approximate design of optimal control for discrete-time systems. Avtomat. i Telemeh., (1976) 1, 90-93.

39. Bakhito R.U., Krapchetkov N.P., Krotov V.F., The synthesis of an approximately optimal control for one class of controlled systems. Avtomat. i Telemeh., (1972) 10, 33-43.

40. Trigub M.B., Approximately optimal stabilization of one class of nonlinear systems. Avtomat. i Telemeh., (1987) 1, 34-47.

41. Lekkerke P.I., Derleman T.W., Calculation method for dynamic optimization using quadratic approximations to minimum surface. Automatica, 7 (1971) 6, 713.

42. Butkovski A.G., Samoilenko Yu I., Control for Quantum-mechanical processes. Nauka. 1984.

43. Krasnov I.V., Shaparev N. Ya., Shkedov I.M., Optimal Control for photoprocesses in gas. - Preprints of Academy of Sci. 1979, N°10.

44. Khorozov O.A., Control of coherent quantum-mechanical processes. - Dissertation, Institute of Cybernetics, Ukrained SSR Academy of Sci., Kiev, 1984.

45. Bunkin F.V., Kirichenko N.A., Lukyonchuk B.S., Optimal modes for heating material by laser radiation. Preprint of 1979, N°146.

46. Vagners J., Neal R., Vlasas G., Optimal laser plasma heating in a solenoidal magnetic field. Physics of fluids, 1975, v.18, 1314-1320.

47. Fain V.M., Photons and nonlinear media "Sov. Radio", 1972.

48. Butylkin V.S., Kaplan A.E., Tronopulo Yu. G., Yakubovich E.I. , Light resonance interactions with substance, Publ. Nauka, 1977.

49. Artamonova N.D., Platonenko V.T., Khokhlov R.V., On control of chemical reactions by resonance photoeffects on molecules. Publ., 1970, vol.58, 2195-2201.

50. Keldysh L.V., Ionization in the field of a strong light wave, 1964, vol.47, 1945-1957.

51. Voronov G.S., Probability of atoms ionization and its dependence on the intensity of photons flow. 1966, vol.51, 1496-1498.

52. Krotov V.F., Feldman I.N., Iteration method of solving the problemsof optimal control. USSR Academy of Sci. Technical Cybernetics, N°2, 1983, 160-169.

53. Krotov V.F., Bukreev V.Z., Gurman V.I., New technique of variational computation in flight dynamics. Publ. Mashinostroyenie, 1969.

54. Somoilenko Yu. I., Optimal Control of Quantum Statistical set. Automatics and Telemechanics.

55. Delone N.B., Krainov V.P., Atom in strong light field. Publ. Energoatomizdat, 1984.

56. Salomaa R., Stendoholm S., Field optimization aspects in two-photon excitations. Appl. Phys., 1978, vol.17, 309-316.

57. Gurman V.I., Rasina I.V., On practical application sufficient conditions for strong relative minimum. Automatics and Telemechanics, 1979, N°10.

Nongaussian linear filtering, identification of linear systems, and the symplectic group

Michiel Hazewinkel
Centre for Mathematics and Computer Science
P.O. Box 4079, 1009 AB Amsterdam, The Netherlands

ABSTRACT

Consider stochastic linear dynamical systems, $dx = Axdt + Bdw, dy = Cxdt + dv, y(0) = 0, x(0)$ a given initial random variable independent of the standard independent Wiener noise processes w, v. The matrices A, B, C are supposed to be constant. In this paper I consider two problems. For the first one A, B and C are supposed known and the question is how to calculate the conditional probability density of x at time t given the observations $y(s), 0 \leq s \leq t$ in the case that $x(0)$ is not necessarily gaussian. (In the gaussian case the answer is given by the Kalman-Bucy filter). The second problem concerns identification, i.e. the A, B, C are unknown (but assumed constant so that $dA = 0, dB = 0, dC = 0$), and one wants to calculate the joint conditional probability density at time t of (x, A, B, C), again given the observations $y(s), 0 \leq s \leq t$. The methods used rely on Wei-Norman theory, the Duncan-Mortensen-Zakai equation and a "real form" of the Segal-Shale-Weil representation of the symplectic group $Sp_n(\mathbf{R})$.

AMS classification: 93E11, 93B30, 17B99, 93C10, 93B35, 93E12
Key words and phrases: nongaussian distribution, identification, non-linear filtering, DMZ equation, Duncan-Mortensen-Zakai equation, propagation of nongaussian initials, Wei-Norman theory, Segal-Shale-Weil representation, reference probability approach, unnormalized density, Kalman-Bucy filter, Lie algebra approach to nonlinear filtering.

1. Introduction

Consider a general nonlinear filtering problem of the following type:

$$dx = f(x)dt + G(x)dw \;,\quad x \in \mathbf{R}^n,\ w \in \mathbf{R}^m \tag{1.1}$$

$$dy = h(x)dt + dv \;,\quad y \in \mathbf{R}^p,\ v \in \mathbf{R}^p \tag{1.2}$$

where f, G, h are vector and matrix valued functions of the appropriate dimensions, and the w, v are standard Wiener processes independent of each other and also independent of the initial random variable $x(0)$. One takes $y(0) = 0$.

The general non-linear filtering problem is this setting asks for (effective) ways to calculate and/or approximate the conditional density $\pi(x,t)$ of x given the observations $y(s), 0 \leq s \leq t$; i.e. $\pi(x,t)$ is the density of $\hat{x} = E[x(t)|y(s), 0 \leq s \leq t]$ the conditional expectation of the state $x(t)$.

One approach to this problem proceeds via the socalled DMZ equation which is an equation of a rather nice form for an unnormalized version $\rho(x,t)$ of $\pi(x,t)$. Here unnormalized means that $\rho(x,t) = r(t)\pi(x,t)$ for some function $r(t)$ of time alone. A capsule description of this approach is given in section 2 below. Using this approach was strongly advocated by Brockett and Mitter (cf. e.g. their contributions in [6]), and initially the approach had a number of non-trivial successes, both in terms of positive and negative results (cf. e.g. the surveys [9] and [4]). Subsequently, the approach became less popular; perhaps because a number of rather formidable mathematical problems arose, and because the number of systems to which the theory can be directly applied appears to be quite small. Cf [4] for a discussion of some aspects of these two points.

It is the purpose of this paper to apply this approach to two problems concerning linear systems, which do not fall within the compass of the usual Kalman-Bucy linear filtering theory. More precisely, consider a linear stochastic dynamical system

Various subselections of the material in this article have formed the subject of various talks at different conferences; e.g. the 2nd conference on the road-vehicle system in Torino in June 1987, the 24-th Winter school on theoretical physics in Karpacz in January 1988, the 3rd meeting of the Bellman continuum in Valbonne in June 1988, the present one, and the special program on signal processing of the IMA in Minneapolis in the summer of 1988. As a result this article may also appear in the proceedings of these meetings.

$$dx = Axdt + Bdw, \quad x \in \mathbf{R}^n, \; w \in \mathbf{R}^m \tag{1.3}$$

$$dy = Cxdt + dv, \quad y, v \in \mathbf{R}^p \tag{1.4}$$

where the A,B,C are matrices of the appropriate sizes. The first problem I want to consider is the filtering of (1.3)-(1.4) in the case that the initial condition $x(0)$ is a non-gaussian random variable. The second problem concerns the identification of (1.3)-(1.4); i.e. one assumes that the matrices A,B,C are constant but unknown and it is desired to calculate the conditional density $\pi(x,A,B,C,t)$ of the (enlarged) state (x,A,B,C) at time t. Technically this means that one adds to (1.3)-(1.4) the equations

$$dA = 0, \; dB = 0, \; dC = 0 \tag{1.5}$$

and one considers the filtering problem for the nonlinear system (1.3)-(1.5). Strictly speaking this problem is not well posed. Simply because A,B,C can not be uniquely identified on the basis of the observations alone. In the DMZ equation approach this shows up only at the very end in the form that $\rho(x,A,B,C,t)$ will be degenerate in the sense that $\rho(Sx,SAS^{-1},SB,CS^{-1},t) = \rho(x,A,B,C,t)$ for all constant invertible real matrices S. As a result the normalization factor $\int \rho(x,A,B,C,t)dxdAdBdC$ does not exist, and in fact $\pi(x,A,B,C,t)$ is also degenerate. One gets rid of this by passing to the quotient space (finite moduli space) $\{(x,A,B,C)\}/GL_n(\mathbf{R})$ for the action just given and/or by considering (local) canonical forms. The normalization factor can be calculated by integrating over this quotient space.

Besides the DMZ-equation, already mentioned, the tools used to tackle the two problems described above are Wei-Norman theory and something which could be called a real form of the Segal-Shale-Weil representation of the symplectic Lie group $Sp_n(\mathbf{R})$. These two topics are discussed in sections 3 and 4 below.

2. The DMZ approach to nonlinear filtering

Consider again the general nonlinear system (1.1)-(1.2). These stochastic differential equations are to be considered as Ito equations. Let $\pi(x,t)$ be the probability density of $E[x(t)|y(s), 0 \le s \le t]$, the conditional expectation of $x(t)$. (Given sufficiently nice f,G and h if can be shown that $\pi(x,t)$ exists.) Then the Duncan-Mortensen-Zakai result [1, 10, 12] is that there exists an unnormalized version $\rho(x,t)$ of $\pi(x,t)$, i.e. $\rho(x,t) = r(t)\pi(x,t)$, which satisfies an evolution equation

$$d\rho = \mathcal{L}\rho dt + \Sigma h_k \rho dy_k(t), \; \rho(x,0) = \psi(x) \tag{2.1}$$

where $\psi(x)$ is the distribution of the initial random variable $x(0)$ and where $\mathcal{L}$ is the second-order partial differential equation

$$\mathcal{L}\phi = \frac{1}{2}\sum_{i,j} \frac{\partial^2}{\partial x_i \partial x_j}(GG^T)_{ij}\phi - \sum_i \frac{\partial}{\partial x_i} f_i \phi - \frac{1}{2}\sum_k h_k^2 \phi \tag{2.2}$$

Here $h_k, y_k(t), f_i$ are components of $h, y(t)$ and f respectively and $(GG^T)_{ij}$ is the (i,j)-entry of the product GG^T of the matrix G and its transpose.

Equation (2.1) is a Fisk-Stratonovič stochastic differential equation. The corresponding Ito differential equation is obtained by removing the $-\frac{1}{2}\Sigma h_k^2 \phi$ term from (2.2).

As it stands (2.1) is a stochastic partial differential equation. However the transformation

$$\tilde{\rho}(x,t) = \exp(\Sigma h_k(x) y_k(t))\rho(x,t) \tag{2.3}$$

turns it into the equation

$$d\tilde{\rho} = (\mathcal{L}\tilde{\rho} + \Sigma \mathcal{L}_i \tilde{\rho} y_i + \frac{1}{2}\Sigma \mathcal{L}_{i,j}\tilde{\rho} y_i y_j)dt \tag{2.4}$$

where $\mathcal{L}_i$ is the operator commutator $\mathcal{L}_i = [h_i, \mathcal{L}] = h_i \mathcal{L} - \mathcal{L} h_i$ and $\mathcal{L}_{ij} = [h_i, [h_j, \mathcal{L}]]$. Cf. [4] for more details. In (2.4) I have explicity indicated the dependence of the various quantities on x,t to stress that here $h(x)$ should simply be seen as a known function of x and not as the time function $h(x(t))$. Equation (2.4) does not involve the derivatives dy_i anymore; it makes sense for all possible paths $y(t)$, and can be regarded as a family of PDE parametrized by the possible observation paths $y(t)$. Thus there is a robust version of (2.1) and we can work with (2.1) as a parametrized family of PDE parametrized by the $y(t)$. Note that knowledge of $\tilde{\rho}(x,t)$ (and $y(t)$) immediately gives $\rho(x,t)$ and that the conditional expectation of any function $\phi(x(t))$ of the state at time t can be calculated by

$$E[\phi(x(t))|y(s), 0 \le s \le t] = \left(\int \rho(x,t)dx\right)^{-1} \int \phi(x)\rho(x,t)\, dx \tag{2.5}$$

Possibly the simplest example of a filtering problem is provided by one-dimensional Wiener noise linearly observed:

$$dx = dw, \; x, \; w \in \mathbf{R} \tag{2.6}$$

$$dy = xdt + dv, \quad y, v \in \mathbf{R}. \tag{2.7}$$

In this case the corresponding DMZ equation is

$$d\rho = \left(\frac{1}{2}\frac{d^2}{dx^2} - \frac{1}{2}x^2\right)\rho dt + x\rho\, dy \tag{2.8}$$

an Euclidean Schrödinger equation for a forced harmonic oscillator.

3. Wei-Norman theory

Wei-Norman theory is concerned with solving partial differential equations of the form

$$\frac{\partial p}{\partial t} = u_1 A_1 p + \cdots + u_m A_m p \tag{3.1}$$

where the $A_i, i=1,...,m$ are linear partial differential operators in the space variables $x_1,...,x_n$, and the $u_i, i=1,...,m$ are given functions of time, in terms of solutions of the simpler equations

$$\frac{\partial p}{\partial t} = A_i p \ , \ i=1,...,m \tag{3.2}$$

which we write as

$$p(x,t) = e^{A_i t}\psi(x) \, , \, \psi(x) = p(x,0) \tag{3.3}$$

Originally, the theory was developed for the finite dimensional case, i.e. for systems of ordinary differential equations

$$\dot{z} = u_1 A_1 z + \cdots + u_m A_m z \tag{3.4}$$

where $z \in \mathbf{R}^k$, and the A_i are $k \times k$ matrices. Both in the finite dimensional case (3.4) and the infinite dimensional case (3.1) it is well known that besides in the given directions $A_1 p,...,A_m p$, the to be determined function or vector can also move (infinitesimally) in the directions given by the commutators $[A_i, A_j]p = (A_i A_j - A_j A_i)p$, and in the directions given by repeated commutators $[[A_i, A_j], A_k]$, $[[A_i, A_j], [A_k, A_l]]$, etc. etc.

Let $\mathrm{Lie}(A_1,...,A_m)$ be the Lie algebra of operators generated by the operators $A_1,...,A_m$. This is the smallest vector space L of operators containing $A_1,...,A_n$ and such that if $A,B \in L$ then also $[A,B] := AB - BA \in L$. In the finite dimensional case (3.4) L is always finite dimensional, a subvector space of $gl_k(\mathbf{R})$, the vectorspace (Lie algebra) of all $k \times k$ matrices. In the infinite dimensional case the Lie algebra generated by the operators $A_1,...,A_m$ in (3.1) can easily be infinite dimensional and it often is; also in the cases coming from filtering problems via the DMZ equation. Cf. [5] for a number of examples.

This is the essential difference between (3.1) and (3.4). Accordingly, here I shall assume that the Lie algebra $L = \mathrm{Lie}(A_1,...,A_m)$ generated by the operators $A_1,...,A_m$ in (3.1) is finite dimensional. For a discussion of various infinite dimensional versions of Wei-Norman theory cf. [4]. Hence, granting this finite dimensionality property, by setting, if necessary, some of the $u_i(t)$ equal to zero, and by combining other $u_j(t)$ in the case of linear dependence among the operators on the RHS of (3.1), without loss of generality, we can assume that we are dealing with an equation

$$\frac{\partial p}{\partial t} = u_1 A_1 p + \cdots + u_n A_n p \tag{3.5}$$

with the additional property that

$$[A_i, A_j] = \sum_k \gamma_{ij}^k A_k \quad ; i,j = 1,...,n \tag{3.7}$$

for suitable real constants γ_{ij}^k; $i,j,k = 1,...,n$.

The central idea of Wei-Norman theory is now to try for a solution of the form

$$p(t) = e^{g_1(t)A_1} e^{g_2(t)A_2} \cdots e^{g_n(t)A_n}\psi \tag{3.8}$$

where the g_i are still to be determined functions of time. The next step is to insert the Ansatz (3.8) into (3.5), to obtain

$$\dot{p} = \dot{g}_1 A \, e^{g_1 A_1} \cdots e^{g_n A_n}\psi + e^{g_1 A_1}\dot{g}_2 A_2 e^{g_2 A_2} \cdots e^{g_n A_n}\psi + \cdots \tag{3.9}$$
$$+ \, e^{g_1 A} \cdots e^{g_{n-1}A_{n-1}}\dot{g}_n A_n e^{g_n A_n}\psi$$

Now, for $i=2,...,n$ insert a term

$$e^{-g_{i-1}A_{i-1}} \cdots e^{-g_1 A_1} e^{g_1 A_1} \cdots e^{g_{i-1}A_{i-1}}$$

just behind $\dot{g}_i A_i$ in the i-th term of (3.9). Then use the adjoint representation formula

$$e^A B e^{-A} = B + [A,B] + \frac{1}{2!}[A,[A,B]] + \frac{1}{3!}[A,[A,B]]] + \cdots \tag{3.10}$$

and (3.7)) repeatedly, and use the linear independence of the $A_1,...,A_n$ to obtain a system of ordinary differential equations for the $g_1,...,g_n$ (with initial conditions $g_1(0)=0=g_2(0)=...=g_n(0)$).

These equations are always solvable for small time. However they may not be solvable for all time, meaning that finite escape time phenomena can occur.

Let's consider an example, viz. the example afforded by the DMZ equation (2.8). One calculates that

$$[\frac{1}{2}\frac{d^2}{dz^2} - \frac{1}{2}x^2, x] = \frac{d}{dx} \, , \, [\frac{1}{2}\frac{d^2}{dx^2} - \frac{1}{2}x^2, \frac{d}{dx}] = x$$

$$[\frac{d}{dz}, x] = 1, \ [A,1] = 0$$

where A is any linear combination of the four operators $\frac{1}{2}\frac{d^2}{dx^2} - \frac{1}{2}x^2, x, \frac{d}{dx}, 1$. Applying the recipe sketched above to the equation

$$\dot{\rho} = (\frac{1}{2}\frac{d^2}{dx^2} - \frac{1}{2}x^2)\rho + x\rho u(t) + \frac{d}{dx}\rho 0 + 1\rho 0 \tag{3.11}$$

one finds the equations

$$\dot{g}_1 = 0\ , \ \cosh(g_1)\dot{g}_2 + \sinh(g_1)\dot{g}_3 = u(t), \tag{3.12}$$
$$\sinh(g_1)\dot{g}_2 + \cosh(g_1)\dot{g}_3 = 0, \ \dot{g}_4 = \dot{g}_3 g_2$$

which are solvable for all time.

This fact and the form of the resulting equations: straightforward quadratures and one set of linear equations $B(t)g = b(t)$, with $B(t), b(t)$ known and $B(t)$ invertible, is typical for the case that the Lie algebra $L = \oplus \mathbf{R} A_i$ spanned by the $A_1,...,A_n$ is solvable. This means the following. Let $[L,L]$ be the subvectorspace of L spanned by all the operators of the form $[A,B], A,B \in L$. It is easily seen that this is again a Lie algebra. Inductively let $L^{(n)} = [L, L^{(n-1)}]$ be the subvectorspace of L spanned by all operators of the form $[A,B]$, $A \in L, B \in L^{(n-1)}$, $L^{(0)} = L$. These are all sub Lie algebras of L.

The Lie algebra of L is called nilpotent if $L^{(n)} = 0$ for n large enough. It is called solvable if $[L,L]$ is nilpotent. The phenomenon alluded to above, i.e. solvability of the Wei-Norman equations for all time, always happens in case L is solvable [11]. (And it is no accident that these algebras have been called solvable. Though this is not the result which gave them that name.)

Note that the DMZ equation (2.1) corresponding to a nonlinear filtering problem (1.1)-(1.2) is of the type (3.1) (with $u_k(t) = dy_k(t)$). Thus the Lie Algebra generated by the operators $\pounds$, $h_1(x),...,h_p(x)$ occuring in (2.1) clearly has much to say about how difficult the filtering problem is. This Lie algebra is called the *estimation Lie algebra* of the system (1.1)-(1.2) and it can be used to prove a variety of positive and negative results about the filtering problem [4, 5, 9].

4. The Segal-Shale-Weil representation and a 'real form'

Let J be the standard symplectic matrix $J = \begin{pmatrix} 0 & I_n \\ -I_n & 0 \end{pmatrix}$, where I_n the $n \times n$ unit matrix. Consider the vector space of $2n \times 2n$ real matrices defined by

$$sp_n(\mathbf{R}) = \{M: JM + M^T J = 0\}. \tag{4.1}$$

Writing M as a 2×2 block matrix, $M = \begin{pmatrix} A & B \\ C & D \end{pmatrix}$, the conditions on the $n \times n$ blocks A,B,C,D become

$$B^T = B, \ C^T = C, \ D = -A^T. \tag{4.2}$$

As we shall see shortly below this set of matrices occurs naturally for filtering problems coming from linear systems (1.1)-(1.2).

The corresponding Lie group to $Sp_n(\mathbf{R})$ is the group of invertible $2n \times 2n$ matrices defined by

$$Sp_n(\mathbf{R}) = \{S \in \mathbf{R}^{2n \times 2n}: S^T J S = J\} \tag{4.3}$$

(This is a *group* of matrices in that if $S_1, S_2 \in Sp_n(\mathbf{R})$ then also $S_1 S_2 \in Sp_n(\mathbf{R})$ and $S_1^{-1} \in Sp_n(\mathbf{R})$ as is easily verified.)

There is a famous representation of $Sp_n(\mathbf{R})$ (or more precisely of its two-field covering group $\tilde{Sp}_n(\mathbf{R})$) in the Hilbert space $L^2(\mathbf{R}^n)$ called the Segal-Shale-Weil representation or the oscillator representation; cf. [8]. Here the word 'representation' means that to each $S \in Sp_n(\mathbf{R})$ there is associated a unitary operator U_S such that $U_{S_1 S_2} = U_{S_1} U_{S_2}$ for all $S_1, S_2 \in Sp_n(\mathbf{R})$.

For the purposes of this paper a modification of it is of importance. It can be described as follows by explicit operators associated to certain specific kinds of elements of $Sp_n(\mathbf{R})$:

(i) Let P be a symmetric $n \times n$ matrix; then to the element

$$\begin{pmatrix} I & P \\ 0 & I \end{pmatrix} \in Sp_n(\mathbf{R})$$

there is associated the operator $f(x) \mapsto \exp(x^T P x) f(x)$

(ii) Let $A \in GL_n(\mathbf{R})$ be an invertible $n \times n$ matrix. Then to the element

$$\begin{bmatrix} A & 0 \\ 0 & (A^{-1})^T \end{bmatrix} \in Sp_n(\mathbf{R})$$

there is associated the operator

$$f(x) \mapsto |\det(A)|^{1/2} f(A^T x)$$

(iii) let Q be a symmetric $n \times n$ matrix. Then to the element

$$\begin{bmatrix} I & 0 \\ Q & I \end{bmatrix} \in Sp_n(\mathbf{R})$$

there is associated the operator

$$f(x) \mapsto \mathcal{F}^{-1}(\exp(x^T Q x)\mathcal{F}f(x))$$

where $\mathcal{F}$ denotes the Fourier transform.

(The operator corresponding to the element

$$\begin{bmatrix} 0 & I \\ -I & 0 \end{bmatrix} \in Sp_n(\mathbf{R})$$

is in fact the Fourier transform itself).

Except for one snag to be discussed below, this suffices to describe the operator which should be associated to any element $S \in Sp_n(\mathbf{R})$. Indeed let

$$S = \begin{bmatrix} S_1 & S_2 \\ S_3 & S_4 \end{bmatrix} \in Sp_n(\mathbf{R}) \tag{4.4}$$

then there is an $s>0, s \in \mathbf{R}$ such that $S_1 + sS_2$ is invertible and we have a factorisation

$$\begin{bmatrix} S_1 & S_2 \\ S_3 & S_4 \end{bmatrix} = \begin{bmatrix} I & 0 \\ (S_3+sS_4)(S_1+sS_2)^{-1} & I \end{bmatrix} \begin{bmatrix} I & S_2(S_1+sS_2)^T \\ 0 & I \end{bmatrix} \begin{bmatrix} S_1+sS_2 & 0 \\ 0 & (S_1^T+sS_2^T)^{-1} \end{bmatrix} \begin{bmatrix} I & 0 \\ -s & I \end{bmatrix} \tag{4.5}$$

(It is easily verified that all four factors on the right are in fact in $Sp_n(\mathbf{R})$.

Now assign to the operator S the product of the four operators corresponding to the factors on the RHS of (4.5) according to the recipe (i)-(iii) given above. There is a conceivable second snag here in that it seems a priori possible that different factorisations could give different operators. This in fact does not happen precisely because the 'representation' described by (i)-(iii) is a 'real form' of the oscillator representation $Sp_n(\mathbf{R}) \to \mathrm{Aut}(L^2(\mathbf{R}^n))$. The relation between the oscillator representation and (i)-(iii) above is given by the substitution $x_k \mapsto \sqrt{i}\, x_k$ where $i = \sqrt{-1}$. (The possible sign ambiguity which could come from the fact that the oscillator representation is really a representation of the covering $\tilde{Sp}_n(\mathbf{R})$ rather than $Sp_n(\mathbf{R})$ itself also seems not to happen; if would in any case be irrelevant for the applications dicussed below.)

It remains to discuss the first snag mentioned just above (5.4) and why the words 'representation' and 'real form' above have been placed in quotation marks. The trouble lies in part (iii) of the recipe. Taking a Fourier transform and than multiplying with a quadratic exponential may well take one out of the class of functions which are inverse Fourier transformable. Another way to see this is to observe that the operator described in (iii) assigns to a function ψ the value in $t=1$ of the solution of the evolution equation

$$\frac{\partial \rho}{\partial t} = ((\frac{\partial}{\partial x})^T Q \frac{\partial}{\partial x})\rho\,,\ \rho(x,0) = \psi(x) \tag{4.6}$$

and if Q is not nonnegative definite this involves anti-diffusion components for which the solution at $t=1$ may not exist. Additionally, - but this is really the same snag - applying recipe (i) to a function may well result in a function that is not Fourier transformable.

What we have in fact is not a representation of all of $Sp_n(\mathbf{R})$ but only a representation of a certain sub-semi-group cone in $Sp_n(\mathbf{R})$.

For the applications to be described below this means that we must be careful to take factorizations such that applying the various operators successively continues to make sense. The factorization (5.5) does not seem optimal in that respect and we shall for the special elements of $Sp_n(\mathbf{R})$ which come from filtering problems use a different one.

Incidentally, one says that two structures over $\mathbf{R}$ are real forms of one another if after tensoring with $\mathbf{C}$ (= extending scalars to $\mathbf{C}$) they become isomorphic (over $\mathbf{C}$). It is in this sense that the 'representation' described by the recipe (i)-(iii) is a 'real form' of the oscillator representation.

5. Propagation of non-gaussian initials

Now, finally, after all this preparation, consider a known linear dynamical system

$$dx = Axdt + Bdw,\ Cxdt + dv_i,\ x \in \mathbf{R}^n, w \in \mathbf{R}^m,\ y,v \in \mathbf{R}^p. \tag{5.1}$$

with a known, not necessarily Gaussian, initial random variable $x(0)$ with probability distribution $\psi(x)$.

The *DMZ* equations in this case is as follows

$$d\rho = \pounds\rho dt + \sum_{j=1}^{p}(Cx)_j dy_j(t) \tag{5.2}$$

where $(Cx)_j$ is the j-th component of the p-vector Cx. The operator $\pounds$ in this case has the form

$$\pounds = \frac{1}{2}\sum_{i,j}(BB^T)_{i,j}\frac{\partial^2}{\partial x_i \partial x_j} - \sum_{i,j}A_{ji}x_j\frac{\partial}{\partial x_i} - \mathrm{Tr}(A) - \frac{1}{2}\sum_j (Cx)_j^2 \tag{5.3}$$

Taking brackets of the multiplication operators $(Cx)_j$ with $\pounds$ yields a linear combination of the operators

$$x_1,\ldots,x_n;\ \frac{\partial}{\partial x_1},\ldots,\frac{\partial}{\partial x_n};\ 1. \tag{5.4}$$

This is a straightforward calculation to check. Moreover, the bracket (= commutator product) of $\pounds$ with any of the operators in (5.4) again yields a linear combination of the operators listed in (5.4). It follows that for linear stochastic dynamical systems (5.1) the associated estimation Lie algebra (= the Lie algebra generated by $\pounds$, $(Cx)_1,\ldots,(Cx)_p$) is always solvable of dimension $\leq 2n+2$.

As a mather of fact it is quite simple to prove that the system (5.1) is completely reachable and completely observable if and only if the dimension of the estimation Lie algebra is precisely $2n+2$ so that a basis of the algebra is formed by the $(2n+1)$ operators of (5.4) and $\pounds$ itself.

In all cases Wei-Norman theory is applicable (working perhaps with a slightly larger Lie algebra than strictly necessarily makes no real difference).

Thus we can calculate effectively the solutions of the unnormalized density equation (5.2) provided we have good ways of calculating the expressions.

$$e^{t\pounds}\psi,\ e^{tx_i}\psi,\ e^{t\frac{\partial}{\partial x_i}}\psi,\ e^{t}\psi \tag{5.5}$$

for arbitrary initial data ψ. The last three expressions of (5.5) cause absolutely zero difficulties $(\exp(t\frac{\partial}{\partial x_i})\psi = \psi(x_1,\ldots,x_{i-1},x_i+t,x_{i+1},\ldots,x_n))$. Thus it remains to calculate the $e^{t\pounds}\psi$ where $\pounds$ is an operator of the form (5.3). It is at this point that the business of the Segal-Shale-Weil representation of the previous section comes in. As a matter of fact the Segal-Shale-Weil representation itself, not the 'real form' described in section 4 above, is a representation of the Lie algebra spanned by the operators

$$i\,x_k x_j,\ x_k\frac{\partial}{\partial x_j} + \frac{1}{2}\delta_{k,j},\ i\frac{\partial^2}{\partial x_k \partial x_j},\ i = \sqrt{-1} \tag{5.6}$$

and apart form multiples of the identity (which hardly matter) and the occurence of $\sqrt{-1}$ these are the constituents of the operators $\pounds$ in (5.3). It is to remove the factors $\sqrt{-1}$ that we have to go to a real form. Cf. [3] for more details on the Segal-Shale-Weil representation itself, and what it, and its real form, have to do with Kalman-Bucy filters.

It is convenient not to have to worry about multiples of the identity. To this end note that if $\pounds' = \pounds + aI$ then $\exp(t\pounds')\psi = \exp(ta)\exp(t\pounds)\psi$, so that neglecting multiples of the identity indeed matters hardly.

The first observation is now that, modulo multiples of the identity operator, if $\pounds$ and $\pounds'$ are two operators of the form (5.3) then their commutator difference $[\pounds,\pounds'] = \pounds\pounds' - \pounds'\pounds$ is again of the same form. (To make this exact replace $\pounds$ in (5.3) by $\pounds + \frac{1}{2}\mathrm{Tr}(A)$ and similarly for $\pounds'$.) Thus these operators actually form a finite dimensional Lie algebra and this is, of course, the symplectic Lie algebra $sp_n(\mathbf{R})$. The correspondence is given by assigning to $\pounds(=\pounds(A,B,C))$ the $2n \times 2n$ matrix

$$\pounds(A,B,C) \rightarrow \begin{pmatrix} -A^T & -C^TC \\ -BB^T & A \end{pmatrix} \tag{5.7}$$

(If you want to be finicky it is the operator $\pounds(A,B,C) + \frac{1}{2}\mathrm{Tr}(A)$ which corresponds to the matrix on the right of (5.7).)

In terms of a basis on the left and right side the correspondence (i.e. the isomorphism of Lie algebras) is given as follows. Let E_{ij} be the $n \times n$ matrix with a 1 in spot (i,j) and zero everywhere else. Then

$$\frac{\partial^2}{\partial x_i \partial x_j} \mapsto \begin{pmatrix} 0 & 0 \\ -E_{ij}-E_{ji} & 0 \end{pmatrix} \tag{5.8}$$

$$x_i \frac{\partial}{\partial x_j} + \frac{1}{2}\delta_{i,j} \mapsto \begin{bmatrix} E_{ij} & 0 \\ 0 & -E_{ji} \end{bmatrix} \tag{5.9}$$

$$x_i x_j \mapsto \begin{bmatrix} 0 & E_{ij}+E_{ji} \\ 0 & 0 \end{bmatrix} \tag{5.10}$$

It is now straightforward to check that this does indeed define an isomorphism of Lie algebras from the Lie algebra of all operators $£(A,B,C)+\frac{1}{2}\mathrm{Tr}(A)$ where $£$ is as in (5.3) and the algebra $sp_n(\mathbf{R})$ described and discussed in section 4 above. For example one has

$$[\frac{\partial^2}{\partial x_1 \partial x_2}, x_2 x_3] = x_3 \frac{\partial}{\partial x_1} \tag{5.11}$$

which fits perfectly with

$$[\begin{bmatrix} 0 & 0 \\ -E_{12}-E_{21} & 0 \end{bmatrix}, \begin{bmatrix} 0 & E_{23}+E_{32} \\ 0 & 0 \end{bmatrix}] = \begin{bmatrix} E_{31} & 0 \\ 0 & -E_{13} \end{bmatrix} \tag{5.12}$$

It is precisely the correspondence (5.8) - (5.10) or, modulo multiples of the identity, (5.7), plus the fact that 'real form' described in section 4 of the SSW representation is precisely the way to remove the $\sqrt{-1}$ factors, plus, again, the fact that the SSW is really a representation, which makes it possible to use finite dimensional calculations to obtain expressions for

$$\exp(t(£(A,B,C)+\frac{1}{2}\mathrm{Tr}(A))\psi \tag{5.13}$$

for arbitrary initial conditions.

Basically the recipe is as follows. Take $£(A,B,C)+\frac{1}{2}\mathrm{Tr}(A)$. Let $M\in sp_n(\mathbf{R})$ be its associated matrix as defined by the RHS of (5.7). Calculate $\exp(tM)=S(t)$. Write $S(t)$ as a product of matrices as in (i), (ii), (iii) in section 4. Apply successively the operators associated to the factors. The result, if defined, will be an expression for (5.13). One factorisation which can be used is that of (4.5) above. It does not, however, seem to be very optimal and it is difficult to show that everything is well defined.

It is better and more efficient to use a preliminary reduction. Consider the algebraic Riccati equation

$$A^T P + PA - PBB^T P + C^T C = 0 \tag{5.14}$$

determined by the triple of matrices (A,B,C). It is easy to check that for any solution P one has

$$\begin{bmatrix} I & -P \\ 0 & I \end{bmatrix} M \begin{bmatrix} I & P \\ 0 & I \end{bmatrix} = \begin{bmatrix} -\tilde{A}^T & 0 \\ -BB^T & \tilde{A} \end{bmatrix} \tag{5.15}$$

where $\tilde{A}=A-BB^TP$. Given this it becomes useful to know when (5.14) has a solution and to know some properties of the solutions. These will also be important for the next section. In fact the function $\mathrm{rc}(A,B,C)$ that assigns to the triple (A,B,C) under suitable conditions the unique positive definite solution of (5.14) is important enough to be considered a standard named function which should be available in accurate tabulated form much as say the Airy function or Bessel functions. I know of no such tables. The symbol 'rc' of course stands for Riccati.

Let A^* be the adjoint of the complex $n\times n$ matrix A, i.e. the conjugated transpose of A, so, if A is real, $A^*=A^T$. Consider the equation (algebraic Riccati equation)

$$A^*P + PA = PBB^*P - C^*C \tag{5.16}$$

(Here A is an $n\times n$ matrix, B an $n\times m$ matrix, C an $p\times n$ matrix.) Some facts about (5.16) are then as follows:

(5.17) If (A,B) is stabilizable, i.e. if there exists an F such that $A-BF$ has all eigenvalues with negative real part, then there is a solution of (5.16) which is positive semidefinite ($P\geqslant 0$) (and for this solution $\tilde{A}=A-BB^*P$ is stable).

(So in particular if (A,B) is completely reachable there is a solution of (5.14).)

(5.18) Suppose (5.16) has a solution $P\geqslant 0$ and suppose that (A,C) is completely observable. Then P is the only nonnegative definite solution of (5.16) and $P>0$.

(5.19) If (A,B,C) is *co* and *cr* then there is a unique $P>0$ which solves (5.16).

This last property is the essential one for this section. For the next one we need something better. Let $L^{co,cr}_{m,n,p}(\mathbf{R})$ be the space of all triples of real matrices (A,B,C) such that (A,B) is completely reachable and (A,C) is completely observable. Let $\mathrm{rc}(A,B,C) := P$ be the unique solution P of (5.16) such that $P>0$ (the matrix P is positive definite and selfadjoint). Then

(5.20)The function $rc(A,B,C)$ from $L^{co,cr}_{m,n,p}(\mathbf{R})$ to the space of selfadjoint matrices is real analytic (and so in particular C^∞ (= smooth)

Moreover

(5.21)$rc(TAT^{-1}, TB, CT^{-1}) = (T^*)^{-1}rc(A,B,C)T^{-1}$

(5.22)$rc(-A^*), \pm C^*, \pm B^*) = rc(A,B,C)^{-1}$

Property (5.21) is important in section 6; more precisely it will be important when these results are really implemented for multi-input multi-output systems. The point is that the matrices (A,B,C) are not determinable from the observations alone, simply because the systems (A,B,C) and (TAT^{-1}, TB, CT^{-1}) for $T \in GL_n(\mathbf{R})$ produce exactly the same input-output behaviour. For completely reachable and completely observable systems this is also the only indeterminacy. Property (5.21) guarantees that the whole analysis of these two section 5 and 6 'descends' to the moduli space (quotient manifold) $L^{co,cr}_{m,n,p}(\mathbf{R})/GL_n(\mathbf{R})$.

Having all this available it is tempting (and natural) to play the trick embodied by (5.15) again, this time using conjugation by a 2×2 block matrix with identities on the diagonal, a zero in the upper right hand corner and a Riccati equation solution Q in the lower left hand corner. This, however, is no particular good because this will introduce both the two factors

$$\begin{bmatrix} I & 0 \\ -Q & I \end{bmatrix}, \begin{bmatrix} I & 0 \\ Q & I \end{bmatrix}$$

in the factorisation of $S(t)=\exp(tM)$, and at least one will cause difficulties with inverse and direct Fourier transforms; cf. part (iii) of the recipe of section 4.

Instead, writing

$$\exp(t\begin{bmatrix} -\tilde{A}^T & 0 \\ -BB^T & \tilde{A} \end{bmatrix}) = \begin{bmatrix} \exp(-t\tilde{A}^T) & 0 \\ -R & \exp(t\tilde{A}) \end{bmatrix} \tag{5.23}$$

one uses the factorisation

$$\exp(t\begin{bmatrix} -\tilde{A}^T & 0 \\ -BB^T & \tilde{A} \end{bmatrix}) = \begin{bmatrix} I & 0 \\ -R\exp(t\tilde{A}^T) & I \end{bmatrix} \begin{bmatrix} \exp(-t\tilde{A}^T) & 0 \\ 0 & \exp(t\tilde{A}) \end{bmatrix} \tag{5.24}$$

giving the following total factorisation for $S(t)=\exp(tM)$

$$S(t) = \begin{bmatrix} I & P \\ 0 & I \end{bmatrix} \begin{bmatrix} I & 0 \\ -R\exp(t\tilde{A}^T) & I \end{bmatrix} \begin{bmatrix} \exp(-t\tilde{A}^T) & 0 \\ 0 & \exp(t\tilde{A}) \end{bmatrix} \begin{bmatrix} I & -P \\ 0 & I \end{bmatrix} \tag{5.25}$$

Except for possibly the second factor on the right hand side of (5.25) applying the recipe of section 4 is a total triviality.

As to that second factor observe that

$$\frac{d}{dt}(\exp(t\begin{bmatrix} -\tilde{A}^T & 0 \\ -BB^T & \tilde{A} \end{bmatrix}) = \begin{bmatrix} -\exp(-t\tilde{A})A^T & 0 \\ -\frac{d}{dt}R & \exp(t\tilde{A})\tilde{A} \end{bmatrix}$$

$$= \begin{bmatrix} \exp(-t\tilde{A}^T) & 0 \\ -R & \exp(t\tilde{A}) \end{bmatrix} \begin{bmatrix} -\tilde{A}^T & 0 \\ -BB^T & \tilde{A} \end{bmatrix} \tag{5.26}$$

from which it follows that

$$\frac{dR}{dt} = -R\tilde{A}^T + \exp(t\tilde{A})BB^T. \tag{5.27}$$

As a result

$$\frac{d}{dt}(R\exp(t\tilde{A}^T) = -R\tilde{A}^T\exp(t\tilde{A}^T) + \exp(t\tilde{A})BB^T\exp(t\tilde{A}^T) + R\tilde{A}^T\exp(t\tilde{A}^T) \tag{5.28}$$

$$= \exp(t\tilde{A})BB^T\exp(t\tilde{A}^T) \geq 0$$

and it follows that

$$R\exp(t\tilde{A}^T) \geq 0 \quad \text{all } t \tag{5.29}$$

which means that applying part (iii) of the recipe of section 4 (= part (iii) of the definition of the real form of the SSW representation) just involves solving a diffusion equation (no anti diffusion component); or, in other words that the inverse Fourier transformation involved will exist. Note also that if the initial condition ψ is Fourier transformable then, if P is nonnegative definite, the result of applying the parts of the recipe corresponding to the third and fourth factors on the RHS of 5.25 will still be a Fourier transformable function.

This concludes the description of the algorithm for propagating non-gaussian initial densities.

6. Identification

Given all that has been said above, this section can be mercifully short. The problem is the following. Given a linear system

$$dx = Axdt + Bdw,\ dy = Cxdt + dv \tag{6.1}$$

with *unknown* A,B,C, but constant A,B,C, we want to calculate the joint conditional density (given the observations $y(s), 0 \leq s \leq t$) for A,B,C,x. This can be approached as a nonlinear filtering problem by adding the equations

$$dA = 0,\ dB = 0,\ dC = 0 \tag{6.2}$$

or, more precisely, the equations stating (locally) that the free parameters remaining after specifying a local canonical form are constant but unknown. More generally one has the same setup and problem when, say, part of the parameters of (A,B,C) are known (or, generalizing a bit more, imperfectly known).

The approach, of course, will be the calculate the DMZ unnormalized version of the conditional density $\rho(x,A,B,t)$ given the observations $y(s)$, $0 \leq s \leq t$. Writing down the DMZ equation for the system (6.1)-(6.2) gives

$$d\rho = \mathcal{L}\rho dt + \sum_{j=1}^{p}(Cx)_j dy_j(t) \tag{6.3}$$

with $\mathcal{L}$ given by (5.3); i.e. exactly the same equation as occurred in section 5 for the case of known A,B,C. And, indeed the only difference is that in section 5 the A,B,C are known, while (6.3) should be seen as a family of equations parametrized by (the unknown parameters in) the A,B,C. Thus if $\rho(x,t|A,B,C)$ denotes the solution of (5.2) and $\rho(x,A,B,C,t)$ denotes the solution of (6.3) then

$$\rho(x,t|A,B,C) = \rho(x,A,B,C,t) \tag{6.4}$$

Now the bank of Kalman-Bucy filters for $\hat{x}$ parametrized by $(A,B,C) \in L^{co,cr}_{m,n,p}$ gives the probability density

$$\pi(x,t|A,B,C) = r(t,A,B,C)^{-1}\rho(x,t|A,B,C) \tag{6.5}$$

so that the normalization factor $r(t,A,B,C)$ can be calculated as $\int \rho(x,t,A,B,C)dx$.
By Bayes

$$\pi(x,A,B,C,t) = \pi(x,t|A,B,C)\pi(A,B,C,t)) \tag{6.6}$$

so that the normalization factor $r(t,A,B,C)$ is, so to speak, precisely equal to the difference between the solution of the DMZ equation (6.3) (or (5.2)) and the bank of Kalman filters producing $\pi(x,t|A,B,C)$. I.e. the marginal conditional density

$$\pi(A,B,C,t) = \int \pi(x,A,B,C,t)dx = \int \rho(x,A,B,C,t)dx \,/ \int \rho(x,A,B,C,t)dxdAdBdC \tag{6.7}$$

is obtainable from the unnormalized version of the bank of Kalman-Bucy filters parametrized by (A,B,C). Given the relations between this bank of filters described in [13] and briefly recalled in section 7 below this may offer further opportunities.

Be that as it may the marginal density $\pi(A,B,C,t)$ which up to a normalization factor is equal to $\int \rho(x,A,B,C,t)dx$ can be effectively calculated by the procedure of section 5 above with the only difference that $P = \mathrm{rc}(A,B,C)$ now has to be treated as a function. Once $\pi(A,B,C,t)$ (or in various cases some unnormalized version $\rho(A,B,C,t)$ is available a host of well known techniques such as maximum likelyhood become available.

If it is possible (as it will be in many cases) to work with a $\rho(A,B,C,t) = r(t)\pi(A,B,C,t)$ there is no (immediate) need to descend to the quotient manifold $L^{co,cr}_{m,n,p}(\mathbf{R}) / GL_n(\mathbf{R})$.

7. On the relation between the 'real form' of the SSW representation and the Kalman-Bucy filter

We have seen that the essential difficulty in obtaining the (unnormalized) conditional density $\rho(x,t)$ lies in 'solving' $\exp(t\mathcal{L})\psi$ where $\mathcal{L}$ is the second order differential operator (5.3). Now $\mathcal{L}$ corresponds in a fundamental way with the $2n \times 2n$ matrix

$$\begin{bmatrix} -A^T & -C^TC \\ -BB^T & A \end{bmatrix} \tag{7.1}$$

Not very surprisingly this matrix in turn is very much related to the matrix Riccati equation part of the Kalman-Bucy filter. Indeed, consider the matrix differential equation

$$\begin{bmatrix} \dot{X} \\ \dot{Y} \end{bmatrix} = \begin{bmatrix} -A^T & -C^TC \\ -BB^T & A \end{bmatrix} \begin{bmatrix} X \\ Y \end{bmatrix} \tag{7.2}$$

and, assuming that $X(t)$ is invertible, let

$$-P = YX^{-1}. \tag{7.3}$$

Then

$$\dot{P} = -\dot{Y}X^{-1} + YX^{-1}\dot{X}X^{-1} = (+BB^TX - AY)X^{-1} + YX^{-1}(-A^TX - C^TCY)X^{-1}$$
$$= +BB^T + AP + PA^T - PC^TCP$$

which is the covariance equation of the Kalman-Bucy filter.

References

1. **T.E. Duncan**, Probability densities for diffusion processes with applications to nonlinear filtering theory, PhD thesis, Stanford, 1967.
2. **M. Hazewinkel**, (Fine) moduli (spaces) for linear systems: what are they and what are they good for. In: C.I. Byrnes, C.F. Martin (eds), Geometric methods for linear system theory (Harvard, June 1979), Reidel, 1980, 125-193.
3. **M. Hazewinkel**, The linear systems Lie algebra, the Segal-Shale-Weil representation and all Kalman-Bucy filters, J. Syst. Th. & Math. Sci. **5** (1985), 94-106.
4. **M. Hazewinkel**, Lectures on linear and nonlinear filtering. In: W. Schiehlen, W. Wedig (eds), Analysis and estimation of stochastic mechanical systems, CISM course June 1987, Springer (Wien), 1988, 103-135.
5. **M. Hazewinkel, S.I. Marcus**, On Lie algebras and finite dimensional filtering, Stochastics **7** (1982), 29-62.
6. **M. Hazewinkel, J.C. Willems** (eds), Stochastic systems: the mathematics of filtering and identification and applications, Reidel, 1981.
7. **O. Hijab**, Stabilization of control systems, Springer, 1987.
8. **R.E. Howe**, On the role of the Heisenberg group in harmonic analysis, Bull. Amer. Math. Soc. **3** (1980), 821-844.
9. **S.I. Marcus**, Algebraic and geometric methods in nonlinear filtering, SIAM J. Control and Opt. **22** (1984), 817-844.
10. **R.E. Mortensen**, Optimal control of continuous time stochastic systems, PhD thesis, Berkeley, 1966.
11. **J. Wei, E. Norman**, On the global representation of the solutions of linear differential equations as products of exponentials, Proc. Amer. Math. Soc. **15** (1964), 327-334.
12. **M. Zakai**, On the optimal filtering of diffusion processes, Z. Wahrsch. verw. Geb. **11** (1969), 230-243.

Logarithmic Transformations with Applications in Probability and Stochastic Control

Wendell H. Fleming
Division of Applied Mathematics, Brown University
Providence, Rhode Island 02912, USA

Abstract

We are concerned with a class of problems described in a somewhat imprecise way as follows. Consider a linear operator of the form $L + V(x)$, where L is the generator of some Markov process x_t and the "potential" $V(x)$ is some real-valued function on the state space of x_t. We are interested in probabilistic representations for solutions $u(t,x)$ of the evolution equation

(1) $$\frac{\partial u}{\partial t} = Lu + V(x)u, \quad t > 0$$

with initial data at $t = 0$. The Feynman-Kac formula gives a well-known stochastic representation for $u(t,x)$. We seek a different probabilistic representation for $I = -\log u$, if $u(t,x)$ is a positive solution to (1). In this representation the operator L is replaced by another generator $\tilde{L}_t$ (perhaps time dependent), chosen to solve a certain stochastic control problem. The dynamic programming equation for this stochastic control problem is

(2) $$\frac{\partial I}{\partial t} = H(I) - V(x), \text{ where}$$
$$H(I) = -e^{I}L(e^{-I}).$$

Another way to view the change of generator from L to $\tilde{L}_t$ is by change of probability measure through conditioning.

Next suppose that the state space of x_t is euclidean R^n, that

$$L = L_\epsilon,\ u = u^\epsilon,\ I^\epsilon = -\epsilon \log u^\epsilon \text{ and}$$
$$H_\epsilon(I) = -e^{I}L_\epsilon(e^{-I}).$$

Under various assumptions it turns out that $I^\epsilon \to I^0$ as $\epsilon \to 0$,

$$\lim_{\epsilon \to 0} \epsilon H_\epsilon(\epsilon^{-1}I) = H_0(x,I_x)$$

where I_x is the gradient, and that $I(t,x)$ is a viscosity solution of the first-order partial differential equation

$$\frac{\partial I}{\partial t} = H_0(x,I_x).$$

When x_t is a nondegenerate diffusion on R^n, then L is a second order elliptic partial differential operator. In this case, the logarithmic transmation provides an analytical approach to large deviations questions of Ventsel-Freidlin type, and for more precise results in the form of asymptotic series expansions of I^ϵ in powers of ϵ. The logarithmic transformation technique is also of use to study certain asymptotic problems in which $u^\epsilon(t,x)$ obeys a nonlinear parabolic partial differential equation.

References

1. G. Barles and B. Perthame, Exit time problems in optimal control and vanishing viscosity, to appear in SIAM J. Control Optimiz.

2. P. DuPuis, Minimizing exit probabilities: a large deviations approximation, Brown Univ. LCDS Report, (1987).

3. L.C. Evans and H. Ishii, A PDE approach to some asymptotic problems concerning random differential equations with small noise intensities, Ann. Inst. H. Poincare Analyse NonLineaire 2 (1985) 1-20.

4. L.C. Evans and P.E. Souganidis, A PDE approach to geometric optics for certain semilinear parabolic equations, preprint.

5. W.H. Fleming, Exit probabilities and optimal stochastic control, Appl. Math. Optimiz. 4 (1978) 329-346.

6. W.H. Fleming, Stochastic calculus of variations and mechanics, J. Optimiz. Th. Appl. 41 (1983) 55-74.

7. W.H. Fleming, Logarithmic transformations and stochastic control, in Springer Lecture Notes in Control and Info. Sci. No. 42 (eds. W.H. Fleming and L.G. Gorostiza) Springer-Verlag (1982) 131-141.

8. W.H. Fleming, A stochastic control approach to some large deviations problems, in Springer Lecture Notes in Math no. 1119 (eds. I. Capuzzo Dolcetta, W.H. Fleming and T. Zolezzi), (1984) 52-66.

9. W.H. Fleming, Controlled Markov processes and viscosity solutions of nonlinear evolution equations, Lezione Fermiane (1986), Accademia Nazionale dei Lincei, Scuola Normale Superiore, Pisa.

10. W.H. Fleming and S.K. Mitter, Optimal control and nonlinear filtering for nondegenerate diffusion processes, Stochastics 8 (1982) 63-77.

11. W.H. Fleming and H.M. Soner, Asymptotic expansions for Markov processes with Levy generators, to appear in Applied Math. and Optimiz.

12. W H. Fleming and P.E. Souganidis, PDE-Viscosity solution approach to some problems of large deviations, Annali Scuola Normale Sup. Pisa Classe Sci., Ser IV 23 (1986) 171-192.

13. W.H. Fleming and P.E. Souganidis, Asymptotic series and the method of vanishing viscosity, Indiana Univ. Math. J. 35 (1986) 425-447.

14. C.J. Holland, A new energy characterization of the smallest eigenvalue of the Schrödinger equation, Comm. Pure Appl. Math. 30 (1977).

15. S.-J. Sheu, Optimal control and its application to large deviation theory, Brown Univ. Ph.D. Thesis 1983.

16. S.-J. Sheu, Stochastic control and principal eigenvalue, Stochastics 11, (1984) 191-211.

MODELISATION ET COMMANDE EN ECONOMIE

MODELS AND CONTROL POLICIES IN ECONOMICS

DYNAMIC OPTIMIZATION OF SOME FORWARD-LOOKING STOCHASTIC MODELS†

Tamer Başar

Decision and Control Laboratory
Coordinated Science Laboratory and the
Department of Electrical and Computer Engineering
University of Illinois
1101 W. Springfield Avenue
Urbana, Illinois 61801 / USA

Abstract

A dynamic decision model is said to be *forward-looking* if the evolution of the underlying system depends explicitly on the expectations the agents form on the future evolution itself. Such models lead to nonstandard stochastic dynamic optimization problems where one has to take into account the fact that there is a circular (closed) relationship between future forecasts and future system behavior. In this paper we study a class of such problems where there is an additional control input designed to make the system track a given trajectory. This leads to a game-theoretic formulation in which context we consider both finite and infinite horizon formulations. It is shown that for the finite horizon problem the unique Nash equilibrium solution requires (fixed size) memory for both agents because of spillover across stages, whereas for the infinite horizon version no memory is needed.

1. An Introduction to Forward-Looking Models

We refer to a dynamic stochastic model as *forward-looking* if one of its inputs involves future expectations of the system trajectory, using (possibly noisy) measurements on the past realizations. Such decision models find wide-spread use in economics, where they are more commonly known as *rational expectations* models. A few representative papers in this area are the works of *Lucas (1975)*, *Sargent and Wallace (1975)*, *Barro (1976)*, *Taylor (1977)*, *Shiller (1978)*, *Blanchard (1979)*, and *Blanchard and Kahn (1980)*. Perhaps the simplest such model that captures the salient features of *forward-looking* behavior is described by the scalar difference equation

$$y_{t+1} = ay_t + bv_t + \epsilon_{t+1}, \tag{1a}$$

where a and b are constant parameters, $\{\epsilon_t\}$ is a sequence of independent zero-mean random variables with finite (positive) variance, and v_t is the decision variable chosen at time t under some "expectation" of the future behavior of the system based on information available at time t. If the forecast of interest is n steps into the future, for example, one possibility is to replace v_t in (1a) by $E_t y_{t+n}$, the conditional mean of y_{t+n} based on the information available at time t. This information, which we denote by η_t, could involve a direct measurement of all the past values of the system trajectory, that is $\{y_t, y_{t-1}, ...\} =: y^t$, or involve some *noisy* measurement on the state trajectory, $\eta_t = z^t$, where $\{z_t\}$ is a measurement sequence defined by

---

† *This work was performed while the author was spending a sabbatical year at INRIA, Sophia Antipolis, France, and it was also partially supported by the Air Force Office of Scientific Research under Grant No. AFOSR 084-0056, through the University of Illinois.*

$$z_t = y_t + \xi_t, \tag{1b}$$

with $\{\xi_t\}$ being another sequence of independent, zero-mean random variables with finite variance.

A basic question addressed in the literature over the years has been the existence of a (unique) stochastic process $\{y_t\}$ that satisfies (1a) whenever $v_t = E_t y_{t+n}$ and the time interval is infinite. The answer to this question is that there is, in general, more than one such solution even in the class of *stationary* processes. However, as we have recently argued in *Başar (1987)*, a better approach towards policy determination in these *forward-looking* models would involve the optimization of an appropriate loss function, by carefully taking into account the informational dependence as well as the correlation of policies across stages. One such criterion would be

$$J_s^T := \min_{\{\gamma_t\}} \sum_{t=s}^{T} E\{[\gamma_t(\eta_t) - y_{t+n}]^2\}\rho^{t-s}, \tag{2}$$

where minimization is subject to the dynamic constraint (1a), with $v_t = \gamma_t(\eta_t)$, and uses the boundary condition $v_t \equiv 0$ for $t > T$. In the above, $[s, T]$ stands for the time horizon, which could also be infinite, and ρ denotes a positive discount factor ($0 < \rho \leq 1$). It has been shown in *Başar (1987)* that the dynamic policy optimization problem admits the solution $v_t = E_t y_{t+1}$ when $n = 1$, but for $n \geq 2$ the unique solution for the finite-horizon version is different from $E_t y_{t+n}$. For $n = 2$, for example, the best forecast into the future (by *two* time steps), under the criterion (2) and using the information $\{\eta_t = y^t\}$, is given by

$$v_t^* = \gamma_t^*(y^t) = \alpha_t y_t + \beta_t v_{t-1}, \tag{3a}$$

for $2 \leq t \leq T$, where the sequences $\{\alpha_t\}$ and $\{\beta_t\}$ are determined recursively *off-line*. For the *noisy* measurement case, $\{\eta_t = z^t\}$, the solution is again unique and is given by

$$v_t^* = \gamma_t^*(z^t) = \alpha_t \hat{y}_t + \beta_t v_{t-1}, \tag{3b}$$

for $2 \leq t \leq T$, where the sequences $\{\alpha_t\}$ and $\{\beta_t\}$ are the same as in (3a), and $\hat{y}_t$ is a sequence of estimates generated recursively by a Kalman filter, under the assumption that the underlying statistics are Gaussian. An interesting feature of the solution is that for the infinite-horizon version (that is as $T \to \infty$) the coefficient sequence $\{\beta_t\}$ vanishes for all finite t, and the solution becomes $v_t^* = E_t y_{t+2}$,thus eliminating the correlation across stages.

In the present paper, we consider a more general formulation than that above, where now two separate agents, say A and B, have influence on the system trajectory, one of them (A) again making a two-step ahead forecast of the trajectory, whereas the other one (B) trying to drive the trajectory as close to a specific target as possible. For such a scenario, the system equation would be replaced by

$$y_{t+1} = a y_t + b v_t + c w_t + \epsilon_{t+1}, \tag{4}$$

where $v_t = \gamma_t(y^t)$ is controlled by agent A and $w_t = \mu_t(y^t)$ by agent B. Taking the time horizon as $[0, T+1]$, the two cost functions to be minimized by A and B, respectively, are

$$J_A(\gamma, \mu) = \sum_{t=0}^{T} E\{[\gamma_t(y^t) - y_{t+2}]^2\}\rho_A^t, \tag{5a}$$

and

$$J_B(\gamma,\mu)=\sum_{t=0}^{T+1}E\{[y_{t+1}-\bar{y}_{t+1}]^2+kw_t^2\}\rho_B^t, \tag{5b}$$

where $\{\bar{y}_t, 2\le t\le T+2\}$ is the desired trajectory, k is a positive weight on agent B's control, ρ_A, ρ_B are the corresponding discount factors, $\gamma:=\{\gamma_T,\gamma_{T-1},..,\gamma_0\}$, $\mu:=\{\mu_{T+1},\mu_T,..,\mu_0\}$, and $v_{T+1}\equiv 0$, the last identity reflecting the fact that no forecast is made at time $t=T+1$. Furthermore, we assume that the independent random variables ϵ_t $(1\le t\le T+2)$ each have zero mean and a probability distribution that assigns positive probability to every measurable open subset of the real line. One such distribution would be the normal (Gaussian) distribution with positive variance.

Since this is a problem with multiple objectives, several equilibrium solution concepts would be applicable, with the one adopted here being the noncooperative *Nash* equilibrium solution. Therefore, we seek a pair (γ^*,μ^*), preferably unique, satisfying the pair of inequalities

$$J_A(\gamma^*,\mu^*)\le J_A(\gamma,\mu^*);\quad J_B(\gamma^*,\mu^*)\le J_B(\gamma^*,\mu), \tag{6}$$

for all admissible γ and μ. Other possibilities would have been the Stackelberg solution with either agent acting as the leader and the Pareto-optimal solution, which, however, we do not discuss here because of space limitations.

The first question we attack, in *section 2*, is a "simpler" version of the above, where agent A's policy is fixed as $v_t=E_ty_{t+2}, t\le T$, which is in general a suboptimal policy for A. We obtain the best policy for B under this additional structural restriction, and derive the corresponding expression for $\{\mu_t\}$ (see *Theorem 1*). Furthermore, we study the limiting behavior of the two policies, for the infinite-horizon problem. Subsequently, in *section 3*, we derive the unrestricted Nash solution and prove its (generic) existence and uniqueness (see *Theorem 2*), with details of the derivation provided in the *Appendix*. We also study the limiting behavior of the solution as $T\to\infty$, and analyze the discrepancies that exist between the two stationary solutions of *Theorem 1* and *Theorem 2*. The paper concludes with a discussion of the "noisy measurement" case and some other possible extensions, in *section 4*. Throughout the analysis, we take the reference trajectory (to be tracked) as the *zero* trajectory, an assumption that does not bring in much loss of conceptual generality but leads to considerable simplifications in the resulting expressions.

2. The Optimal Tracking Strategy Under Perfect Myopic Forecast

With v_t taken as E_ty_{t+2} (which myopically minimizes each term of (5*a*)), and $\{\bar{y}_t\}$ taken as the *zero* trajectory, the dynamic policy optimization problem faced by agent B is the minimization of F_0^{T+1}, where

$$F_s^{T+1}=\sum_{t=s}^{T+1}E\{y_{t+1}^2+kw_t^2\}\rho_B^{t-s}, \tag{7a}$$

the dynamic constraint is

$$\begin{aligned} y_{T+2}&=ay_{T+1}+cw_{T+1}+\epsilon_{T+2},\\ y_{t+1}&=ay_t+bE_ty_{t+2}+cw_t+\epsilon_{t+1},\quad t\le T,\end{aligned} \tag{7b}$$

and the information constraint is $w_t=\mu_t(y^t)$. Note that this is not a standard linear-quadratic stochastic control problem because of the presence of the conditional expectations term in (7b), which could even make the dynamic constraint nonlinear in the past values of the trajectory. We will show below, however, that the optimal control is still linear, thus making the corresponding forecast also linear in the available information. The derivation entails a recursive approach where the structure of v_t is determined alongside the optimal control at each step of the iterative minimization.

Before presenting the main result of this section, we first introduce two sequences $\{p_t\}$ and $\{\nu_t\}$ which are defined recursively by

$$p_t = 1 + \frac{\rho_B a^2 k p_{t+1}}{c^2 p_{t+1} + k\nu_t^2}; \quad p_{T+2} = 1, \tag{8a}$$

$$\nu_{t-1} = 1 - \frac{abk\nu_t}{c^2 p_{t+1} + k\nu_t^2}; \quad \nu_{T+1} = 1. \tag{8b}$$

Next we define a third sequence $\{g_t\}$ in terms of the other two, according to

$$g_t = -cap_{t+1}/(c^2 p_{t+1} + k\nu_t^2), \quad t \leq T+1. \tag{8c}$$

We are now in a position to state the main result, after invoking a condition which generically holds.

Condition 1. *The sequence $\{\nu_t\}$ generated by (8b) does not vanish for any $t \leq T+1$.*

Theorem 1. *Let Condition 1 be satisfied. Then, the dynamic policy optimization problem with myopic forecast admits the unique solution*

$$w_t = \hat{\mu}_t(y_t) = g_t y_t, \quad 0 \leq t \leq T+1, \tag{9a}$$

with the corresponding forecast policy given by

$$v_t = E_t y_{t+2} = \frac{(a + cg_{t+1})(a + cg_t)}{\nu_{t+1}\nu_t} y_t := h_t y_t. \tag{9b}$$

The minimum value of F_0^{T+1} in (6a) is

$$\hat{F}_0^{T+1} = p_0 E\{y_0^2\} + \lambda_0, \tag{9c}$$

where λ_0 is the last step of the backward recursion

$$\lambda_{T+1} = var(\epsilon_{T+2}),$$
$$\lambda_{t-1} = \rho_B \lambda_t + p_t var(\epsilon_t).$$

Proof. The proof proceeds by recursively showing that the minimum value of F_s^{T+1} is given, for each $s \leq T+1$, by

$$\hat{F}_s^{T+1} = [(p_s - 1)/\rho_B] E\{y_s^2\} + \lambda_s.$$

The result is trivially true for $s = T+2$, where we take $\lambda_{T+2} = 0$. Let us therefore assume its validity, along with (9a) and (9b), up to $s+1$, and verify the expression, as well as (9a) and (9b), for s. The minimization problem faced by agent **B** at time s is

$$\min_{\mu_s} [\rho_B \hat{F}_{s+1}^{T+1} + E\{y_{s+1}^2 + k\mu_s(y^s)^2\}], \tag{$\bullet$}$$

which is equivalent to

$$\min_w E\{p_{s+1}y_{s+1}^2 + kw^2 \mid y^s\},$$

which uses the dynamic constraint

$$y_{s+1} = ay_s + bE_s y_{s+2} + cw + \epsilon_{s+1}. \qquad (\star)$$

We also have the relationship

$$y_{s+2} = (a + cg_{s+1})y_{s+1} + b[(a + cg_{s+2})(a + cg_{s+1})/\nu_{s+2}\nu_{s+1}]y_{s+1} + \epsilon_{s+2},$$

where we have explicitly used (9a) and (9b), with t replaced by $s+1$. (Of course, if $s = T+1$, the last relationship would not be needed since the conditional expectation term in ($\star$) would be missing.) Now, taking the conditional expectation of the last expression with respect to y^s, substituting this into ($\star$), taking the conditional expectation of the resulting expression again with respect to y^s, and solving for the resulting $E_s y_{s+1}$ in terms of y_s and w, we arrive at the expression

$$E_s y_{s+1} = \frac{1}{\nu_s}[ay_s + cw].$$

Using this, $E_s y_{s+2}$ can easily be evaluated to be

$$E_s y_{s+2} = \frac{a + cg_{s+1} + bh_{s+1}}{\nu_s}[ay_s + cw], \qquad (\star\star)$$

under which the dynamic constraint becomes equivalent to

$$y_{s+1} = \frac{1}{\nu_s}[ay_s + cw] + \epsilon_{s+1}.$$

This makes the minimization problem a standard *linear-quadratic* one, and hence it readily follows that the minimizing control is uniquely given by (9a) with $t = s$. Substitution of this solution into ($\bullet$) and ($\star\star$) finally verifies the asserted form for $\hat{F}_s^{T+1}$ and the structure of the forecast policy as given by (9b). We should note that *Condition 1* has explicitly been used in the proof, to make sure that one can solve uniquely for $E_s y_{s+1}$ and $E_s y_{s+2}$. $\diamond$

Condition 1, under which the existence and uniqueness of the optimal control (9a) is valid, holds whenever a and b have opposite signs, regardless of the magnitudes of the parameters of the problem. The result follows by inspection, since with $ab < 0$ and $\nu_{T+1} = 1$, we have $\nu_t > 0$ for all $t \le T+1$. For $ab < 0$, however, there may exist isolated values for the parameters for which the condition does not hold for some t. [A more precise statement here would be that with all but one of the parameters fixed (and $ab > 0$), there will exist at most a finite number of different values of that parameter for which *Condition 1* is violated. This follows since for each t, ν_t is a rational function of the quintuplet (a, b, c, ρ_B, k).] For example, for the parameter values $a = c = k = 1$, $b = 2$, we have $\nu_T = 0$, which shows that *Condition 1* may fail even for a two-stage problem. However, if we perturb the value of b to $b = 2.1$, and take $\rho_B = 1$, then *Condition 1* holds for all values of t. In fact, running the coupled recursive equations (8a)-(8b) in retrograde time, we find that (for these parameter values) the pair (ν_t, p_t) converges to $(0.504147, 1.880960)$ in 29 steps, within the accuracy of six decimal places. Hence, in this case, the infinite-horizon version (even with no discounting) admits a unique optimal stationary control, given by $w_t = \hat{\mu}(y_t) = -1.135124\, y_t$. If, in the above, b is instead taken to be 1, again *Condition 1* holds, the pair (ν_t, p_t) converges to $(0.694146, 1.787692)$ in 9 iterations, and the optimal control policy converges to $w_t = \hat{\mu}(y_t) = -0.787692\ y_t$. As a final numerical experimentation, reflecting a different set of parameter values, we consider the case of $a = 2, b = -3, \rho_B = 0.8, c = k = 1$. For this set, we already know that *Condition 1* holds, since $ab < 0$. Studying the convergence of the optimal policy to a stationary control, we find that the pair (ν_t, p_t) converges to $(2.796267, 1.521150)$ in 26 iterations, with the resulting stationary policy being $w_t = \hat{\mu}(y_t) = -0.325719\ y_t$.

3. The Nash Equilibrium Solution

We now remove the restriction that agent **A**'s input to the system is a myopic forecast, and allow him to determine the "best" choice for v_t by minimizing the cost function J_A. As we have discussed in *section 1*, this joint optimization problem can best be treated as a noncooperative game, and hence we study in this section the Nash equilibrium of the underlying game, as defined by (6).

There are two general approaches to the derivation of Nash equilibria in such dynamic games. One would be first to guess (or propose) a structure for the solution in terms of some parameters, and then to validate the equilibrium property of the asserted structure and to obtain the corresponding values of the parameters so that the resulting policies are in Nash equilibrium. A second approach would be to obtain the Nash solution recursively (by employing the definition of *stagewise* or *feedback* equilibrium; see, for example, *Başar and Olsder (1982)*) by solving static games conditioned on the available (common) information, at each step of the iteration. Note that this would be applicable only if both agents have identical information (which is the case here), since otherwise stagewise decomposition would not be possible. Now, two disadvantages of the first method are that (i) one has to guess the structure of the solution correctly, and (ii) even if the initial guess is correct there is no way to show (using this method) that the validated Nash solution is unique. The second method, on the other hand, is capable of answering the uniqueness question, but it only produces candidate solutions which subsequently have to be checked for their equilibrium property. What we will, therefore, choose to do in the sequel is to use an appropriate combination of the two approaches, to generate candidate solutions and verify their existence and uniqueness. We should note in passing that even though the problem may look, at the outset, as a standard *linear-quadratic* one, the presence of the two-step delay in the cost function of agent **A** makes the game a nonstandard one, thus eliminating the possibility of direct application of results available on linear-quadratic feedback Nash games (as, for example, covered in *Başar and Olsder (1982)*).

Before presenting the solution in *Theorem 2* below, we first introduce some sequences which will be needed in the characterization of the equilibrium policies. Towards this end, let $\{m_t\}$, $\{\tilde{m}_t\}$, $\{n_t\}$ be three sequences generated by

$$m_{t-1} = \frac{1}{1 - bm_t}[a + c\tilde{\alpha}_t + b\alpha_t\tilde{m}_t]; \quad m_T = \frac{ak}{k + c^2}, \tag{10a}$$

$$\tilde{m}_{t-1} = \frac{1}{1 - bm_t}[c\tilde{\beta}_t + b\beta_t\tilde{m}_t]; \quad \tilde{m}_T = 0, \tag{10b}$$

$$n_{t-1} = \frac{\rho_A n_t(1 - bm_t)^2}{b^2 + \rho_A n_t(1 - bm_t)^2}; \quad n_T = 1, \tag{10c}$$

where α_t, $\tilde{\alpha}_t$, β_t, $\tilde{\beta}_t$ are defined, for $t \le T$, by

$$\alpha_t := \frac{(1 - bm_t)\rho_A n_t m_t - b}{b^2 + \rho_A n_t(1 - bm_t)(1 - bm_t - \tilde{m}_t)}[a + c\tilde{\alpha}_t], \quad t \le T, \tag{11a}$$

$$\widetilde{\alpha}_{t-1} := -\frac{[ak_{11,t} + (bk_{11,t} + k_{22,t})\alpha_{t-1}]c}{k + c^2 k_{11,t}}, \qquad t \le T+1, \tag{11b}$$

$$\beta_t := \frac{(1 - bm_t)\rho_A n_t m_t c\widetilde{\beta}_t + b(1 - c\widetilde{\beta}_t)}{b^2 + \rho_A n_t(1 - bm_t)(1 - bm_t - \widetilde{m}_t)}, \quad t \le T, \tag{12a}$$

$$\widetilde{\beta}_{t-1} := -\frac{(bk_{11,t} + k_{12,t})c\beta_{t-1}}{k + c^2 k_{11,t}}, \qquad t \le T+1, \tag{12b}$$

and

$$K_t := \begin{pmatrix} k_{11,t} & k_{12,t} \\ k_{12,t} & k_{22,t} \end{pmatrix}$$

is a 2×2 matrix sequence generated by the discrete time Riccati equation

$$\begin{aligned} K_t &= \rho_B A_t'[K_{t+1} - K_{t+1}C(C'K_{t+1}C + k)^{-1}C'K_{t+1}]A_t + Q, \\ K_{T+1} &= [ka^2 \rho_B/(k + c^2)]Q, \end{aligned} \tag{13a}$$

with

$$A_t := \begin{pmatrix} a + b\alpha_t & b\beta_t \\ \alpha_t & \beta_t \end{pmatrix}, \quad Q := \begin{pmatrix} 1 & 0 \\ 0 & 0 \end{pmatrix}, \quad C := \begin{pmatrix} c \\ 0 \end{pmatrix}. \tag{13b}$$

Finally, let $r_{\alpha t}$, $r_{\tilde{\alpha} t}$, $r_{\beta t}$, $r_{\tilde{\beta} t}$ be defined by

$$r_{\alpha t} := \frac{c[(1 - bm_t)\rho_A m_t n_t - b]}{b^2 + \rho_A n_t(1 - bm_t)(1 - bm_t - \widetilde{m}_t)}, \tag{14a}$$

$$r_{\tilde{\alpha} t} := -\frac{c(bk_{11,t+1} + k_{22,t+1})}{k + c^2 k_{11,t+1}}, \tag{14b}$$

$$r_{\beta t} := \frac{c(1 - bm_t)\rho_A n_t m_t - bc}{b^2 + \rho_A n_t(1 - bm_t)(1 - bm_t - \widetilde{m}_t)}, \tag{15a}$$

$$r_{\tilde{\beta} t} := -\frac{c(bk_{11,t+1} + k_{12,t+1})}{k + c^2 k_{11,t+1}}. \tag{15b}$$

The last four expressions are the coefficient terms in (11*a*)-(12*b*), indicating the dependence of α_t, $\widetilde{\alpha}_t$, β_t and $\widetilde{\beta}_t$ on $\widetilde{\alpha}_t$, α_t, $\widetilde{\beta}_t$ and β_t, respectively. A certain relationship between these coefficient terms in fact determines the existence of a unique Nash equilibrium solution, as to be elucidated below.

Condition 2. *For all* $t \le T$,

$$r_{\alpha t} r_{\tilde{\alpha} t} \ne 1, \quad r_{\beta t} r_{\tilde{\beta} t} \ne 1, \tag{16a}$$

$$bm_t \ne 1, \tag{16b}$$

$$\rho_A n_t(1 - bm_t)(1 - bm_t - \widetilde{m}_t) \ne -b^2. \tag{16c}$$

Theorem 2. *Let Condition 2 be satisfied. Then, the forward-looking tracking model* (4)-(5) *admits a unique Nash equilibrium solution* $\{\gamma_t^*, \mu_t^*\}$, *where agent* **A***'s (best forecast) policy is*

$$\begin{aligned} v_t = \gamma_t^*(y^t) &= \alpha_t y_t + \beta_t \widetilde{v}_{t-1}, \quad & t \ge 1 \\ &= \alpha_0 y_0, & t = 0, \end{aligned} \tag{17}$$

and agent B's (best tracking) policy is

$$\begin{aligned} w_t = \mu_t^*(y^t) &= \tilde{\alpha}_t y_t + \tilde{\beta}_t \tilde{v}_{t-1}, & 1 \le t \le T \\ &= -[ac/(k+c^2)]y_{T+1}, & t = T+1 \\ &= \tilde{\alpha}_0 y_0, & t = 0 \quad , \end{aligned} \tag{18a}$$

where the sequence $\{\tilde{v}_t\}$ *is generated by*

$$\begin{aligned} \tilde{v}_t = &= \alpha_t y_t + \beta_t \tilde{v}_{t-1}, & t \ge 1 \\ &= \alpha_0 y_0, & t = 0. \end{aligned} \tag{18b}$$

Proof. We will first verify the structural consistency of the solution (17)-(18) under the Nash inequalities (6), and then discuss the existence of the solution. Some details of the derivation, as well as a proof for the uniqueness of the solution will be given in the *Appendix*.

Towards verifying the validity of (6), first consider the second inequality, where agent A's policy has been fixed as given by (17). Then, agent B faces a stochastic control problem with cost function J_B (given by (5*b*) with *zero* reference trajectory) and state dynamics

$$\begin{aligned} y_{t+1} &= (a + b\alpha_t)y_t + b\beta_t\tilde{v}_{t-1} + cw_t + \epsilon_{t+1} \quad , & t \le T \\ &= ay_{T+1} + cw_{T+1} + \epsilon_{T+2} \quad , & t = T+1, \end{aligned}$$

where the sequence $\{\tilde{v}_t\}$ is generated by (18*b*) in view of (17). The optimal control at time $T+1$, w_{T+1}, can readily be obtained, to be given by the second line in (18*a*). To obtain the remaining controls, we introduce a new state vector, $x_t := (y_t, \tilde{v}_{t-1})$, and reformulate the problem as one of minimizing J_B under the dynamic constraint

$$x_{t+1} = A_t x_t + C w_t + D\epsilon_{t+1} \quad ; \qquad D := (1 \quad 0)',$$

where control w_t is allowed to depend on x^t, $t \le T$. [Note that even though v^{t-1} is not available to agent B, $\tilde{v}^{t-1}$ is since it is generated by y^{t-1}.] This is the familiar linear-quadratic optimal control problem, whose unique solution is

$$w_t = -(k + C'K_{t+1}C)^{-1}C'K_{t+1}A_t x_t, \tag{$\star$}$$

where $\{K_t\}$ is generated by (13*a*). [Note that the terminal constraint on K_t at $t = T+1$ is not Q because we have already substituted for the optimal w_{T+1} and have reduced the cost function J_B to the one where the leading term is now y_{T+1}^2 instead of y_{T+2}^2.] Now, the optimal control ($\star$) is clearly linear in y_t and $\tilde{v}_{t-1}$, at time t, and a little algebra shows that it can be expressed in the form (18*a*).

We now focus attention on the first inequality of (6), where agent B's policy is fixed as given by (18*a*). Then the problem faced by agent A is one of optimal forecast, where the cost function is J_A (given by (5*a*)) and the dynamic constraint is

$$\begin{aligned} y_{t+1} &= (a + c\tilde{\alpha}_t)y_t + c\tilde{\beta}_t\tilde{v}_{t-1} + bv_t + \epsilon_{t+1}, & 1 \le t \le T \\ &= -[ak/(k+c^2)]y_{T+1} + \epsilon_{T+2}, & t = T+1 \\ &= (a + c\tilde{\alpha}_0)y_0 + bv_0 + \epsilon_1, & t = 0. \end{aligned}$$

Because of the form of the cost function, the available linear-quadratic theory cannot be directly applied to this problem; nevertheless, a one can employ a *dynamic programming* type argument to construct the optimal solution in retrograde time, as in the proof of *Theorem 2.1* of *Başar (1987)*. It has been shown in the *Appendix* that the optimal solution is unique (under some conditions which will be specified later), and the optimal policy at time t is a function of three variables, y_t, $\tilde{v}_{t-1}$ and v_{t-1}. The precise expression is

$$\begin{aligned} v_t = \gamma_t(y^t) &= \hat{\alpha}_t y_t + \hat{\beta}_t v_{t-1} + \bar{\beta}_t \tilde{v}_{t-1}, \quad 1 \le t \le T \\ &= \hat{\alpha}_0 y_0, \qquad\qquad t = 0, \end{aligned} \tag{$**$}$$

where

$$\hat{\alpha}_t = \frac{1}{b^2 + (1-bm_t)^2 \rho_A n_t}[\rho_A n_t(1-bm_t)(\tilde{m}_t\alpha_t + (a + c\tilde{\alpha}_t)m_t) - b(a + c\tilde{\alpha}_t)] \tag{$\bullet$}$$

$$\hat{\beta}_t = \frac{b}{b^2 + (1-bm_t)^2 \rho_A n_t}$$

$$\bar{\beta}_t = \frac{1}{b^2 + (1-bm_t)^2 \rho_A n_t}[\rho_A n_t(1-bm_t)(\tilde{m}_t\beta + c\tilde{\beta}_t m_t) - bc\tilde{\beta}_t], \tag{$\bullet\bullet$}$$

and $\{m_t\}$, $\{\tilde{m}_t\}$, $\{n_t\}$ are generated by (10*a*)-(10*c*). In writing down these expressions, we have already assumed the validity of (16*a*) and (16*c*), since otherwise m_t and $\tilde{m}_t$ would not have been well defined. We should note, however, that even in the *pure* forecast problem discussed in *Başar (1987)*, a condition similar to (16*b*) was required for the well-posedness of the problem.

Now, to complete the derivation, we substitute for α_t and β_t in ($\bullet$) and ($\bullet\bullet$) from (11*a*) and (12*a*), respectively, and observe that the resulting expression for $\hat{\alpha}_t$ is identical with that of α_t, and also when the resulting expression for $\bar{\beta}_t$ is added to $\hat{\beta}_t$, the outcome is identical to β_t; in other words,

$$\hat{\alpha}_t \equiv \alpha_t \quad , \quad \hat{\beta}_t + \bar{\beta} \equiv \beta_t .$$

When the latter is used in ($**$) recursively, it follows that $\{v_t\}$ is generated by the same sequence (of y_s's) as $\{\tilde{v}_t\}$, and hence that ($**$) admits the simpler representation (17).

This then completes the verification of the existence part of the theorem; more precisely, of the fact that the policies (17)-(18) constitute a Nash equilibrium pair under *Condition 2*. Note that (16*a*) in *Condition 2* simply guarantees that there is a unique solution to the two pairs of coupled equations (11) and (12), for all t, and it may also be referred to as the *Nash condition*.

As we have indicated earlier, the uniqueness part of the theorem has been verified separately in the *Appendix*. ◇

Several observations and remarks would be in order here. Firstly, we note that, as opposed to the *memoryless* solution of *Theorem 1* (obtained under myopic forecast), the unique Nash equilibrium solution incorporates memory, for both agents. For agent A, the "best" forecast policy is a linear function of the most recent measurement and the most recent decision taken by that agent. [This is true since $\tilde{v}_{t-1}$ in (17) can be replaced by v_{t-1}, without affecting the solution.] For agent B, on the other hand, the "best" tracking policy is a linear function of the most recent measurement and a linear aggregate of all past measurements, weighted in an appropriate manner. The solution is characterized in terms of four gain coefficients $(\alpha_t, \tilde{\alpha}_t, \beta_t, \tilde{\beta}_t)$, which can be computed recursively. Hence, the solution does not change structurally over time, which makes it feasible to obtain *stationary* Nash policies for the infinite-horizon version, provided that the sequences $\{\alpha_t^T\}$, $\{\tilde{\alpha}_t^T\}$, $\{\beta_t^T\}$, $\{\tilde{\beta}_t^T\}$ converge for all finite t as $T \to \infty$, where the superscript T in the sequences denotes the dependence of each sequence on the terminal time, taken as a parameter. Even though the computation of the four critical quantities $(\alpha_t, \widetilde{\alpha}_t, \beta_t, \widetilde{\beta}_t)$ may look complicated at the outset, the iterations are in fact quite straightforward, requiring simple algebraic manipulations at each step. The order one has to follow in the computation is as follows:

> *Starting at $t=T$, first compute the quadruple $(\alpha_T, \widetilde{\alpha}_T, \beta_T, \widetilde{\beta}_T)$ from (11a)-(12b), using the given boundary conditions on K_{T+1}, m_T, $\widetilde{m}_T$ and n_T. Note that this computation involves the solution of two pairs of coupled linear equations, at which point we invoke the Nash condition (16b) to obtain a unique solution. At this stage also condition (16c) is invoked, so that (11a) and (12a) are well defined. The next step would be to obtain the new values for k_{t+1}, m_t, $\widetilde{m}_t$, n_t at $t=T+1$, using the iterations (13a), (10a), (10b) and (10c), respectively. At this stage, condition (16a) is invoked so that (10a) and (10b) are well-defined relationships. These new values for K, m, $\widetilde{m}$, n are then used again in (11a)-(12b) to update the values of the gain coefficients, and this procedure is repeated until the initial stage $t=0$ is reached.*

We should point out that similar to *Condition 1* in *Section 2*, *Condition 2* also generically holds, in the sense that if all but one of the parameter values are fixed, then there is only a finite number of values for that parameter for which the condition fails.

Even though it is not our intention to provide here a general convergence analysis for the infinite-horizon problem (this would in fact be quite a challenging task), it would nevertheless be instructive to study some properties of the stationary solution, assuming that such a solution exists and *Condition 2* holds for all t of interest. Accordingly, letting

$$\alpha^* := \lim_{T\to\infty} \alpha_t^T, \quad \widetilde{\alpha}^* := \lim_{T\to\infty} \widetilde{\alpha}_t^T, \quad \beta^* := \lim_{T\to\infty} \beta_t^T, \quad \widetilde{\beta}^* := \lim_{T\to\infty} \widetilde{\beta}_t^T, \quad n^* := \lim_{T\to\infty} n_t^T,$$

it readily follows that $n^*=0$. In view of this, we arrive at the stationary Nash policies

$$v_t = \gamma^*(y^t) = \alpha^* y_t + \beta^* v_{t-1}, \tag{19a}$$

$$w_t = \mu^*(y^t) = \widetilde{\alpha}^* y_t + \widetilde{\beta}^* \widetilde{v}_{t-1}, \tag{19b}$$

where $\{\widetilde{v}_t\}$ is generated by

$$\widetilde{v}_t = \alpha^* y_t + \beta^* \widetilde{v}_{t-1}, \tag{19c}$$

and the following relationship holds:

$$\alpha^* = -\frac{a + c\widetilde{\alpha}^*}{b} \quad , \quad \beta^* = \frac{1 - c\widetilde{\beta}^*}{b} \quad . \tag{20}$$

Now, using these stationary policies in the system equation (4), we arrive at the result that the equilibrium trajectory $\{y_t^*\}$ is generated by

$$y_{t+1}^* = (a + b\alpha^* + c\widetilde{\alpha}^*) y_t + b\beta^* v_{t-1}^* + c\widetilde{\beta}^* \widetilde{v}_{t-1}^* + \epsilon_{t+1},$$

where $\{v_t^*\}$ and $\{\widetilde{v}_t^*\}$ denote the discrete-time stochastic processes generated by (19a) and (19b), respectively, when $y_t = y_t^*$, $t \geq 0$. Note that, as stochastic processes, they are identical *almost surely*, and hence, by also using (20), it can be shown that the equilibrium trajectory $\{y_t^*\}$ is generated by the simpler dynamics

$$\begin{aligned} y_{t+1}^* &= v_{t-1}^* + \epsilon_{t+1} \\ v_t^* &= \alpha^* y_t^* + \beta^* v_{t-1}^* \quad , \end{aligned}$$

which admits the *ARMA* representation

$$y^\star_{t+1} + \beta^\star y^\star_t - \alpha^\star y^\star_{t-1} = \epsilon_{t+1} + \beta^\star \epsilon_t. \tag{21}$$

An important observation that can be made here is that the relationship

$$E_{t-1} y^\star_{t+1} = v^\star_{t-1} \quad ,$$

holds, that is we have *perfect foresight.* Said differently, the stationary Nash solution satisfies the side condition of myopic foresight introduced in *section 2*, in spite of the fact that the two solutions (of *Theorem 1* and *Theorem 2*) are structurally different. [Compare (20) (or its stationary version) with (17)-(18).] This clearly implies that the Nash solution is disadvantageous to agent B (at least in the limit as $T \to \infty$), since it does not yield the best (optimum) solution obtainable under the side condition induced by the equilibrium solution itself. A reason for this *inefficient* behavior on the part of B is that in the analysis of *section 3* agent A is also an active player, whereas in *section 2* he was passive. Such features can be observed even in finite-horizon problems, as the following example demonstrates.

Numerical example 1. In our general formulation, let T=0, $E[y_0^2] =: \sigma_0$, and all other parameter values be unity. Then, the two solutions given in *Theorem 1* and *Theorem 2* and the corresponding values of expected costs and trajectory sequences can be computed to be as follows:

Theorem 1:

$$w_1 = \hat{\mu}_1(y_1) = -\frac{1}{2} y_1, \quad w_0 = \hat{\mu}_0(y_0) = -\frac{6}{7} y_0, \quad v_0 = \frac{1}{7} y_0, \tag{22a}$$

$$\hat{J}_A = \frac{5}{4}; \quad \hat{J}_B = \frac{5}{2} + \frac{6}{7}\sigma_0,$$

$$\hat{y}_1 = \frac{2}{7} y_0 + \epsilon_1; \quad \hat{y}_2 = \frac{1}{2} \hat{y}_1 + \epsilon_2.$$

Theorem 2:

$$w_1 = \mu_1^\star(y_1) = -\frac{1}{2} y_1, \quad w_0 = \mu_0^\star(y_0) = -\frac{3}{4} y_0, \quad v_0 = \gamma_0^\star(y_0) = \frac{1}{4} y_0, \tag{22b}$$

$$J_A^\star = \frac{5}{4}; \quad J_B^\star = \frac{5}{2} + \frac{15}{16}\sigma_0,$$

$$y_1^\star = \frac{1}{2} y_0 + \epsilon_1; \quad y_2^\star = \frac{1}{2} y_1^\star + \epsilon_2.$$

A number of observations can be made in connection with this example:

1. In both cases above, we obtain *perfect foresight* (i.e. $v_0 = E_0 y_2$), but the corresponding trajectories are different. Even though (as we have seen earlier) the Nash solution does not generally enjoy perfect foresight for the finite-horizon case, here it does, mainly because the problem involves basically a single stage, thus eliminating the effect of spillover across consecutive periods.
2. Agent A incurs equal expected costs in both cases, whereas agent B does worse with the Nash solution. This is, of course, consistent with our earlier comments just preceding this example, which, even though were made in the context of the infinite-horizon problem, are equally valid here since the Nash solution satisfies the boundary condition (i.e. perfect foresight) of the myopic solution.

3. Since $\hat{\mu}_1 = \mu_1^\star$ is a *universally optimal* policy for agent **B** at stage t=1, whichever equilibrium solution is adopted (even outside the two considered here) the trajectory will be given by

$$y_2 = \frac{1}{2}y_1 + \epsilon_2$$
$$y_1 = y_0 + v_0 + w_0.$$

Now, if we let $v_0 = E_0 y_2$, and attempt to solve for v_0 from the above equations, we first obtain (since $w_0 = \mu_0(y_0)$ is known to B for each fixed μ_0)

$$v_0 = E_0 y_2 = \frac{1}{2}E_0 y_1 = \frac{1}{2}y_0 + \frac{1}{2}v_0 + \frac{1}{2}w_0,$$

from which v_0 can be solved uniquely to give

$$v_0 = \gamma_0(y_0) = y_0 + w_0; \quad , w_0 = \mu_0(y_0). \tag{$\circ$}$$

This shows that the actual choice for $v_0 = \gamma_0(y_0)$ (under perfect foresight) depends explicitly on B's policy μ_0, and the two solutions given above are two different manifestations of this dependence. Both (22*a*) and (22*b*) use ($\circ$) as a constraint, but while in (22*a*) J_B is minimized subject to ($\circ$), in (22*b*) the choices are determined by the Nash solution of a game played between the two agents at time t=0. One could envision other scenarios between the two agents which would lead to still different choices for μ_0 (and thereby γ_0), but in all cases the resulting expected cost to **A** will be the constant 5/4, independent of μ_0 and σ_0. ◇

We now conclude this section with a second example, which is an extended version of the previous example with an additional stage. It will serve to demonstrate some additional features of the solution given in *Theorem 2*.

Numerical example 2. In the general formulation, let T=1, and all parameter values be unity. Then, the unique Nash equilibrium solution (as presented in *Theorem 2*) can be computed to be as follows:

$$\begin{aligned} &w_2 = \mu_2^\star(y_2) = -\frac{1}{2}y_2, \quad w_1 = \mu_1^\star(y^1) = -\frac{3}{8}y_1 - 0.190476y_0, \\ &w_0 = \mu_0^\star(y_0) = -0.746032y_0; \\ &v_1 = \gamma_1^\star(y^1) = -\frac{3}{8}y_1 + 0.31746y_0, \quad v_0 = \gamma_0^\star(y_0) = 0.253968y_0. \end{aligned} \tag{23}$$

The corresponding equilibrium trajectory is generated by

$$\begin{aligned} y_1^\star &= 0.5079366y_0 + \epsilon_1 \\ y_2^\star &= 0.25y_1^\star + 0.126984y_0 + \epsilon_2 \\ y_3^\star &= 0.5y_2^\star + \epsilon_3 \quad , \end{aligned}$$

from which it follows that $E_1 y_3^\star = 0.125y_1^\star + 0.063492y_0 \not\equiv \gamma_1^\star(y_1^\star, y_0)$; that is, the solution does not lead to perfect foresight at time t=1. However, $E_0 y_2^\star = 0.253968y_0 \equiv \gamma_0^\star(y_0)$; that is, there is perfect foresight at t=0. This latter result is not a feature of this example only, but holds for the general solution of *Theorem 2* (even though it may be rather difficult to prove algebraically). Through an indirect reasoning that follows the proof of *Theorem 2*, as given in the *Appendix*, one can conclude that $E_0 y_2^\star = \gamma_0^\star(y_0)$ is a genuine property of the general Nash solution since, at the initial stage, the variable v_0 minimizes an expression that is a perfect square in y_2 (see $(A.1)$) and there is no spillover effect. $\diamond$

4. Some Extensions

A first extension of the results presented in *section 2* and *section 3* would be to the more general case where the reference trajectory is not *zero* and the cost function (5*b*) contains additional (time-varying) weights on the deviation from the desired trajectory (*i.e.* the first term). The reason why we have not included this in our presentation here is because such an extension does not entail anything conceptually new, while requiring some additional notation which would have complicated the resulting expressions considerably. The gist of the results for the nonzero reference trajectory case is that the statements of both *Theorem 1* and *Theorem 2* remain essentially intact, with the only difference being that now each policy includes an additive (bias) term which depends linearly on the desired reference trajectory. The existence conditions in both cases are identical to the earlier ones. For the case when there is a time-varying weight in the first term of (5*b*), the results again remain intact, with only the additive term 1 in (8*a*) replaced by this new weight and Q in (13*a*) adjusted accordingly.

A second extension would be to the class of problems where the agents do not have direct access to the trajectory $\{y_t\}$, but rather acquire common noisy measurements $\{z_t\}$, as defined by (1*b*), where now $\eta_t = z^t$. Towards studying this extension, let us assume that $\{\epsilon_t\}$ and $\{\xi_t\}$ are sequences of independent Gaussian zero-mean random variables with variances $var(\epsilon_t) =: \varphi > 0$, $var(\xi_t) =: \varsigma_t > 0$, and that they are independent of y_0 which is also a Gaussian zero-mean random variable, with variance σ_0. Then, in the formulation of *section 2*, we interpret the operator E_t as the conditional expectation $E\{\cdot|z^t, w^{t-1}\}$. Note that here we have replaced $\eta_t = z^t$ with $\tilde{\eta}_t := (z^t, w^{t-1})$, without any loss of generality, since w^{t-1} is measurable with respect to z^{t-1}. Now, letting $\hat{y}_t := E_t y_t$, it is a standard result (see, for example, *Bertsekas (1987)* or *Kumar and Varaiya (1986)*) that $\hat{y}_t$ is generated by the Kalman filter equations:

$$\begin{aligned}\hat{y}_{T+2} &= a\hat{y}_{T+1} + cw_{T+1} + [\hat{\sigma}_{T+2}/(\hat{\sigma}_{T+2} + \varsigma_{T+2})]r_{T+2},\\ \hat{y}_{t+1} &= a\hat{y}_t + bE_t y_{t+2} + cw_t + [\hat{\sigma}_{t+1}/(\hat{\sigma}_{t+1} + \varsigma_{t+1})]r_{t+1}, \quad t \le T; \quad \hat{y}_{-1} = 0,\end{aligned} \tag{24a}$$

$$r_{t+1} := z_{t+1} - a\hat{y}_t - bE_t y_{t+2} - cw_t, \tag{24b}$$

$$\hat{\sigma}_{t+1} = [a^2\varsigma_t/(\hat{\sigma}_t + \varsigma_t)]\hat{\sigma}_t + \varphi_{t+1}, \quad \hat{\sigma}_0 = \sigma_0, \tag{24c}$$

where $\{r_t\}$ is a sequence of independent Gaussian random variables, known as the *innovation* sequence. In writing down these relationships, we have made explicit use of the fact that both $E_t y_{t+2}$ and w_t are z^t-measurable.

Now note that the error sequence $\{e_t\}$, $e_t := y_t - \hat{y}_t$, is generated by

$$e_{t+1} = ae_t + \epsilon_{t+1} - [\hat{\sigma}_{t+1}/(\hat{\sigma}_{t+1} + \varsigma_{t+1})]r_{t+1}; \quad e_0 = 0, \tag{Δ}$$

and that $E_t e_{t+n} = 0$ for all $n \ge 1$. In view of this last property,

$$E_t y_{t+2} = E_t\hat{y}_{t+2} + E_t e_{t+2} = E_t\hat{y}_{t+2},$$

and hence (24*a*) can be rewritten as

$$\begin{aligned}\hat{y}_{T+2} &= a\hat{y}_{T+1} + cw_{T+1} + [\hat{\sigma}_{T+2}/(\hat{\sigma}_{T+2} + \varsigma_{T+2})]r_{T+2},\\ \hat{y}_{t+1} &= a\hat{y}_t + bE_t\hat{y}_{t+2} + cw_t + [\hat{\sigma}_{t+1}/(\hat{\sigma}_{t+1} + \varsigma_{t+1})]r_{t+1}, \quad t \le T; \quad \hat{y}_{-1} = 0.\end{aligned} \tag{25}$$

Furthermore, since $y_t = \hat{y}_t + e_t$, and $\hat{y}_t$ is orthogonal to e_t, the counterpart of (7*a*) for the noisy case would be:

$$F_s^{T+1} = \sum_{t=s}^{T+1} E\{\hat{y}_{t+1}^2 + kw_t^2\}\rho_B^{t-s} + \sum_{t=s}^{T+1} E\{e_{t+1}^2\}\rho_B^{t-s}, \tag{26}$$

where the second summation term does not enter the optimization, since the sequence $\{e_t\}$ generated by (Δ) is independent of the control sequence $\{w_t\}$. Hence, the problem faced by agent B is the minimization of the first term of (26) subject to the dynamics (25), where $w_t = \mu_t(\hat{y}^t)$, which is compatible with the original information $\tilde{\eta}_t = (y^t, w^{t-1})$ since $\hat{y}_t$ is generated by $(\tilde{\eta}_{t-1}, w_{t-1})$. Then, the problem is identical with the perfect information case (apart from a change of notation), in view of the fact that $\{r_t\}$ is a zero-mean independent sequence, playing the role of $\{\epsilon_t\}$ in (7*b*). This shows that the problem (with myopic forecast) features *certainty equivalence*, making the statement of *Theorem 1* valid also in the noisy case, with only y_t replaced by $\hat{y}_t$, and (9*c*) including an additional positive term due to the second term of (26). The following theorem summarizes this result.

Theorem 3. *Let Condition 1 be satisfied. Then, the dynamic policy optimization problem with myopic forecast, as formulated in section 2 but with common noisy measurements* (1*b*) *for both agents, admits the unique solution*

$$w_t = \hat{\mu}_t(y_t) = g_t\hat{y}_t, \quad 0 \le t \le T+1, \tag{27a}$$

with the corresponding forecast policy given by

$$v_t = E_t y_{t+2} = \frac{(a + cg_{t+1})(a + cg_t)}{\nu_{t+1}\nu_t}\hat{y}_t, \tag{27b}$$

where $\{\hat{y}_t\}$ *is generated by* (25), *and* $\{g_t\}$, $\{\nu_t\}$ *are as defined by* (8*c*) *and* (8*b*), *respectively.*
◇

Hence, for the noisy case, certainty equivalence holds under myopic forecast, and the statement of *Theorem 1* basically remains intact. For *Theorem 2*, however, there is no direct counterpart, and derivation of the Nash equilibrium solution is quite a nontrivial task. We will not pursue this extension here, since presenting the full details of the derivation of the Nash equilibrium solution would at least double the length of the present paper. What we can say at this point, however, is that (guided by the results presented in *Başar (1978b)* for a linear-quadratic nonzero-sum dynamic game with a different type of an information pattern and a different type of a cost function for one of the agents) the problem will generically admit a unique Nash equilibrium solution, linear in the available common information. This solution will not satisfy the certainty equivalence or separation principle of stochastic control, and thus will have no relationship with the solution presented in *Theorem 2.* The following numerical example (which is the "noisy" version of the second example of *section 3*) should serve to corroborate this claim and to give some indication as to the intricacies involved in the derivation of the general solution.

Numerical example 3. Consider the second numerical example of *section 3*, but with noisy measurement (1*b*) for both agents, and with all parameter values (including the noise variances) equal to unity. Hence, the cost functions are

$$\begin{aligned} J_A &= E\{(v_1 - y_3)^2 + (v_0 - y_2)^2\} \\ J_B &= E\{y_3^2 + w_2^2 + y_2^2 + w_1^2 + y_1^2 + w_0^2\}, \end{aligned} \tag{28}$$

and the dynamic constraints are

$$\begin{aligned} y_3 &= y_2 + w_2 + \epsilon_3 \\ y_2 &= y_1 + v_1 + w_1 + \epsilon_2 \\ y_1 &= y_0 + v_0 + w_0 + \epsilon_1, \end{aligned} \tag{29}$$

where $w_2 = \mu_2(z^2)$, $w_1 = \mu_1(z^1)$, $w_0 = \mu_0(z_0)$, $v_1 = \gamma_1(z^1)$, $v_0 = \gamma_0(z_0)$; $z_0 = y_0 + \xi_0$, $z_1 = y_1 + \xi_1$ and $z_2 = y_2 + \xi_2$. The first significant difference between the perfect and the noisy measurement cases appears in the construction of the best μ_2, which now depends explicitly on (μ_1, μ_0) and (γ_1, γ_0). [Recall that in the perfect measurement case covered by *Theorem 2*, there was a universally optimal policy for agent B at the terminal stage of the game.] With the quadruple $(\gamma_1, \gamma_0; \mu_1, \mu_0)$ fixed, say at (γ, μ), the minimization of J_B with respect to μ_2 becomes a standard quadratic optimization problem,

$$\min_{w_2} E\{(y_2 + w_2 + \epsilon_3)^2 + w_2^2 | z^2, \gamma, \mu\},$$

whose unique solution is

$$w_2 = \mu_2(z^2, \gamma, \mu) = -\frac{1}{2} E[y_2 | z^2, \gamma, \mu] =: \frac{1}{2}\hat{y}_{2\gamma\mu}. \tag{30}$$

Here $\hat{y}_{2\gamma\mu}$ is generated by the Kalman filter:

$$\begin{aligned} \hat{y}_{2\gamma\mu} &= \hat{y}_{1\gamma_0\mu_0} + \gamma_1(z^1) + \mu_1(z^1) + \frac{8}{13}(z_2 - \hat{y}_{1\gamma_0\mu_0} - \gamma_1(z^1) - \mu_1(z^1)) \\ \hat{y}_{1\gamma_0\mu_0} &= \hat{y}_0 + \gamma_0(z_0) + \mu_0(z_0) + \frac{3}{5}(z_1 - \hat{y}_0 - \gamma_0(z_0) - \mu_0(z_0)) \\ \hat{y}_0 &= \frac{1}{2} z_0, \end{aligned} \tag{31}$$

which depends on (γ, μ) partly directly and partly through $\hat{y}_{1\gamma_0\mu_0} := E[y_1 | z^1, \gamma_0, \mu_0]$. To obtain the pair (γ, μ) that is in Nash equilibrium with (30), we follow a procedure quite analogous (in principle) to the one followed in the proof of uniqueness (for *Theorem 2*) in the *Appendix*, geared towards obtaining a (unique) stagewise equilibrium. Accordingly, the derivation involves the solution of two static games, one at t=1 and the other one at t=0. To characterize the static game at t=1, we substitute (31) into (28), eliminate the intermediate variables and take expectation over the statistics of ϵ_3, ϵ_2 and ξ_2, to arrive at the reduced conditional (on z^1) cost functions:

$$\begin{aligned} J_A^1 = E\{\frac{1}{169}[4v_1 - 9(y_1 + w_1) + \frac{5}{2}\hat{y}_1 + \frac{5}{2}(\gamma_1 + \mu_1)]^2 \\ + [v_0 - y_1 - v_1 - w_1]^2 | z^1\}, \end{aligned}$$

$$\begin{aligned} J_B^1 = E\{&\frac{1}{169}[9(y_1 + v_1 + w_1) - \frac{5}{2}\hat{y}_1 - \frac{5}{2}(\gamma_1 + \mu_1)]^2 \\ &+ \frac{1}{676}[5(\hat{y}_1 + \gamma_1 + \mu_1) + 8(y_1 + v_1 + w_1)]^2 + (y_1 + v_1 + w_1)^2 + w_1^2 | z^1\}. \end{aligned}$$

In the above, we have made notational simplifications by suppressing the (γ_0, μ_0)-dependence of $\hat{y}_1$ and the arguments of (γ_1, μ_1). This is clearly a static game in the pair (v_1, w_1), and its Nash solution can be obtained for each fixed (γ_1, μ_1) and (γ_0, μ_0), where we take $v_0 = \gamma_0(z_0)$.

Differentiating J_A^1 with respect to v_1 and J_B^1 with respect to w_1, and setting the resulting expressions equal to zero after conditioning on z^1, we find that the Nash condition is satisfied and there exists a unique solution to the pair of equations, linear in γ_1, μ_1, $\hat{y}_1$ and γ_0. Now requiring consistency in the solution (as in the proof of uniqueness for *Theorem 2* in the *Appendix*), we set $v_1 = \gamma_1(z^1)$, $w_1 = \mu_1(z^1)$, and solve the resulting pair of linear equations (in v_1 and w_1) uniquely, to arrive at the policies:

$$v_1 = \gamma_1(z^1, \gamma_0, \mu_0) = -0.523810\hat{y}_1 + 1.547619\gamma_0(z_0), \tag{32a}$$

$$w_1 = \mu_1(z^1, \gamma_0, \mu_0) = -0.285714\hat{y}_1 - 0.928571\gamma_0(z_0). \tag{32b}$$

Note that here γ_0 is yet to be determined.

To complete the solution, we next formulate the game at t=0, by substituting (30) and (32) into (28), again eliminating the intermediate variables and averaging over the statistics of the random variables involved, to obtain the reduced conditional (on z_0) cost functions:

$$\begin{aligned} J_A^0 = E\{&(0.619048\hat{y}_0 + 1.208790\gamma_0 - 0.029304\mu_0 + 0.648352(v_0 + w_0))^2 \\ &+ (0.485714v_0 + 0.295238\hat{y}_0 - 0.514286w_0 + 0.190477\gamma_0 + 0.809524\mu_0)^2 | z_0\}, \end{aligned}$$

$$\begin{aligned} J_B^0 = E\{&\frac{1}{4}(0.619048\gamma_0 - 1.194139\hat{y}_1 + 1.384615y_1)^2 \\ &+ \frac{1}{4}(0.619048\gamma_0 - 0.424909\hat{y}_1 + 0.615385y_1)^2 + (0.619048\gamma_0 - 0.809524\hat{y}_1 + y_1)^2 \\ &+ (0.285714\hat{y}_1 + 0.928571\gamma_0)^2 + y_1^2 + w_0^2 | z_0\}, \end{aligned}$$

where both y_1 and $\hat{y}_1$ depend on (v_0, w_0), the latter through z_1, as given in (31).

The procedure here is the same as at t=1: First obtain the Nash solution of (J_A^0, J_B^0) in terms of (γ_0, μ_0), then require consistency (v_0=$\gamma_0(z_0)$,w_0=$\mu_0(z_0)$) and solve for (v_0, w_0) from the resulting equations, which will lead to policies whose argument is z_0. At each step the uniqueness condition is met, and thus the procedure yields the unique Nash equilibrium policies (at t=0):

$$v_0 = \gamma_0^*(z_0) = +0.248227\hat{y}_0, \tag{33a}$$

$$w_0 = \mu_0^*(z_0) = -0.751773\hat{y}_0. \tag{33b}$$

These policies are finally used in (32) and (31) to complete the characterization of the Nash equilibrium policies:

$$v_1 = \gamma_1^*(z^1) = -0.523810\hat{y}_1^* + 0.384161\hat{y}_0, \tag{34a}$$

$$w_1 = \mu_1^*(z^1) = -0.285714\hat{y}_1^* - 0.230496\hat{y}_0, \tag{34b}$$

$$w_2 = \mu_2^*(z^2) = -0.5\hat{y}_2^*, \tag{34c}$$

where

$$\begin{aligned} \hat{y}_2^* &= 0.059102\hat{y}_0 + 0.073260\hat{y}_1^* + 0.615385z_2 \\ \hat{y}_1^* &= 0.198582\hat{y}_0 + 0.6z_1 \\ \hat{y}_0 &= 0.5z_0. \end{aligned} \tag{35}$$

An equivalent representation for γ_1^* in (34a) would be

$$v_1 = \gamma_1^*(z^1) = -0.523810\hat{y}_1^* + 1.547619v_0,$$

which shows explicit dependence on v_0. Note that the policies (34) are different from their counterparts in the noise-free case (*i.e.* (23)), thus corroborating our earlier remark that the "noisy version" does not feature certainty equivalence.

The equilibrium trajectory corresponding to the unique Nash solution is generated by

$$y_3^* = y_2^* - 0.5\hat{y}_2^* + \epsilon_3$$
$$y_2^* = y_1^* - 0.809524\hat{y}_1^* + 0.153664\hat{y}_0 + \epsilon_2$$
$$y_1^* = y_0 - 0.503546\hat{y}_0 + \epsilon_1.$$

Using these, it is easy to check that, as in the second example of *section 3*, $E_1 y_3^* \not\equiv \gamma_1^*(z^1)$, while $E_0 y_2^* \equiv \gamma_0^*(z_0)$, which shows that the Nash solution could lead to perfect foresight at the initial stage, even in the noisy case. As we will discuss in a companion paper, this turns out to be a general property of the Nash solution for the "noisy version" of the problem of *section 3*. ◇

Appendix

In this appendix, we first complete the proof of the existence part of *Theorem 2* by showing that the policy (∗∗) given there indeed solves agent A's optimization problem. Subsequently, we establish the uniqueness of the Nash solution presented in *Theorem 2*.

Existence. The optimization problem faced by agent A is the minimization of J_0^T, where

$$J_s^T = \sum_{t=s}^{T} E\{(v_t - y_{t+2})^2\}\rho_A^{t-s},$$

under the constraints

$$y_{T+2} = [ak/(k+c^2)]y_{T+1} + \epsilon_{T+2}$$
$$y_{t+1} = (a + c\tilde{\alpha}_t)y_t + c\tilde{\beta}_t\tilde{v}_{t-1} + bv_t + \epsilon_{t+1}, \quad 1 \le t \le T$$
$$y_1 = (a + c\tilde{\alpha}_0)y_0 + \epsilon_1;$$
$$\tilde{v}_t = \alpha_t y_t + \beta_t \tilde{v}_{t-1}, \quad v_t = \gamma_t(y^t).$$

We now claim that, for a general t,

$$\min_{\{\gamma_s\}_{s=t}^T} J_t^T = \min_{\gamma_{t+1},\gamma_t} E\{\rho_A n_{t+1}(v_{t+1} - m_{t+1}y_{t+2} - \tilde{m}_{t+1}\tilde{v}_{t+1})^2 + (v_t - y_{t+2})^2\} + q_t, \quad (A.1)$$

where $\{q_t\}$ is a sequence depending only on the variances of the additive stochastic terms ϵ_t, $t \le T+2$. Under the validity of this assertion, the optimal policy at time t is obtained by minimizing the following quantity with respect to the scalar variable $v_{t+1}=:v$, for each fixed y^{t+1}:

$$E\{\rho_A n_{t+1}[v - m_{t+1}(a + c\tilde{\alpha}_{t+1})y_{t+1} - m_{t+1}c\tilde{\beta}_{t+1}\tilde{v}_t - m_{t+1}bv - \tilde{m}_{t+1}\alpha_{t+1}y_{t+1} - \tilde{m}_{t+1}\beta_{t+1}\tilde{v}_t]^2 + [(a + c\tilde{\alpha}_{t+1})y_{t+1} + c\tilde{\beta}_{t+1}\tilde{v}_t + bv - v_t]^2 | y^{t+1}\}. \quad (A.2)$$

Being quadratic and strictly convex (in v), this optimization problem admits a unique solution (for each fixed $y_{t+1}, \tilde{v}_t, v_t$), given by

$$v_{t+1} = \gamma_{t+1}(y^{t+1}) = \hat{\alpha}_{t+1}y_t + \hat{\beta}_{t+1}v_t + \bar{\beta}_{t+1}\tilde{v}_t, \quad (A.3a)$$

for $0 \leq t \leq T-1$, and at the initial stage by

$$v_0 = \gamma_0(y_0) = \hat{\alpha}_0 y_0, \tag{A.3b}$$

where

$$\hat{\alpha}_t = \frac{1}{b^2 + (1-bm_t)^2 \rho_A n_t}[\rho_A n_t(1-bm_t)(\tilde{m}_t \alpha_t + (a + c\tilde{\alpha}_t)m_t) - b(a + c\tilde{\alpha}_t)]$$

$$\hat{\beta}_t = \frac{b}{b^2 + (1-bm_t)^2 \rho_A n_t}$$

$$\bar{\beta}_t = \frac{1}{b^2 + (1-bm_t)^2 \rho_A n_t}[\rho_A n_t(1-bm_t)(\tilde{m}_t \beta + c\tilde{\beta}_t m_t) - bc\tilde{\beta}_t].$$

As we have discussed earlier (in the proof of the existence part of *Theorem 2*), substitution for α_t and β_t (from (11*a*) and (12*a*), respectively) into the three expressions above, leads to the equivalences $\hat{\alpha}_t \equiv \alpha_t$ and $\hat{\beta}_t + \bar{\beta}_t \equiv \beta_t$. Hence, the optimal solution (*A*.3) admits the equivalent representation

$$\begin{aligned} v_t &= \alpha_t y_t + \hat{\beta}_t v_{t-1} + (\beta_t - \hat{\beta}_t)\tilde{v}_{t-1}, & 1 \leq t \leq T \\ &= \alpha_0 y_0 \qquad\qquad , & t = 0. \end{aligned} \tag{A.4}$$

We now turn to verification of the structural form (*A*.1). The result trivially holds for $t{=}T$, with $m_T = ak/(k+c^2)$, $\tilde{m}_T = 0$. Let us therefore assume the validity of the assertion for $t{+}1$ and prove it for t. Towards this end, we substitute (*A*.4), with t replaced by $t{+}1$, into (*A*.2), and arrive (after some rather tedious algebra) at an expression which is a perfect square in v_t, y_{t+1} and $\tilde{v}_t$:

$$E\{n_t(v_t - m_t y_{t+1} - \tilde{m}_t \tilde{v}_t)^2 | y^{t+1}\}. \tag{A.5}$$

Here m_t, $\tilde{m}_t$ and n_t are defined in terms of m_{t+1}, $\tilde{m}_{t+1}$ and n_{t+1} as in (10*a*) through (10*c*). [In fact, it is not difficult to see that the resulting cost should be a perfect square, because (*A*.2) can be made equal to zero by appropriately choosing v_t and v_{t+1}. With this observation, it then remains to find the three coefficients n_t, m_t and $\tilde{m}_t$.] Now, since the minimum of (*A*.2) over v_{t+1} is equal to (*A*.5), we have

$$\begin{aligned} \min_{\{\gamma_s\}_{s=t-1}^{T}} J_{t-1}^T &= \min_{\gamma_t, \gamma_{t-1}} E\{(v_{t-1} - y_{t+1})^2 + \rho_A \min_{\{\gamma_s\}_{s=t+1}^{T}} J_t^T\} \\ &= \min_{\gamma_t, \gamma_{t-1}} E\{(v_{t-1} - y_{t+1})^2 \rho_A n_t (v_t - m_t y_{t+1} - \tilde{m}_t \tilde{v}_t)^2\} \\ &\quad + \rho_A[q_t + (1 + \rho_A n_{t+1} m_{t+1}^2) var(\epsilon_{t+2})], \end{aligned}$$

which is in the same form as (*A*.1), with

$$q_{t-1} := \rho_A[q_t + (1 + \rho_A n_{t+1} m_{t+1}^2) var(\epsilon_{t+2})].$$

This then completes the proof of optimality of (⋆⋆) in the proof of the existence part of *Theorem 2*.

Uniqueness. It is a well-known fact that dynamic games could admit nonunique Nash equilibria, with each such equilibrium leading to a different cost pair which are in general incomparable (see, for example, *Başar and Olsder (1982)*). Thus, "uniqueness" is an important question to pose, if the proposed equilibrium is to be of value. As we have discussed extensively in earlier papers (for example, *Başar(1976)*, *Başar(1977)*), the main source of nonuniqueness in Nash equilibria is the so-called *informational nonuniqueness* which arises if each agent, in his one-sided optimization, has the freedom of choosing different *representations* of the same policy. What we prove in the sequel is that for the game problem covered by *Theorem 2* there is no informational nonuniqueness, and the structural form (17)-(18) is the only form in which a Nash equilibrium can exist. Furthermore, we show that structural uniqueness is guaranteed under *Condition 2*. In the proof, we will not explicitly derive the expressions for this unique Nash solution, since we have already shown in the first part of the proof that (17)-(18) exists as a Nash equilibrium.

Towards devising a proof for uniqueness, we first introduce two *generic* functions $quad(\cdot)$ and $lin(\cdot)$, where

$$quad(\cdot) = \text{a quadratic function of its arguments}$$
$$lin(\cdot) = \text{a linear function of its arguments.}$$

Furthermore, we introduce a class of *nested subgames* $\{G_s\}$, parameterized by s, each one being a replica of the original game but defined on a shorter time interval, $[s, T+1]$, $0 \le s \le T+1$. More precisely, for the subgame G_s, the cost functions are defined by (5a)-(5b) with the lower limits changed to $t=s-1$, and with the action variables being $v_s^T := (v_T, .., v_{s+1}, v_s)$ for A, and $w_s^{T+1} := (w_{T+1}, .., w_{s+1}, w_s)$ for B, where $v_t = \mu_t(y^t)$, $w_t = \gamma_t(y^t)$, and a similar convention as above applying to the policy variables μ_s^{T+1},γ_s^T. To be consistent with this convention, for s=0 we extend the limit of the summation to $t = -1$ in both J_A and J_B, by adding *zero* as the incremental cost term at $t = -1$. Now let $(\tilde{\gamma} := \tilde{\gamma}_0^T, \tilde{\mu} := \tilde{\mu}_0^{T+1})$ be a Nash equilibrium solution for the original game (G_0), such as the one given in *Theorem 2*. Then, it is a well-known property of the Nash solution (called the *stagewise equilibrium* property) that for any s, the truncated version of these policies, $(\tilde{\gamma}_s^T, \tilde{\mu}_s^{T+1})$, constitutes a Nash equilibrium solution for G_s, with the past policies $(\gamma_0^{s-1}, \mu_0^{s-1})$ fixed at $(\tilde{\gamma}_0^{s-1}, \tilde{\mu}_0^{s-1})$.

We now develop a procedure for studying the uniqueness of the solutions of these individual subgames. First consider the case $s = T+1$, where G_{T+1} is not really a game but a one-sided optimization problem for agent B, since only B is active at $t = T+1$. Then, clearly the solution is unique, and is given by the second line in (18a). Note that this solution is both informationally and structurally unique (regardless of the past policy choices), the former being due to our assumption in *section 1* on the structure of the probability distribution of the additive system noise. Hence, in the study of the second game in the sequence, G_T, we can take μ_{T+1} as in (18a), without any loss of generality. Accordingly, substituting this μ_{T+1}, say μ_{T+1}^*, into both J_A and J_B, eliminating the intermediate variables using the evolution equation (4) and averaging over the statistics of the random variables by employing their independence property, we arrive at the structural forms

$$\begin{aligned} cost_A(G_T) &= quad(y_T, v_T, w_T, v_{T-1}) \\ cost_B(G_T) &= quad(y_T, v_T, w_T) + quad(w_{T-1}^2), \end{aligned} \tag{A.6}$$

which are the costs incurred to A and B, respectively, conditioned on the information available at time T, *i.e.* $\eta_T = y^T$. Since the first cost shows explicit dependence on v_{T-1}, we fix $v_{T-1} = \gamma_{T-1}(y^{T-1})$, and solve for the Nash equilibrium of the resulting static game. Because of the quadratic structure of the cost functions, the Nash solution, if it exists, will be linear in the pair (y_T, v_{T-1}); furthermore it will be (structurally) unique under conditions not depending on y_T and v_{T-1}, and *Condition 2* precisely serves this purpose. Hence, the static game defined by (*A*.6) admits a unique Nash solution, for each fixed γ_{T-1}, given by

$$v_T = \tilde{\gamma}_T(y^T) \equiv lin(y_T, v_{T-1}) \tag{A.7a}$$

$$w_T = \tilde{\mu}_T(y^T) \equiv lin(y_T, v_{T-1}), \tag{A.7b}$$

where $v_{T-1} = \gamma_{T-1}(y^{T-1})$. The linear functions here are precisely the ones given in (17) and (18), with v_{T-1} in the latter case replaced by $\tilde{v}_{T-1}$. The solution is also unique *representationwise* since, because of our nonsingular statistics assumption on the probability distributions of the random variables involved, y_T cannot be expressed in terms of the past values of the trajectory *almost surely* (which would have been possible in a purely deterministic problem). We now note that the complete (unique) solution to subgame G_T is (*A*.7) along with μ^*_{T+1} which was the unique solution (for agent B) to subgame G_{T+1}.

The next game in the sequence, G_{T-1}, involves the the action variables (v_T, v_{T-1}) for agent A and (w_{T+1}, w_T, w_{T-1}) for agent B. Since every Nash equilibrium is necessarily a stagewise equilibrium and since the unique (linear) Nash solution of G_T does not depend structurally on v_{T-1} and w_{T-1}, it follows that every Nash equilibrium for G_{T-1} should match with that of G_T for policies μ_{T+1}, μ_T and γ_T. Hence, the equilibrium solution of G_{T-1} will be nonunique only if the last components of the policy sequences, $(\gamma_{T-1}, \mu_{T-1})$, are nonunique at equilibrium. Towards a study of this, we substitute the solution of G_T into J_A and J_B, with v_{T-1} in (*A*.7*b*) replaced by a general function of y^{T-1}, say $\psi_{T-1}(y^{T-1})$, since B does not have direct access to v_{T-1}. [It is important to note at this point that if B had direct access to v_{T-1}, the solution would have been informationally nonunique, for reasons discussed extensively in *Başar (1978a)* for a different class of such games.] Now, after eliminating the intermediate variables and averaging over the stochastic variables, we arrive at the following *reduced* costs for G_{T-1}, conditioned on the common information available at time $T-1$, *i.e.* y^{T-1}:

$$cost_A(G_{T-1}) = quad(y_{T-1}, v_{T-1}, w_{T-1}, v_{T-2}, \psi_{T-1}(y^{T-1}))$$
$$cost_B(G_{T-1}) = quad(y_{T-1}, v_{T-1}, w_{T-1}, \psi_{T-1}(y^{T-1})).$$

Here, in addition to the unknown (but fixed) function ψ_{T-1}, we also have $v_{T-2} = \gamma_{T-2}(y^{T-2})$ fixed by an arbitrary choice of γ_{T-2}. Under an appropriate condition which is independent of ψ_{T-1} and γ_{T-2} (which is also guaranteed by *Condition 2*), this static game admits a unique equilibrium for each fixed ψ_{T-1} and γ_{T-2}:

$$v_{T-1} = \bar{\gamma}_{T-1}(y_{T-1}, v_{T-2}, \psi_{T-1}(y_{T-1})) \equiv lin(y_{T-1}, v_{T-2}, \psi_{T-1}(y_{T-1})) \tag{A.8a}$$

$$w_{T-1} = \bar{\mu}_{T-1}(y_{T-1}, v_{T-2}, \psi_{T-1}(y_{T-1})) \equiv lin(y_{T-1}, v_{T-2}, \psi_{T-1}(y_{T-1})), \tag{A.8b}$$

where $v_{T-2}=\gamma_{T-2}(y^{T-2})$. Next, we impose consistency in the solution for each fixed γ_{T-2}, which requires that $\bar{\gamma}_{T-1} \equiv \psi_{T-1}$. Using this side condition in (*A*.8*a*), we arrive at

$$v_{T-1} = lin(y_{T-1}, v_{T-2}, v_{T-1})$$

which, being linear, admits the unique solution (for each fixed y^{T-1} and γ_{T-2})

$$v_{T-1} = \tilde{\gamma}_{T-1}(y_{T-1}, v_{T-2}) \equiv lin(y_{T-1}, v_{T-2}), \tag{A.9a}$$

under a nonsingularity condition which is met under *Condition 2*. Letting $\psi_{T-1} = \tilde{\gamma}_{T-1}$ in (*A*.8*b*), we finally obtain for w_{T-1} (for each fixed γ_{T-2}):

$$w_{T-1} = \tilde{\mu}_{T-1}(y_{T-1}, v_{T-2}) \equiv lin(y_{T-1}, v_{T-2}) \tag{A.9b}$$

This then completes the verification of the uniqueness of the solution of G_{T-1}, for each fixed γ_{T-2}. Note that the complete solution to G_{T-1} is given by μ^*_{T+1}, (*A*.7) and (*A*.9), with

v_{T-1} in (*A.7b*) replaced by the expression in (*A.9a*). Here we could also have expressed (*A.7a*) in terms of y^T, instead of (y_T, v_{T-1}), by substituting for v_{T-1} from (*A.9a*), but this is not necessary since agent **A** does have access to his past decision value, and enriching his information set by also including past decision values does not lead to informational nonuniqueness.

The important observation here is that, for each fixed γ_{T-2}, the solution of subgame G_{T-1} (to be denoted $(\tilde{\gamma}_T, \tilde{\gamma}_{T-1}; \tilde{\mu}_{T+1}, \tilde{\mu}_T, \tilde{\mu}_{T-1})$) is structurally unique, with each strategy being linear in its arguments. More precisely, we have $\tilde{\gamma}_T$ linear in (y_T, v_{T-1}), $\tilde{\gamma}_{T-1}$ linear in (y_{T-1}, v_{T-2}), $\tilde{\mu}_{T+1}$ linear in y_{T+1}, $\tilde{\mu}_T$ linear in $(y_T, y_{T-1}, \gamma_{T-2}(y^{T-2}))$ and $\tilde{\mu}_{T-1}$ linear in $(y_{T-1}, \gamma_{T-2}(y^{T-2}))$. Furthermore, the solution is informationally unique because of the nonsingular statistics of the additive noise in the dynamics (4). Then, in the construction of the Nash solution for subgame G_{T-2}, we first substitute for $(\gamma_T, \gamma_{T-1}; \mu_{T+1}, \mu_T, \mu_{T-1})$ from the unique solution of G_{T-1}, with γ_{T-2} replaced by a general function ψ_{T-2}, as in the construction of the solution for G_{T-1}. Repeating the same procedure as in G_{T-1}, we can obtain a linear stagewise Nash solution for G_{T-2} for each fixed γ_{T-3}, whose uniqueness is again guaranteed by *Condition 2*. Following this procedure in retrograde time, we find that for each s, the subgame G_s admits a unique stagewise equilibrium (for each fixed γ_{s-1}), linear in the available information as well as in γ_{s-1}. Since γ_{-1} is trivially zero, the process halts at s=0, leading to the conclusion that the game G_0 admits a unique stagewise equilibrium, linear in the common information available to the agents. This then establishes uniqueness of the Nash solution of the original problem (which is identical with G_0), since every Nash equilibrium is a stagewise equilibrium and we have already proven that the game admits at least one Nash equilibrium.

We conclude this *Appendix* by pointing to the fact that the above procedure would have been an alternative method for the construction of the Nash solution given in *Theorem 2*, but alone it would not be sufficient, since a stagewise equilibrium need not be a Nash equilibrium. ◇

References

Barro, R. J. (1976) , "Rational Expectations and the Role of Monetary Policy", *Journal of Monetary Economics*, vol.2, pp.1-33.

Başar, T. (1976) , "On the Uniqueness of the Nash Solution in Linear-Quadratic Differential Games", *International Journal of Game Theory*, vol.5, no.2/3, pp.65-90.

Başar, T. (1977) , "Informationally Nonunique Equilibrium Solutions in Differential Games", *SIAM Journal on Control*, vol.15, no.4, pp.636-660.

Başar, T. (1978a) , "Decentralized Multicriteria Optimization of Linear Stochastic Systems", *IEEE Transactions on Automatic Control*, vol.AC-23, no.2, pp.233-243.

Başar, T. (1978b) , "Two-Criteria LQG Decision Problems with One-Step Delay Observation Sharing Pattern", *Information and Control*, vol.32, no.1, pp.21-50.

Başar, T. (1987) , "Some Thoughts on Rational Expectations Models, and Alternate Formulations", invited contribution to a special issue of *Computer and Mathematics with Applications*, to appear in 1988/1989.

Başar, T. and G. J. Olsder (1982) , *Dynamic Noncooperative Game Theory*, Academic Press, London/New York.

Bertsekas, D. P. (1987) , *Dynamic Programming: Deterministic and Stochastic Models*, Prentice Hall, Englewood Cliffs, New Jersey.

Blanchard, O. (1979) , "Backward and Forward Solutions for Economies with Rational Expectations", *American Economic Review*, vol.69, pp.114-118.

Blanchard, O. and C. M. Kahn (1980) , "The Solution of Linear Difference Models under Rational Expectations", *Econometrica*, vol.48, no.5, pp.1305-1311.

Kumar, P. R. and P. P. Varaiya (1986) , *Stochastic Systems: Estimation, Identification and Adaptive Control*, Prentice Hall, Englewood Cliffs, New Jersey.

Lucas, R. (1975) , "An Equilibrium Model of the Business Cycle", *Journal of Political Economy*, vol.83, pp.1113-1144.

Sargent, T. J. and N. Wallace (1975) , "Rational Expectations, the Optimal Monetary Instrument, and the Optimal Money Supply", *Journal of Political Economy*, vol.83, pp.241-254.

Shiller, R. (1978) , "Rational Expectations and the Dynamic Structure of Macroeconomic Models: A Critical Review", *Journal of Monetary Economics*, vol.4, pp.1-44.

Taylor, J. (1977) , "Condition for Unique Solutions in Stochastic Macroeconomic Models with Rational Expectations", *Econometrica*, vol.45, pp.1377-1385.

Victory and Defeat in Differential Games

Jean-Pierre Aubin

CEREMADE, UNIVERSITÉ DE PARIS-DAUPHINE
F-75775, Paris cx(16) France

Abstract

We construct the set-valued feedback *map wich allow players in a differential game the possiblity of winning, separately or colletively, or the certainty of winning or loosing and we characterize the indicator functions of their graphs as solutions to (contingent) partial differential equations. Decisions are defined to be the derivatives of the controls of players, and we provide decision rules for each of these set-valued feedback maps allowing the players to abide by them as time elapses.*

Résumé

Nous construisons des correspondances de rétroaction qui procurent aux joueurs d'un jeu différentiel la possibilité de gagner (séparément ou collectivement), ou la certitude de gagner ou de perdre, et nous caractérisons les fonctions indicatrices de leurs graphes comme solutions d'équations aux dérivées partielles (contingentes). Les décisions sont définies comme les dérivées des contrôles des joueurs, et nous procurons des règles de décision associées à chacune de ces corespondances de rétroaction permettant de les respecter au cours du temps.

1 Description of the Game

Let our two players Xavier and Yves act on the evolution of the state $z(t) \in \mathbf{R}^n$ of the differential game governed by the differential equation

(1) $$z'(t) = h(z(t, u(t), v(t)))$$

by choosing Xavier's controls

(2) $$\forall t \geq 0,\ u(t) \in U(z(t))$$

and by choosing Yves's controls

(3) $$\forall t \geq 0,\ v(t) \in V(t)$$

Here, h, describing the dynamics of the game, maps continuously $\mathbf{R}^n \times \mathbf{R}^p \times \mathbf{R}^q$ into $\mathbf{R}^n$, and $U : \mathbf{R}^n \rightsquigarrow \mathbf{R}^p$ and $V : \mathbf{R}^n \rightsquigarrow \mathbf{R}^q$ are closed[1] set-valued maps describing the state-dependent constraints bearing on the players.

We shall assume that the open-loop controls $u(\cdot)$ and $v(\cdot)$ are absolutely continuous and obey a growth condition of the type[2]

(4) $$\begin{cases} i) & \|u'(t)\| \leq \rho(\|u(t)\| + 1) \\ ii) & \|v'(t)\| \leq \sigma(\|v(t)\| + 1) \end{cases}$$

We shall refer to them as "smooth open-loop controls", the non negative parameters[3] ρ and σ being fixed once and for all. The domain K of the game is the subset of

[1] This means that the graph of the set-valued map is closed. Upper semicontinuous set-valued maps with compact values are closed, and thus, closedness can be regarded as a weak continuity requirement.

[2] one can replace $\rho(\|u\| + 1)$ by any continuous function $\phi(u)$ with linear growth.

[3] or any other linear growth condition $\phi(\cdot)$ or $\psi(\cdot)$ which makes sense in the framework of a game under investigation.

$$(5) \qquad \begin{cases} (z,u,v) \in \mathbf{R}^n \times \mathbf{R}^p \in \mathbf{R}^q \text{ such that} \\ u \in U(z) \;\&\; v \in V(z) \end{cases}$$

Roughly speaking, Xavier may win as long as its opponent allows him to choose at each instant $t \geq 0$ controls $u(t)$ in the subset $U(z(t))$, and must loose if for any choice of open-loop controls, there exists a time $T > 0$ such that $u(T) \notin U(z(T))$.

Definition 1.1 *Let (u_0, v_0, z_0) be an initial situation such that initial controls $u_0 \in U(z_0)$ and $v_0 \in V(z_0)$ of the two players are consistent with the initial state z_0.*

We shall say that

— *Xavier* must win *if and only if for all smooth open-loop controls $u(\cdot)$ and $v(\cdot)$ starting at u_0 and v_0, there exists a solution $z(\cdot)$ to (1) starting at z_0 such that (2) is satisfied.*

— *Xavier* may win *if and only if there exist smooth open-loop controls $u(\cdot)$ and $v(\cdot)$ starting at u_0 and v_0 and a solution $z(\cdot)$ to (1) starting at z_0 such that (2) is satisfied.*

— *Xavier* must loose *if and only if for all smooth open-loop control $u(\cdot)$ and $v(\cdot)$ starting at u_0 and v_0 and solution $z(\cdot)$ to (1) starting at z_0, there exists a time $T > 0$ such that*

$$u(T) \notin U(z(T))$$

— *The initial situation is* playable *if and only if there exist open-loop controls $u(\cdot)$ and $v(\cdot)$ starting at u_0 and v_0 and a solution $z(\cdot)$ to (1) starting at z_0 satisfying* both *relations (2)and (3).*

Naturally, if both Xavier and Yves must win, then both relations (2)and (3) are satisfied. This is not necessarily the case when both Xavier and Yves may win, and this is the reason why we are led to introduce the concept of playability.

2 The Main Theorems

Theorem 2.1 *Let us assume that h is continuous with linear growth and that the graphs of U and V are closed. Let the growth rates ρ and σ be fixed.*

There exist five (possibly empty) closed set-valued feedback maps from $\mathbf{R}^n$ to $\mathbf{R}^p \times \mathbf{R}^q$ having the following properties:

— *$R_U \subset U$ is such that whenever $(u_0, v_0) \in R_U(z_0)$, Xavier* may win *and that whenever $(u_0, v_0) \notin R_U(z_0)$, Xavier* must loose

— *If h is lipschitzean, $S_U \subset R_U$ is the largest closed set-valued map such that whenever $(u_0, v_0) \in S_U(z_0)$, Xavier must win.*

— *$S_V \subset R_V \subset V$, which have analogous properties.*

— *$R_{UV} \subset R_U \cap R_V$ is the largest closed set-valued map such that any initial situation satisfying $(u_0, v_0) \in R_{UV}(z_0)$ is playable.*

Knowing these five set-valued feedback maps, we can split the domain K of initial situations into ten areas which describe the behavior of the differential game from the position of the initial situation.

$(z_0, u_0, v_0) \in$	Graph(S_U)	Graph(R_U)	$K\backslash$Graph(R_U)
Graph(S_V)	Xavier must win Yves must win	Xavier may win Yves must win	Xavier must loose Yves must win
Graph(R_V)	Xavier must win Yves may win	? ? ? ? PLAYABILITY ? ? ? ?	Xavier must loose Yves may win
$K\backslash$Graph(R_V)	Xavier must win Yves must loose	Xavier may win Yves must loose	Xavier must loose Yves must loose

The 10 areas of the domain of the differential game

In particular, the complement of the graph of R_{UV} in the intersection of the graphs of R_U and R_V is the instability region, where either Xavier or Yves may win, but not both together.

The problem is to characterize these five set-valued maps, the existence of which is now guaranteed, by solving the "contingent extension" of the partial differential equation[4]

$$\frac{\partial \Phi}{\partial z} \cdot h(z,u,v) - \rho(\|u\|+1)\left\|\frac{\partial \Phi}{\partial u}\right\| - \sigma(\|v\|+1)\left\|\frac{\partial \Phi}{\partial v}\right\| = 0 \tag{6}$$

which can be written in the following way:

$$\frac{\partial \Phi}{\partial z} \cdot h(z,u,v) + \inf_{\|u'\| \leq \rho(\|u\|+1)} \frac{\partial \Phi}{\partial u} \cdot u' + \inf_{\|v'\| \leq \sigma(\|v\|+1)} \frac{\partial \Phi}{\partial v} \cdot v' = 0$$

We shall also introduce the partial differential equation[5]

$$\frac{\partial \Phi}{\partial z} \cdot h(z,u,v) + \rho(\|u\|+1)\left\|\frac{\partial \Phi}{\partial u}\right\| + \sigma(\|v\|+1)\left\|\frac{\partial \Phi}{\partial v}\right\| = 0 \tag{7}$$

which can be written in the following way:

$$\frac{\partial \Phi}{\partial z} \cdot h(z,u,v) + \sup_{\|u'\| \leq \rho(\|u\|+1)} \frac{\partial \Phi}{\partial u} \cdot u' + \sup_{\|v'\| \leq \sigma(\|v\|+1)} \frac{\partial \Phi}{\partial v} \cdot v' = 0$$

The link between the feedback maps and the solutions to the solutions to these partial differential equations is provided by the indicators of the graphs: we associate with the set-valued maps S_U, R_U and R_{UV} the functions Φ_U, Ψ_U and Ψ from $\mathbf{R}^n \times \mathbf{R}^p \times \mathbf{R}^q$ to $\mathbf{R}_+ \cup \{+\infty\}$ defined by

[4]If Φ is a solution to this partial differential equation, one can check that for any initial situation $(z_0, u_0, v_0) \in \mathrm{Dom}(\Phi)$, there exists a smooth solution $(z(\cdot), u(\cdot), v(\cdot))$ such that

$$t \to \Phi(z(t), u(t), v(t)) \text{ is non increasing}$$

This property remains true for the solutions to the contingent partial differential equation (9).

[5]One can check that if f is lipschitzean and Φ is a solution to this partial differential equation, for any initial situation $(z_0, u_0, v_0) \in \mathrm{Dom}(\Phi)$, any smooth solution $(z(\cdot), u(\cdot), v(\cdot))$ satisfies

$$t \to \Phi(z(t), u(t), v(t)) \text{ is non increasing}$$

This property remains true for the solutions to the contingent partial differential equation (10).

$$(8)\qquad \begin{cases} i) & \Phi_U(z,u,v) := \begin{cases} 0 & \text{if } (u,v) \in S_U(z) \\ +\infty & \text{if } (u,v) \notin S_U(z) \end{cases} \\ ii) & \Psi_U(z,u,v) := \begin{cases} 0 & \text{if } (u,v) \in R_U(z) \\ +\infty & \text{if } (u,v) \notin R_U(z) \end{cases} \\ iii) & \Psi(z,u,v) := \begin{cases} 0 & \text{if } (u,v) \in R_{UV}(z) \\ +\infty & \text{if } (u,v) \notin R_{UV}(z) \end{cases} \end{cases}$$

and the functions Ψ_V and Φ_V associated to the set-valued map R_V and S_V in an analogous way.

These functions being only lower semicontinuous, but not differentiable, cannot be solutions to either partial differential equations (6) and (7). But we can define the *contingent epiderivatives* of any function $\Phi : \mathbf{R}^n \times \mathbf{R}^p \times \mathbf{R}^q \to \mathbf{R} \cup \{+\infty\}$ and replace the partial differential equations (6)and (7) by the contingent partial differential equations

$$(9)\qquad \inf_{\substack{\|u'\| \leq \rho(\|u\|+1) \\ \|v'\| \leq \sigma(\|v\|+1)}} D_\uparrow\Phi(z,u,v)(h(z,u,v),u',v')$$

and

$$(10)\qquad \sup_{\substack{\|u'\| \leq \rho(\|u\|+1) \\ \|v'\| \leq \sigma(\|v\|+1)}} D_\uparrow\Phi(z,u,v)(h(z,u,v),u',v')$$

respectively.

Let Ω_U and Ω_V be the indicators of the graphs of the set-valued maps U and V defined by

$$(11)\qquad \begin{cases} i) & \Omega_U(z,u,v) := \begin{cases} 0 & \text{if } u \in U(z) \\ +\infty & \text{if } u \notin U(z) \end{cases} \\ ii) & \Omega_V(z,u,v) := \begin{cases} 0 & \text{if } v \in V(z) \\ +\infty & \text{if } v \notin V(z) \end{cases} \end{cases}$$

Theorem 2.2 *We posit the assumptions of Theorem 2.1. Then*

— Ψ_U *is the smallest lower semicontinuous solution to the contingent partial differential equation (9) larger than or equal to* Ω_U

— Ψ_V *is the smallest lower semicontinuous solution to the contingent partial differential equation (9) larger than or equal to* Ω_V

— Ψ *is the smallest lower semicontinuous solution to the contingent partial differential equation (9) larger than or equal to* $\max(\Omega_U,\Omega_V)$

— *If h is lipschitzean,* Φ_U *is the smallest lower semicontinuous solution to the contingent partial differential equation (10) larger than or equal to* Ω_U

— *If h is lipschitzean,* Φ_V *is the smallest lower semicontinuous solution to the contingent partial differential equation (10) larger than or equal to* Ω_V

If any of the above solutions is the constant $+\infty$, the corresponding feedback map is empty.

Proof of Theorem 2.1 — Let us denote by B the unit ball and introduce the set-valued map F defined by

$$H(z,u,v) := \{h(z,u,v)\} \times \rho(\|u\|+1)B \times \sigma(\|v\|+1)B$$

The evolution of the differential game described by the equations (1) and (4) is governed by the differential inclusion

$$(z'(t),u'(t),v'(t)) \in H(z(t),u(t),v(t))$$

— Since the graph of U is closed, we know that there exists a largest closed viability domain contained in Graph$(U) \times \mathbf{R}^q$, which is the set of initial situations (z_0, u_0, v_0) such that there exists a solution $(z(\cdot), u(\cdot), v(\cdot))$ to this differential inclusion remaining in this closed set. This is the graph of R_U. Indeed, if $(u_0, v_0) \in R_U(z_0)$, there exists a solution to the differential inclusion remaining in the graph of U, i.e., Xavier may win. If not, all solutions starting at (z_0, u_0, v_0) must leave this domain in finite time.

The set-valued feedback map is defined in an analogous way.

— For the same reasons, the graph of the set-valued feedback map R_{UV} is the largest closed viability domain of the set K of initial situations.

— When h is lipschitzean, so is F. Then the solution-map $S(z_0, u_0, v_0)$ is also lipschitzean thanks to Filippov's Theorem[6], so that the subset of initial situations such that all the functions of $S(z_0, u_0, v_0)$ remain in a closed subset is also closed. This is the largest closed invariant domain by F of this closed subset. Then the largest closed invariant domain contained in Graph$(U) \times \mathbf{R}^q$ is the graph of the set-valued feedback map S_U. □

Proof of Theorem 2.2 — We recall that thanks to Haddad's viability Theorem, a subset $L \subset \mathbf{R}^n \times \mathbf{R}^p \times \mathbf{R}^q$ is a viability domain of F if and only if

$$\forall\, (z,u,v) \in L,\ T_L(z,u,v) \cap H(z,u,v) \neq \emptyset$$

Let Ψ_L denote the indicator of L. We know that the epigraph of the contingent epiderivative $D_\uparrow \Psi_L(z,u,v)$ of Ψ_L is the contingent cone to the epigraph of Ψ_L at $((z,u,v),0)$. Since the latter subset is equal to $L \times \mathbf{R}_+$, its contingent cone is equal to $T_L(z,u,v) \times \mathbf{R}_+$, and coincides with the epigraph of the indicator of $T_L(z,u,v)$. Hence *the indicator of the contingent cone* $T_L(z,u,v)$ *is the contingent epiderivative* $D_\uparrow \Psi_L(z,u,v)$ *of the indicator* Ψ_L *of* L *at* (z,u,v).

Therefore, the above tangential condition can be reformulated in the following way:

$$\forall\, (z,u,v) \in L,\ \exists\, w \in H(z,u,v) \text{ such that } D_\uparrow \Psi_L(z,u,v)(w) = \Psi_{T_L(z,u,v)}(w) = 0$$

Since the epiderivative is lower semicontinuous and the images of F are compact, this is equivalent to say that

$$\forall\, (z,u,v) \in L, \quad \inf_{w \in H(z,u,v)} D_\uparrow \Psi_L(z,u,v)(w) = 0$$

By the very definition of the set-valued map F, we have proved that L is a closed viability domain if and only if its indicator function Ψ_L is a solution to the contingent partial differential equation (9).

— Hence to say that the graph of R_U is the largest closed viability domain contained in the graph of U amounts to saying that its indicator Ψ_U is the smallest lower semicontinuous solution to the contingent partial differential equation (9) larger than or equal to the indicator Ω_U of Graph$(U) \times \mathbf{R}^q$. The same reasoning shows that indicator Ψ_V of R_V is the smallest lower semicontinuous solution to the contingent partial differential equation (9) larger than or equal to Ω_V and that the indicator Ψ of the graph of R_{UV} is the smallest lower semicontinuous solution to the contingent partial differential equation (9) larger than or equal to the indicator of K, which is equal to $\max(\Omega_U, \Omega_V)$.

— We know that the a closed subset $L \subset \mathbf{R}^n \times \mathbf{R}^p \times \mathbf{R}^q$ is "invariant" by a lipschitzean set-valued map F if and only if

$$\forall\, (z,u,v) \in L,\ T_L(z,u,v) \subset H(z,u,v)$$

This condition can be reformulated in terms of contingent epiderivative of the indicator function Ψ_L of L by saying that

$$\forall\, (z,u,v) \in L, \quad \sup_{w \in H(z,u,v)} D_\uparrow \Psi_L(z,u,v)(w) = 0$$

[6] See [3, p.120]

Hence to say that the graph of S_U is the largest closed invariance domain contained in the graph of U amounts to saying that its indicator Φ_U is the smallest lower semicontinuous solution to the contingent partial differential equation (10) larger than or equal to the indicator Ω_U of Graph$(U) \times \mathbf{R}^q$. □

3 Closed-Loop Decision Rules

When the initial situation (z_0, u_0, v_0) belongs to one of the following subsets:

(12) $$\text{Graph}(S_U) \cap \text{Graph}(S_V) \text{ or } K \backslash (\text{Graph}(R_U) \cup \text{Graph}(R_V))$$

then the players has nothing to worry about because both of them must either win or loose whatever the choice of their control.

In the other areas, at least one of the players may win, but for achieving victory, he has to find open-loop or closed-loop controls which remain in the appropriate set-valued feedback map.

Let us denote by R one of the feedback maps R_U, R_V, R_{UV} and assume that the initial situation belongs to the graph of the set-valued feedback map R (when it is not empty). The theorem states only that there exists at least a solution $(z(\cdot), u(\cdot), v(\cdot))$ to the differential game such that

$$\forall\, t \geq 0,\ (u(t), v(t)) \in R(z(t))$$

To implement these strategy, players have *to make decisions, i.e., to choose velocities of controls* in an adequate way:

We observe that playable solutions

Proposition 3.1 *The solutions to the game satisfying*

$$\forall\, t \geq 0,\ (u(t), v(t)) \in R(z(t))$$

are the solutions to the system of differential inclusions

(13) $$\begin{cases} i) & z'(t) = h(z(t), u(t), v(t)) \\ ii) & (u'(t), v'(t)) \in G_R(z(t), u(t), v(t)) \end{cases}$$

*where we have denoted by G_R the R-*decision map *defined by*

(14) $$G_R(z, u, v) := DR_R(z, u, v)(h(z, u, v))$$

For simplicity, we shall set $G := G_R$ whenever there is no ambiguity.

Proof — Indeed, since the absolutely continuous function $(z(\cdot), u(\cdot), v(\cdot))$ takes its values into Graph(R), then its derivative $(z'(\cdot), u'(\cdot), v'(\cdot))$ belongs almost everywhere to the contingent cone

$$T_{\text{Graph}(R)}(z(t), u(t), v(t)) := \text{Graph}(DR(z(t), u(t), v(t)))$$

We then replace $z'(t)$ by $h(z(t), u(t), v(t))$.

The converse holds true because equation (13) makes sense only if $(z(t), u(t), v(t))$ belongs to the graph of R. □

The question arises whether we can construct selection procedures of the decision components of this system of differential inclusions. It is convenient for this purpose to introduce the following definition.

Definition 3.2 () *We shall say that a selection $(\tilde{c}, \tilde{d})$ of the contingent derivative of the smooth regulation map R in the direction h defined by*

(15) $$\forall\ (z, u, v) \in \text{Graph}(R),\ \tilde{c}(z, u, v) \in DR(z, u, v)(h(z, u, v))$$

is a closed-loop decision rule.

The system of differential equations

$$(16)\qquad \begin{cases} i) & z'(t) = h(z(t), u(t), v(t)) \\ ii) & u'(t) = c(z(t), u(t), v(t)) \\ iii) & v'(t) = d(z(t), u(t), v(t)) \end{cases}$$

is called the associated closed-loop decision game.

Therefore, closed-loop decision rules being given for each player, the closed-loop decision system is just a system of ordinary differential equations.

It has solutions whenever the maps c and d are continuous (and if such is the case, they will be continuously differentiable).

But they also may exist when c or d or both are no longer continuous. This is the case when the decision map is lower semicontinuous thanks to Michael's Theorem:

Theorem 3.3 *Let us assume that the decision map $G := G_R$ is lower semicontinuous with non empty closed convex values on the graph of R. Then there exist continuous decision rules c and d, so that the decision system 16 has a solution whenever the initial situation $(u_0, v_0) \in R(z_0)$*

But we can obtain explicit decision rules which are not necessarily continuous, but for which the decision system 16 has a still solution.

It is useful for that purpose to introduce the following definition:

Definition 3.4 (Selection Procedure) *A* selection procedure of the regulation map $G : \mathbf{R}^n \rightsquigarrow \mathbf{R}^p \times \mathbf{R}^q$ *is a set-valued map* $S_G : \mathbf{R}^n \rightsquigarrow \mathbf{R}^p \times \mathbf{R}^q$

$$(17)\qquad \begin{cases} i) & \forall z \in K,\ \ S(G(z)) := S_G(z) \cap G(z) \neq \emptyset \\ ii) & \text{the graph of } S_G \text{ is closed} \end{cases}$$

and the set-valued map $S(G) : z \rightsquigarrow S(G(z))$ is called the selection *of G.*

It is said convex-valued *or simply,* convex *if its values are convex and* strict *if moreover*

$$(18)\qquad \forall z \in \mathrm{Dom}(G),\ S_G(z) \cap G(z) = \{\tilde{d}(z)), \tilde{c}(z)\}$$

is a singleton.

Hence, we obtain also the following existence theorem for closed-loop decision rules obtained through sharp convex selection procedures.

Theorem 3.5 *Let S_G be a convex selection of the set-valued map G. Then, for any initial state $(z_0, u_0, v_0) \in$ graph(R), there exists a starting at (z_0, u_0, v_0) to the associated system of differential inclusions*

$$(19)\qquad \begin{cases} i) & z'(t) = h(z(t), u(t), v(t)) \\ ii) & (u'(t), v'(t)) \in S(DR(z(t), u(t), v(t))h(z(t), u(t), v(t))) \\ & := G(z(t), u(t), v(t)) \cap S_G(z(t), u(t), v(t)) \end{cases}$$

In particular, if we assume further that the selection procedure S_G is sharp, then the single-valued map

$$(\tilde{c}(z, u, v), \tilde{d}(z, u, v)) := S(G)(z, u, v)$$

is closed-loop decision rule, for which decision system 16 has a solution for any initial state $(z_0, u_0, v_0) \in$ graph(R).

Proof— We shall replace the system of differential inclusions (13) by the system of differential inclusions

$$(20)\qquad \begin{cases} i) & z'(t) = h(z(t),u(t),v(t)) \\ ii) & (u'(t),v'(t)) \in S_G(z(t),u(t),v(t)) \end{cases}$$

Since the convex selection procedure S_G has a closed graph and convex values, the right-hand side is upper semicontinuous set-valued map with nonempty compact convex images and with linear growth. It remains to check that Graph$R$ is still a viability domain for this new system of differential inclusions. Indeed, by construction, we know that there exists an element w in the intersection of $G(z,u,v)$ and $S_G(z,u,v)$. This means that the pair $(h(z,u,v),w)$ belongs to $h(z,u,v)\times S_G(z,u,v)$ and that it also belongs to

$$\text{Graph}(G) := T_{\text{Graph}R}(z,u)$$

Therefore, we can apply Haddad's Viability Theorem. For any initial situation (z_0,u_0,v_0), there exists a solution $(z(\cdot),u(\cdot),v(\cdot))$ to the new system of differential inclusions (20) which is viable in Graph(R). Consequently, for almost all $t>0$, the pair $(z'(t),u'(t),v'(t))$ belongs to the contingent cone to the graph of R at $(z(t),u(t),v(t))$, which is the graph of the contingent derivative $DR(z(t),u(t),v(t))$. In other words,

$$\text{for almost all } t>0,\ (u'(t),v'(t)) \in G(z(t),u(t),v(t))$$

We thus deduce that for almost all $t>0$, $(u'(t),v'(t))$ belongs to the selection $S(G)(z(t),u(t),v(t))$ of the set-valued map $G(z(t),u(t),v(t))$. Hence, we have found a solution to the system of differential inclusions (19). □

We can now multiply the possible corollaries, since we have given several instances of selection procedures of set-valued maps.

Example— COOPERATIVE BEHAVIOR

Let σ : Graph$(G)\mapsto$ G be continuous.

Corollary 3.6 *Let us assume that the set-valued map G is lower semicontinuous with nonempty closed convex images on* Graph(R). *Let σ be continuous on* Graph(G) *and convex with respect to the pair (u,v). Then, for all initial situation $(u_0,v_0)\in R(z_0)$, there exist a solution starting at (z_0,u_0,v_0) and to the differential game (1)-(4) which are regulated by:*

$$(21)\qquad \begin{cases} \text{for almost all } \geq 0,\ (u'(t),v'(t)) \in G(z(t),u(t),v(t)) \text{ and} \\ \sigma(z(t),u(t),v(t),u'(t),v'(t)) \\ = \inf_{u',v'\in G(z(t),u(t),v(t))} \sigma(z(t),u(t),v(t),u',v') \end{cases}$$

In particular, the game can be played by the heavy decision *of minimal norm:*

$$\begin{cases} (c^\circ(z,u,v),d^\circ(z,u,v)) \in G(z,u,v) \\ \|c^\circ(z,u,v)\|^2 + \|d^\circ(z,u,v)\|^2) = \min_{(u',v')\in G(z,u,v)}(\|u'\|^2+\|v'\|^2) \end{cases}$$

Proof — We introduce the set-valued map S_G defined by:

$$S_G(z) := \{(c,d)\in Y \mid \sigma(z,u,v,c,d) \leq \inf_{(u',v')\in G(z,u,v)} \sigma(z,u,v,u',v')\}$$

It is a convex *selection procedure* of G. Indeed, since G is lower semicontinuous, the function

$$(z,u,v,c,d) \mapsto \sigma(z,u,v,c,d) + \sup_{(u',v')\in G(z,u,v)} (-\sigma(z,u,v,u',v'))$$

is lower semicontinuous thanks to the Maximum Theorem. Then the graph of S_G is closed because

$$\text{Graph}(S_G) =$$
$$\{(z,u,v) \mid \sigma(z,u,v,c,d) + \sup_{(u',v')\in G(z,u,v)}(-\sigma(z,u,v,u',v')) \leq 0\}$$

The images are obviously convex. Consequently, the graph of G being also closed, so is the selection $S(G)$ equal to:

$$S(G)(z,u,v) = \{(c,d) \in G(z,u,v) \mid \sigma(z,u,v,c,d) \leq \inf_{(u',v')\in G(z,u,v)} \sigma(z,u,v,u',v'))\}$$

We then apply Theorem 3.5. We observe that when we take

$$\sigma(z,u,v,c,d) \quad := \quad \|c\|^2 + \|d\|^2$$

the selection procedure is strict and yields the decisions of minimal norm. □

Example— NONCOOPERATIVE BEHAVIOR

We can also choose controls in the regulation sets $G(z,u,v)$ in a non cooperative way, as saddle points of a function $a(z,u,v,\cdot,\cdot)$.

Corollary 3.7 *Let us assume that the set-valued map G is lower semicontinuous with nonempty closed convex images on* Graph(R) *and that $a : \mathbf{R}^n \times \mathbf{R}^p \times \mathbf{R}^q \to \mathbf{R}$ satisfies*

$$(22) \qquad \begin{cases} i) & a \text{ is continuous} \\ ii) & \forall(z,u,v,d), \;\; c \mapsto a(z,u,v,c,d) \text{ is convex} \\ iii) & \forall(z,u,v,c), \;\; d \mapsto a(z,u,v,c,d) \text{ is concave} \end{cases}$$

Then, for all initial situation $(u_0,v_0) \in R(z_0)$, there exist a solution starting at (z_0,u_0,v_0) and to the differential game (1)-(4) which are regulated by:

$$\text{for almost all } t \geq 0, \quad \begin{cases} i) & (u'(t),v'(t)) \in G(z(t),u(t),v(t)) \\ ii) & \forall\,(u',v') \in G(z(t),u(t),v(t)), \\ & a(z(t),u(t),v(t),u'(t),v') \\ & \leq a(z(t),u(t),v(t),u'(t),v'(t)) \\ & \leq a(z(t),u(t),v(t),u',v'(t)) \end{cases}$$

Proof — We prove that the set-valued map S_G associating to any $(z,u,v) \in$ Graph(R) the subset

$$\begin{array}{l} S_G(z,u,v) \; := \; \{\, (c,d) \text{ such that} \\ \forall(u',v') \in G(z,u,v), \;\; a(z,u,v,c,v') \leq a(z,u,v,u',d) \,\} \end{array}$$

is a convex selection procedure of G. The associated selection map $S(G(\cdot))$ associates with any (z,u,v) the subset

$$\begin{array}{l} S(G(z,u,v)) \; := \; \{\, (c,d) \in G(z,u,v) \text{ such that} \\ \forall(u',v') \in G(z,u,v), \;\; a(z,u,v,c,v') \leq a(z,u,v,u',d) \,\} \end{array}$$

of saddle-points of $a(z,u,v,\cdot,\cdot)$ in $G(z,u,v)$. Von Neumann' Minimax Theorem states that the subsets $S(G(z,u,v))$ of saddle-points are not empty since $G(z,u,v)$ are convex and compact. The graph of S_G is closed thanks to the assumptions and the Maximum Theorem because it is equal to the lower section of a lower semicontinuous function:

$$\text{Graph}(S_G) = \{(z,u,v,c,d) \mid \sup_{(u',v')\in G(z,u,v)} (a(z,u,v,c,v') - a(z,u,v,u',d)) \leq 0\}$$

We then apply Theorem 3.5. □

Remark — Whenever the subset $R_{UV}(z(t))\backslash R_V(z(t))$ is not empty, Xavier may be tempted to choose a control $u(t)$ such that

$$(u(t),v(t)) \in R_{UV}(z(t))\backslash R_V(z(t))$$

because in this case, Xavier may win and Yves is sure to loose eventually. Naturally, Yves will use the opposite behavior.

Hence we can attach to the game two functions

$$(23)\quad \begin{cases} i) & a_U(z,u,v) := d((u,v), R_{UV}(z)\backslash R_V(z)) \\ ii) & b_V(z,u,v) := d((u,v), R_{UV}(z)\backslash R_U(z)) \end{cases}$$

and look for closed-loop controls $(\hat{u}(z), \hat{v}(z))$ which are Nash equilibria of this game:

$$(24)\quad \begin{cases} i) & (\hat{u}(z,\hat{v}(z))) \in R_{UV}(z) \\ ii) & \forall\,(u,v) \in R_{UV}(z),\ \ a_U(z,\hat{u}(z),\hat{v}(z)) \leq a_U(z,u,\hat{v}(z)) \\ & \&\ b_V(z,\hat{u}(z),\hat{v}(z)) \leq b_V(z,\hat{u}(z),v)\ \ \square \end{cases}$$

Unfortunately, the selection procedure which could yield such behavior are not convex. The answer to this question remains unknown for the time.

References

[1] AUBIN J.-P. (1988) *Qualitative Differential Games: a Viability Approach.* Annales de l'Institut Henri-Poincaré, Analyse Non Linéaire.

[2] AUBIN J.-P. (1988) *Contingent Isaac's Equations of a Differential Game. Proceedings of the Third International Meeting on Differential Games, INRIA Sophia-Antipolis.*

[3] AUBIN J.-P. & CELLINA A. (1984) DIFFERENTIAL INCLUSIONS. Springer-Verlag (Grundlehren der Math. Wissenschaften, Vol.264, 1-342)

[4] AUBIN J.-P. & EKELAND I. (1984) APPLIED NONLINEAR ANALYSIS. Wiley-Interscience

[5] BERKOWITZ L. (1988) *This volume*

[6] BERNHARD P. (1979) Contribution à l'étude des jeux différentiels à somme nulle et information parfaite. Thèse Université de Paris VI

[7] BERNHARD P. (1980) *Exact controllability of perturbed continuous-time linear systems.* Trans. Automatic Control, 25, 89-96

[8] BERNHARD P. (1987) In Singh M. G. Ed. SYSTEMS & CONTROL ENCYCLOPEDIA, Pergamon Press

[9] FLEMMING W. & RISHEL R.W. (1975) DETERMINISTIC AND SOCHASTIC OPTIMAL CONTROL Springer-Verlag

[10] FRANKOWSKA H. (1987) *L'équation d'Hamilton-Jacobi contingente.* Comptes Rendus de l'Académie des Sciences, PARIS,

[11] FRANKOWSKA H. *Optimal trajectories associated to a solution of contingent Hamilton-Jacobi Equation* Appl. Math. Opt.

[12] FRANKOWSKA H. (to appear) *Hamilton-Jacobi Equations:viscosity solutions and generalized gradients.* J. Math.Anal. Appli.

[13] GUSEINOV H. G. , SUBBOTIN A. I. & USHAKOV V. N. (1985) *Derivatives for multivalued mappings with applications to game theoretical problems of control.* Problems of Control and Information Theory, Vol.14, 155-167

[14] ISSACS R. (1965) DIFFERENTIAL GAMES. Wiley, New York

[15] KRASOVSKI N. N. & SUBBOTIN A. I. (1974) POSITIONAL DIFFERENTIAL GAMES. Nauka, Moscow

[16] LEITMANN G. (1980) *Guaranteed avoidance strategies.* Journal of Optimization Theory and Applications, Vol.32, 569-576

[17] LIONS P. -L. (1982) GENERALIZED SOLUTIONS OF HAMILTON-JACOBI EQUATIONS. Pitman

[18] LIONS P. -L. & SOUGANIDIS P.E. (1985) *Differential games, optimal control and directional derivatives of viscosity solutions of Bellman and Isaacs' equations.* SIAM J. Control. Optimization, 23

[19] SUBBOTIN A. I. (1985) *Conditions for optimality of a guaranteed outcome in game problems of control.* Proceedings of the Steklov Institute of Mathematics, 167, 291-304

[20] SUBBOTIN A. I. & SUBBOTINA N. N. (1983) *Differentiability properties of the value function of a differential game with integral terminal costs.* Problems of Control and Information Theory, 12, 153-166

[21] SUBBOTIN A. I. & TARASYEV A. M. (1986) *Stability properties of the value function of a differential game and viscosity solutions of Hamilton-Jacobi equations.* Problems of Control and Information Theory, 15, 451-463

BARGAINING WITH DYNAMIC INFORMATION

Jukka Ruusunen, Harri Ehtamo and Raimo P. Hämäläinen
Espoo, Finland

Abstract: The Nash bargaining solution in a sequential cooperative game with dynamic information is studied. The bargaining scheme is applied at each stage, where the gains from the cooperation are measured by considering the past, the current and the future gains. It is shown that at each stage the cooperative policy can be selected as if the current contract would remain in effect over the remainder of the time horizon of cooperation.

1. Introduction

Let us consider a group of decision makers (DMs), who are in a position to choose their policies cooperatively. It is assumed that there are outcomes that all the DMs prefer to the status quo outcome. Since the DMs do not in general have the same choice set for the cooperative outcomes, there is a conflict of interests between the DMs in cooperation. The cooperative outcome is required to be individually rational such that all the DMs gain from the cooperation. However, depending on the contract chosen an individual DM can gain more or less.

The Nash bargaining solution, see e.g. Roth (1979), has been developed for games in normal form. The cooperative outcome is selected by maximizing the product of the DMs' gains from cooperation. The gains are measured as differences between the payoffs from a cooperative outcome and the status quo outcome. In the present study bargaining takes place in a dynamic framework under future uncertainty. Decision making is described by a cooperative game in an extensive form. Thus an extension of the Nash bargaining scheme to extensive form games has to be made.

Haurie (1976) has studied a dynamic bargaining game, where at each stage only future gains are considered in bargaining, and the past is ignored. This kind of an extension was shown to lead to the sustainability problem: reopening of the negotiations at an intermediate stage would change the cooperative policy. The solution concept proposed by Tolwinski (1982) is based on the assumption that rebargaining actually takes place at every stage. A cooperative policy related to a given stage is constructed by considering only current and future gains relative to the stage. The policies are constructed according to the principle of dynamic programming assuming that rebargaining will take place at every future stage. The solution satisfies the axioms of the bargaining scheme in all of the resulting subgames. However, the axioms are not satisfied in the original cooperative game.

In this paper we propose a bargaining mechanism that satisfies the rules of fairness of the Nash bargaining scheme over the *whole* time interval of cooperation. At each stage the cooperative policies are functions of the current information. The past history of bargaining relative to a given stage is described by the DMs' gains over the past stages. A procedure to construct the bargaining solution is given. An attractive feature of this procedure is that at each stage the DMs can select the cooperative policy as if the contract defined would remain in effect over the remainder of the time interval.

2. A bargaining problem with dynamic information

Consider a group $\theta_1 = \{1, \dots, M\}$ of M ($M \geq 2$) DMs who have joined together to cooperate. The bargaining takes place over N time periods in a stochastic environment. The uncertainty related to the environment in period $k \in \theta_2 = \{0, \dots, N-1\}$ is described by the random variable $\xi_k \in \Xi_k$ with probability measure $p_k(\cdot | I_{k-1})$, where $I_{k-1} = (\xi_{-1}, \xi_0, \dots, \xi_{k-1})$ is the information related to period $k-1$. The first term in the information vector, ξ_{-1}, describes the past behaviour of the random variable before the initial period $k = 0$. It is assumed that the actual value taken by ξ_k is known before the decision in period k is made.

The admissible policies are functions of the information vector, i.e. $u_k = u_k(I_k) \in U_k \subset R^m$, $k \in \theta_2$. Furthermore, u_k is of the form $u_k = (u_k^1, \dots, u_k^M)$, where $u_k^i \in U_k^i \subset R^{m_i}$ is the admissible policy for DMi, $i \in \theta_1$. The decision u_k and the sequence of random variables described by the information I_k yield a payoff $g_k^i(u_k, I_k)$ for DMi in period k. The objective of DMi is the maximization of the sum of these payoffs over the time interval of cooperation,

$$J^i = \sum_{l=0}^{N-1} g_l^i(u_l, I_l). \tag{1}$$

The benefits of cooperation are measured from a status quo, which would prevail in the noncooperative situation. The status quo policies are denoted by $v_k = \delta_k(I_k)$, $k \in \theta_2$.

The rules of fairness of the Nash bargaining scheme are used to determine the cooperative solution over the whole time interval of cooperation. Consequently, the time interval is partitioned into the past and the future relative to each period $k \in \theta_2$. For given information I_{k-1}, the past periods are described by the cumulative benefits for the DMs up to period k

$$\begin{aligned} y_k^i(u_0, \dots, u_{k-1}; I_{k-1}) &= \sum_{l=0}^{k-1} \left[g_l^i(u_l, I_l) - g_l^i(\delta_l(I_l), I_l) \right], \\ y_0^i(I_0) &= 0. \end{aligned} \tag{2}$$

The expected future payoffs from period k on are

$$J_k^i(u_k, \dots, u_{N-1}; I_k) = g_k^i(u_k(I_k), I_k) + E_{\xi_{k+1}, \dots, \xi_{N-1}} \{ \sum_{l=k+1}^{N-1} g_l^i(u_l(I_l), I_l) \}. \tag{3}$$

If the status quo policies were used, these expected future payoffs would become

$$D_k^i(I_k) = J_k^i(v_k, \ldots, v_{N-1}; I_k). \tag{4}$$

According to the Nash bargaining scheme, the contract definition problem related to period $k \in \theta_2$ can be stated as follows: For given $y_k^i(u_0^*, \ldots, u_{k-1}^*; I_{k-1})$, I_k find an admissible policy $\{\bar{u}_k, \ldots, \bar{u}_{N-1}\}$ such that

$$\begin{aligned} &\prod_{i\in\theta_1} \left[y_k^i(u_0^*, \ldots, u_{k-1}^*; I_{k-1}) + J_k^i(\bar{u}_k, \ldots, \bar{u}_{N-1}; I_k) - D_k^i(I_k)\right] \\ &\quad \geq \prod_{i\in\theta_1} \left[y_k^i(u_0^*, \ldots, u_{k-1}^*; I_{k-1}) + J_k^i(u_k, \ldots, u_{N-1}; I_k) - D_k^i(I_k)\right] \end{aligned} \tag{5}$$

for all admissible policies $\{u_k, \ldots, u_{N-1}\}$ for which

$$y_k^i(u_0^*, \ldots, u_{k-1}^*; I_{k-1}) + J_k^i(u_k, \ldots, u_{N-1}; I_k) - D_k^i(I_k) \geq 0, \quad \forall i \in \theta_1. \tag{6}$$

3. Computation of the contract

To solve the problem defined by (5) and (6) we need two results; for the proofs of these results see Ehtamo *et al.* (1987). Let U be a set, and $J^i : U \to R$, $i \in \theta_1$. Define

$$F^i(u) = J^i(u) - D^i, \quad i \in \theta_1, \tag{7}$$

$$F(u) = \prod_{i\in\theta_1} F^i(u), \quad u \in U, \tag{8}$$

where $D^i \in R$, $i \in \theta_1$, are fixed, and consider

$$\max_{u\in U} F(u) \tag{9}$$

$$\text{subject to } F^i(u) \geq 0, \quad i \in \theta_1. \tag{10}$$

Lemma 1: Let $u^* \in U$ be such that

$$\sum_{i\in\theta_1} \mu^i F^i(u^*) \geq \sum_{i\in\theta_1} \mu^i F^i(u) \qquad \forall u \in U, \tag{11}$$

where

$$\mu^i = \prod_{j\neq i} F^j(u^*), \quad F^i(u^*) > 0, \quad \forall i \in \theta_1. \tag{12}$$

Then u^* solves (9), (10).

Lemma 2: Suppose U is a convex set in a vector space X and J^i , $i \in \theta_1$ are concave functionals on U. Suppose $u^* \in U$ solves (9), (10) with $F^i(u^*) > 0$ for all i. Then (11) holds, where μ^i is as in (12).

Lemma 1 is now applied to the contract definition problem (5) and (6). Denote $\Lambda^> = \{\mu \in R^M | \mu^i > 0,\ \forall i\}$. For given l, I_l and $\mu \in \Lambda^>$ let $u_l = \beta_l(I_l, \mu)$ solve the maximization problem

$$\max_{u_l \in U_l} \sum_{i \in \theta_1} \mu^i g_l^i(u_l, I_l), \tag{13}$$

and define

$$V_k^i(I_k, \mu) = E_{\xi_{k+1},\ldots,\xi_{N-1}} \{ \sum_{l=k+1}^{N-1} g_l^i(\beta_l(I_l, \mu), I_l) \}, \quad \forall k \in \theta_2. \tag{14}$$

The bargaining solution $\{\bar{u}_k, \ldots, \bar{u}_{N-1}\}$ in period $k \in \theta_2$ is defined as follows. Let $\bar{\mu} \in \Lambda^>$ satisfy

$$\bar{\mu}^i = \prod_{\substack{j \in \theta_1 \\ j \neq i}} [y_k^j(u_0^*, \ldots, u_{k-1}^*; I_{k-1}) + g_k^j(\beta_k(I_k, \bar{\mu}), I_k) + V_k^j(I_k, \bar{\mu}) - D_k^j(I_k)]. \tag{15}$$

Set $\bar{u}_l(I_l) = \beta_l(I_l, \bar{\mu})$ for $l \geq k$. Then we have

Proposition 1: Let $k \in \theta_2$. For given I_k and $y_k(u_0^*, \ldots, u_{k-1}^*; I_{k-1})$ let $\{\bar{u}_k, \ldots, \bar{u}_{N-1}\}$ be defined as above. Then $\{\bar{u}_k, \ldots, \bar{u}_{N-1}\}$ solves (5) and (6).

Proof: Since $E\{\max[\cdot]\} \geq \max[E\{\cdot\}]$, we have, using the defininition of the functions $\beta_l(I_l, \mu)$,

$$\sum_{i \in \theta_1} \bar{\mu}^i J_k^i(\bar{u}_k, \ldots, \bar{u}_{N-1}; I_k) \geq \sum_{i \in \theta_1} \bar{\mu}^i J_k^i(u_k, \ldots, u_{N-1}; I_k) \tag{16}$$

for all admissible $\{u_k, \ldots, u_{N-1}\}$. The result is then implied by (15), (16) and Lemma 1. □

According to Proposition 1 the cooperative policy for the present stage can be selected by assuming that the current contract $\bar{\mu}$ will remain in effect over the remainder of the time interval of cooperation. At the next stage rebargaining takes place and cooperative actions for that stage are selected, assuming that the updated contract will remain in effect over the remainder of the time interval of cooperation, and so forth. Hence, the actual cooperative policy $\{u_0^*, \ldots, u_{N-1}^*\}$, where u_k^* is the policy to be applied at stage k, $k \in \theta_2$, is chosen as follows:

(i) Set $k = 0$, $\nu_0(I_0) = \mu$, where μ satisfies

$$\mu^i = \prod_{\substack{j \in \theta_1 \\ j \neq i}} [y_0^j(I_0) + g_0^j(\beta_0(I_0, \mu), I_0) + V_0^j(I_0, \mu) - D_0^j(I_0)]. \tag{17}$$

Set $u_0^*(I_0) = \beta_0(I_0, \nu_0(I_0))$.

(ii) Suppose $u^*_{k-1}(I_{k-1})$ has been chosen. Set $\nu_k(I_k) = \mu$, where μ satisfies

$$\mu^i = \prod_{\substack{j \in \theta_1 \\ j \neq i}} \left[y^j_k(u^*_0, \ldots, u^*_{k-1}; I_{k-1}) + g^j_k(\beta_k(I_k,\mu), I_k) + V^j_k(I_k,\mu) - D^j_k(I_k)\right]. \tag{18}$$

Set $u^*_k(I_k) = \beta_k(I_k, \nu_k(I_k))$.

4. Conclusion

A mechanism for sequential bargaining under uncertainty is presented. The cooperative solution satisfies the rationality axioms of the Nash bargaining scheme over the whole time interval of cooperation. At each stage the past history of the disturbances is known. It is also assumed that the actual value taken by the disturbance related to the present stage is known before the decisions are made. It has been shown that the expected future gains contributing to the current contract can be computed as if no further contracts will be made in the future.

5. References

Ehtamo, H., J. Ruusunen, and R. P. Hämäläinen (1987), "On the Computation of the Nash Bargaining Solution with an Energy Management Example," in Proc. 26th IEEE CDC, Los Angeles, CA, pp. 263–266.

Haurie, A. (1976), "A note on nonzero-sum differential games with bargaining solution," Journal of Optimization Theory and Applications, Vol. 18, pp. 31–39.

Roth, A. E. (1979), "Axiomatic Models of Bargaining," Springer-Verlag, Berlin Heidelberg.

Tolwinski, B. (1982), "A concept of cooperative equilibrium for dynamic games," Automatica, Vol. 18, pp. 431–441.

OPTIMAL BAYESIAN CONTROL OF A NONLINEAR REGRESSION PROCESS WITH UNKNOWN PARAMETERS

by

Nicholas M. Kiefer
Department of Economics
Cornell University
Ithaca, New York 14853

and

Yaw Nyarko
Department of Economics
Brown University
Providence, RI 02912

1. Introduction

Economic Agents operating in uncertain, stochastic environments can face a tradeoff between current period expected reward and accumulation of information of uncertain value. For example, a firm producing to meet uncertain demand might produce at the expected current reward maximizing output, based on his current beliefs about the form of the demand curve, or it might choose to experiment by varying output, thus taking short term losses in order to sharpen beliefs about the form of the demand curve. A parametric representation of the agent's problem is made by considering the utility function $u(x,y)$ and the conditional density $f(y|x,\theta)$. Here the random variable y is what the agent is trying to control (e.g., current period profits) and x is the control variable. The parameters θ of the conditional density of y given x are unknown, but the agent has opinions about θ given by a distribution μ. The agent attempts to minimize the present discounted value of the stream of expected losses, $E\Sigma\delta^t u(x_t,y_t)$, where the expectation is taken with respect to current beliefs. The problem is complicated by the fact that beliefs are updated from period to period using Bayes Rule; consequently current period actions can be expected to influence future period beliefs. This introduces stochastic dynamics into the model.

This paper considers the problem in the case in which the density $f(y|x,\theta)$ is a location family. In this case the model can be written $y = g(x,\beta) + \varepsilon$, where ε is an i.i.d. random variable whose distribution may involve unknown parameters. When $g(x,\beta) = x'\beta$ the problem is one of controlling a linear regression process with unknown parameters over an infinite horizon. Many approximate control rules for this problem have been proposed, for example sequential least-squares estimation combined with one-period optimization conditioning on the current estimates. The analogous policy for the nonlinear model is clear. In practice several policies can work "well," though it is possible to compose examples in which the policy mentioned, for example, is easily improved. From an economic modelling point of view, however, we are interested in the optimal policy, and in the consequences for convergence of beliefs and policies of following the optimal policy. Will it be optimal for an agent to learn the parameters (and thus converge to "rational expectations")?

This paper gives general conditions under which the sequence of beliefs converges to a limit and the sequence of optimal policies converges to a limit. Under further conditions the limit policy is the optimal one-period policy for limit beliefs. Conditions under which the limit belief is point mass at true parameter values, corresponding to consistent parameter estimates are more stringent and are still under investigation.

Least-squares control rules in the linear regression model have been widely discussed and studied analytically by Taylor (1974) and Jordan (1985) and experimentally by Anderson and Taylor (1976). Improvements using a Bayesian approach were suggested by Zellner (1971) and studied by Harkema (1975). The optimal policy in the linear regression case has been studied by Kiefer and Nyarko (1987), who obtain results on convergence of beliefs and policies. convergence in a different class of models has been studied by Easley and Kiefer (1986). Results on optimal learning while controlling a stochastic process are collected along with an example in Kiefer (1988).

2. The Decision Problem: Uncertainty, Policies and Rewards

In this section we sketch the general framework we wish to study.

Let Ω' be a complete and separable metric space, let $\mathcal{F}'$ be its Borel field, and $(\Omega', \mathcal{F}', P')$ a probability space. Define the stochastic process $\{\epsilon_t\}^{\infty}$ on $(\Omega', \mathcal{F}', P')$. The ϵ_t are assumed to be independent and identically distributed, with the common marginal distribution $p(\epsilon_t|\xi)$ depending on some parameter, ξ in R^h, which is unknown to the agent. We assume that the set of probability measures, $\{p(\cdot|\xi)\}$, is continuous in the parameter ξ (in the weak topology of measures); and that for any ξ, $\int \epsilon\, p(d\epsilon|\xi) = 0$. Let $\bar{X}$, the action space, be a compact subset of R^k. Define $\Theta = R^m \times R^h$ to be the parameter space. If the "true parameter" is $\theta = (\beta,\xi) \in \Theta$, and the agent chooses an action $x_t \in \bar{X}$ at date t, then the agent observes y_t, where,

$$y_t = g(x_t,\beta) + \epsilon_t \tag{2.1}$$

and ϵ is chosen according to $p(\cdot|\xi)$. The function g is assumed measurable;

further restrictions are introduced implicitly through assumptions on the updating equation (2.2) and the reward function (2.3).

One example is the simple linear regression model with unknown slope and intercept and with the ϵ_t independent draws from the normal distribution with mean zero and variance σ^2. In that example Ω' is R^∞, $\mathcal{F}'$ is the collection of Borel sets on R^∞, and P' is the infinite product of independent univariate normal distributions with means zero and common variance σ^2. The parameter ξ is the variance of ϵ, σ^2. The action space $\bar{X}$ is a closed interval in R^1. The parameter $\beta \epsilon R^2$ consists of the slope and intercept of the regression. The space Θ is $R^2 \times R^1_+$.

Let $\mathcal{F}$ be the Borel field of Θ, and let $P(\Theta)$ be the set of all probability measures on $(\Theta, \mathcal{F})$. Endow $P(\Theta)$ with its weak topology, and note that $P(\Theta)$ is then a complete and separable metric space (see e.g., Parthasarathy (1967, Ch. II, Theorems 6.2 and 6.5)). Let $\mu_0 \epsilon P(\Theta)$ be the prior probability on the parameter space, with finite first moment.

The agent is assumed to use Bayes rules to update the prior probability at each date after any observation of (x_t, y_t). For example, in the initial period, date 1, the prior distribution is updated after the agent chooses an action x_1, and observes the value of y_1. The updated prior, i.e., the posterior, is then $\mu_1 = \Gamma(x_1, y_1, \mu_0)$, where $\Gamma:\bar{X} \times R^1 \times P(\Theta) \rightarrow P(\Theta)$ represents the Bayes rule operator. If the prior, μ_0, has a density function, then the posterior may be easily computed. In general, the Bayes rule operator may be defined by appealing to the existence of certain conditional probabilities, although some care is needed (see Diaconis and Freedman (1986)). Under some conditions the operator Γ is continuous in its arguments, and we assume this throughout. Any $\{x_t, y_t\}$ process will therefore result in a posterior process, $\{\mu_t\}$, where for all $t = 1,2,\ldots,$

$$\mu_t = \Gamma(x_t, y_t, \mu_{t-1}) \tag{2.2}$$

Let $\bar{H}_n = P(\Theta) \times \prod_{i=1}^{n-1} [\bar{X} \times R^1 \times P(\Theta)]$. A <u>partial history</u>, h_n, at date n is any element $h_n = (\mu_0,(x_1, y_1, \mu_1)\ldots,(x_{n-1}, y_{n-1},\mu_{n-1})) \epsilon \bar{H}$; h_n is said to be admissible if (2.2) holds for all $t = 1,2,\ldots, n-1$. Let H_n be the subset of $\bar{H}_n$ consisting of all admissible partial histories at date n. A <u>policy</u> is a sequence $\pi = \{\pi_t\}_{t=1}^\infty$, where for each $t \geq 1$, the policy function $\pi_t:H_t \rightarrow \bar{X}$ specifies the date t action $x_t = x_t(h_t)$, as a Borel function of the partial history, h_t in

H_t, at that date. A policy function is <u>stationary</u> if $\pi_t(h_t) = g(\mu_t)$ for each t, where the function $g(\cdot)$ maps $P(\Theta)$ into $\tilde{X}$.

Define $(\Omega, \mathcal{F}, P) = (\Theta, \mathcal{F}, \mu_0) \times (\Omega', \mathcal{F}', P')$. Any policy, π, then generates a sequence of random variables $\{(x_t(\omega), y_t(\omega), \mu_t(\omega)\}_{t=1}^{\infty}$ on $(\Omega, \mathcal{F}, P)$ as described above, using (2.1) and (2.2). See Kiefer and Nyarko (1987) for technical details.

For any $n = 1,2,\ldots$, let $\mathcal{F}_n$ be the sub-field of $\mathcal{F}$, generated by the random variables (h_n, x_n). Notice that x_n is $\mathcal{F}_n$ -measurable but y_n and μ_n are not $\mathcal{F}_n$ - measurable. Next define $\mathcal{F}_\infty = \vee_{n=0}^{\infty} \mathcal{F}_n$.

Let $u:\tilde{X} \times R^1 \to R^1$ be the utility function, so $u(x_t, y_t)$ is the utility to the agent when action x_t is chosen at date t and the observation y_t is made. The reward function $r:\tilde{X} \times P(\Theta) \to R^1$, is defined by

$$r(x_t, \mu_{t-1}) = \int_\Theta \int_R u(x_t, y_t) p(d\epsilon_t | \xi) \mu_{t-1}(d\theta) \tag{2.3}$$

The inner integration marginalizes with respect to ϵ, given the parameter ξ, the outer integration is with respect to parameters. Assume that the reward function is uniformly bounded, continuously, and concave in x for given μ. Note that this assumption restricts $g(\cdot,\cdot)$, $U(\cdot,\cdot)$ and $p(\cdot|\cdot)$.

Let δ in $[0,1)$ be the discount factor. Any policy π generates a sum of expected discounted rewards equal to

$$V_\pi(\mu_0) = \int \sum_{t=1}^{\infty} \delta^{t-1} r(x_t(\omega), \mu_{t-1}(\omega)) P(d\omega) \tag{2.4}$$

where the (x_t, μ_t) processes are those obtained using the policy π. A policy π^* is said to be an <u>optimal policy</u> if for all policies π and all priors μ_0 in in $P(\Theta)$, $V_{\pi^*}(\mu_0) \geq V(\mu_0)$. Even though the optimal policy, π^* (when it exists) may not be unique, the value function $V(\mu_0) = V_{\pi^*}(\mu_0)$ is always well-defined.

3. <u>Existence of a Stationary Optimal Policy</u>

Straightforward dynamic programming arguments can be used to show that stationary optimal policies exist and the value function is continuous.

<u>Theorem 3.1</u>: A stationary optimal policy $g:P(\Theta) \to \tilde{X}$ exists. The value function, V, is continuous on $P(\Theta)$, and the following functional equation holds:

$$V(\mu) = \max \{r(x, \mu) + \delta\int V(\tilde{\mu})p(d\epsilon|\xi)\mu(d\theta)\} \tag{3.1}$$

where $\tilde{\mu} = \Gamma(x, y, \mu)$ and $y = g(x, \beta) + \epsilon$, and where the integral is taken over $R^1 \times \Theta$.

Proof: Let $S = \{f:P(\Theta) \to R \mid f$ is continuous and bounded$\}$.

Define $T:S \to S$ by

$$Tw(\mu) = \max_{x\epsilon\bar{X}} \{r(x,\tilde{\mu}) + \delta\int V(\mu)p(d\epsilon|\phi)\mu(d\theta)\} \tag{3.2}$$

One can easily show that for $w\epsilon S$, $Tw\epsilon S$; and that T is a contraction mapping. Hence there exists a $v\epsilon S$ such that $v=Tv$. Replacing w with v in (3.2) then results in (3.1); and since $v\epsilon S$, v is continuous. Finally, it is immediate that the solution to the maximization exercise in (3.2) (replacing w with v) results in a stationary optimal policy function (see Blackwell (1965) or Maitra (1968) for the details of the above arguments).

4. Convergence of the Process $\{\mu_t\}$.

In this section we prove that the posterior process converges for P-a.e ω in Ω, to a well-defined probability measure (with the convergence taking place in a weak topology).

Note that for any Borel subset, D, of the parameter space Θ, if we suppress the ω's and let, for some fixed ω, $\mu_t(D)$ represent the mass that measure $\mu_t(\omega)$ assigns to the set D, then

$$\mu_t(D) = E[1_{\{\theta\epsilon D\}}|\mathcal{F}_t] \tag{4.1}$$

Define a measure μ_∞ on Θ by setting, for each Borel set D in Θ,

$$\mu_\infty(D) = E[1_{\{\theta\epsilon D\}}|\mathcal{F}_\infty] \tag{4.2}$$

The measure μ_∞ is the limiting posterior distribution and is indeed a well-defined probability measure.

Theorem 4.1. The posterior process $\{\mu_t\}$ converges, for P-a.e. ω in Ω, in the weak topology, to the probability measure μ_∞.

Summary of Proof: Use (4.1) above to show that for any Borel set D in Θ, $\mu_t(D)$ is a Martingale measure, establish that the sequence of probability measures, $\mu_t(\omega)$, for fixed ω, is tight using the assumption that the first moment of μ_∞ is finite, then apply Prohorov's Theorem (e.g., Billingsley (1968, Theorem 6.1)) to deduce that μ_∞ is a probability measure.

Note that this result on convergence of beliefs is quite different from the standard consistency result looked for in econometrics. The Martingale Convergence Theorem allows us to establish convergence, but the limit measure μ_∞ is a random variable, in the sense that it depends on the particular sequence of shocks realized. In a standard estimation problem, the limit result is that beliefs converge and the limit belief is independent of sample paths, and the limit belief is correct in the sense that μ_∞ assigns point mass to the true parameter value. Standard results do not hold here because along any sample path for which beliefs converge, the sequence of actions $\{x_t\}$ may also be converging. But if actions converge too rapidly, they may not generate enough information to identify all the unknown parameters. One can construct examples in related problems in which this phenomenon occurs (see e.g., Kiefer (1988)).

5. Optimization and Limit Beliefs and Actions

In Theorem 4.1, convergence of beliefs was established for an arbitrary $\{x_t\}$ sequence (i.e., without taking into account the underlying maximization problem). In this section we ask what action (or actions) $\bar{x}$ corresponds to the limiting beliefs μ_∞.

Theorem 5.1 establishes that the limit action is the action which maximizes single period reward for limit beliefs.

Theorem 5.1: The limit action $\bar{x} = \lim_{t\to\infty} x$ exists, is unique for given μ) and maximizes the one-period reward, $r(x,\mu_\infty)$, for limit beliefs μ_∞.

Proof of Theorem 5.1: Recall from Theorem 4.1 that $\lim_{t\to\infty} \mu_t = \mu_\infty$ exists for all sample paths. The sequence $\{x_t\}$ and $\{\mu_t\}$ satisfies for each t (simultaneously, a.e.) the functional equation

$$V(\mu_t) = r(x_t,\mu_t) + \delta\int V(\Gamma(x_t,y_t,\mu_t))p(d\varepsilon|\xi)\mu_t(d\theta). \quad (5.1)$$

Taking limits along any convergent subsequence gives

$$V(\mu_\infty) = r(\bar{x},\mu_\infty) + \delta\int V(\Gamma(\bar{x},y,\mu_\infty))p(d\epsilon|\xi)\mu_\infty(d\theta)$$

where $\bar{x}$ is a limit point of the $\{x_t\}$ sequence. (In taking the limits one uses the fact that V is bounded and the integral in (5.1) is $E[V(\mu_t)|\mathcal{F}_{t-1}]$ to apply Chung (1974, Theorem 9.4.8).) However, from convergence of beliefs $(\bar{x},y)$ yields no information so $\Gamma(\bar{x},y,\mu_\infty) = \mu_\infty$, and (5.1) becomes $V(\mu_\infty) = r(\bar{x},\mu_\infty) + \delta V(\mu_\infty)$.

Now we show that $\bar{x}$ solves the problem

$$\max_{x\epsilon\bar{X}} r(x,\mu_\infty) \tag{5.2}$$

Suppose on the contrary that there is an $\hat{x}\epsilon\bar{X}$ such that $r(\hat{x},\mu\ ) > r(\bar{x},\mu\ )$. Then by the functional equation

$$V(\mu_\infty) \geq r(\hat{x},\mu_\infty) + \delta\int V(\Gamma(\hat{x},\hat{y},\mu_\infty))p(d\epsilon|\theta)\mu_\infty(d\theta). \tag{5.3}$$

But by Blackwell's Theorem (see e.g., Kihlstrom (1984, Lemma 1, p. 18)), since the experiment "observe $(\hat{x},\hat{y})$" is trivially sufficient for the experiment "make no observations," we obtain,

$$\int V(\Gamma(x,y,\mu_\infty))p(d\epsilon|\phi)\mu_\infty(d\theta) \geq V(\mu\ ) \tag{5.4}$$

Hence, from (5.3) and (5.4) $V(\mu\ ) > r(\bar{x},\mu\ ) + \delta V(\mu\ )$, which is a contradiction. So $\bar{x}$ solves problem (5.2); that is, $\bar{x}$ maximizes the one-period reward $r(x,\mu)$ for limit beliefs, μ . Since $r(\cdot,\mu_\infty)$ is strictly concave in x, $\bar{x}$ must be unique.

6. Conclusion

We have considered the decision problem facing an agent controlling a nonlinear regression process when parameters in the mean function and in the error distribution are unknown. The agent faces a tradeoff between accumulating information by varying the values of the regressors and accumulating one-period reward by following the one-period expected reward maximizing policy. We show that the problem can be brought into the dynamic programming framework and that the value function satisfies the usual functional equation. The sequence of beliefs about the unknown parameters is shown to converge almost surely. Further, the optimal action process converges to the one-period optimal action under limit beliefs.

7. Acknowledgements

This research is supported in part by the National Science Foundation.

REFERENCES

Anderson, T.W. and J. Taylor, (1976), "some Experimental Results on and Statistical Properties of Least Squares Estimates in Control Problems," Econometrica, 44:1289-1302.

Billingsley, P., (1968), Convergence of Probability Measures, Wiley, New York.

Blackwell, D., (1965), "Discounted Dynamic Programming," Annals of Mathematical Statistics, 36, pp. 2226-235.

Chung, K.L., (1974), A Course in Probability Theory, 2nd edition, Academic Press, New York.

Diaconis, P. and D. Freedman, (1986), "On The Consistency Of Bayes Estimates," Annals of Statistics, 14, 1-26 (discussion and rejoinder 26-27).

Easley, D. and N.M. Kiefer, (1986), "Controlling a Stochastic Process with Unknown Parameters," Cornell University working paper, forthcoming in Econometrica.

Harkema, R., (1975), "An Analytical Comparison of Certainty Equivalence and Sequential Updating," JASA, 70, 348-350.

Kiefer, N.M. and Y. Nyarko, "Control of a Linear Regression Process with Unknown Parameters" in W. Barnett, E. Berndt and H. White (eds.), Dynamic Econometric Modelling, New York: Cambridge University Press, 1987.

Kiefer, N.M., "Optimal Collection of Information by Partially Informed Agents," Cornell working paper, 1988.

Kihlstrom, R.E., (1984), "A 'Bayesian' Exposition of Blackwell's Theorem on the Comparison of Experiments," in Bayesian Models in Economic Theory, eds. M. Boyer and R.E. Kihlstrom, Elsevier Science Publishers B.V.

Jordan, J.S., (1985), "The Strong Consistency of the Least Squares Control Rule and Parameter Estimates," manuscript.

Maitra, A., (1968), "Discounted Dynamic Programming in Compact Metric Spaces," Sankhya, Ser A, 30, pp. 211-216.

Parthasarathy, K., (1967), Probability Measures on Metric Spaces, Academic Press, New York.

Taylor, J.B., (1974), "Asymptotic Properties of Multiperiod Control Rules in the Linear Regression Model," International Economic Review, 15, 472-484.

Zellner, A., (1981), An Introduction to Bayesian Inference in Econometrics, Wiley: New York.

/

A THREE-MIRROR PROBLEM ON DYNAMIC PROGRAMMING

Seiichi Iwamoto

Department of Economic Engineering
Faculty of Economics
Kyushu University 27, Fukuoka 812, Japan

1. INTRODUCTION

The essence of dynamic programming states that a simultaneous optimization of real-valued two-variable functions is assured by the two-stage optimization under both separability and monotonicity [15, 16]. We call these two properties the recusiveness with monotonicity —— dynamic programming structure —— [8, 11]. This structure yields what we call dynamic programmable function [11].

In this paper we focus our attention on both dynamic programming structure and quasililearization for a class of objective functions. Given a differentiable strictly increasing convex function $f : R^1 \longrightarrow R^1$, we approximate $f(x)$ by its linear approximation $f(x;h)$ $R^1 \times R^1 \longrightarrow R^1$, which is strictly increasing in h for $x \in R^1$. Thus, $f(x)$ is a quasilinearization of $f(x;h)$. The N-times composition of $f(x_n;\cdot)$ generates a dynamic programmable function $F(x;h) : R^N \times R^1 \longrightarrow R^1$. Similarly, inverse function $f^{-1}(y)$, reverse function $f_{-1}(x;k)$ which is the inverse function of $f(x;h)$ with respect to h for fixed x, and conjugate function $f^*(y)$ also generate dynamic programmable functions $F^{-1}(y;k)$, $F_{-1}(x;k)$, and $F^*(y;h) : R^N \times R^1 \longrightarrow R^1$, respectively. Thus, the function f yields four —— main, inverse, reverse, and conjugate —— optimization problems on R^N. These problems are solved through dynamic programming approach. Some relations between them are established. Finally we illustrate two interesting examples from Bellman [1].

2. PROBLEMS

First of all let us consider the following famous problem [1, p. 102; 8, p.101; 10, p.18]:

$$\text{Max} \quad e^{x_1}(1-x_1) + e^{x_1+x_2}(1-x_2) + \dots + e^{x_1+\dots+x_N}(1-x_N) + e^{x_1+\dots+x_N}\times h$$

$$\text{s.t.} \quad -\infty < x_n < \infty \qquad 1 \leq n \leq N$$

where h is a real constant. We remark that the N-times iteration of

$$f(x;h) = e^x(1-x+h)$$

yields the objective function

$$f(x_1;f(x_2;\dots;f(x_N;h)\dots))$$

$$= e^{x_1}(1-x_1) + e^{x_1}\Big[e^{x_2}(1-x_2) + e^{x_2}\big[\dots + e^{x_{N-1}} \times\big[e^{x_N}(1-x_N) + e^{x_N}\times h\big]\dots\big]\Big]$$

(See also [11, p.278; 12, p.285]).

Second we consider the following maximization problem:

$$\text{Max} \quad (1-2x_1^2)\exp(x_1^2) + 2x_1(1-2x_2^2)\exp(x_1^2+x_2^2) + 4x_1x_2 \times(1-2x_3^2)\exp(x_1^2+x_2^2+x_3^2) + 8x_1x_2x_3(1-2x_3^2) \times\exp(x_1^2+x_2^2+x_3^2)h$$

$$\text{s.t.} \quad x_1 \geq 0,\ x_2 \geq 0,\ x_3 \geq 0$$

where $h \geq 0$. The three-times iteration of

$$f(x;h) = (1 - 2x^2 + 2xh)\exp(x^2)$$

generates

$$f(x_1;f(x_2;f(x_3;h)))$$

$$= (1-2x_1^2)\exp(x_1^2) + 2x_1\exp(x_1^2)\left[(1-2x_2^2)\exp(x_2^2) + 2x_2\exp(x_2^2)\times\right.$$

$$\left.\left[(1-2x_3^2)\exp(x_3^2) + 2x_3\exp(x_3^2)h\right]\right].$$

These two functions are called *recursive functions on* R^N *(resp.* R^3_+*) with strict increasingness* ([10, 11]). A function $F : R^N \times R^1 \longrightarrow R^1$ is called *dynamic programmable function on* R^N if it is expressed as follows

$$F(x_1,x_2,\ldots,x_N;h)$$

$$= f_1(x_1;f_2(x_1,x_2;\ldots;f_N(x_1,x_2,\ldots,x_N;h)\ldots))$$

where $f_n: R^n \times R^1 \longrightarrow R^1$ and $f_n(x_1,x_2,\ldots,x_n;\ \cdot): R^1 \longrightarrow R^1$ is non-decreasing for $1 \leq n \leq N$, $(x_1,x_2,\ldots,x_n) \in R^n$. Therefore, any recursive function with strict increasingness is a dynamic programmable function. In the following we are mainly concerned with a class of recusive functions on $X(\subset R^N)$ with strict increasingness.

3. MAIN RESULT

First, we prepare the following fundamental lemma. Let X and Y be two nonempty sets. For each $x \in X$ let $Y(x)$ be a nonempty subset of Y. That is, $Y(\cdot) : X \longrightarrow 2^Y$ is a point-to-set-valued mapping, where 2^Y denotes the set of all nonempty subsets of Y. Let

$$G_r(Y) = \{(x,y) \mid y \in Y(x),\ x \in X\} \subset X \times Y$$

be the graph of the mapping $Y(\cdot)$. In the following it will be clear from the context whether a notation Y is considered the set or the mapping.

LEMMA 1 (Maximax Theorem [11; p.268]) Let $f : X \times R^1 \longrightarrow R^1$ be a function such that $f(x;\cdot) : R^1 \longrightarrow R^1$ is nondecreasing for $x \in X$. Let $g : G_r(Y) \longrightarrow R^1$ be a function. If $\underset{x\in X}{\mathrm{Max}}\, f(x; \underset{y\in Y(x)}{\mathrm{Max}}\, g(x,y))$ exists, then $\underset{(x,y)\in G_r(Y)}{\mathrm{Max}}\, f(x; g(x,y))$ exists and both are equal:

$$\underset{x\in X}{\mathrm{Max}}\, f(x; \underset{y\in Y(x)}{\mathrm{Max}}\, g(x,y)) = \underset{(x,y)\in G_r(Y)}{\mathrm{Max}}\, f(x; g(x,y)).$$

REMARK This equality remains valid even if the operator Max is replaced by the operator min under the same condition as stated above. Furthermore, as a special case we have

$$\underset{-\infty<x<\infty}{\mathrm{Max}}\, f(x; \underset{-\infty<y<\infty}{\mathrm{Max}}\, g(y)) = \underset{-\infty<x,y<\infty}{\mathrm{Max}}\, f(x; g(y)).$$

In general we have for any differentiable convex function $f : R^1 \longrightarrow R^1$

$$f(h) = \underset{-\infty<x<\infty}{\mathrm{Max}}\, f(x;h) \qquad (1)$$

where

$$f(x;h) = F(x) + f'(x)h$$
$$F(x) = f(x) - xf'(x). \qquad (2)$$

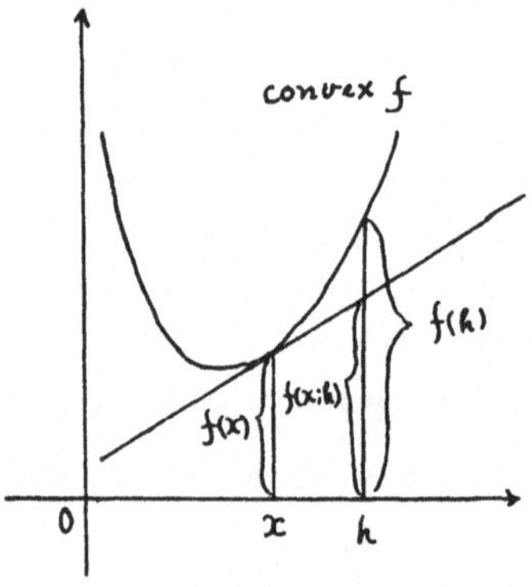

Fig. 1

Thus, $f(x;h)$ is the *linear approximation of* $f(x)$ *at* h:

$$f(x;h) = f(x) + (h - x)f'(x). \tag{3}$$

The expression (1) is called a *quasilinearization of f(x)* ([1; p.135 ; 13; 14]).

Furthermore, from Lemma 1, we obtain under $f'(x) \geq 0$, $-\infty < x < \infty$

$$\begin{aligned} f(f(h)) &= \operatorname*{Max}_{-\infty<x_1<\infty} f(x_1; \operatorname*{Max}_{-\infty<x_2<\infty} f(x_2; h)) \\ &= \operatorname*{Max}_{-\infty<x_1,x_2<\infty} f(x_1; f(x_2;h)). \end{aligned} \tag{4}$$

that is

$$\begin{aligned} f(f(h)) &= \operatorname*{Max}_{-\infty<x_1<\infty} [F(x_1) + f'(x_1)(\operatorname*{Max}_{-\infty<x_2<\infty} [F(x_2) + f'(x_2)h])] \\ &= \operatorname*{Max}_{-\infty<x_1,x_2<\infty} [F(x_1) + f'(x_1)F(x_2) + f'(x_1)f'(x_2)h]. \end{aligned} \tag{5}$$

DEFINITION Let $f : R^1 \longrightarrow R^1$ be a differentiable increasing (resp. strictly increasing) convex function. Then we define $F : R^N \times R^1 \longrightarrow R^1$ by

$$\begin{aligned} F(x;h) &= f(x_1; f(x_2; \ldots; f(x_N;h)\ldots)) \\ &= F(x_1) + f'(x_1)F(x_2) + \ldots + f'(x_1)f'(x_2)\ldots f'(x_{N-1}) \\ &\quad \times F(x_N) + f'(x_1)f'(x_2)\ldots f'(x_N)h \end{aligned} \tag{6}$$

where $f(x;h)$ and $F(x)$ are defined in (2), and $x = (x_1, x_2, \ldots, x_N)$. The function $F : R^N \times R^1 \longrightarrow R^1$ is the *recursive function with increasingness (resp. strict increasingness) generated by f* or simply *dynamic programmable function generated by f.*

In the following, it will be clear from the context a function f (resp. F) is considered $f(x)$ or $f(x;h)$ (resp. $F(x)$ or $F(x;h)$).

REMARK The equalities (1) and (4) (or (5)) remain valid if we replace 'Max' and 'convex' with 'min' and 'concave', respectively. Similarly, a differentiable increasing (resp. strictly increasing) *concave* function $g : R^1 \longrightarrow R^1$ *generates the recursive function* $G : R^N \times R^1 \longrightarrow R^1$ *with increasingness (resp. strict increasingness)*, which is also called *dynamic programmable function generated by g:*

$$\begin{aligned} G(y;k) &= g(y_1;\ g(y_2;\ \ldots;\ g(y_N;k)\ldots)) \\ &= G(y_1) + g'(y_1)G(y_2) + \ldots + g'(y_1)g'(y_2)\ldots.g'(y_{N-1}) \\ &\quad \times G(y_N) + g'(y_1)g'(y_2)\ldots g'(y_N)k \end{aligned} \tag{7}$$

where

$$y = (y_1, y_2, \ldots, y_N),$$

$$\begin{aligned} g(y;k) &= g(y) + (k - y)g'(y) \\ &= G(y) + g'(y)k, \end{aligned} \tag{8}$$

$$G(y) = g(y) - yg'(y).$$

Therefore we have the following main result:

THEOREM 1. (i) Let $f: R^1 \longrightarrow R^1$ be a differentiable increasing convex function. Then for $h \in R^1$

$$f^N(h) = \operatorname*{Max}_{x\varepsilon R^N} F(x;h) \tag{9}$$

and $x_1^* = f^{N-1}(h)$, $x_2^* = f^{N-2}(h)$, ..., $x_{N-1}^* = f(h)$, $x_N^* = h$ attains the maximum, here and in the following $f^n(h)$ is the n-times composition of $f(x)$:

$$f^n(x) = f(f(\ldots f(x)\ldots)).$$

(ii) Let $g : R^1 \longrightarrow R^1$ be a differentiable increasing concave function. Then for $k \varepsilon R^1$

$$g^N(k) = \min_{y\varepsilon R^N} G(y;k) \tag{10}$$

and $\hat{y}_1 = g^{N-1}(k)$, $\hat{y}_2 = g^{N-2}(k)$, ..., $\hat{y}_{N-1} = g(k)$, $\hat{y}_N = k$ attains the minimum.

4. INVERSION, REVERSION AND CONJUGATION

First we consider the inverse function f^{-1} to a continuous strictly increasing function f. We remark that $f : R^1 \longrightarrow R^1$ is an onto differentiable strictly increasing convex function iff $f^{-1} : R^1 \longrightarrow R^1$ is the onto differentiable strictly increasing concave function. Then we have

COROLLARY (i) Let $f : R^1 \longrightarrow R^1$ be an onto differentiable strictly increasing convex function. Then for $k \varepsilon R^1$

$$f^{-N}(k) = \min_{y\varepsilon R^N} F^{-1}(y;k) \tag{11}$$

and $\hat{y}_1 = f^{-N+1}(k)$, $\hat{y}_2 = f^{-N+2}(k)$, ..., $\hat{y}_{N-1} = f^{-1}(k)$, $\hat{y}_N = k$ attains the minimum, where $F^{-1}(y;k)$ is the dynamic programmable function generated by f^{-1} and $f^{-n}(y)$ is the n-time composition of f^{-1}:

$$f^{-n}(y) = f^{-1}(f^{-1}(\ldots f^{-1}(y)\ldots)).$$

(ii) Let $g : R^1 \longrightarrow R^1$ be an onto differentiable strictly increasing concave function. Then for $h \varepsilon R^1$

$$g^{-N}(h) = \underset{x \varepsilon R^N}{\mathrm{Max}_N}\, G^{-1}(x;h) \tag{12}$$

and $x_1^* = g^{-N+1}(h)$, $x_2^* = g^{-N+2}(h)$, ..., $x_{N-1}^* = g^{-1}(h)$, $x_N^* = h$ attains the maximum, where $G^{-1}(x;h)$ is the dynamic programmable function generated by g^{-1}.

Here we remark that

$$F^{-1}(y;k) = F^{-1}(y_1) + f^{-1\prime}(y_1)F^{-1}(y_2) + \ldots + f^{-1\prime}(y_1)f^{-1\prime}(y_2) \times \ldots f^{-1\prime}(y_{N-1})F^{-1}(y_N) + f^{-1\prime}(y_1)f^{-1\prime}(y_2)\ldots f^{-1\prime}(y_N)k \tag{13}$$

where

$$F^{-1}(y) = f^{-1}(y) - yf^{-1\prime}(y) \tag{14}$$

and $f^{-1\prime}$ is the derivative of the inverse function f^{-1}. Similarly, $G^{-1}(x;h)$ is defined and omitted.

Second we consider the reversion of the linear approximation $f(x;h)$ of $f(x)$ —— not the reversion of $f(x)$ itself —— as follows. For any onto differentiable strictly increasing convex function $f : R^1 \longrightarrow R^1$, its linear approximation $f : R^1 \times R^1 \longrightarrow R^1$

defined by (2) or (3) is continuous strictly increasing and linear in h for $x \in R^1$. Therefore, $f(x;\cdot) : R^1 \longrightarrow R^1$ is invertible for $x \in R^1$. Its inverse function $f_{-1}(x;\cdot) : R^1 \longrightarrow R^1$ becomes

$$f_{-1}(x;k) = F_{-1}(x) + \frac{k}{f'(x)} \tag{15}$$

where

$$F_{-1}(x) = x - \frac{f(x)}{f'(x)}. \tag{16}$$

We call $f_{-1} = f_{-1}(x;k)$ the *reverse function* of $f = f(x;h)$. As we noted in (1), we have

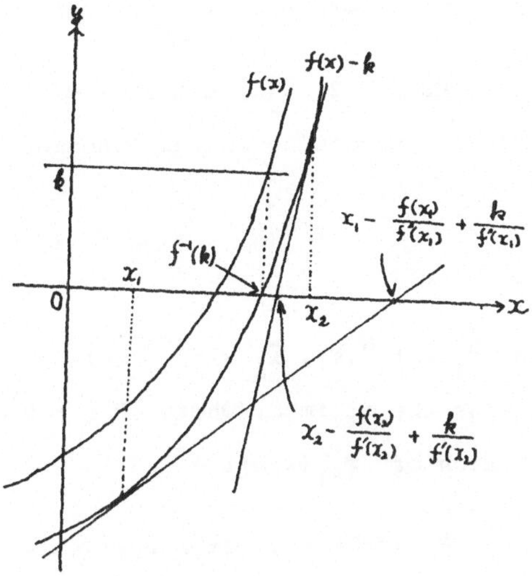

Fig. 2

$$\begin{aligned} f(h) &= \operatorname*{Max}_{-\infty<x<\infty} f(x;h) \\ &= \operatorname*{Max}_{-\infty<x<\infty} [F(x) + f'(x)h] \\ &= \operatorname*{Max}_{-\infty<x<\infty} [f(x) + (h - x)f'(x)] \end{aligned} \tag{17}$$

and $x^* = h$ attains the maximum. This fact is equivalently transformed to

$$\begin{aligned} f^{-1}(k) &= \min_{-\infty<x<\infty} f_{-1}(x;k) \\ &= \min_{-\infty<x<\infty} \left[F_{-1}(x) + \frac{k}{f'(x)}\right] \end{aligned} \tag{18}$$

$$= \min_{-\infty<x<\infty} \left[x + \frac{k - f(x)}{f'(x)}\right]$$

and $x = f^{-1}(k)$ attains the minimum (see Fig.2). This fact reflects also the main idear of Newton method from a viewpoint of optimization. Therefore, we have the following reversed form of (9):

THEOREM 2. (i) Let $f : R^1 \longrightarrow R^1$ be an onto differentiable strictly increasing convex function. Then for $k \varepsilon R^1$

$$f^{-N}(k) = \min_{x \varepsilon R^N} F_{-1}(x;k) \tag{19}$$

and $\hat{x}_1 = f^{-N}(k)$, $\hat{x}_2 = f^{-N+1}(k)$, ..., $\hat{x}_{N-1} = f^{-2}(k)$, $\hat{x}_N = f^{-1}(k)$ attains the minimum, where $F_{-1} : R^N \times R^1 \longrightarrow R^1$ is the N-times composition of $f_{-1}(x;k)$:

$$F_{-1}(x;k) = f_{-1}(x_1;\ f_{-1}(x_2;\ \ldots;\ f_{-1}(x_N;k)\ldots)). \tag{20}$$

(ii) Let $g : R^1 \longrightarrow R^1$ be an onto differentiable strictly increasing concave function. Then for $h \varepsilon R^1$

$$g^{-N}(h) = \operatorname{Max}_{y \varepsilon R^N} G_{-1}(y;h) \tag{21}$$

and $y_1^* = g^{-N}(h)$, $y_2^* = g^{-N+1}(h)$, ..., $y_{N-1}^* = g^{-2}(h)$, $y_N^* = g^{-1}(h)$ attains the maximum, where $G_{-1} : R^N \times R^1 \longrightarrow R^1$ is the N-times composition of $g_{-1}(y;h)$:

$$G^{-1}(y;h) = g_{-1}(y_1;\ g_{-1}(y_2;\ \ldots;\ g_{-1}(x_N;h)\ldots)). \tag{22}$$

Here we remark that

$$F_{-1}(x;k) = F_{-1}(x_1) + \frac{F_{-1}(x_2)}{f'(x_1)} + \dots + \frac{F_{-1}(x_N)}{f'(x_1)f'(x_2)\dots f'(x_{N-1})}$$

$$+ \frac{k}{f'(x_1)f'(x_2)\dots f'(x_N)} \qquad (23)$$

where $F_{-1}(x)$ is defined in (16). Similarly, $G_{-1}(y;h)$ is defined from $G_{-1}(y_n)$, $g'(y_n)$ and h. We call $F_{-1}(x;k)$, $G_{-1}(y;h)$ the *dynamic programmable function generated by reverse function* $f_{-1}(x;k)$, $g_{-1}(y;h)$), respectively.

We have the following relation between $F^{-1}(y;k)$ and $F_{-1}(x;h)$:

THEOREM 3. (i) Let $f : R^1 \longrightarrow R^1$ be an onto differentiable strictly increasing convex function. Then we have by the monotone transformation $y = f(x)$

$$f^{-1}(y;k) = f_{-1}(x;k). \qquad (24)$$

Furthemore, the monotone transfomation $y_n = f(x_n)$ $1 \leq n \leq N$ yields

$$F^{-1}(y;k) = F_{-1}(x;k). \qquad (25)$$

(ii) Let $g : R^1 \longrightarrow R^1$ be an onto differentiable strictly increasing concave function. Then we have by the monotone transformation $x = g(y)$

$$g^{-1}(x;h) = g_{-1}(y;h). \qquad (26)$$

Furthermore, the monotone transformation $x_n = g(y_n)$ $1 \leq n \leq N$ yields

$$G^{-1}(x;h) = G_{-1}(y;h). \tag{27}$$

Proof. It is straightforward.

Finally we consider conjugations $*$ and $\wedge$. For any convex function $f : R^1 \longrightarrow R^1$, we define its *conjugate function* $f^* : R^1 \longrightarrow R^1$

$$f^*(y) = \sup_{-\infty<x<\infty} [xy - f(x)]. \tag{28}$$

On the other hand, for any concave function $g : R^1 \longrightarrow R^1$, we denote its *conjugate function* $\hat{g} : R^1 \longrightarrow R^1$ by

$$\hat{g}(x) = \inf_{-\infty<y<\infty} [yx - g(y)]. \tag{29}$$

If both operations $*$ and $\wedge$ are well defined, they are dual in the following sense:

$$\widehat{(-f)}(y) = -f^*(-y) \qquad y \in R^1.$$

LEMMA 2. Let $f : R^1 \longrightarrow R^1$ be a twice differentiable strictly increasing strictly convex function. Then we have for $f'(-\infty) < y < f'(\infty)$

(i) $f^*(y) = xy - f(x)$

(ii) $f^{*\prime}(y) = x$ and in particular $f^{*\prime}(y) > 0$ for $f'(0)<y<f'(\infty)$

and

(iii) $f^{*\prime\prime}(y) = \dfrac{1}{f''(x)} > 0$

where x satisfies uniquely $f'(x) = y$. Therefore, $f^* : (f'(0),$

$f'(\infty)) \longrightarrow R^1$ is strictly increasing strictly convex. Thus we have the following result for f^*:

THEOREM 4. Let $f : R^1 \longrightarrow R^1$ be a twice differentiable strictly increasing strictly convex function. Then we have for $f'(0) < f^{*n}(h) < f'(\infty) \quad 0 \leq n \leq N-1$

$$f^{*N}(h) = \underset{f'(0)<y_n<f'(\infty) \quad 1\leq n\leq N}{\text{Max}} F^*(y;h) \tag{30}$$

and $y_1^* = f^{*(N-1)}(h)$, $y_2^* = f^{*(N-2)}(h)$, ..., $y_{N-1}^* = f^*(h)$, $y_N^* = h$ attains the maximum, where $F^*(y;h)$ is the dynamic programmable function generated by f^* and f^{*n} is the n-time composition of f^*.

Similarly, for concave function g, we have the following:

LEMMA 3. Let $g : R^1 \longrightarrow R^1$ be a twice differentiable strictly increasing strictly concave function. Then we have for $g'(\infty) < x < g'(-\infty)$

(i) $\hat{g}(x) = yx - g(y)$

(ii) $\hat{g}'(x) = y$ and in particular $\hat{g}'(x) > 0$ for $g'(\infty)<x<g'(0)$

and

(iii) $\hat{g}''(x) = \frac{1}{g''(y)} < 0$

where y satisfies uniquely $g'(y) = x$. Therefore, $\hat{g} : (g'(\infty), g'(0)) \longrightarrow R^1$ is strictly increasing strictly concave.

THEOREM 5. Let $g : R^1 \longrightarrow R^1$ be a twice differentiable strictly increasing strictly concave function. Then we have for $g'(\infty) < g^n(k) < g'(0) \quad 0 \le n \le N-1$

$$\hat{g}^N(k) = \min_{g'(\infty)<x_n<g'(0) \quad 1\le n\le N} \hat{G}(x;k) \tag{31}$$

and $\hat{x}_1 = g^{N-1}(k)$, $\hat{x}_2 = g^{N-2}(k)$, ..., $\hat{x}_{N-1} = g(k)$, $\hat{x}_N = k$ attains the minimum, where $\hat{G}(x;k)$ is the dynamic programmable function generated by $\hat{g}$ and $\hat{g}^n$ is the n-times composition of $\hat{g}$.

Here we remark that

$$F^*(y;h) = F^*(y_1) + f^{*\prime}(y_1)F^*(y_2) + \dots + f^{*\prime}(y_1)f^{*\prime}(y_2)\dots \times f^*(y_{N-1})F^*(y_N) + f^{*\prime}(y_1)f^{*\prime}(y_2)\dots f^{*\prime}(y_N)h \tag{32}$$

where

$$\begin{aligned} F^*(y) &= f^*(y) - yf^{*\prime}(y) \\ &= -f(x). \end{aligned} \tag{33}$$

Here x satisfies uniquely $f'(x) = y$. Similar expressions for $\hat{G}(x;k)$ and $\hat{G}(x)$ are omitted.

5. EXAMPLES

In this section we illustrate explicit form of $f(x;h)$, $F(x;h)$, $F^{-1}(y;k)$, $f_{-1}(x;k)$, $F_{-1}(x;k)$, $F^*(y;k)$ and others for a given $f(x)$.

5.1 $f(x) = e^x : (-\infty, \infty) \longrightarrow (0, \infty)$

In this case we have the following expressions. First we have from (2),(6)

$$f(x;h) = (1 - x + h)e^{x} \qquad -\infty < x,h < \infty$$

$$F(x;h) = e^{x_1}(1 - x_1) + e^{x_1+x_2}(1 - x_2) + \ldots + e^{x_1+\ldots+x_{N-1}}$$

$$\times(1 - x_N) + e^{x_1+\ldots+x_N}\times h \qquad -\infty < x_n,h < \infty.$$

Second, for inversion, we have from (13),(14)

$$g(y) \equiv f^{-1}(y) = \log y : (0, \infty) \longrightarrow (-\infty, \infty) \tag{34}$$

$$g(y;k) = f^{-1}(y;k) = -1 + \log y + \frac{k}{y} \qquad 0 < y,k < \infty$$

$$G(y;k) = F^{-1}(y;k) = -1 + \log y_1 + (y_1)^{-1}(-1 + \log y_2)$$

$$+ (y_1 \ldots y_{N-1})^{-1}(-1 + \log y_N) + (y_1 \ldots y_N)^{-1}k$$

$$y_n > 0, \quad k \gg 0$$

where $k \gg 0$ means that k is sufficiently large that log...logk (N-times log operation) becomes well defined. That is, in this case,

$$k > e^{e^{\cdot^{\cdot^{\cdot^{e}}}}} \quad ((N-1)\text{'s } e).$$

Third, for reversion, we have from (15),(16),(20)

$$f_{-1}(x;k) = x - 1 + e^{-x}k \qquad -\infty < x < \infty, \quad k > 0$$

$$F_{-1}(x;k) = x_1 - 1 + e^{-x_1}(x_2 - 1) + \ldots + e^{-x_1-\ldots-x_{N-1}}(x_{N-1}-1)$$

$$+ e^{-x_1-\ldots-x_N}\times k$$

$$-\infty < x_n < \infty, \; k \gg 0.$$

Moreover, the reversion of $g = g(y)$ defined in (34) becomes

$$g_{-1}(x;h) = y(1 - \log y) + yh \qquad y > 0, \quad -\infty < h < \infty$$

$$G_{-1}(x;h) = y_1(1 - \log y_1) + y_1y_2(1 - \log y_2) + \dots + y_1\dots y_N$$

$$\times(1 - \log y_N) + y_1\dots y_N h \qquad y_n > 0, \quad -\infty < h < \infty.$$

Fourth, for conjugation, we have from (28), (29), (32), (33)

$$f^*(y) = (-1 + \log y)y : (0, \infty) \longrightarrow [-1, \infty)$$

$$f^{*\prime}(y) = \log y > 0 \qquad \text{on} \quad (1, \infty)$$

$$f^{*\prime\prime}(y) = 1/y > 0$$

$$f^*(y;k) = -y + k\times\log y \qquad y > 1, \quad k > 1$$

$$F^*(y;k) = -y_1 - y_2\log y_1 - \dots - y_N\log y_1\dots\log y_{N-1}$$

$$+ k\times\log y_1\dots\log y_N \qquad y_n > 1, \quad k > e^2$$

$$\hat{g}(x) = 1 + \log x : (0, \infty) \longrightarrow (-\infty, \infty)$$

$$\hat{g}(x;h) = \log x + x^{-1}h \qquad 0 < x,h < \infty$$

$$\hat{G}(x;h) = \log x_1 + (x_1)^{-1}\log x_2 + \dots + (x_1\dots x_{N-1})^{-1}\log x_N$$

$$+ (x_1\dots x_N)^{-1}h \qquad x_n > 0, \quad h \gg 0$$

where $h \gg 0$ in this case means that

$$h > e^{-1+e^{-1+e^{\cdot^{\cdot^{-1+e^{-1}}}}}} \qquad (N\text{'s } e).$$

Finally, for reversion of $\hat{g}(y;k)$, we have

$$\hat{g}_{-1}(x;k) = -x\log x + xk \qquad x > 0, \quad -\infty < k < \infty$$

$$\hat{G}_{-1}(x;k) = -x_1\log x_1 - x_1x_2\log x_2 - \ldots - x_1\ldots x_N\log x_N$$
$$+ x_1\ldots x_N k \qquad x_n > 0, \quad -\infty < k < \infty.$$

5.2 $\quad f(x) = x^2 : [0, \infty) \longrightarrow [0, \infty)$

In this case we have the following result. First, we get

$$f(x;h) = -x^2 + 2xh \qquad x,h \geq 0$$

$$F(x;h) = -x_1^2 - 2x_1x_2^2 - \ldots - 2^{N-1}x_1\ldots x_{N-1}x_N^2$$
$$+ 2^N x_1\ldots x_N h \qquad x_n \geq 0, \; h \geq 0.$$

In particular Theorem 1 for case $N = 1$ implies

$$\max_{-\infty<x<\infty} \left[2xh - x^2 \right] = h^2 \qquad -\infty < h < \infty.$$

This is one of the simplest quasilinearization [1; p.134].

Second, the inversion becomes

$$g(y) \equiv f^{-1}(y) = \sqrt{y} : (0, \infty) \longrightarrow (0, \infty)$$

$$g(y;k) = f^{-1}(y;k) = \frac{1}{2}\left(y + \frac{k}{\sqrt{y}}\right) \qquad y,k > 0$$

$$G(y;k) = F^{-1}(y;k)$$
$$= \frac{1}{2}(y_1)^{1/2} + \frac{1}{2^2}(y_2/y_1)^{1/2} + \ldots + \frac{1}{2^N}(y_N/y_1\ldots y_{N-1})^{1/2}$$
$$+ \frac{k}{2^N}(y_1\ldots y_N)^{-1/2} \qquad y_n > 0, \; k > 0.$$

Therefore, Corollary (ii) for case $N = 1$ reduces

$$\min_{x>0} \left[\frac{1}{2}\sqrt{x} + \frac{1}{2\sqrt{x}}\right] = \sqrt{k} \qquad k > 0$$

(see also [1; p.134]).

Third, for reversion, we have

$$f_{-1}(x;k) = \frac{1}{2}(x + \frac{k}{x}) \qquad x,k > 0$$

$$F_{-1}(x;k) = \frac{1}{2}x_1 + \frac{1}{2^2}(x_2/x_1) + \dots + \frac{1}{2^N}(x_N/x_1\dots x_{N-1})$$

$$+ \frac{k}{2^N}(x_1\dots x_N)^{-1} \qquad x_n > 0, \quad k > 0.$$

Finally, the conjugation yields

$$f^*(y) = \frac{1}{4}y^2 : [0, \infty) \longrightarrow [0, \infty)$$

$$f^*(y;k) = -\frac{1}{4}y^2 + \frac{1}{2}yk \qquad y,k \geq 0$$

$$F^*(y;k) = -\frac{1}{4}y_1^2 - \frac{1}{4\cdot 2}y_1y_2^2 - \dots - \frac{1}{4\cdot 2^{N-1}}y_1\dots y_{N-1}y_N^2$$

$$+ \frac{1}{2^N}y_1\dots y_N k \qquad y_n \geq 0, \quad k \geq 0$$

$$\hat{g}(x) = -\frac{1}{4x} : (0, \infty) \longrightarrow (-\infty, 0)$$

$$\hat{g}(x) = -\frac{1}{2x} + \frac{h}{4x^2} \qquad x,h > 0.$$

Therefore we get

$$\hat{g}(x) = \min_{0<x<\infty} \hat{g}(x;h) .$$

However if $N \geq 2$, then it does not hold that

$$\hat{g}^N(h) = \min_{0<x_n<\infty} \hat{G}(x;h) \qquad h > 0,$$

because of $\hat{g}(h) < 0$.

References

1. R. Bellman, Dynamic Programming, Princeton Univ. Press, Prinston N.J., 1957.
2. R. Bellman and R. Kalaba, Quasilinearization and Nonlinear Boundary-value Problems, American Elsevier, N.Y., 1965
3. R. Bellman and Wm. Karush, On a new functional transform in analysis : the maximum transform, Bull. Amer. Math. Soc. 67 (1961), 501-503.
4. R. Bellman and Wm. Karush, Mathematical programming and the maximum transform, J. SIAM Appl. Math. 10(1962), 550-567.
5. R. Bellman and Wm. KARUSH, On the maximum transform and semi-groups of transformations, Bull. Amer. Math. Soc. 68(1962), 516-518.
6. R. Bellman and Wm. Karush, Functional equations in the theory of dynamic programming - XII: an application of the maximum transform, J. Math. Anal. Appl. 6(1963), 155-157
7. R. Bellman and Wm. Karush, On the maximum transform, J. Math. Anal. Appl. 6(1963), 67-74.
8. N. Furukawa and S. Iwamoto, Dynamic programming on recursive reward systems, Bull. Math. Statist. 17(1976) 103-126.
9. S. Iwamoto, Some operations on dynamic programmings with one-dimensional state space, J. Math. Anal. Appl. 69(1979), 263-282.
10. S. Iwamoto, Reverse function, reverse program and reverse theorem in mathematical programming, J. Math. Anal. Appl. 95 (1983), 1-19.
11. S. Iwamoto, Sequential minimaximization under dynamic programming structure, J. Math. Anal. Appl. 108(1985), 267-282.
12. S. Iwamoto, R.J. Tomkins and C.-L. Wang, Some theorems on reverse inequalities, J. Math. Anal. Appl. 119(1986), 282-299.
13. E. Stanley Lee, Dynamic Programming, quasilinearization and dimensiomality difficulty, J. Math. Anal. Appl. 27(1968), 303-322.

14. E. Stanley Lee, Quasilinearization and Invariant Imbedding, Academic Press, N.Y., 1968.

15. L.G. Mitten, Composition principle for synthesis of optimal multistage process, Operations Res. 12(1964), 601-619.

16. G.L. Nemhauser, Introduction to Dynamic Programming, John Wiely and Sons, 1966.

MODELISATION ET COMMANDE DES SYSTEMES BIOLOGIQUES ET DES ECOSYSTEMES

MODELS AND CONTROL POLICIES FOR BIOLOGICAL SYSTEMS AND ECOSYSTEMS

CONTROL THEORY AND BIOLOGICAL REGULATIONS: BIPOLAR CONTROLS

Daniel CLAUDE

Laboratoire des Signaux et Systèmes,
C.N.R.S.- E.S.E.,
Plateau du Moulon, 91190 Gif-sur-Yvette, France.

Abstract : In memory of Richard Bellman, we present bipolar controls in biology. From its therapeutic application in the field of cerebral tumors and cancerology alone, Richard Bellman would have certainly been in favour of this methodology which links control theory to biological regulation. We show all its richness in opening other prospects that entirely justify the link between mathematics and medicine which interested him so much.

I. INTRODUCTION

For several decades, numerous research workers have thought of making a link between Mathematics and Medicine (cf. recent books by Winfree [26] and Swan [25]), particularly in attempts in the modelization of certain biological phenomena and, for example, in cancerology, in the search for medicinal procedures (chemotherapy) or the placing of protocols of emission of specific active particles (radiotherapy). In this way they wished to combine mathematical theory with medical practice. Control theory, applied to some biological regulations, answers this demand and this hope.

In biology, numerous regulations demand many factors with coupled actions. This is the case in the regulation of cellular hydration or the control of the mitosis in which corticoids and vasopressin act respectively, just as insulin and glucagon regulate glycemic activity. The failure, in some pathologies, of therapies which give only one hormone results in the fact that the reaction of the other hormone, caused by a subtle game of crossed feedbacks, has been neglected. Moreover, biology is a strongly nonlinear field where the principle of superposition of actions does not work.

Every measurable therapeutic action should thus go through a multivariable nonlinear modelization complete enough to take care of the preponderant aspects of the phenomena studied, and simple enough to be able reasonably to control those systems and deduce therapeutical actions. Because of couplings, the solutions proposed, in character falsely paradoxical, can surprise, disturb, and even arouse hostility. Nevertheless, clinical results authenticated by radiography and C.T. scans exist, and it is hoped that the two examples we will treat convince the reader of the necessity for rapid improvement in the field of bipolar therapy, in which Bernard-Weil is a pioneer.

II. THE ADRENAL-POSTPITUITARY SYSTEM AND VASOPRESSINO-CORTICOTHERAPY

Within the framework of the application of control theory to cancer chemotherapy, Sundareshan and Fundakowski [24] pose questions on the dual character of the object of therapies and try to find agents which are able to destroy malignant cells while protecting healthy cells. In fact, at the heart of the organism exists an important system which ensures the regulation of cellular growth both in cell division and in cell hydration. This is the adrenal-postpituitary hormonal system.

The adrenal-postpituitary system, which is formed on the one hand by the corticoadrenal glands and on the other hand by the neuropostpituitary gland, plays a leading part in clinical manifestations seen in neurosurgical patients. This system is responsible for certain cerebral oedemas, figures in some cases of cerebral collapse, thus endangering the possible continuation of operations for subdural haemotoma, and intervenes in the evolution of malignant cerebral tumours.

The recognition of the coupling between these two glands dates from the 30s (cf. [23]), and the system, which has agonistic-antagonistic actions (cf. [4, 6, 7]), ensures regulations of major importance. Thus cortisone, secreted by the corticoadrenal glands, is a marvellous agent, not only against the hyperhydration of cells, but also as an antimitogenic agent. This has been proved *in vitro* both in the case of malignant cerebral tumours in culture and in the case of any other cancerous cell cultures. As for vasopressin, secreted by the neuropostpituitary gland, it is responsible for the reabsorption of water by the renal tube and is an important growth factor. This first polypeptide growth factor was discovered in 1968 by Bernard-Weil, Dalage, Olivier and Piette [9] and their result has been subsequently taken up by the American authors, Rozengurt et al. [20] in 1979, and Monaco et al. [18] in 1982. We refer to Pawlikowski [19] for a recent evocation of the mitrogenic influence of neuropeptides. The imbalance between corticoids and vasopressin, with an excess of vasopressin promoting tumoral growth, has been measured again recently in the cancerology of the digestive system (cf. [11]), but it has been ascertained in many other cases. Because of coupling between these hormones, some cerebral oedemas are resistant to cortisone, and cancerous tumours are only really influenced by corticoids for a short space of time and by very high doses of these hormones. Worse still, the diseased system takes a position of "pathological homeostasis" (cf.

Bernard-Weil [4]) and this controlled imbalance benefits from potent biological safeguards which tend to maintain it as if it were in a state of "normal" physiological functioning.

In this way the vasopressin-corticoid imbalance is preserved in cancer patients with the administering of corticoids having the effect of increasing the ratio of vasopressin, which is already abnormally high (cf. [3]). The solution thus consists of considering the simultaneous administering of vasopressin and corticoids (cf. [5]), a multivariable nonlinear model supporting the first intuitions of the physician (cf. [4, 6, 7]).

This model is made up of a nonlinear differential system with two inputs, e_1 and e_2, two disturbances, p and q, and two outputs, z_1 and z_2. It can be written in the following way (cf. [15]):

$$\dot{\mathcal{X}} = \sum_{i=1}^{3} [k_i (u+p)^i + c_i(v+q)^i] + e_1$$

$$\dot{\mathcal{Y}} = \sum_{i=1}^{3} [k_i' (u+p)^i + c_i'(v+q)^i] + e_2$$

$$\dot{X} = e_1 \qquad (2.1)$$

$$\dot{Y} = e_2$$

$$z_1 = \mathcal{X} - \mathcal{Y}$$

$$z_2 = \mathcal{X} + \mathcal{Y} - m$$

with $\mathcal{X} = x+X$; $\mathcal{Y} = y+Y$, where x is an endogenous secretion of the adrenocortical hormones, y an endogenous secretion of vasopressin, and X and Y the same type of hormones as x, y, but exogenous (therapeutic).

The right-hand side of the first two equations are the development in a series in function of the antagonistic expression $u = \mathcal{X} - \mathcal{Y}$ and the agonistic expression $v = m \operatorname{Log}[1 + (\mathcal{X} + \mathcal{Y} - m)/m] + \theta(t)$, with $\theta(t) = A + B \sin(\omega t) + C \cos(\omega t)$, where the constants A, B, C and ω ($\omega = 2\pi / 24$ in the case of the circadian rhythm) bring about the synchronizer $\theta(t)$ linked to biological rhythms. The introduction of the cubic power is justified by the conditions of stability of the system (cf. [4]); p(t) represents a possible osmotic stimulus; q(t) corresponds to an eventual volemic stimulus (haemorrhage, for instance) or stress; k_i, c_i, k_i', c_i' ($i = 1,2,3$) are constant parameters; m is a generally constant parameter ($m = 0.8$) but can also be regarded as a time-variable parameter. Thus when q has positive values through a sharp increase in volemia, for instance, and in such a way that x and y become negative, the possibility is forseen to leave the value 0.8 at m during the necessary transient period.

The system is written in a system of common unit (c.u.) such that 0.4 c.u. = 77 ng/ml of plasmatic cortisol (F) = 1.1 µU/ml of plasmatic vasopressin (VP), values which correspond to the mean experimental values of circadian rhythms of these hormones. The values x, y, X, Y can be compared to hormonal concentrations and are thus liable to constraints of positivity. In the physiological case ($X = 0, Y = 0 ; p = 0, q = 0$), the equilibration is simulated with a parametric field of (2.1) giving a limit cycle such that the pair (u, v) admits the origin (0, 0) as a critical point. The equilibration ($X = 0, Y = 0$) becomes pathological if an alteration of the field (2.1) allows a new limit cycle to appear.

The parameters $k_i, c_i, k_i^{'}, c_i^{'}$ ($i = 1, 2, 3$) for the system simulating the pathology, and $\bar{k}_i, \bar{c}_i, \bar{k}_i^{'}, \bar{c}_i^{'}$ for the system simulating the physiological cycle, are identified from clinical and physiological data, by means of the Davidon-Fletcher-Powell method of numerical integration with constraints (cf. [1]). The criterion to minimize, $\mathfrak{J}(k_i, c_i, k_i^{'}, c_i^{'}, T)$, is given by:

$$\mathfrak{J}\ (k_i, c_i, k_i^{'}, c_i^{'}, T) = \sum_j \left[(\bar{x}_j - x_j)^2 + (\bar{y}_j - y_j)^2 \right] \qquad (2.2)$$

where $\bar{x}$ et $\bar{y}$ designate experimental values and x and y the solutions of system (2.1) with $X = 0, Y = 0$, $p = 0, q = 0$. The quantity T corresponds to three cycles, here equal to 72 hours.

In the pathological case, the "therapeutical simulation" consists of determining exogenous hormones X et Y so as to bring the system back to a physiological situation. One method (cf. [6, 7]) consists of writing the inputs e_1 and e_2 in a form similar to that of the endogenous hormones, i.e.:

$$\begin{aligned} e_1 &= \sum_{i=1}^{3} \left[k_{3+i}\ (u + p)^i + c_{3+i}\ (v + q)^i \right] + \sum_{i=1}^{3} \lambda_i\ (X - \alpha_1)^i \\ e_2 &= \sum_{i=1}^{3} \left[k_{3+i}^{'}\ (u + p)^i + c_{3+i}^{'}\ (v + q)^i \right] + \sum_{i=1}^{3} \lambda_i^{'}\ (X - \alpha_1^{'})^i \end{aligned} \qquad (2.3)$$

with $\lambda_1, \lambda_2, \lambda_3, \lambda_1^{'}, \lambda_2^{'}, \lambda_3^{'}, \alpha_1, \alpha_1^{'}$ constant parameters with the role of avoiding the drift of the limit cycle of dimension 4 which the four states of the system follows. The parameters of relations (2.3) are then identified by means of the Davidon-Fletcher-Powell method.

Remark 1. For inputs e_1 and e_2, the temptation to take the difference between the physiological state equations and the pathological state equations leads to a control which could neither satisfy the conditions of positivity of the variables x, y, X, Y, nor ensure the existence of a limit cycle (cf. [6]).

A second method, based firstly (cf. [15]) on the decoupling and linearization of nonlinear systems (cf. [12, 13, 14] and references therein), consists in fact of inversing the system (2.1) (cf. [16, 17]). From system (2.1), the following relations are then considered:

$$\begin{aligned} H &= 1/2\,(z_1 + z_2 + m) \\ V &= 1/2\,(- z_1 + z_2 + m) \\ X &= H - x \\ Y &= V - y \end{aligned} \tag{2.4}$$

x et y being the solutions of differential equations:

$$\begin{aligned} \dot{x} &= \sum_{i=1}^{3} [\, k_i\,(z_1 + p)^i + c_i\,(m \operatorname{Log}(1 + z_2/m) + \theta + q)^i \,] \\ \dot{y} &= \sum_{i=1}^{3} [\, k_i'\,(z_1 + p)^i + c_i'\,(m \operatorname{Log}(1 + z_2/m) + \theta + q)^i \,] \end{aligned} \tag{2.5}$$

It is a matter of allowing the outputs z_1 and z_2 of system (2.1) to pass from the pathological state, given by differential equations:

$$\begin{aligned} \dot{\psi}_1 &= \sum_{i=1}^{3} [\, (k_i - k_i')(\psi_1 + p)^i + (c_i - c_i')(v + q)^i \,] \\ \dot{\psi}_2 &= \sum_{i=1}^{3} [\, (k_i + k_i')(\psi_1 + p)^i + (c_i + c_i')(v + q)^i \,]) \end{aligned} \tag{2.6}$$

with $v = m \operatorname{Log}(1 + \psi_2/m) + A + B \sin(\omega t) + C \cos(\omega t)$ and $\omega = 2\pi / 24$, to the physiological balance described by the differential equations obtained from experimental data:

$$\begin{aligned} \dot{\varphi}_1 &= \sum_{i=1}^{3} [\, (\bar{k}_i - \bar{k}_i')\, \varphi_1^i + (\bar{c}_i - \bar{c}_i'\, v^i \,] \\ \dot{\varphi}_2 &= \sum_{i=1}^{3} [\, (\bar{k}_i + \bar{k}_i')\, \varphi_1^i + (\bar{c}_i + \bar{c}_i')\, v^i \,]) \end{aligned} \tag{2.7}$$

with $v = m \operatorname{Log}(1 + \varphi_2/m) + A + B \sin(\omega t) + C \cos(\omega t)$ and $\omega = 2\pi / 24$. We write z_1 and z_2 in the form

$$z_1 = \delta_1 + \varphi_1$$

$$z_2 = \delta_2 + \varphi_2 \qquad (2.8)$$

The wish of the therapist is then to find functions δ_1 and δ_2 which first permit the definition of a transient way leading initial pathological curves, depicted by x and y, towards physiological curves which the variables $\mathbf{H}$ and $\mathbf{Y}$, sum of the actions of endogenous and exogenous hormones, should follow. Secondly, after the transient period (two or three days), the therapist wishes, at one and the same time, to see installed a permanent regime as close as possible to the physiological circadian rhythm for variables $\mathbf{H}$ and $\mathbf{Y}$, and to realize, for numerous reasons easy to guess, a periodic therapeutic action - depicted by X and Y - of a period equal here to 24 hours.

Nevertheless, the immediate analysis of equations (2.5) shows that with pathological coefficients, there is no reason that the insertion of physiological rhythms into these equations would lead to the appearance of a limit cycle. On the contrary, as numeric solutions confirm, we note an affine drift of the cycle and the demonstration of this phenomenon is evident.

Thus the only possibility, in a permanent regime, is to deform the physiological rhythm as little as possible to ensure the periodicity of the therapy presented by X et Y, the functions δ_1 and δ_2 also being periodic. This leads us to make an optimization under the following constraints: $x \geq 0$, $y \geq 0$, $X \geq 0$, $Y \geq 0$. Finally, we need to ensure that the limit cycle obtained is stable and that in addition the system is structurally stable.

It should be observed that the use of optimization is judicious in comparison to the notion of mean physiological rhythm which is used, and also in relation to the uncertainty brought about by modelization. Thus the conditions of positivity of x, y, X, Y variables can be satisfied by the functions of δ_1 and δ_2 and, after a transient period, these functions should ensure the existence of a permanent cyclic regime founded on the circadian rhythm. It is then a matter of finding a class of functions wide enough to include the solutions sought. The 4-parametric family of functions given by

$$f(x) = \int_0^{x+a} \frac{bd}{1 + d^2 - \cos(bt)} \cos(e^t)\, dt + c \quad \text{with } d > 0 \qquad (2.9)$$

is dense in the set of continuous functions on any compact interval of **R** and gives an idea of the minimal number of necessary parameters. This class of functions is used by Boshernitzan [2] in the search for universal differential equations (cf. [2, 22]).

Remark 2. System (2.8) always admits at least the mathematical solution $\delta_1 = \psi_1 - \varphi_1$, $\delta_2 = \psi_2 - \varphi_2$, but this solution can of course in no way be the therapeutical solution!

For the present, we envisage a nonlinear control in the following form:

$$\dot{z}_1 = \sum_{i=1}^{3} [\, k_{3+i} (z_1 + p)^i + c_{3+i} (m \operatorname{Log} (1 + z_2/m) + \theta + q)^i \,] + \sum_{i=1}^{3} \lambda_i (z_1 - \alpha_1)^i$$

$$(2.10)$$

$$\dot{z}_2 = \sum_{i=1}^{3} [\, k'_{3+i} (z_1 + p)^i + c'_{3+i} (m \operatorname{Log} (1 + z_2/m) + \theta + q)^i \,] + \sum_{i=1}^{3} \lambda'_i (z_2 - \alpha'_1)^i$$

The parameters of equations (2.10) are then determined by the minimization of the gap between the solutions of equations (2.7) and (2.10).

This second method is under study.

Remark 3. We may be uneasy about the impossibility of finding a therapeutic control capable of restoring physiological rhythms, but we should not forget that in reality the parameters which cause the behaviour of the system are variables. If they have gone from the physiological state to the pathological situation, the therapist postulates in the case of reversibility that a forced maintaining of a rhythm close to the physiological rhythm, during an adequate period, will allow the parameters to readjust onto the physiological homeostasis.

III. THE INSULIN-GLUCAGON COUPLING AND DIABETES

Glycemic activity can be considered as resulting from antagonistic action of glucagon which increases glycemia and of insulin which reduces glycemia, these two hormones acting in a coupled way. In relation to the adrenal-postpituitary system, this system presents a remarkable particularity from the anatomical point of view. In the case of the response of glycemia, Nature has installed the control mechanism in the same place - the islets of Langerhans - within the pancreas. In these cellular clusters is found simultaneously the manufacture of insulin and glucagon under the coordinated effect of somatostatin. With clinical results obtained by vasopressino-corticotherapy, it seemed interesting, considering the stakes in diabetology, to propose a modelization of the insulin-glucagon systems from the bipolar point of view of agonistic-antagonistic systems defined by Bernard-Weil.

The modelization proposed takes the form of a nonlinear differential system with three inputs e_1, e_2

and p, and three outputs, z_1, z_2 et z_3, defined as follows (cf. [8]):

$$\dot{\mathcal{X}} = \sum_{i=1}^{3} [\, k_i(\mathcal{X} - \mathcal{Y} + p))^i + c_i(\mathcal{X} + \mathcal{Y} - m)^i \,] + e_1$$

$$\dot{\mathcal{Y}} = \sum_{i=1}^{3} [\, k_i'(\mathcal{X} - \mathcal{Y} + p)^i + c_i'(\mathcal{X} + \mathcal{Y} - m)^i \,] + e_2$$

$$\dot{X} = e_1$$

$$\dot{Y} = e_2 \qquad (3.1)$$

$$\dot{G} = g_1(G_0 - G) + g_2[g_3[\mathrm{th}(g_4(\mathcal{X} - \mathcal{Y} + Y - X + g_5)) + \mathrm{th}(g_4(X - Y + g_5)) - 2\mathrm{th}(g_4 g_5)] + p]$$

$$z_1 = \mathcal{X} - \mathcal{Y}$$

$$z_2 = \mathcal{X} + \mathcal{Y} - m$$

$$z_3 = G$$

with $\mathcal{X} = x + X$, $\mathcal{Y} = y + Y$, x and y the endogenous actions of glucagon and insulin respectively and X and Y the actions of the exogenous hormones (therapeutic). $G_0 = 0.78$, indicates the physiological load of G(t) glycemia and m = 2.1. A common unit is given by: :10 μU/ml of insulin = 100 pg/ml of glucagon.

In the case of the study of the oral glucose tolerance test, the input p(t) which is linked with the oral absorption of 100g of glucose, is given by the function:

$$p(t) = (\, p_1 / (p_1 - p_2)\,) \,.\, 100 \,.\, p_3 \,[\, \exp(-p_2 t) - \exp(-p_1 t) \,] \qquad (3.2)$$

In the same way as for the model of the adrenal-postpituitary system, the parameters of equations (3.1) and (3.2) were identified by means of the Davidon-Fletcher-Powell method of nonlinear optimization from experimental curves. The parameters defining the function p(t) were adjusted once only because the conditions of intestinal absorption of glucose are less influenced by hormonal anomalies than other processes of glucose metabolism. On the other hand, the parameters of equation (3.1) are to be identified in both the physiological and the pathological cases. The search for control (therapeutic) which aims at correcting the anomalies of glycemia in diabetics was obtained at first (cf.[8]) by taking the inputs e_1 and e_2 in the form:

$$e_1 = \sum_{i=1}^{3} \left[k_{3+i} (\mathbf{X} - \mathbf{Y} + p)^i + c_{3+i} (\mathbf{X} + \mathbf{Y} - m)^i \right]$$

$$e_2 = \sum_{i=1}^{3} \left[k'_{3+i} (\mathbf{X} - \mathbf{Y} + p)^i + c'_{3+i} (\mathbf{X} + \mathbf{Y} - m)^i \right] \qquad (3.3)$$

They allow the setting in place of an asymptotical control which tends to bring the pathological limit position back to the physiological mean value of glycemia (1 g/l). The initial glucagon-insulin imbalance before the dose in glucose, are, like the physiological balance, stable critical points of the physiological and pathological models. We can also operate as for the adrenal-postpituitary system and consider the following relations:

$$\mathbf{X} = 1/2\,(z_1 + z_2 + m)$$

$$\mathbf{Y} = 1/2\,(- z_1 + z_2 + m)$$

$$X = \mathbf{X} - x \qquad (3.4)$$

$$Y = \mathbf{Y} - y$$

x and y being solutions of the differential equations:

$$\dot{x} = \sum_{i=1}^{3} \left[k_i (z_1 + p)^i + c_i (z_2)^i \right]$$

$$\dot{y} = \sum_{i=1}^{3} \left[k'_i (z_1 + p)^i + c'_i (z_2)^i \right] \qquad (3.5)$$

Our concern here is to allow outputs z_1, z_2 and z_3 of system (3.1) to pass from the pathological position:

$$z_1(0)\,;\, z_2(0)\,;\, G(0) \qquad (3.6)$$

to the asymptotical physiological balance:

$$z_1 = 0\,;\, z_2 = 0\,;\, G = 1 \qquad (3.7)$$

The physiological balance should of course be attained before the next ingestion, that is to say, within a period of about 5 hours.

To determine the "therapy" - X, Y - to be applied to the "pathological" system (3.1) we can, for example, use the same method given by formulas (2.10) and execute an optimization under constraints $x \geq 0$, $y \geq 0$, $X \geq 0$, $Y \geq 0$, by minimizing the gap between the three inputs z_1, z_2 and z_3 of the controlled "pathological" system (3.1) and the three outputs φ_1, φ_2 and φ_3 of the "physiological" system (3.1) bound to the inputs $e_1 = 0$, $e_2 = 0$ et p(t). This will be the object of our next study, but the simulations executed with the inputs e_1 and e_2 in the form (3.3) (cf. [8]) already show that a better approximation of the glycemia curve is obtained with the simultaneous intervention of two actions X et Y (insulin and glucagon) rather than with the insulin action alone.

IV. CONCLUSION

We have presented, and illustrated with two examples, a new method of research which links control theory to biology. This method, of which Bernard-Weil is the initiator, opens a vast field of investigation and allows us, by way of an original modelization which is connected to Rosen's dynamical metaphors [21], to take into account the agonistic-antagonistic aspect occurring in many biological regulations. This modelization, which is able to simulate both the pathology and the physiology, proposes bipolar controls with therapeutical repercussions which are at times surprising. There is no question that the specialist in control theory provide precise medical details, as he does not have the ability to do this, but he can all the same indicate (as a number of medical publications (cf. [5, 10, 11]) have already done) that the practice of bipolar therapy enlarges its field of application step by step. There is no doubt that at some future date, which we should try to make as near as possible, this therapy will lead to the alleviation of suffering of many human beings.

REFERENCES

[1] M.S. BAZARAA and C.M. SHETTY, Nonlinear Programming, Theory and Algorithms, Wiley, New York, 1979.

[2] M. BOSHERNITZAN, Universal formulae and universal differential equations, Annals of Mathematics, 124, 1986, pp. 273-291.

[3] E. BERNARD-WEIL, Effects of a week of ACTH or corticosteroid treatment on the neuropostpituitary response to corticosteroid load, Steroids Lip. Res., 3 , 1972, pp.24-29.

[4] E. BERNARD-WEIL, Formalisation et contrôle du système endocrinien surréno-posthypophysaire par le modèle mathématique de la régulation des couples ago-antagonistes, Thèse d'Etat, Universté Paris VI, France, 1979.

[5] E. BERNARD-WEIL, Lack of response to a drug: a system theory approach, Kybernetes, 14, 1985,pp. 25-30.

[6] E. BERNARD-WEIL, Interactions entre les modèles empirique et mathématique dans la vasopressino-corticothérapie de certaines affections cancéreuses dans "Régulations physiologiques : Modèles récents", G. Chauvet et J.A. Jacquez, éd., Masson, Paris, 1986, pp. 133-155.

[7] E. BERNARD-WEIL, A general model for the simulation of balance, imbalance and control by agonistic-antagonistic biological couples, Mathem. Modelling, 7, 1986, pp. 1587-1600.

[8] E. BERNARD-WEIL and D. CLAUDE, Simulation du test de tolérance au glucose par le modèle de la régulation des couples ago-antagonistes. Contrôle bipolaire, C.R. Acad. Sci. Paris, 305, série I, 1987, pp. 303-306.

[9] E. BERNARD-WEIL, C. DALAGE, C. PIETTE and L. OLIVIER, Action of lysine-vasopressin on the protein content of Hela cell cultures and on the RNA and DNA concentrations of tissueincubation, Experta Med. Internat. Congr. Ser. : Protein and Polypeptide Hormones, 161, 1968, pp. 547-548.

[10] E. BERNARD-WEIL and B. PERTUISET, Mathematical model for hormonal therapy (vasopressin, corticoids) in cerebral collapse and malignant tumors of the brain (36 cases), Neurol. Res., 5, 1983, pp. 19-35.

[11] E. BERNARD-WEIL, J.L. JOST and P. VAYRE, Nouvel aspect des relations hôte-tumeur. Etude du système surréno-post-hypophysaire chez le cancéreux digestif, Chirurgie, 113, 1987, pp. 293-298.

[12] D. CLAUDE, Découplage des systèmes non linéaires, séries génératrices non commutatives et algèbres de Lie, SIAM J. Control Optimiz., 24, 1986, pp. 562-578.

[13] D. CLAUDE, Everything you always wanted to know about linearization but were afraid to ask, in "Algebraic and geometric methods in nonlinear control theory", M. Fliess and M. Hazewinkel, ed., D. Reidel Publishing Company, 1986, pp. 181-226.

[14] D. CLAUDE, Découplage et linéarisation des systèmes non linéaires par bouclages statiques, Thèse d'Etat, Université Paris-Sud, France, 1986.

[15] D. CLAUDE and E. BERNARD-WEIL, Découplage et immersion d'un modèle neuro-endocrinien, C.R. Acad. Sci. Paris, 299, série I, 1984, pp.129-132.

[16] M. FLIESS, Automatique et corps différentiels, Forum mathématiques, 1, 1989.

[17] U. KOTTA, Application of inverse system for linearization and decoupling, Systems Control Lett., 8, 1987, pp. 453-457.

[18] M. MONACO, P.H. KOHN, W.R. KIDWELL, J.S. STROBL and M.E. LIPPMAN, Vasopressin action on WRK-1 rat mammory tumor cells, J.N.C.I., 68, 1982, pp. 267-270.

[19] M. PAWLIKOWSKI, The effect of neuropeptides on cellular proliferation, Materia Medica Polona,Fasc. 1, 61, 1987, pp. 17-20.

[20] E. ROZENGURT, A. LEGG and P. PETTICAN, Vasopressin stimulation of mouse 3T3 cell growth, Proc. Natl. Acad. Sci. USA, 76, 1979, pp. 1284-1287.

[21] R. ROSEN, Dynamical System Theory in Biology, Wiley-Interscience, New York, 1970.

[22] A. RUBEL, A universal differential equation, Bull. Amer. Math. Soc. (N.S.), 4, 1981, pp. 345-349.

[23] A. SILVETTE and S. BRITTON, A theory of corticoadrenal and postpituitary influences on the kidney, Science, 88, 1938, pp. 150-151.

[24] M.K. SUNDARESHAN and R.A. FUNDAKOWSKI, Stability and control of a class of compartmental systems with application to cell proliferation and cancer therapy, IEEE Trans. Automat. Contr., AC-31, 1986, pp. 1022-1032.

[25] G.W. SWAN, Application of Optimal Control Theory in Biomedicine, Dekker, New York, 1984.

[26] A.T. WINFREE, The Geometry of Biological Time, Springer-Verlag, New York, 1980.

COMPUTER MODELS APPLIED TO CANCER RESEARCH

Werner Düchting
Department of Electrical Engineering
University of Siegen
Hölderlinstr. 3, D-5900 Siegen, West Germany

ABSTRACT: The aim of this contribution is to illustrate the impact of computer simulation in the field of biology and medicine. This paper shows how systems analysis, control theory and computer science can stimulate new approaches to interpret cancer, to predict tumor growth and to optimize tumor treatment.

Starting with a review of the current biological knowledge about the origin of cancer a computer model is constructed
- to simulate the time behaviour of disturbed cell growth control circuits
- to predict spatial tumor growth (2-D, 3-D) and
- to simulate different kinds of cancer treatment (surgery, radiation- and chemotherapy).

In the long run the aim of our work is to optimize treatment strategies and schedules in vitro and in vivo by computer simulation prior to clinical therapy.

1. BIOLOGICAL BACKGROUND OF THE CANCER PROBLEM

Cancer is a multistep process with the stages of initiation, promotion and progression. Characteristic features of malignant tumors (1) are uncontrolled proliferation, invasion in adjacent normal tissue, metastases induced to other tissues via lymphatic channels and the ability to evade immune surveillance. Though cancer treatment is concentrated on a prevention of metastases (2) the central question in the background of research is: Which is the initiating event that is responsible for a stepwise transformation of a normal cell into a tumor cell? Recent investigations in the field of molecular biology have focussed on dominant cellular genes called "proto-oncogenes" which can be activated by tumor viruses, gene amplification, gene translocation and genetic mutation. In spite of this progress (3) the main question how genes and the growth of normal and malignant cells are regulated still remains open.

Most of the normal tissues in the body contain some cells that can renew themselves (neurons, liver cells, kidney cells) if a tissue is injured. The division of a cell into two new ones involves four stages: G1 $\rightarrow$ S $\rightarrow$ G2 $\rightarrow$ M (G1 is a gap after stimulation; S is the phase of DNA replication; G2 is a second gap period and M is the stage of mitosis). When the replacement has been completed the repair process stops. Furthermore, at particular stages of the cell cycle the cells may be blocked by drugs or agents, or they may move out of the cell cycle into a resting phase known as G0 (4).

In contrast to the normal cell a tumor cell is theoretically able to divide indefinitely. In addition a different morphology, larger nucleus, abnormal number of chromosomes and the formation of new capillaries (tumor angiogenesis) which is associated with a more rapidly growing tumor (5) can be noticed.

For studying the process of carcinogenesis tumors are induced to animals or to cell cultures (in vitro). Cell cultures are not only used to study the division of tumor cells, but also to determine the effect of chemotherapeutic drugs. During the last years a large progress has been made in experiments gaining hard data about normal and abnormal cell-growth control processes for instance of cell-cycle phase durations.

2. MODELING APPROACHES

Starting from basic biological test results a large body of mathematically oriented work applying mathematics to the field of biology and medicine has been published (6-10). Unfortunately these models which consist of complicated formulae, are in most cases not completely understood by clinicians. In this dilemma the combined application of methods of systems analysis, control theory, automata theory, computer sciences and heuristics is a good link between the diverging areas of medicine and mathematics.

Our own approach developing closed-loop control circuits for tumor growth started in 1968 (11). At that time the subject of consideration was focussed on stability conditions and on the interpretation of cancer as an unstable closed-loop control circuit. Step by step the dynamic behaviour of cell renewal control loops (Fig. 1) was investigated. Blockoriented simulation languages have been used for simulating the macromodels. As a result the number of cells as a function of time has been plotted (12).

Then oncologists advised us to consider not only the time but also the spatial behaviour of tumor growth. In a first approach we developed models at a cellular level which described the 2-D behaviour of a normal cell inoculated into a nutrient medium (in a Petri dish). Next we extended this approach and tried to simulate tumor growth in the tissue of a tobacco leaf (13).

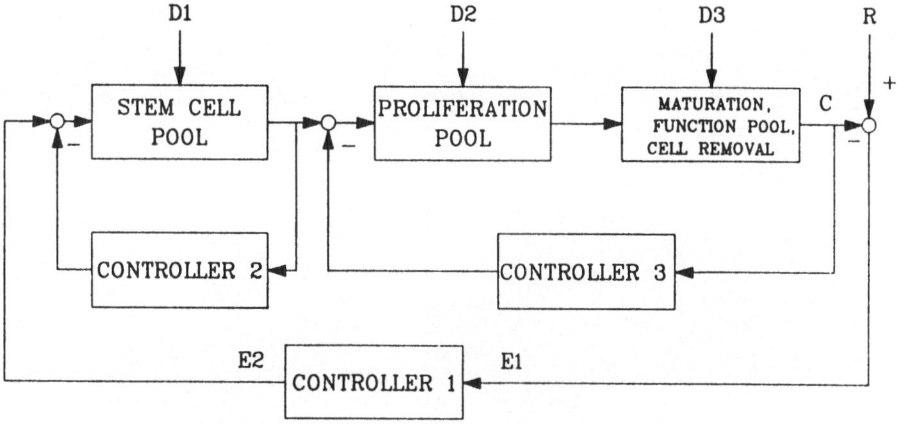

R: Required tissue oxygen (desired number of erythrocytes)

C: Number of red blood cells (erythrocytes)

E2: Production of the erythropoietin hormone

D1, D2, D3: Disturbance

Fig. 1: Multi-loop control circuit of erythropoiesis

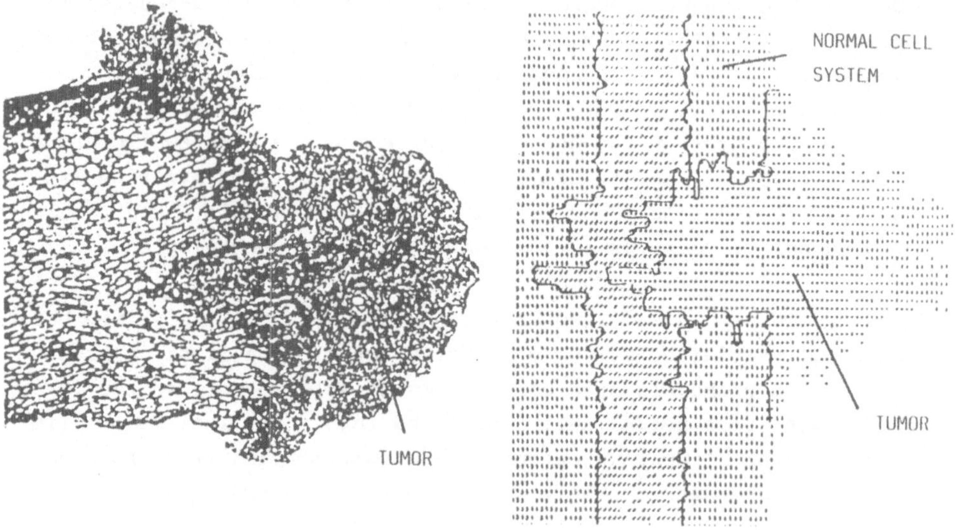

Fig. 2: Simulation of tumor formation in the tissue of a tobacco leaf

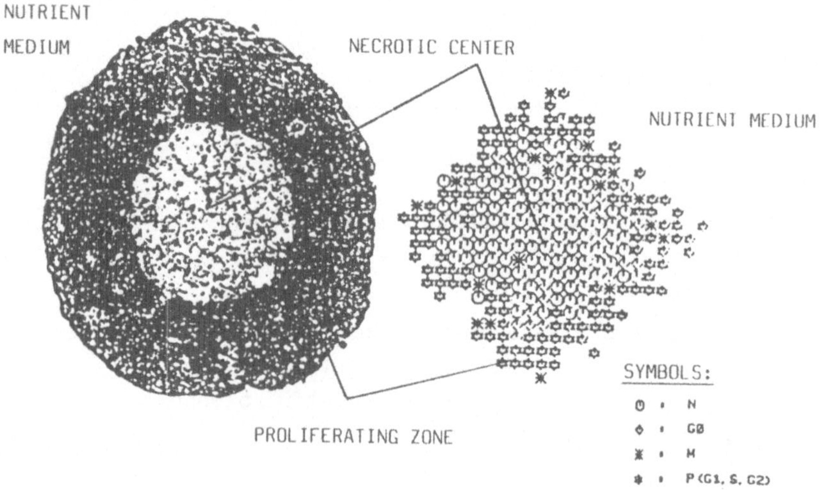

Fig. 3: Simulation of the formation of a tumor spheroid. The initial configuration consisted of a single tumor cell placed in the center of the nutrient medium

After getting the results shown in Figure 2 we improved these models by introducing distinguished cell cycle phases (G1, S, G2, M, GO, N). Thus, we were able to simulate the 3-D growth of a single dividing tumor cell (14) inoculated into the center of the cell space of a nutrient medium at the beginning of the simulation run (Fig. 3).

The introduction of distinguished cell-cycle phases was necessary because chemotherapeutic agents and rays effect only a very particular phase of the cell cycle that means they act phase - specifically.

After simulating in vitro tumor growth the attempt was made to substitute the nutrient medium by static blood vessels (15). However, very soon it was clear that a more realistic structure of capillaries was desirable for simulating in vivo tumor growth.

3. DESIGN STRATEGIES OF A HEURISTIC MODEL

The modeling of complex cell growth requires a considerable simplification. Some of the oversimplifying assumptions are

- constant volume of a cubic cell
- constant phase duration and constant cell loss
- only horizontal and vertical communication between neighboring cells
- a limited tissue volume by computer facilities
- side effects, immunologic reactions, heterogenity, drug resistance and the formation of metastases are neglected.

If you want to construct a model of high order, it is necessary to design a modular concept. In this case it means to design modular structured subsystems.

(i) You need control models (Fig. 4) which describe the cell division of normal and tumor cells at a cellular level including experimentally gained data e.g. of cell-cycle phase durations.

(ii) Heuristic cell-production and interaction rules are required describing the cell-to-cell communication. For instance one rule of the catalogue may say:
All tumor cells residing at a distance larger than 100 μm from the capillaries after the next division step will enter the resting phase G0.

(iii) Cell movement is described by transport equations (diffusion-, Poisson-equation), that means we have to introduce into the model gradients for pressure and metabolic compounds.

(iv) To represent 2-D and 3-D simulation results computer-graphics software packages are necessary.

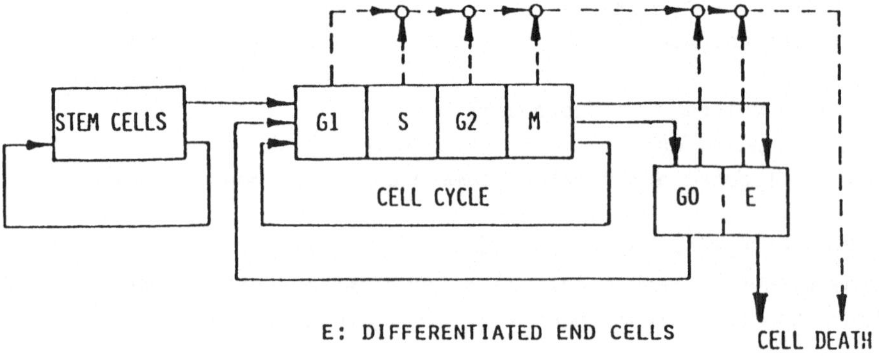

Fig. 4: Simplified cytokinetic model describing the division of a normal cell

The large body of statements, rules and equations has been transformed into algorithms. In addition algorithms considering tumor treatment (surgery, radiation- and chemotherapy) have been developed in subprograms written in FORTRAN IV. To start the simulation program packages the following input data have to be fed into the computer (VAX 730):

- notations about the character of a cell (normal, malignant)
- cell-cycle phase durations
- cell-loss rates
- initial configuration of normal tissue and of tumor cells
- distinguished data about the kind of the planned tumor treatment.

4. SELECTED SIMULATION RESULTS

Numerous simulation runs have been performed by Düchting and Vogelsaenger (15-17) simulating tumor growth and different kinds of treatment. Some special results will be demonstrated now.

4.1 Growth of capillaries

The simulation of in-vivo tumor growth requires a realistic structure of capillaries. Therefore Vogelsaenger (16) investigated the question: Is the formation of capillaries a stochastic or a regulated process? In (16) the assumption is made that each cell of an organ in evolution has a special request for oxygen and glucose. Therefore, parallel to the formation of tissue capillaries are built with a specific structure corresponding to the required oxygen and glucose. That means from the viewpoint of control theory the request for oxygen supply is regulated to a constant level by building a special structure of capillaries. A comparison between Figure 5 and Figure 6 shows that for the cortex of a rat the simulation result is highly similar to the experimental result received by Bär (18).

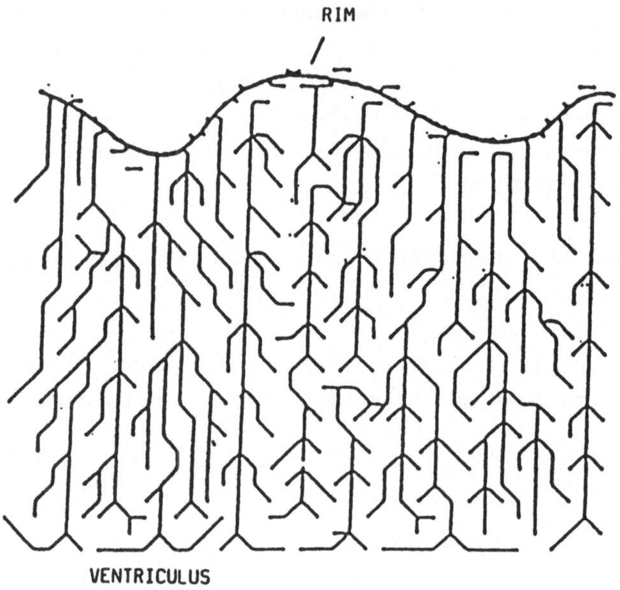

Fig. 5: Capillary network in the cortex (simulation result)

4.2 Spread of tumor cells in the cortex

Now the assumption is made that a single tumor cell is arbitrarily placed in the tissue of the cortex at T=1 unit of time. If this tumor cell resides close to a capillary it will divide and move in accordance with the cell production rules (Fig. 7). Further tumor growth is possible only because tumor cells produce a substance which is called tumor-angiogenesis factor (TAF). This factor stimulates nearby blood vessels to send out new capillaries (Fig. 8) which grow towards the tumor, penetrate it and lead to further rapid tumor growth. Recently great efforts have been made to attack cancer by trying to find a protein which inhibits the production of the tumor-angiogenesis factor.

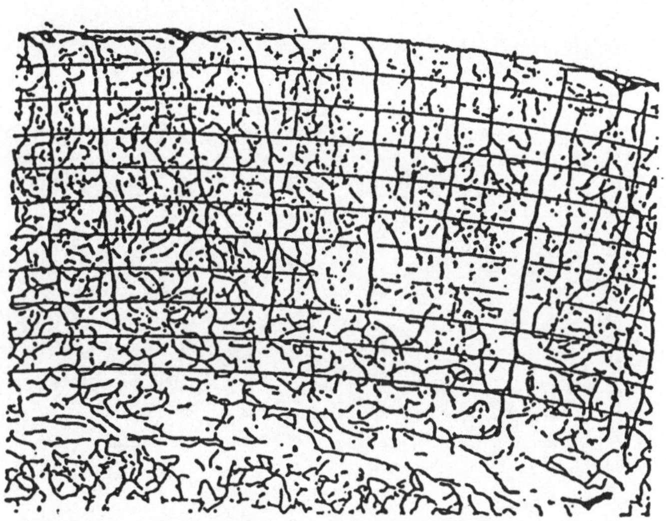

Fig. 6: Vascularization of the cortex (18)

4.3 Chemotherapeutic treatment in vitro

As pointed out in section 1, the cytotoxic effect of chemotherapeutic drugs is tested in cell cultures. These are very good in-vitro systems which can be simulated by a computer model. Figure 9(a) shows a tumor spheroid at T=200 units of time which has grown up from a single tumor cell inoculated into the center of the cell space at T=1 unit of time.

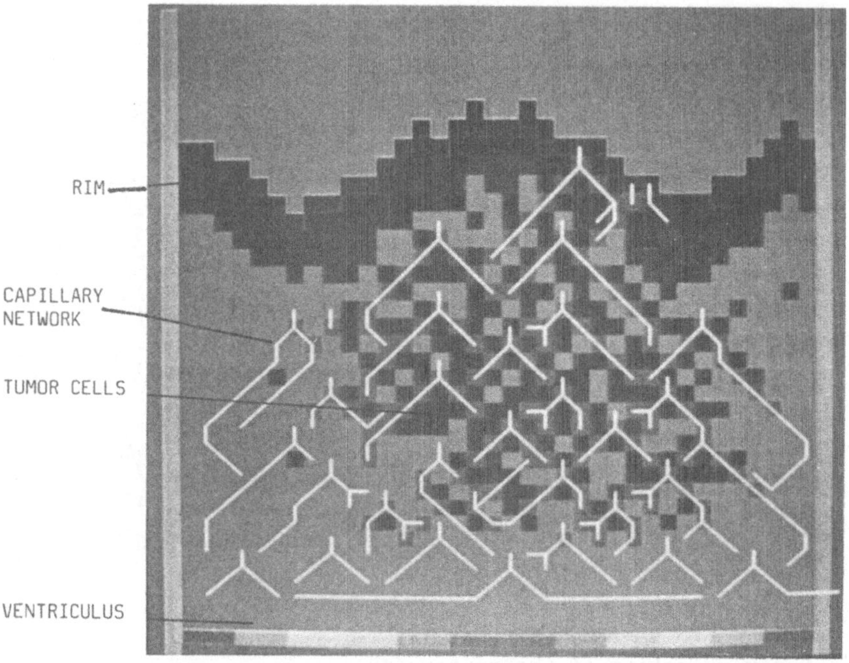

Fig. 7: Spread of tumor cells in the cortex at T=45 units of time

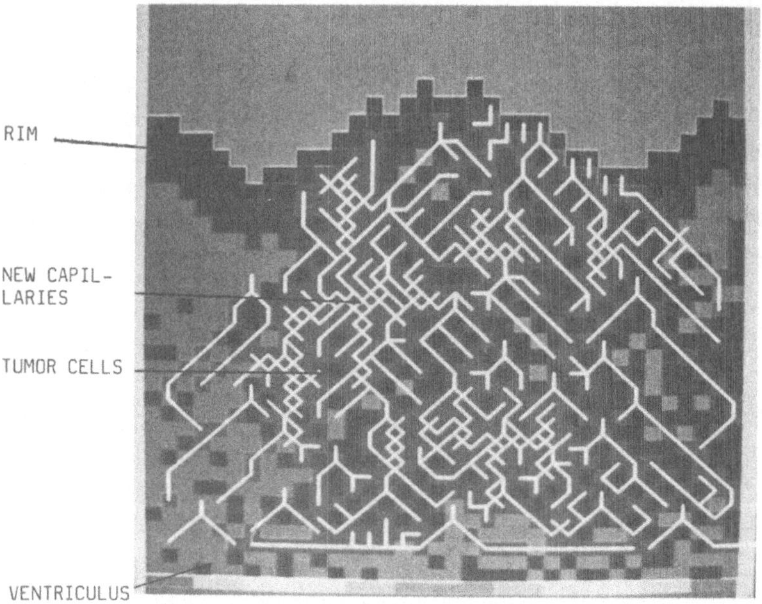

Fig. 8: Formation of new capillaries at T=120 units of time (tumor-angiogenesis effect)

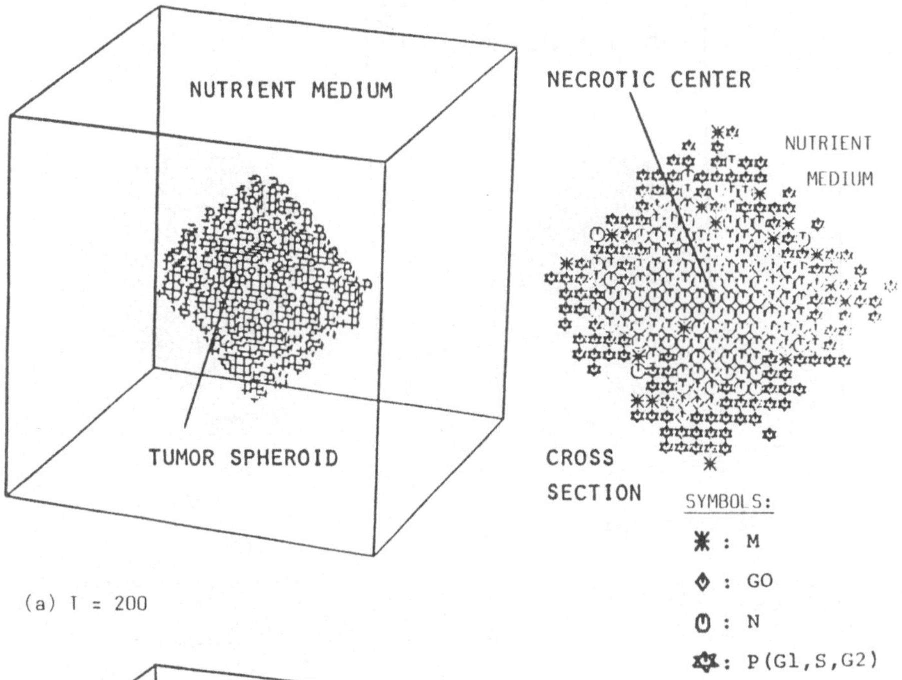
NUTRIENT MEDIUM
TUMOR SPHEROID
NECROTIC CENTER
NUTRIENT
MEDIUM
CROSS
SECTION
SYMBOLS:
: M
: GO
: N
: P(G1,S,G2)

(a) T = 200

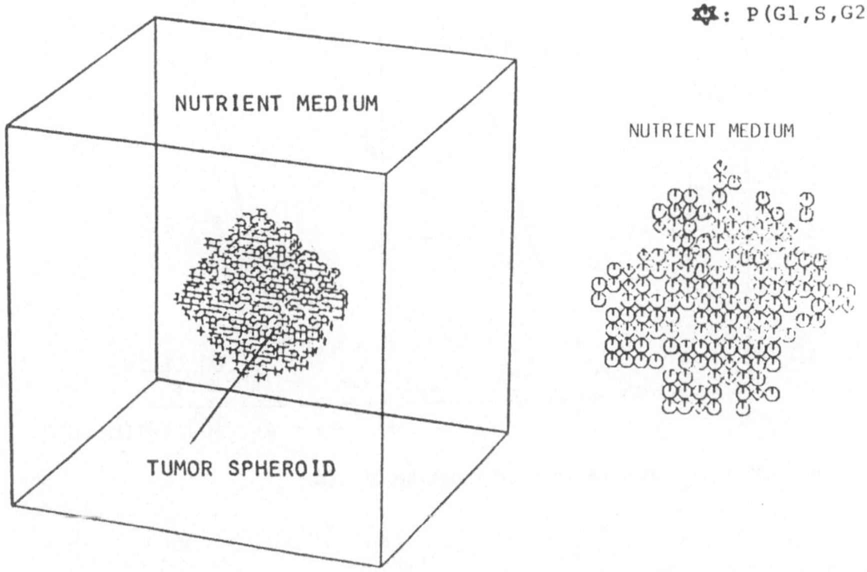
NUTRIENT MEDIUM
TUMOR SPHEROID
NUTRIENT MEDIUM

(b) T = 201

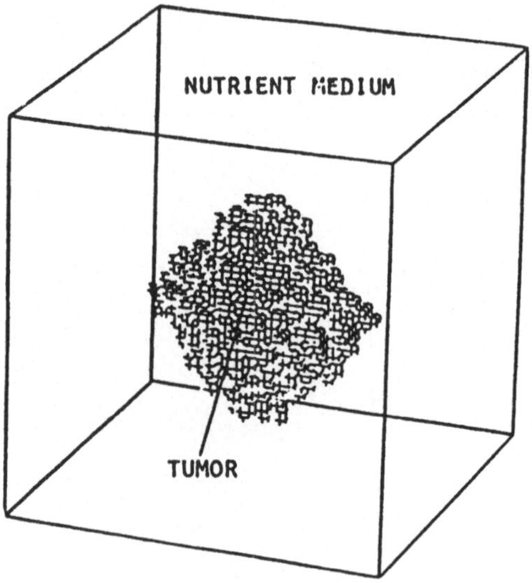

(c) T = 300

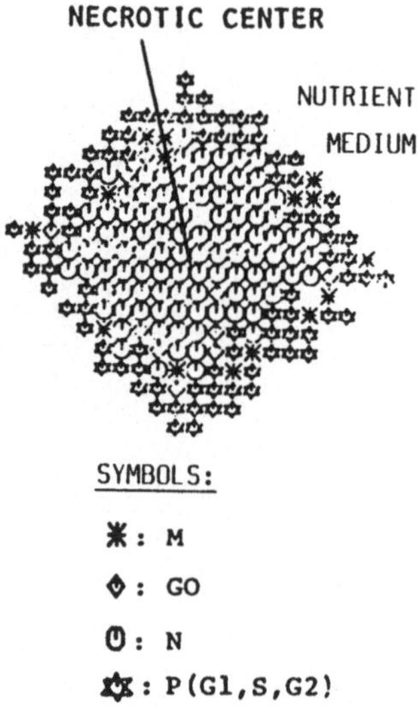

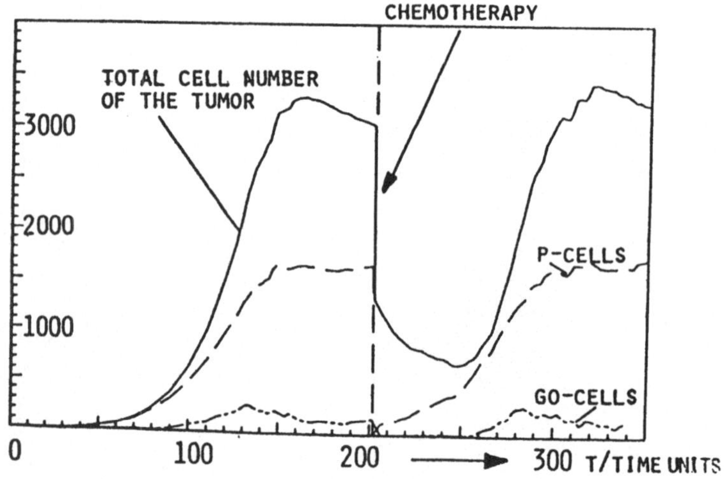

(d) Number of tumor cells as a function of time

Fig. 9 (a)-(d): Simulation of a chemotherapeutic treatment of a tumor spheroid (in vitro)

At T=201 units of time it is assumed that all proliferating tumor cells (i.e. the outside rim) have been killed by a cytotoxic drug (Fig. 9(b)). Now the remaining resting tumor cells (GO-phase) in the neighborhood of the nutrient medium are being recruited into the cell cycle again, and after a short time of remission the tumor spheroid continues to grow (Fig. 9(c)-(d)). Therefore, a second therapeutic attack or a combined approach is recommended. The task which has been solved in (15) is to determine the optimum time at which the drug has to be applied for a second (and more) time(s).

5. FUTURE PROSPECTS

From the voluminous catalogue of unsolved problems in the area of cancer research I think there are three promising avenues of future work in the modeling field:

- Optimization of distinguished methods and schedules of cancer treatment.
- Generation of a more realistic initial configuration of a tumor by combining CT-pictures (Computer Tomography) with predictive models describing tumor growth and last not least
- Consideration of facts which had to be neglected so far (formation of metastases, immunologic reactions, drug resistance, heterogenity, side effects).

6. REFERENCES

(1) Tannock, I.F. and Hill, R.P. (eds.), The Basic Science of Oncology, Pergamon Press, New York 1987.

(2) Sherbet, G.V., The Metastatic Spread of Cancer, MacMillan Press, London 1987.

(3) Poste, G. and Crooke, St.T., New Frontiers in the Study of Gene Functions, Plenum Press, New York 1987.

(4) Baserga, R., Molecular Biology of the Cell Cycle, Int. J. Radiat. Biol., Vol. 49, No. 2 (1986): 219-226.

(5) Folkman, J. and Klagsbrun, M., Angiogenic Factors, Science, Vol. 235 (1987): 442-447.

(6) Cherruault, Y., Mathematical Modelling in Biomedicine, D. Reidel Publishing Company, Dordrecht 1986.

(7) Segel, L., Modeling Dynamic Phenomena in Molecular and Cellular Biology, Cambridge University Press, Cambridge 1984.

(8) Swan, G.W., Applications of Optimal Control Theory in Biomedicine, Marcel Dekker Inc., New York 1984.

(9) Wolfram, St., Cellular Automata as Models of Complexity, Nature, Vol. 311, No. 5985 (1984): 419-424.

(10) Meinhard, H., Models of Biological Pattern Formation, Academic Press, London 1982.

(11) Düchting, W., Krebs, ein instabiler Regelkreis, Versuch einer Systemanalyse, Kybernetik, 5. Band, 2. Heft (1968): 70-77.

(12) Düchting, W., Computer Simulation of Abnormal Erythropiesis - an Example of Cell Renewal Regulating Systems, Biomed. Techn. 21 (1976): 34-43.

(13) Düchting, W. and Dehl, G., Spatial Structure of Tumor Growth: A Simulation Study, IEEE Transactions on Systems, Man and Cybernetics SMC-10, No. 6(1980): 292-296.

(14) Düchting, W. and Vogelsaenger, Th., Three-Dimensional Pattern Generation applied to Spheroidal Tumor Growth in a Nutrient Medium, Int. J. Bio-Medical Computing 12(1981): 377-392.

(15) Düchting, W. and Vogelsaenger, Th., Aspects of Modelling and Simulating Tumor Growth and Treatment, J. Cancer Res. Clin. Oncol. 105(1983): 1-12.

(16) Vogelsaenger, Th., Modellbildung und Simulation von Regelungsmechanismen wachsender Blutgefäßstrukturen in normalen Geweben und malignen Tumoren, Dissertation Siegen, 1986.

(17) Düchting, W., Simulation of 3-D Tumor Growth and Radiation Therapy in "Proceedings of the International Symposium Computer Assisted Radiology" edited by H.U. Lemke, M.L. Rhodes, C.C. Jaffee and R. Felix, Springer-Verlag, Berlin 1987: 335-339.

(18) Bär, Th., Patterns of Vascularization in the Developing Cerebral-Cortex, CIBA Found. Symp. 100(1983): 20-36.

QUASILINEARIZATION IN BIOLOGICAL SYSTEMS MODELING

E. S. Lee* and K. M. Wang**

The estimation of parameters in differential equations is a basic problem in biological systems modeling. However, these parameters cannot be estimated easily when the equations are too complicated and cannot be solved in closed form. Although Dr. Bellman has proposed to use quasilinearization to solve this problem, more numerical experiments are needed to show the effectiveness of this approach. In this paper, quasilinearization is used to estimate the parameters in various biological models. It is shown that this approach is quite effective and converges very fast in most situations. Thus, the quadratic convergence property is preserved.

QUASILINEARIZATION AND THE NONLINEAR ESTIMATION OF PARAMETERS

The algorithm of quasilinearization in estimation is well documented [1-3], only the essential equations will be discussed in the following. Consider a system represented by the following system of nonlinear differential equations

$$\frac{dx}{dt} = f(x, \alpha, t) \tag{1}$$

where x and f are M-dimensional vectors with components $x_1, x_2, \ldots, x_M$ and $f_1, f_2, \ldots, f_M$, respectively and α represents the L dimensional unknown parameters. Let us assume that the L parameters cannot be measured directly and only M_1 of the M variables can be measured. These measured values are

$$x_j^{(exp)}(t_s) = b_s^{(j)}, \qquad S = 1,2,\ldots,m, \qquad j = 1,2,\ldots,M_1 \tag{2}$$

with $t_m = t_f$. The problem is to estimate the parameters $\alpha_\ell(t)$, $\ell = 1,2,\ldots,L$ and the initial conditions

$$x_i(0) = c_i, \qquad i = 1,2,\ldots,M \tag{3}$$

from the given or measured data, Equation (2). It should be emphasized that the measured values $b_s^{(j)}$ do contain noise. Let us establish the vector equation

$$\frac{d\alpha}{dt} = 0 \tag{4}$$

---

* Corresponding author, E. S. Lee, Dept. of Ind. Engg., Kansas State University, Manhattan, KS 66506

** Dept. of Ind. Engg., Tsinghua University, Taiwan, China

The problem can be stated as find the values of the vectors c and α so that the least square expression

$$J = \sum_{j=1}^{M_1} \sum_{s=1}^{m} \left[x_j(t_s) - b_s^{(j)}\right]^2 \tag{5}$$

is minimized subject to the constraints of Equations (1) and (4). This is a multipoint boundary value problem with minimization. It can be solved by the use of quasilinearization . Equations (1) and (4) can be combined to obtain

$$\frac{dy}{dt} = g(y,t) \tag{6}$$

where y and g are M + L dimensional vectors. Equation (6) can be linearized by the use of Taylor Series with second and higher order terms omitted. The resulting vector equation is

$$\frac{dy_{k+1}}{dt} = g(y_k, t) + J(y_k)(y_{k+1} - y_k) \tag{7}$$

where y_k is assumed known and is obtained from the previous iteration and Y_{k+1} is the unknown function. The expression $J(y_k)$ is the Jacobian matrix. Because of the fast convergence rate, Equation (7) with unknown initial conditions can be solved quickly by the use of the superposition principle. In general, less than ten iterations are needed to obtain a very high accuracy.

THE ARTIFICIAL KIDNEY SYSTEM

Consider the following simple model of the artificial kidney system [4, 10].

$$V_1 \frac{dC_1}{dt} = G - K(C_1 - C_2) \tag{8}$$

$$V_2 \frac{dC_2}{dt} = K(C_1 - C_2) - C_k C_2 - C_d C_2 \tag{9}$$

where G = urea (or creatinine) production rate

k = mass transfer parameter

C_k = clearance rate of patient kidney

C_d = dialyzer clearance

C_1 = urea concentration in intracellular cell

C_2 = urea concentration in extracellular cell

V_1 = volume of intracellular cell

V_2 = volume of extracellular cell

In actual experimental situations, the constants or parameters cannot be measured, only C_2 can be measured at the various values of t. Our problem is to estimate k and $C_1(0)$ for Equations (8) and (9) from the experimental data

$$C_2^{(exp)}(t_s) = C_{2s}, \qquad s = 1,2,\ldots,m \tag{10}$$

Notice that the initial condition of $C_2(t=0)$ can be measured, but $C_1(t=0)$ must be estimated. Thus, an equation like Equation (4) can be established for the parameter k.

This problem is solved by quasilinearization with the following experimental data [4]

$$C_2^{(exp)}(t_s=1) = 2.070,$$

$$C_2^{(exp)}(t_s=2) = 1.818$$

$$C_2^{(exp)}(t_s=3) = 1.674$$

and the values of

$$G = 0.031, \quad C_d = 3.6, \quad C_k = 0, \quad \Delta t = 0.01,$$

$$C_2(t=0) = 2.538, \; t_f = 3$$

Four different experiments were carried out with four different sets of initial approximations. The convergence rates are summarized in Table 1. Notice that five digits accuracy are obtained in 6 to 10 iterations. The Runge-Kutta integration technique was used.

GLUCOSE AND INSULIN KINETICS MODELING

Consider the following simple one compartment model of glucose and insulin in plasma [5, 6]

$$\frac{dH}{dt} = - I_1 H + I_3 G + I_2 \tag{11}$$

$$\frac{dG}{dt} = - I_4 G - I_6 H + I_5 \tag{12}$$

where G = plasma glucose concentration

H = plasma IRI concentration

I_i = parameters or constants.

The problem is to estimate I_1, I_3, I_4, I_6, H(t=0) and G(t=0) from experimental data for H and G at various values of t. Again, equations like equation (4) can be established for the four parameters.

The four parameter values and the two initial conditions are estimated by quasilinearization. The numerical values used are

$I_2 = -1.56, \quad I_5 = 6.94, \quad t_f = 180$ minutes

$\Delta t = 0.2.$

The experimental data used are listed in Table 2. Several different sets of initial approximations are used. One of the typical results are listed in Table 3. The initial approximations are obtained by integrating the equations with the values for the Zeroth iteration as the initial conditions. The Runge-Kutta technique is again used. Notice that even with the very extreme initially assumed initial conditions of zero, only nine iterations are needed to obtain a five digits accuracy.

CARDIOVASCULAR INDICATOR DILUTION MODELING

Consider the following four cell cardiovascular indicator dilution model [7, 8].

$$\frac{dC_1}{dt} = B_1C_1 + B_2C_4$$

$$\frac{dC_2}{dt} = B_1(C_1 - C_2)$$

$$\frac{dC_3}{dt} = B_1(C_2 - C_3) \tag{13}$$

$$\frac{dC_4}{dt} = B_1(C_3 - C_4)$$

where $B_1 = F/V$, $B_2 = F_s/V$, with

F = volumetric flow rate

F_s = recycle volumetric flow rate

V = volume of the well-mixed cells

The boundary conditions for Equation (13) are

$$C_1(t=0) = \frac{M}{V} = B_3, \qquad C_2(t=0) = 0$$

$$C_3(t=0) = 0 \qquad C_4(t=0) = 0 \tag{14}$$

where M is the mass of the injection and the C_i's are the concentrations of the corresponding cells.

In actual experiments, only the C's can be measured, the parameters B_1 and B_2 cannot be measured directly and must be estimated indirectly from experimental data.

The values of B_1, B_2 and B_3 are estimated by quasilinearization with the numerical data listed in Table 4. The Runge-Kutta numerical integration formula with $\Delta t = 0.2$ is used. Various different initial approximations for B_1, B_2 and B_3 were used. The convergence rate is again very fast. Three typical convergence results are listed in Table 5 for three different sets of initial approximations.

METHOTREXATE PHARMACOKINETICS MODELING

Consider the following pharmacokinetic model used to predict the detailed distribution and excretion of methotrexate in mammalian species over a wide range of doses [9]. The material balance equations representing the various anatomical compartments are

Plasma: $$V_p \frac{dC_p}{dt} = Q_L \frac{C_L}{R_L} + Q_K \frac{C_K}{R_K} + Q_M \frac{C_M}{R_M} - (Q_L + Q_K + Q_M)\, C_p \tag{15}$$

Muscle: $$V_M \frac{dC_M}{dt} = Q_M \left(C_p - \frac{C_M}{R_M}\right) \tag{16}$$

Kidney: $$V_K \frac{dC_K}{dt} = Q_K \left(C_p - \frac{C_K}{R_K}\right) - k_K \frac{C_K}{R_K} \tag{17}$$

Liver: $$V_L \frac{dC_L}{dt} = (Q_L - Q_G)\left(C_p - \frac{C_L}{R_L}\right) + Q_G \left(\frac{C_G}{R_G} - \frac{C_L}{R_L}\right) - r \tag{18}$$

Gut Tissue: $$V_G \frac{dC_G}{dt} = Q_G \left(C_p - \frac{C_G}{R_G}\right) + 1/4 \sum_{i=1}^{4} \left(\frac{k_G C_i}{K_G + C_i} + b\, C_i\right) \tag{19}$$

Gut Lumen: $$\frac{dC_{GL}}{dt} = 1/4 \sum_{i=1}^{4} \frac{dC_i}{dt} \tag{20}$$

$$\frac{V_{GL}}{4} \frac{dC_1}{dt} = r_3 - k_F\, V_{GL}\, C_1 - 1/4\left(\frac{k_G C_1}{K_G + C_1} + b\, C_1\right) \tag{21}$$

$$\frac{V_{GL}}{4} \frac{dC_i}{dt} = k_F\, V_{GL}\, (C_{i-1} - C_i) - 1/4\left(\frac{K_G C_i}{K_G + C_i} + b\, C_i\right) \tag{22}$$

$$i = 2,3,4$$

where the value of r in Equation (18) can be represented by

$$r = \frac{K_L (C_1/R_L)}{K_L + (C_L/R_L)} \tag{23}$$

which is the secretion rate of methotrexate out of the liver cells into the bile ducts. Using the three compartments model, we have

$$\tau \frac{dr_1}{dt} = r - r_1 \tag{24}$$

$$\tau \frac{dr_2}{dt} = r_1 - r_2 \tag{25}$$

$$\tau \frac{dr_3}{dt} = r_2 - r_3 \tag{26}$$

where C is the drug concentration in the various anatomical compartments, r is the drug transport rate in the bile, V is the volume of the various compartments, b is the rate constant for nonsaturable gut absorption, Q is the plasma flow rate, R is the tissue plasma equilibrium ratio for linear binding and K_k is kidney clearance and is equal to 1.1 mℓ/min for rat. The other numerical values used for rat are:

V_p = 9 mℓ	Q_k = 5 mℓ/min
V_M = 100 mℓ	Q_L = 6.5 mℓ/min
V_k = 1.9 mℓ	Q_G = 5.3 mℓ/min
V_L = 8.3 mℓ	R_M = 0.15
V_G = 11 mℓ	R_k = 3.0
V_{GL} = 11 mℓ	R_L = 3.0
Q_M = 3 mℓ/min	

The body weight for rat is 200 g. Notice that three compartments were assumed for bile secretion and 4 compartments were assumed for gut lumen. Some of the parameters such as R_G, k_G and K_G are not measurable. These parameters for methotrexate in rat will be estimated by quasilinearization using experimental data obtained by Bischoff et al. [9]. These experimental data as a function of time for the drug concentrations in the various compartments are listed in Table 6 and are obtained from the figures of reference [9].

It should be emphasized that the parameters R_g, k_G and K_G cannot be estimated easily. This is because that the systems of differential equations cannot be solved in closed form. thus, quasilinearization forms an ideal and powerful approach.

In addition to the 13 differential equations represented by Equations (15) - (26), 3 additional differential equations in the form of Equation (4) can be formulated for the 3 unknown parameters. Thus, there are a total of 16 differential equations. The initial conditions for the 13 differential equations are all equal to zero except $C_p(t)$ which is

$$C_p(t) = 1200/9 \tag{27}$$

The 16 different equations can be linearized by using Equation (7). The unknown parameters can then be obtained by using Equation (5) and superpositoin principle. The homogeneous and particular solutions can be obtained by numerically integrating the linearized equations. In the present work, the modified Adam-Moulton integration scheme is used with step size as

$$\Delta t = 0.01 \text{ minute for } 0 \leq t \leq 30$$

$$\Delta t = 0.1 \text{ minute for } 30 \leq t \leq 240.$$

The convergence rates for the three parameters are listed in Table 7. Notice the fast convergence rates. Only 5 iterations are needed to obtain 4 digits accuracy.

Since the results of the previous iteration for all t must be stored in the computer, the storage requirement can be quite large. For example, the pharmacokinetic model needs (30/0.01 + 210/0.1 + 1) 16 = 81616 storage spaces. In order to reduce this storage requirement, we can store only the initial conditions of the previous iteration. The complete profile for all t of the previous iteration can be obtained by integrating the equations when we calculate the current iterations. The storage requirements can thus be reduced tremendously. For the pharmacokinetic problem, the storage requirement is reduced from 81616 to 16.

REFERENCES

[1] Bellman, R. and R. Kalaba, Quasilinearization and nonlinear Boundary Value Problems, American Elsevier, NY (1965).

[2] Lee, E. S., Quasilinearization and Invariant Imbedding, Academic Press, NY (1968).

[3] Lee, E. S., "Quasilinearization" The Bellman Continuum, World Scientific Publishing Co. (1986).

[4] Bell, R. L., F. K. Curtis and A. L. Babb, "Analog Simulation of the Patient-Artificial Kidney System" Trans. Am. Soc. Artificial Internal Organs, 11, 183 (1965).

[5] Ackerman, E., L. C. Gatewood, J. W. Rosevear and G. D. Molnar, "Model Studies of Blood Glucose Regulation," Bull. Math. Biophys, 27, 21 (1965).

[6] Norwich, K. H. "Mathematical Models of the Kinetics of Glucose and Insulin in Plasma," Bull. Math. Biophys., 31, 105 (1969).

[7] Harris, T. R., "The identification of recirculating systems in the frequency domain," Bull. Math. Biophys., 30, 87 (1968).

[8] Nicholes, K. K. and H. R. Warner, "Study of dispersion of an indicator in the circulation," Ann. N.Y. Acad. Sci., 115, 721 (1964).

[9] Bischoff, K. B., R. L. Dedrick, D. S. Zaharko and J. A. Longstreth, "Methotrexate Pharmacokinetics," J. Pharmaceutical Science, 60, 1128 (1971).

[10] Abbrecht, P. H. and N. W. Prodany "A Model of the Patient-Artificial Kidney System," IEEE Trans. Bio-Medical Eng., BME-18, 257 (1971).

Table 1 Convergence Rates of the Artificial Kidney Model

Iteration	$C_1(0)$	K	$C_1(0)$	K	$C_1(0)$	K	$C_1(0)$	K
0	2.538	5.	2.538	12.	2.538	19.2	2.538	25.
1	2.9513	6.1718	2.7879	5.2057	3.1695	-35.947	2.4352	18.639
2	2.7675	7.5204	2.8314	7.4735	2.9149	- 4.7906	3.1274	-34.37
3	2.7997	7.5318	2.8023	7.4970	2.9895	6.9627	2.7892	- 7.5045
4	2.8000	7.5279	2.7994	7.5351	2.7776	7.5923	2.6165	5.6438
5	2.7999	7.5288	2.8000	7.5272	2.7991	7.5369	2.8398	7.8420
6	2.7999	7.5286	2.7999	7.5289	2.8000	7.5270	2.8016	7.5273
7	2.7999	7.5286	2.7999	7.5285	2.7999	7.5290	2.7999	7.5292
8			2.7999	7.5286	2.7999	7.5285	2.7999	7.5285
9			2.7999	7.5286	2.7999	7.5286	2.7999	7.5286
10					2.7999	7.5286	2.7999	7.5286

Table 2 Experimental Data for Glucose and Insulin Kinetics Model

t_s	$H^{(exp)}(ts)$	$G^{(exp)}(ts)$
0	177	581
30	155	182
60	40	95
90	26	87
120	20	97
150	24	106
180	28	110

Table 3 Convergence Rates of Glucose and Insulin Kinetics Model

Iteration	I_1	I_2	I_4	I_6	H(0)	G(0)
0	0.	0.	0.	0.	177.	581.
1	0.051076	0.025872	0.048153	0.22224	181.31	576.58
2	0.038405	0.017182	0.020605	0.052089	177.16	580.37
3	0.045445	0.021543	0.028009	0.043957	177.56	580.06
4	0.046151	0.022149	0.028790	0.043174	177.27	580.35
5	0.046411	0.022281	0.028581	0.043500	177.24	580.38
6	0.046408	0.022286	0.028565	0.043510	177.23	580.39
7	0.046423	0.022293	0.028555	0.043523	177.23	580.39
8	0.046421	0.022292	0.028555	0.043523		
9	0.046422	0.022293	0.028555	0.043524		
10	0.046422	0.022293	0.028555	0.043524		

Table 4 Experimental Data for Cardiovascular Model

t_s	$C_1(t_s)$	$C_2(t_s)$	$C_3(t_s)$	$C_4(t_s)$
0.0	0.9997	0.0	0.0	0.0
2.0	0.2289	0.3314	0.2609	0.1387
4.0	0.1327	0.1887	0.2391	0.2366
6.0	0.1141	0.1347	0.1682	0.2009
8.0	0.0909	0.1066	0.1269	0.1528
10.0	0.0702	0.0834	0.0988	0.1175
12.0	0.0543	0.0646	0.0768	0.0912
14.0	0.0421	0.0501	0.0595	0.0707
16.0	0.0327	0.0388	0.0462	0.0549
18.0	0.0253	0.0301	0.0358	0.0425
20.0	0.0196	0.0234	0.0278	0.0329

Table 5 Convergence Rate of Cardiovascular Model

Iteration	B1	B2	B3	B1	B2	B3	B1	B2	B3
0	0.1	0.01	0.1	0.6	0.2	0.8	2.	1.5	3
1	0.4379	0.0725	0.4969	0.7663	0.3755	0.9903	1.7167	1.2619	1.0049
2	0.4896	0.1772	0.8993	0.7966	0.3970	0.9992	0.6846	0.2886	0.9983
3	0.6522	0.2679	0.9658	0.8013	0.4015	0.9996	0.8014	0.4021	0.9992
4	0.7616	0.3635	0.9952	0.8017	0.4018	0.9997	0.8017	0.4018	0.9997
5	0.7974	0.3979	0.9993	0.8017	0.4018	0.9997	0.8017	0.4018	0.9997
6	0.8014	0.4015	0.9997						
7	0.8017	0.4018	0.9997						
8	0.8017	0.4018	0.9997						

Table 6 Experimental Data for Pharmacokinetics Modeling

t_s (min)	$C_P(t_s)$	$C_M(t_s)$	$C_K(t_s)$	$C_L(t_s)$	$C_{GL}(t_s)$
15	7.7	1.5	20.	20.9	23.98
30	4.0	0.75	10.8	11.5	47.00
60	1.5	0.25	4.0	4.97	59.00
90	1.14	0.16	2.8	3.60	45.50
120	0.80	0.13	2.2	2.80	36.00
180	0.45	0.072	1.1	1.45	18.25
240	0.27	0.043	0.67	0.86	8.90

Table 7 Convergence Rates of Pharmacokinetics Model

Iteration	R_G	k_G	K_G
0	1.	20.	200.
1	1.108	22.64	237.2
2	1.112	21.61	224.6
3	1.112	21.97	229.3
4	1.112	21.85	227.7
5	1.112	21.89	228.3
6	1.112	21.89	228.3

CYCLIC CONTROL IN ECOSYSTEMS

Joseph Bentsman
Department of Mechanical and Industrial Engineering

Bruce Hannon
Department of Geography
Affiliate Scientist, Illinois Natural History Survey

University of Illinois at Urbana-Champaign
Urbana, IL 61801

The theory of feedback control as a possible stabilizing mechanism has already been introduced into ecosystem analysis. One problem in the theory is the identification of the informational links by which such controls operate. Cyclic controls, for example, zero–mean sine functions added to certain exchange flows in the system, might also contribute to system stability. Their advantage is that they operate without need for information from the rest of the system. The theory of ecosystem cyclic control is presented and applied to data from an oyster reef ecosystem.

I. INTRODUCTION

To address the problem of ecosystem stability and performance, the previous control studies utilized solely classical control principles, feedback and feedforward (Olsen, 1961; Lowes and Blackwell, 1975; Mulholland and Sims, 1976; Vincent, et.al., 1977; Goh, 1979; Hannon, 1985b,c, 1986; DeAngelis, 1986). If knowledge of the current output is used to modify the inputs to control the system, we have a feedback control situation (Wonham, 1984). Feedforward control uses current knowledge of the disturbance (rather than output) as the basis for a corrective action (Takahashi, et. al., 1970). The major problem with these kinds of controls, however, lies in explaining how the requisite information flows occur.

An alternative approach to ecosystem stability is found in the concept of cyclic (or vibrational) control (Meerkov, 1980; Bellman, et.al.,1986). Basically, cyclic controls are periodic variations (zero–mean) in the flows between components in an ecosystem or between the ecosystem and the surrounding environment. If the amplitudes and frequencies of these variations are within the appropriate range, the ecosystem, unstable without such variations, could under certain conditions be stabilized by their introduction <u>without any information flows</u>.

Oscillations–induced stabilization of ecosystems has been investigated by a number of researchers. Armstrong and McGehee (1976) developed a theory for the coexistence of a variety of species using a smaller number of resources. Their technique involved a the sequential staging of the species in a periodic manner, sharing the resource through time. Kemp and Mitsch (1979) used an empirical model to demonstrate the stable coexistence of three plankton species on the same resource if one of the resource inputs (wave energy) was regularly pulsing. They speculated that only a special range of frequencies and

Supported in part by the Illinois Department of Energy and Natural Resources

pulse amplitudes would produce the needed stability. The pulsing resource appeared to force a sharing between the three species, disadvantaging the species which was the most prolific under steady conditions. Levins (1979) established the sufficient conditions of coexistence by requiring that the resource or the species functions contain externally induced time–varying elements that enter the equations nonlinearly. Nonlinear dynamics in Levins' treatment was essential since it resulted in terms with even powers of zero mean oscillatory functions. The averages of such terms gave rise to the "average" nonzero inputs which acted as effective new resources and under certain conditions ensured stable oscillatory regimes of the system.

The goal of the present paper is to assess cyclic (vibrational) control theory as a tool in ecosystem analysis and management. We show that an unstable linear system can be made asymptotically stable by zero mean parametric excitations as well, and hence, nonlinearities are not necessary for oscillatory stabilization. We also utilize nonzero averages of even powers of zero mean oscillatory functions to obtain stabilizing corrections. However, we average not the original system with oscillations, but some other specially constructed system, the average of which reveals the dynamics of the original cycling system. For the purpose of illustration, we have chosen a modeling technique known as flow analysis (Hannon, 1973, 1985a; Barber, et. al., 1979) from a variety of ecosystem modeling approaches, each valid for certain system classes. First, we briefly review the flow analysis technique and present the theory of linear cyclic control of ecosystems. Then, we apply cyclic control to an oyster reef ecosystem where it acts in only one of the component flows. The extension of the theory to nonlinear systems can be done on the basis of the work of Bellman, et. al., 1986. The theory indicates the range of the amplitude/frequency ratios in which stabilizing cycles should be sought and asserts the existence of stabilizing cycles in this range. The actual stabilizing amplitudes and frequencies are determined via trial and error solutions of the differential equations.

II. FLOW ACCOUNTING

In the analysis of complex dynamic systems, it is necessary to develop consistent definitions and categorize all the identifiable flows. We start with the diagram shown in Figure 1. For more details on the ecological accounting system, see Hannon (1973), Finn (1976), Levine (1977, 1980), Hannon (1979), Patten, et. al., (1976), Herendeen (1981), Ulanowicz (1984) and Hannon (1985a).

In Figure 1, n x n matrix **P** is called the production–consumption matrix[1]. This matrix represents n processes which consume and produce n commodities. By process, we mean an aggregation of similar consumers–producers which is viewed as a single ecosystem component. By commodities, we mean the substances produced and consumed by the components of the ecosystem. The elements of the ith column represent the breakdown of the main part of the consumption of the ith process. The elements of the ith row describe the breakdown of the main part of the production by the same process. Therefore, each element of **P** is the amount of commodity i (row number) which is used by process j (column number) in the given time period. For example, p_{ij} could be the daily amount of algal biomass (commodity i) consumed by a particular class of herbivores (process j). This is a multicommodity system since commodities listed along any of the rows are noncommensurable with commodities in any other row. Therefore, the row sums may be calculated since they are all the same commodity and, we assume, possess the same nutritional qualities for all consumers (The exception to this rule is the nonbasal heat of respiration which by definition has zero value to any component in the ecosystem). But, in general, the column sums cannot be formed because a common measure of a value of each element along the columns may not exist. Commodities of different qualities, even though measured in the same units (e.g.,gms-carbon) cannot be meaningfully added together. The inputs to omnivores and detritivores, for example, are of different qualities, both chemically and in nutritional meaning, to the consumer.

The diagonal elements in **P** are the self-use terms which are for example, own-waste consumption by rabbits and the consumption of decomposers by decomposers and cannibalism.

The full output vector **q'** is the sum of the vector of the nonbasal heat **w** given off by each of the components and the total output vector **q**.

The system in Figure 1 is shown without joint products, that is, each process (column) is assumed to produce a commodity of only one type. The joint product case is discussed in Hannon, 1985a and Costanza and Hannon, 1986.

The relationship to the external environment of the measurable quantities in the ecosystem modeled in Figure 1 is summarized in Table 1. The features of each quantity in this table are identified by the letters in the corresponding boxes. The table shows two vectors: **r** and **e**. The net output vector **r** is composed of three types of flows: exports (A & D), imports (D & E) and the heat of basal metabolism (B). By imports we mean those quantities which can be produced by the ecosystem but enter the system from the external environment. Exports are those quantities which can be produced by the ecosystem but which are not necessarily produced by it, and which leave the ecosystem for the external environment. The letter D in the import and export columns indicates those measured quantities which are passing through the ecosystem in the given time period, therefore, the quantity A – E is the net export. The system is perturbed by the externally induced change of the net export. The heat of basal metabolism (basal respiration) is that given off by the organism at rest. We take the heat of basal metabolism (B) as a surrogate for the commodity flows which are used in rebuilding the stocks metabolized during the given period. By stocks we mean the accumulated output quantities in each of the components in the system.

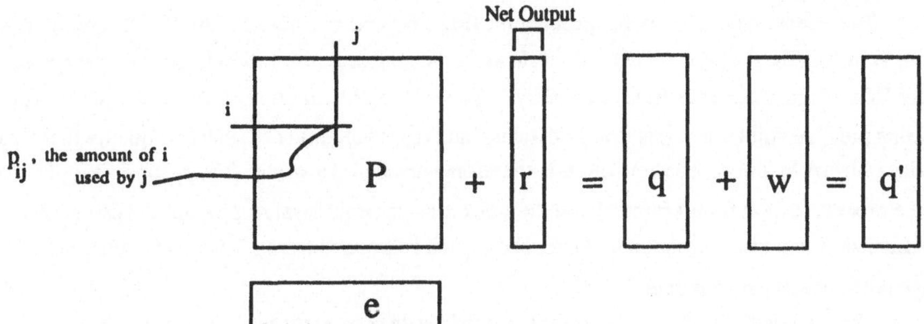

Figure 1. Steady State Ecosystem Flow Accounting Diagram

[1]Matrices are upper case symbols and vectors are lower case; both are in bold type. The elements of either are in plain type with the appropriate subscripting. A dot over a symbol indicates the time derivative.

<table>
<tr><td colspan="3" rowspan="3"></td><td colspan="3">Net Output</td><td>Non–Produced Input</td></tr>
<tr><td colspan="3">**r**</td><td rowspan="2">**e**</td></tr>
<tr><td>Exports</td><td>Imports</td><td>Basal Heat</td></tr>
<tr><td rowspan="4">Commodities that the Ecosystem is capable of Producing</td><td rowspan="2">Produced by the System</td><td>Leaves the System</td><td>A</td><td></td><td>B</td><td></td></tr>
<tr><td>Stays</td><td></td><td></td><td></td><td></td></tr>
<tr><td rowspan="2">Not Produced by the System</td><td>Leaves</td><td>D</td><td>D</td><td rowspan="2"></td><td rowspan="2"></td></tr>
<tr><td>Stays</td><td></td><td>E</td></tr>
<tr><td>Commodities that the Ecosystem is incapable of Producing</td><td></td><td></td><td></td><td></td><td></td><td>F</td></tr>
</table>

Table 1. Description of the Quantities which Form the Net Output and Nonproduced Input.

The stocks are, for example, the amount of biomass of algae which has accumulated in the producer (sun capturing) component of an aquatic ecosystem. The vector **e** stands for those input commodities that the ecosystem is incapable of producing (e.g., sunlight) but that are necessary for ecosystem functioning.

III. FLOW ANALYSIS

Next we combine the flow definitions above with the possibility of a growth in the stock of process j during the given time period Δt. These flows are graphically shown in Figure 2 for the individual process.

The consumption flows p_{ij}, production flows p_{jk} and the storage flow $\Delta s_j/\Delta t$ are internal to the ecosystem boundary, while the net output flows r_j, the nonbasal respiration flow w_j and the nonproduced input flow e_j cross the ecosystem boundary. The nonbasal respiration flow (e. g., the energy used in chasing prey, avoiding predators, food–searching and reproduction) is of such low quality that it cannot be utilized further by the ecosystem, and it is therefore considered a waste. The r_j consists of the net export of the process (export minus import) and the stock replacement (basal respiration). The net input vector e is assumed to cause no restriction to the level of q_j and is dropped from further consideration at the current stage of the model development.

The total outflow q'_j is defined for the steady state ecosystem as

$$q'_j \triangleq \sum_{k=1}^{n} p_{jk} + r_j + w_j \ . \tag{1}$$

To take into account a growth in stock, Δs_j, over the time period Δt when the system is not in the steady state, definition (1) is augmented as

$$q'_j \triangleq \sum_{k=1}^{n} p_{jk} + r_j + w_j + \frac{\Delta s_j}{\Delta t} \ . \tag{2}$$

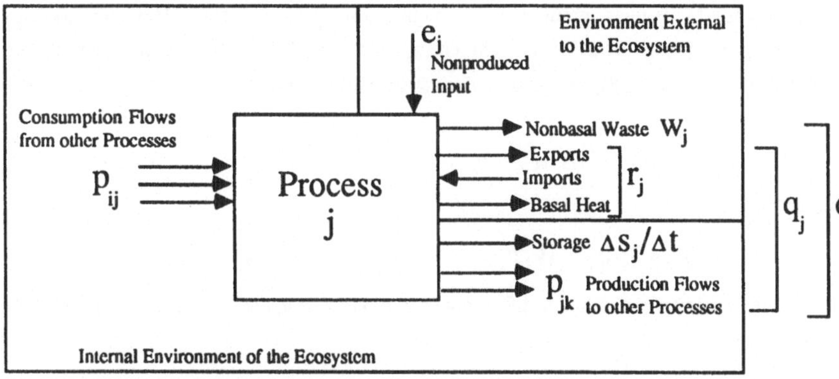

Figure 2.,The Definition of the Input and Output Flows of a Typical Process (j).

Three important simplifying assumptions are now made for the ecosystem shown in Figure 2 with q_j' defined in (2).

i) a commodity weighting or importance factor is assigned to each of the commodities produced in the system. The weight for each commodity is independent of which component consumes this commodity. A weight of zero is given to the nonbasal heat of respiration, and therefore, the vector **w** disappears from the formulation. The element q_j can be then be formed by the simple addition of all the elements along the jth row of matrix **P**, the rate of the jth stock growth and the jth element of vector **r**. For a more complete discussion of the commodity weighting issue, see Hannon, 1985a.

ii) the inputs to process j, p_{ij} form a constant ratio with the output of process j, q_j. Thus, $p_{ij}/q_j = g_{ij}$ = constant. The constants g_{ij} are determined from the data on the ecosystem at its steady state and they are assumed to remain constant for the dynamic form of our model presented below. These constants represent the internal behavior of the jth process. The g_{ij} incorporate the consumption flows into the model by locking them into a constant relationship with the output of the receiving process. Thus, the problem of summing the consumption flows (see Figure 2) is avoided.

iii) the stock (s_j) of any process (j) stays in constant proportion to the total output (q_j) of this process. That is: $b_{jj} = s_j/q_j$ = constant, forming a diagonal matrix $\mathbf{B} = \text{diag}\{b_{11},\ldots,b_{nn}\}$. This assumption allows us to obtain a balance equation using definition (2) since now

$$q_j = s_j/b_{jj} \quad . \tag{3}$$

If the results of assumptions i) and ii) are combined with (2) and (3), and if Δt becomes infinitesimal, we have

$$\frac{s_j}{b_{jj}} = \sum_{k=1}^{n} g_{jk}\frac{s_k}{b_{kk}} + r_j + \dot{s}_j \quad , \tag{4}$$

where $\dot{s}_j = ds_j/dt$.

Equation (4) is the dynamic description of the stock for process j. However, most experimental ecosystem data is presented as flows. Therefore, we change (4) into a dynamic description of the flows for process j. Substituting (3) and its time derivative into (4) yields

$$b_{jj}\,\dot{q}_j = q_j - \sum_{k=1}^{n} g_{jk}\,q_k - r_j \quad \forall\, j,\ 1 \le j \le n,$$

or in matrix form

$$\dot{\mathbf{q}} = \mathbf{A}\mathbf{q} - \mathbf{B}^{-1}\mathbf{r}, \qquad \mathbf{A} \triangleq \mathbf{B}^{-1}(\mathbf{I} - \mathbf{G}). \tag{5}$$

This time invariant ordinary differential equation (5) is in the "standard" form for the flow analysis approach.

IV. STABILITY ANALYSIS

The stability properties of the behavior of **q** when the system is subjected to a step change in **r** depend entirely on the matrix A in (5). If the real parts of all the eigenvalues of A were negative, the system would respond in a stable manner (Luenberger, 1979, p 158). However, in (5) the sum of the eigenvalues of matrix A is always positive. Therefore, the system will always respond to "sufficiently rich" changes in **r** in an unstable manner.

From an ecological viewpoint, a positive **r** represents an output of the ecosystem (for example, the amount of fish caught in the annual season). From the control theory viewpoint however, this output represents an *input* to the system or a *control* action. For example, the amount of fish caught directly affects the rate of (re)production of fish *and* many other quantities produced in the ecosystem, which in turn, also affect the fishing success. If the system (5) is to accurately represent the functioning of an ecosystem, the equations must be judiciously modified to include stabilizing or controlling flows. Equations (5) can be made to respond stably by modifying **r** to include a feedforward or a feedback control. Let us, however, demonstrate the use of cyclic control for ecosystem stabilization through the addition of a cyclic flow to one of the elements in the matrix **G**.

In the flow accounting framework, cyclic control alone cannot guarantee stability of the system. However, only a very simple form of constant feedback is required to make cyclic control effective. Such feedback can be easy to maintain since it need not ensure stability but only "condition" the system for cyclic control. On the other hand, for a broad class of the so-called decentralized systems, no constant or time-varying feedback exists that can stabilize the system (Anderson and Moore, 1981). In these cases, the addition of cyclic control can result in the desired stabilizing effect.

Since equation (5) is still always unstable, several changes must be made to **r** to demonstrate the cyclic control. First, **r** must be broken into two parts: a vector of net outputs which are independent of the output **q**, and another vector which contains the feedback and cyclic control and depends on **q**. The first vector contains the "set point" vector for the system, $\mathbf{r}_s$: the vector of net outputs which in the absence of cyclic control determines the unstable steady state level $\mathbf{q}_s$ of the total output. The introduction of cyclic control converts the unstable steady state $\mathbf{q}_s$ into asymptotically stable T-periodic operating regime, $\mathbf{q}_s(t)$, where T is the period of a cyclic control. A feedback control is needed to convert the trace of the matrix in equation (5) to a negative value (Meerkov, 1980). Assume that this is an internal control that changes the net output from the system in linear proportion to the production flows, a "flow" control (Hannon, 1986). For simplicity, let the linear proportionality be represented by a diagonal matrix of constants, **Q**. In this case, vector **r** in equation (5) is given by:

$$\mathbf{r} = \mathbf{r}_1 + \mathbf{r}_s = \mathbf{r}_s - \mathbf{r}_c + \mathbf{Qq} \ ,$$

where $\mathbf{r}_c = \mathbf{Qq}_s(0)$

Equation (5) then becomes:

$$\begin{aligned} \dot{\mathbf{q}} &= \mathbf{B}^{-1}(\mathbf{I} - \mathbf{G} - \mathbf{Q})\mathbf{q} + \mathbf{B}^{-1}(\mathbf{r}_c - \mathbf{r}_s) \\ &= \mathbf{Nq} + \mathbf{B}^{-1}(\mathbf{r}_c - \mathbf{r}_s), \ \mathbf{N} \triangleq \mathbf{B}^{-1}(\mathbf{I} - \mathbf{G} - \mathbf{Q}). \end{aligned} \tag{6}$$

The constant vector $\mathbf{B}^{-1}(\mathbf{r}_c - \mathbf{r}_s)$ will be dropped because it is independent of $\mathbf{q}$ and therefore does not affect the stability analysis.

Matrix **Q** must have only one non-zero element with sufficiently large absolute value to cause a sign change in the trace of **N**. Therefore, we further assume that matrix **Q** makes the trace of matrix **N** negative, but does not guarantee system stability, i. e., we simulate the circumstances where the feedback controls (like **Q**) are not adequate to make all of the eigenvalues fall in the left-half plane. This situation can arise if the information gathering processes of the system are somehow limited, resulting in lack of controllability and/or observability (Luenberger, 1979), but are sufficient to condition the system for cyclic control.

Let us again augment the vector $\mathbf{r} = \mathbf{Qq} - \mathbf{D}(t)\mathbf{q}$, where **D**(t) is a periodic, zero mean matrix. The periodic input **D**(t), is weighted by the state vector of the system **q**, and therefore **D**(t) appears in the system equation in the form of parametric perturbations or *cyclic control*. In this case equation (6) becomes

$$\dot{\mathbf{q}} = [\mathbf{N} + \mathbf{B}^{-1}\mathbf{D}(t)]\mathbf{q} \ . \tag{7}$$

Because equation (7) is time-varying, eigenvalues can no longer describe its stability. It is possible, however, to associate stability properties of the oscillatory system (7) with a certain constant matrix that describes its <u>average</u> behavior. The stabilizing action of cyclic controls consists in converting the remaining right-half plane eigenvalues of system (6) into "left-half plane on-the-average" ones. In this case, stabilization is achievable without the need for additional information flows, provided that the amplitudes and frequencies of the cyclic controls are within a critical range.

Assume, for simplicity, that the ijth element of the cyclic control matrix **D**(t) is given by $d_{ij}(t) = c_{ij}\cos(\omega_{ij}t)$, where c_{ij} is the amplitude and ω_{ij} is the frequency of the oscillation.

In order to describe the average behavior of system (7), we introduce the parameter ε as

$$\varepsilon \triangleq \max_{ij}(1/\omega_{ij})$$

and define

$$c_{ij} \triangleq \alpha_{ij}/\varepsilon \quad \text{and} \quad \omega \triangleq \beta_{ij}/\varepsilon$$

so that the ijth element of **D**(t) can be rewritten as $d_{ij}(t) = (\alpha_{ij}/\varepsilon)\cos(\beta_{ij}t/\varepsilon)$.

With this notation, the cyclic control matrix **D**(t) takes the form

$$D(t) = \frac{1}{\varepsilon} D'(\frac{t}{\varepsilon}),$$

and system (7) becomes

$$\dot{q} = [N + \frac{1}{\varepsilon} B^{-1} D'(\frac{t}{\varepsilon})]q. \tag{8}$$

Thus, if the α_{ij}'s and β_{ij}'s are assumed constant, the amplitudes c_{ij} and the frequencies ω_{ij} of the zero-mean cyclic terms $d_{ij}(t)$ are parameterized by a positive ε. It has been proven (Bellman, Bentsman and Meerkov, 1985) that there exists an ε_0 = constant > 0, such that for any ε satisfying the inequality $0 < \varepsilon < \varepsilon_0$, the stability properties of system (8) are defined by the eigenvalues of a constant matrix

$$M = \lim_{T \to \infty} \frac{1}{T} \int_0^T \Phi(\tau)^{-1} N \Phi(\tau) \, d\tau \quad , \tag{9}$$

where $\Phi(t)$ is the state transition matrix of

$$\frac{dq}{d\tau} = B^{-1} D'(\tau) q \tag{10}$$

where $t = t/\varepsilon$.

Specifically, for sufficiently small ε, system (8) is asymptotically stable if all the eigenvalues of **M** have negative real parts. As seen from this result, the elements of matrix **M** are defined in terms of the elements of matrices **N**, B^{-1}, "amplitude/frequency" ratios α_{ij}, and "frequency/frequency" ratios β_{ij}. Consequently, **M** provides a link between α_{ij}, β_{ij} and stability of (7): If α_{ij} and β_{ij} are found which place all the eigenvalues of **M** in the left-half plane, then there exists an ε such that oscillations with amplitudes α_{ij}/ε and frequencies β_{ij}/ε guarantee asymptotic stability of system (7). The matrix

$$M' \triangleq M-N$$

can be thought of as a "correction" of **N** induced by oscillations.

In the context of ecological systems, cyclic control is easy to apply. Indeed, ecological systems are usually described by sparse matrices and therefore the cyclic control matrix **D**(t) might often satisfy condition $D^2(t) = 0$ independently of the magnitudes and frequencies of the oscillations. In this case, since **B** is a diagonal matrix, all non-zero elements of matrix **M'** are given as

$$m'_{ij} = -\frac{\gamma_{ij}^2}{2} n_{ji} \quad , \quad \gamma_{ij} \triangleq \frac{1}{b_{ii}} \frac{\alpha_{ij}}{\beta_{ij}}, \tag{11}$$

where n_{ji} denotes the jith element of the matrix **N**. Therefore, the only elements of **D**(t) that will affect the eigenvalues of **M** are those off-diagonal elements that have a corresponding non-zero symmetric element in **N**.

The first step in the search for amplitudes and frequencies of the stabilizing oscillations is to find m'_{ij}'s that move all the eigenvalues of **N+M'** to the left-half plane. A straightforward way to accomplish this is to try only one of the appropriate elements at a time, and let g_{ij} increase from 0 to a sufficiently large number. When the appropriate set of elements m'_{ij}, and, hence, g_{ij} have been identified, we must

Table 2. Oyster reef Input-Output flow matrix (P), along with vectors for net export + stock replacement respiration (r), total output excluding waste heat (q), waste heat (w), and total output including waste heat (q').

	P									
	(1)	(2)	(3)	(4)	(5)	(6)	r	q	w	q'
Oysters 1	0	15.79	0	0	0	0.51	17.80	34.10	7.365	41.47
Detritus 2	0	0	8.17	7.27	0.64	0	6.19	22.27	0	22.27
Microbiota 3	0	0	0	1.21	1.21	0	2.875	5.295	2.875	8.17
Meiofauna 4	0	4.24	0	0	0.66	0	1.75	6.65	1.75	8.4
Deposit Feeders 5	0	1.91	0	0	0	0.17	0.215	2.295	0.215	2.51
Predators 6	0	0.33	0	0	0	0	0.2	0.53	0.15	0.68
Net Input e	41.47	0	0	0	0	0				
Control Q	1.52	2.28	.94	1.26	2.09	1.38				

return to equation (8), placing $(\alpha_{ij}/\varepsilon)\cos(\beta_{ij}/\varepsilon)$ at these locations in $\mathbf{D}(t)$. Then, by changing ε and repeatedly solving equation (8) for stabilizing pairs of (α,β), the areas of stabilizing amplitudes and frequencies can be found. The search for stabilizing oscillations becomes complicated when the stabilizing matrices that satisfy $\mathbf{D}^2(t) = 0$ do not exist (see for example, Wu, 1975).

Cyclic control could naturally arise in an ecosystem as i) an oscillation of the flows between various components or ii) a part of the net output, a cyclical export (import) from (to) a particular component, the interpretation used in this paper.

What follows is a simple example of ecosystem stabilization by a cyclic control.

V. APPLICATION TO THE OYSTER REEF ECOSYSTEM

In this section, we apply the theory presented above to the oyster reef ecosystem (Dame and Patten, 1981). This compact but complex system is shown at steady state (i.e., for constant flows) in Figure 3.

The data from Figure 3 have been arranged in the proposed accounting framework (Figure 1) in Table 2. In this arrangement, estimates of the basal metabolism or structural-rebuilding respiration are included in the net output.

From the data in Table 2, we constructed **G** for use in the **N** matrix. With the feedback control elements of diagonal matrix **Q**, shown in Table 2, the trace of **N** is negative and its eigenvalues are: 0.0726 ±0.0371i, -0.1753, -.0089, -0.0994 and -0.0028. Because the complex pair has positive real parts, the system is unstable. Let us demonstrate that a cyclic control can be found to stabilize the system at the given steady state.

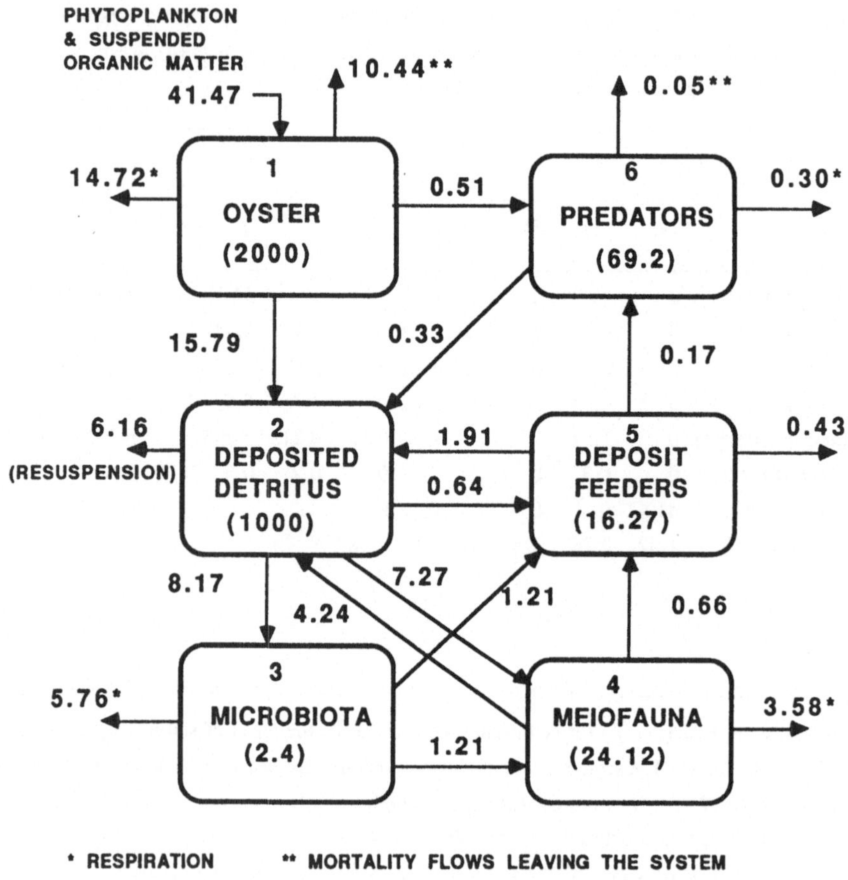

Figure 3. The Oyster Reef Ecosystem. Flow units are kcal/m^2-day. Stock unit: kcal/m^2.

Let $m'_{5,3}$ be the only non-zero element of matrix **M'**, indicating a cyclic net input to deposit feeders and a cycle in the flow 5–3. Then by experiment, for $m'_{5,3} > 0.0346$, all the eigenvalues of matrix N + M' are in the left-half plane. Choosing $\beta_{5,3} = 1.0$, from equation 11 we obtain

$$\alpha_{5,3} = b_{5,5}\left[\frac{-2m'_{5,3}}{n_{3,5}}\right]^{\frac{1}{2}} = 1.7298,$$

where $b_{5,5}= 7.0893$ and $n_{3,5}=-1.1632$. Thus, according to the theory of Section IV, oscillations of the form $d_{5,3}(t) = \alpha_{5,3}\omega \mathrm{Sin}(\omega t)$, $\alpha_{5,3} > 1.73$, should stabilize the system for sufficiently large ω. The asymptotic nature of the theory implies however, that condition $\alpha_{5,3} >1.73$ should be partially observed for smaller ω as well. It is precisely this insight that motivates the numerical search for the actual parameters of stabilizing cycles at low frequencies. In Figure 4, we demonstrate that condition, $\alpha_{5,3} >1.73$, is partially observed for $\omega/2\pi > 0.08$. The amplitudes are $q_3 d_{5,3}(t)/b_{5,5}$. The cross-hatched region in Figure 4 corresponds to the actual stabilizing amplitudes and frequencies of the cycles $d_{5,3}(t)$.

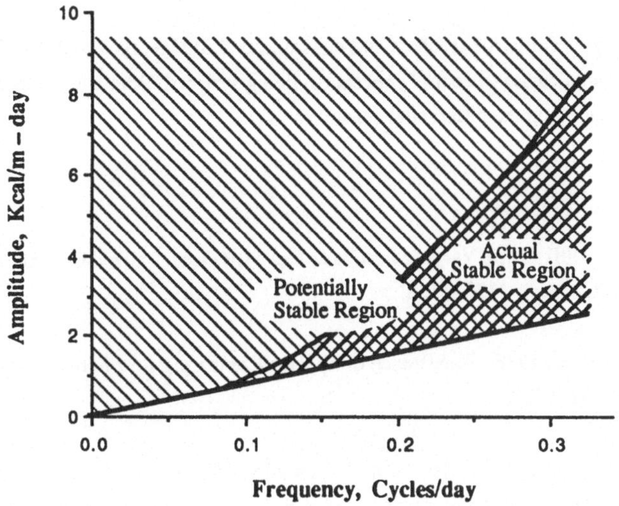

Figure 4. Cyclic Control in the Oyster Reef Ecosystem. The Range of the Parameters of the Stabilizing Oscillations of the Net Input to the Deposit Feeders 5 and of the Connection to the Microbiota 3.

While our choice of Q was largely arbitrary, we find the data in Figure 4 interesting. They show, for example, that a cyclic net input to the Deposit Feeders (which in turn allows them to cycle their feeding on the microbiota) can stabilize this ecosystem (given the above Q). With a cycle frequency of once in seven days, the stabilizing amplitude would range from about 1.1 to 1.7 kcals/m^2–day, encompassing the average value of the flow from 3 to 5 of 1.2 kcals/m^2–day (see Figure 3). It seems possible that such a cyclic flow could occur. No data on the variation of flows in this oyster reef ecosystem were given (Dame, 1976, 1979; Dame and Patten, 1981). From Figure 4, we also see that smaller stabilizing amplitudes are associated with lower frequencies. This application to the oyster reef system is expected to convey a biological possibility of ecosystem stabilization by already existing or intentionally introduced oscillations.

VI. CONCLUSION

The material presented above demonstrates that cyclic control is a biologically feasible stabilizing mechanism that could either develop in the course of evolution or be introduced by an ecosystem manager.

The important point about cyclic control is that stabilization can be provided without any information exchange. Therefore, the components that can establish a balanced cyclic exchange of materials or energy with the external environment and/or with other components might bring stability to the whole system without the cost of building and maintaining additonal information links. Thus, since cycles often occur in ecosystems naturally or can be introduced intentionally, cyclic control theory constitutes a viable tool for the ecosystem analysis and management.

ACKNOWLEDGMENT

The first author is grateful to his teachers and co–workers S.M. Meerkov and R.E. Bellman for arousing his interest in the application of control concepts to living systems. Both authors thank Salvatore Cusumano for his help in setting up the differential equation solver.

REFERENCES

Anderson, B. and J. Moore, 1981, Time Varying Feedback Laws for Decentralized Control, IEEE Trans. Autom. Control, **AC–26**, 5, 1133–1138.

Armstrong, R. and R. McGehee, 1976, Coexistence of Species Competing for Shared Resources, Theo. Population Biol., **9**, 317–328.

Barber, M., B. Patten and Finn, J., 1979, Review and Evaluation of I-O Flow Analysis for Ecological Systems, in: Compartmental Analysis of Ecosystem Models, Vol. **10** of Statistical Ecology, Matis, J., Patten, B. and White, G., eds., International Cooperative Publishing House, Fairland, Md.

Bellman, R., J. Bentsman and S. M. Meerkov, 1985, Stability of Fast Periodic Systems, IEEE Trans. on Automatic Control, **AC-30**, 3, 289-291.

________, 1986, Vibrational Control of Nonlinear Systems, Vibrational Stabilizability, IEEE Trans. Autom. Control, **AC–31**, 8, 710–716.

________, 1986, Vibrational Control of Nonlinear Systems, Vibrational Controllability and Transient Behavior, IEEE Trans. Autom. Control, **AC–31**, 8, 717–724.

Costanza, R., C. Neill, S. Leibowitz, J. Fruci, L. Bahr, and J. Day, 1983, Ecological Models of the Mississippi Deltaic Plain Region: Data Collection and Presentation, U. S. Fish and Wildlife Service, Washington, DC., FWS/OBS–82/86.

Costanza, R. and B. Hannon, , 1987, Multicommodity Ecosystem Analysis: Dealing with Apples and Oranges in Flow and Compartmental Analysis, to appear in: Progress in Systems Ecology: Mid–1980's Issues and Perspectives, B. Patten and S. Jørgensen, Editors.

Dame, R. and B. Patten, 1981, Analysis of Energy Flows in an Intertidal Oyster Reef, Marine Ecology Progress Series, **5**:115-24. See also: Dame, R., 1976, Energy Flow in An Intertidal Oyster Population, Estuarine and Coastal Marine Science, **4**, 243-253. See also: (1979), The Abundance, Diversity and Biomass of Macrobenthos on North Inlet , South Carolina, Intertidal Oyster Reefs, Proceedings of the National Shellfisheries Association, **69**, 6-10.

DeAngelis, D., W. Post and C. Travis, 1986, Positive Feedback in Natural Systems, Biomathematics, Springer–Verlag, Berlin, **15**, 233.

Finn, J. 1976, Measures of Ecosystem Structure and Function Derived from Analysis of Flows, J. Theo. Biol., **56**, 363-380.

Goh, B. 1979, The Usefulness of Optimal Control Theory to Ecological Problems, in: Theoretical Systems Ecology, E. Halfon, ed., Academic Press, N.Y., 385-399.

Hannon, B. 1973, The Structure of Ecosystems, J. Theo. Biol., **41**, 535-46.

Hannon, B 1979, Total Energy Costs in Ecosystems, J. Theo Biol., **80**, 271-293.

Hannon, B. 1985a, Ecosystem Flow Analysis, Canadian Journal of Fisheries and Aquatic Sciences **213**, Ecological Theory for Biological Oceanography, R. Ulanowicz and T. Platt, eds., 97-118.

Hannon, B. 1985b, Conditioning the Ecosystem, Mathematical Biosciences, **75**, 1, 23-42.

Hannon, B. 1985c, Linear Dynamic Ecosystems, J. Theo. Biol., **116**, 1, 89-98.

Hannon, B. 1986, Ecosystem Control Theory, J. Theo. Biol., **121**, 417–437.

Herendeen, R. 1981, Energy Intensities In Ecological and Economic Systems, J. Theo. Biol. **91**, 607-620.

Kemp, W. and W. Mitsch, 1979, Turbulence and Phytoplankton Diversity: A General Model of the "Paradox of Plankton", Ecol. Model., **7**, 201–222.

Luenberger, D. 1979, Introduction to Dynamic Systems: Theory, Models, and Application, Wiley, NY.

Levine,S. 1977, Exploitation Interactions and the Structure of Ecosystems, J. Theo. Biol., **69**, 345-355.

Levine, S. 1980, Several Measures of Trophic Structure Applicable to Complex Food Food Webs, J. Theo. Biol., **83**, 195-207.

Levines, R., 1979, Coexistence in a Variable Environment, Amer. Natur., **114**, 6, 765–783.

Lowes, A. and C. Blackwell, 1975, Applications of Modern Control Theory to Ecological Systems, in: Ecosystems Analysis and Prediction, S. Levin, ed., SIAM, Phil., Pa., 299-305.

Meerkov, S. M., 1980, Principle of Vibrational Control: Theory and Applications, IEEE Trans. Auto. Control, **AC-25**, 4, 755-762.

Mulholland, R. and C. Sims, 1976, Control Theory and Regulation of Ecosystems, in: Systems Analysis and Simulation in Ecology, Patten, B., ed., Academic Press, NY., 373-390.

Olsen, J. 1961, Analog Computer Models for Movement of Nuclides Through Ecosystems, Radioecology, Proc. First National Symp., Colo. State University, Schultz and Klement, A. eds., Rheinhold Publishing Co., N.Y., 121-125

Patten, B., R. Bosserman, J. Finn, and W. Cale, 1976, Systems Analysis and Simulation in Ecology, 4, 457-574.

Takahashi, Y., M. Rabins and D. Auslander, 1970, Control and Dynamic Systems, Addison–Wesley Publishing, Reading, MA, 348.

Ulanowicz, R. 1984, Growth and Development: A Phenomenological Perspective, Center for Environmental And Estuarine Studies, Chesapeake Biological Laboratory, University of Maryland, Solomons, MD.

Vincent, T.L., C.S. Lee and B.S. Goh, 1977, Control Targets for the Management of Biological Systems, Ecological Modeling, 3, 285–300.

Wonham, W., 1984, Regulation, Feedback and Internal Models, in: Adaptive Control of Ill–Defined Systems, O. Selfridge and E. Rissland, eds., Plenum Press, NY., 75–88.

Wu, M., 1975, Some Results in Linear Time Varying Systems, IEEE AC–20, 159–161.

SELF CONTROLLED GROWTH POLICY
FOR A FOOD CHAIN SYSTEM

George Bojadziev
Department of Mathematics and Statistics
Simon Fraser University
Burnaby, B.C. V5A 1S6
Canada

Abstract. A behavioural policy of controlled growth for a food chain model of length $2n$ is considered. The highest trophic level population controls its own growth in order to restrain the growth of the other $2n-1$ populations in the system so as to avoid undesirable outcomes.

1. INTRODUCTION

The present research concerning control policies for biological systems in population dynamics mainly deals with human control added to models of interacting populations. Various pest management programs provide typical examples of this kind of *external* control [1,2]. However in reality there are also situations in which one or more populations participating in the system are the controllers. Such systems change behaviour abruptly in response to changes of the size of the interacting populations, climatic conditions, diseases, etc. We call this type of control *internal*. The classical models in population dynamics usually do not reflect either the external nor the internal control. The controlling populations can apply the internal control to their own members (*self control*) or to all or some of the other participating populations in the model. In this paper the attention is focused on the concept of self control.

Generalizing a previous paper (Bojadziev and Skowronski [3]) here we study a food chain system of size $2n$ involving a controlling factor $u(t)$ which adjusts the number of the highest trophic level population so that a reasonable size of all populations is maintained. Making use of a methodology developed by Leitmann and Skowronski [4] (see also Blaquière, Gerard, and Leitmann [5]) for dynamical systems, we derive conditions under which the designed control policy results in avoidance of a prescribed region in R^{2n} so that undesirable outcomes are avoided.

2. THE FOOD CHAIN MODEL

Consider the food chain model with control

$$\bar{x}'(t) = f(\bar{x}(t),\ u(t)) \tag{1}$$

where $t \in R_+$ is the time variable, $\bar{x}(t) = (x_1,\ldots,x_{2n})^T$ is the population vector, $u(t)$ is the control, and the components of the vector function $\bar{f}(x,u) = (f_1,\ldots,f_{2n})^T$ are given by

$$f_1(\bar{x},u) = x_1(\alpha_1 - \frac{\beta_1}{\gamma_1} x_2) \ ,$$

$$f_{2k}(\bar{x},u) = x_{2k}\left(-\alpha_{2k} + \frac{\beta_{2k-1}}{\gamma_{2k}} x_{2k-1} - \frac{\beta_{2k}}{\gamma_{2k}} x_{2k+1}\right),$$

$$f_{2k+1}(\bar{x},u) = x_{2k+1}\left(-\alpha_{2k+1} + \frac{\beta_{2k}}{\gamma_{2k+1}} x_{2k} - \frac{\beta_{2k+1}}{\gamma_{2k+1}} x_{2k+2}\right), \tag{2}$$

$$f_{2n}(\bar{x},u) = x_{2n}\left(-\alpha_{2n} + \frac{\beta_{2n-1}}{\gamma_{2n}} x_{2n-1}\right) + ux_{2n}^2 \ ,$$

$$k=1,\ldots,n-1, \qquad f_i(\bar{x},u) = f_i(\bar{x},0), \qquad i=1,\ldots,2n-1 \ .$$

For $u=0$ the model (1) reduces to the uncontrolled food chain model

$$\bar{x}'(t) = \bar{f}(\bar{x}(t),0) \ . \tag{3}$$

In (1) $x_i, i=1,\ldots,2n$, is the size of the i-th population; α_i (growth rate coefficient), β_i (interaction coefficient), and γ_i (trophic weight factor) are positive constants; γ_j/γ_i expresses the gain-loss ratio when population i interacts with population j. The control $u(t) \in \mathcal{U}[t_o,\hat{t}] = \{u(t): u(t) \in U$ and $u(t)$ measurable on $[t_o,\hat{t}]\}$, $0 \le t_o < \hat{t} < \infty$, $U \subset R$ is a compact set to be specified later in accordance to a growth restriction policy.

The biological meaning of the control term ux_{2n}^2 in the last expression (2) which takes part in (1) is that for $u > 0$ the population with size x_{2n} (the highest trophic level population in the food chain) is enhanced by increasing the population density (increasing returns) and for $u < 0$ it dampers its own growth (diminishing returns). The 2n-th population can be considered as a consumer or predator of a higher level in terms of organization and brain capability in comparison to the other 2n-1 populations or resources. The self controlled growth of the consumer (predator) will affect the growth of all populations in the food chain system.

Each choice of control, say $u(t_o) = c_o \in U$ on some time interval starting at $t = t_o$, generates a solution or response $k[t] = k(\bar{x}(t_o), c_o, t)$ of the system (1) with initial state $\bar{x}(t_o) \in R_+^{2n}$ which geometrically is represented by an orbit ℓ_o in the phase space R^{2n}. If $c_o = 0$ (no control, hence (1) reduces to (3)) the response $k(x(t_o), 0, t)$ of (3) can exhibit large variation and may endanger the existence of an acceptable size of some populations. In order to avoid such undesirable outcomes, the consumer population with size x_{2n} may opt to self control its own growth which will affect the growth of the other populations in the food chain. This can be accomplished by selecting a suitable control value $u(t_1) = c_1 \in U$ at a point $\bar{x}(t_1) \in R_+^{2n}$ (switching point) on some time inverval starting at $t = t_1$, $t_1 > t_o$. The control value $u(t_1) = c_1$ will generate a response $k(\bar{x}(t_1), c_1, t)$ along a new orbit ℓ_1, $\ell_0 \cap \ell_1 = \bar{x}(t_1)$.

Using a Liapunov function for the uncontrolled model (3) we define for the response of (1) an avoidance region A, a security zone S which safeguards the response of entering A, and design a control policy for avoidance.

3. THE LIAPUNOV FUNCTION

The coordinates of the nontrivial equilibrium $E^0(\bar{x}^0)$, $\bar{x}^0 = (x_1^0, \ldots, x_{2n}^0)^T \in R^{2n}$, of (3) are

$$x_2^0 = \frac{\alpha_1\gamma_1}{\beta_1}, \qquad x_{2n-1}^0 = \frac{\alpha_{2n}\gamma_{2n}}{\beta_{2n-1}},$$

$$x_{2k-1}^0 = \frac{\alpha_{2k}\gamma_{2k} + \beta_{2k}x_{2k+1}^0}{\beta_{2k-1}}, \qquad k=1,\ldots,n-2, \tag{4}$$

$$x_{2k+2}^0 = \frac{-\alpha_{2k+1}\gamma_{2k+1} + \beta_{2k}x_{2k}^0}{\beta_{2k+1}}, \qquad k=1,\ldots,n-1.$$

We require that $E^0 \in \text{Int } R_+^{2n}$, the interior of the closed positive cone, so that E^o has biological meaning. Since $x_{2n-1}^0 > 0$, it follows from (4) that $x_{2k-1}^0 > 0$, k-1,...,n-1. Also from (4) we see that $x_2^0 > 0$. However, in order to secure that $x_{2k}^0 > 0$, k=2,...,n-1, we assume that $x_{2k}^0 > \alpha_{2k+1}\gamma_{2k+1}/\beta_{2k}$.

The model (3) has the Volterra function (Huang and Morowitz [6])

$$V(\bar{x}) = \sum_{i=1}^{2n} \gamma_i x_i^0 \left(\frac{x_i}{x_i^0} - \ln \frac{x_i}{x_i^0} - 1 \right), \tag{5}$$

continuous on Int R_+^{2n}, which is actually a Liapunov function with the following properties.

(i) The minimum of $V(\bar{x})$ is attained at the equilibrium $E^0(\bar{x}^0)$ given by (4); $\min V(\bar{x}^0) = 0$;

(ii) $V(\bar{x})$ is monotone increasing about E^0 (has the nesting property);

$$\text{(iii)} \quad \frac{dV(\bar{x})}{dt} = \sum_{i=1}^{2n} \frac{\partial V}{\partial x_i} f_i(\bar{x},0) = 0 \,, \tag{6}$$

where f_i are given by (2). From here follows that the equilibrium $E^0(\bar{x}^0)$ is stable.

The model (3) has a first integral

$$V(\bar{x}) = h, \qquad h = \text{const} > 0 \,, \tag{7}$$

which represents a family of level surfaces V_h in R^{2n+1}. The orthogonal projection of V_h onto R^{2n} generates 2n dimensional hypersurfaces H_h in R^{2n} which are closed, do not intersect, contain inside the equilibrium E^0, and accommodate orbits of (3). Further, if $h_1 < h_2$, the hypersurface H_{h_1} is inside the hypersurface H_{h_2}.

4. AVOIDANCE CONTROL

Here, marking use of a Liapunov design technique [4], we introduce definitions and prove a theorem concerning the food chain model (1).

Definition 1 (Avoidance set A). Given $\bar{\varepsilon} = (\varepsilon_1,\ldots,\varepsilon_{2n})^T \in \text{Int } R_+^{2n}$ and the Liapunov function $V(\bar{x})$ by (5),

$$A \triangleq \{\bar{x} \in R^{2n}: V(\bar{x}) \geq V(\bar{\varepsilon}) = h_\varepsilon\} \,, \tag{8}$$

where ε_i (avoidance parameters), i=1,...,2n, are small as desired for a particular study. The boundary of A is

$$\partial A = H_{h_\varepsilon} \triangleq \{\bar{x} \in R^{2n}: V(\bar{x}) = h_\varepsilon\} \,. \tag{9}$$

Definition 2 (Security zone S). Given $\bar{\delta} = (\delta_1,\ldots,\delta_{2n})^T \in \text{Int } R^{2n}$, $\delta_i > \varepsilon_i$, and $V(\bar{x})$ by (5),

$$S \triangleq \{\bar{x} \in R^{2n}: V(\bar{x}) \geq V(\bar{\delta}) = h_\delta\} - A \,, \tag{10}$$

δ_i, i=1,...,2n, are security parameters. The boundary of S is given by

$$\partial S = H_{h_\delta} \triangleq \{\bar{x} \in R^{2n}: V(\bar{x}) = h_\delta\} \,. \tag{11}$$

From the nesting property of $V(\bar{x})$ it follows that $h_\delta < h_\varepsilon$, hence in

R_+^{2n} the hypersurface (9) encloses the hypersurface (11).

Definition 3 The set A defined by (8) is avoidable if there is a set S defined by (10) and a control $u \in U$ such that for all $\bar{x}^s(t_s) \in S$, the response $k(\bar{x}^s(t_s), u(t_s), t)$ of (1) cannot enter A, i.e.

$$k(\bar{x}^s(t_s),\ u(t_s),\ t) \cap A = \phi \ \forall\ t\ . \tag{12}$$

Now we establish sufficient conditions for the avoidance of A.

Theorem The food chain model (1) is controllable for avoidance of A if there is a control $u(t) \in U$ and a Liapunov function $V(\bar{x})$ defined by (5) so that

$$\frac{dV(\bar{x})}{dt} = \sum_{i=1}^{2n} \frac{\partial V}{\partial x_i} f_i(\bar{x},u) \le 0\ , \tag{13}$$

where $f_i(\bar{x},u)$ are given by (2).

Proof. Assume that A is not avoidable, i.e. (12) is violated. Hence for some $\bar{x}^s(t_s) \in S$, the response $k(\bar{x}^s(t_s),\ u(t_s), t)$ enters A, $t > t_s$. Then there is a $t_a > t_s$ for which $\bar{x}^a(t_a) = k(\bar{x}^s(t_s), u(t_s), t_a) \in \partial A$. From the nesting property of $V(\bar{x})$ it follows that $V(\bar{x}^s(t_s)) < V(\bar{x}^a(t_a))$, meaning that the function $V(\bar{x})$ is increasing. This contradicts (13) which states that $V(\bar{x})$ is non-increasing along every response of (1).

5. THE CONTROL POLICY

To design a policy for avoidance the region A by the response of (1) we use the theorem in the previous section. Substituting $f_i(\bar{x},u)$ from (2) into (13) with (5) gives

$$\frac{dV(\bar{x})}{dt} = \sum_{i=1}^{2n} \frac{\partial V}{\partial x_i} f_i(\bar{x},0) + \frac{\partial V}{\partial x_{2n}} u x_{2n}^2 \le 0\ .$$

According to (6) the summation term above is zero; the second term gives

$$\gamma_{2n} x_{2n}^0 \left(\frac{1}{x_{2n}^0} - \frac{1}{x_{2n}} \right) u x_{2n}^2 \le 0$$

which can be written as

$$\left(\frac{1}{x_{2n}^0} - \frac{1}{x_{2n}} \right) u \le 0\ . \tag{14}$$

The inequality (14) establishes a relationship between the control u

and the controlling population x_{2n}. It requires that

$$u \le 0 \ \forall \ x_{2n} > x_{2n}^0 \,,$$
$$u \ge 0 \ \forall \ x_{2n} < x_{2n}^0 \,. \qquad (15)$$

According to (15) we specify that

$$u(t) \in U = [-r,r] \subset R, \qquad r=\text{const.} \qquad (16)$$

On the basis of (15) we formulate the following behavioural policy.

Avoidance control policy: If the response $k[t] = k(\bar{x}(t_o),u(t_o),t)$ of the food chain model (1) with initial state $\bar{x}(t_o)$ and fixed control $u(t_o) \in U$, U specified by (16), enters the security zone S given by (10), in order to prevent $k[t]$ of entering into A defined by (8), a new control value $u(t_s)$ should be selected from U at a switching point $\bar{x}(t_s) \in S$ with corresponding response $k(\bar{x}(t_s),u(t_s),t)$, $t_s > t_o$. If $x_{2n} > x_{2n}^0$, the new control value $u(t_s)$ should be negative and if $x_{2n} < x_{2n}^0$, it should be positive.

Note 1. The control $u=0$ satisfies (15) but then the response will be accommodated on a hypersurface H_{h_s} enclosed in the security zone S, $H_{h_\delta} < H_{h_s} < H_{h_\varepsilon}$, which may not be satisfactory since large population fluctuations occur.

Note 2. The particular situation $x_{2n} = x_{2n}^0$ at $\bar{x}(t_s) \in S$ satisfies (14), hence any value $u \in U$ can be selected temporarily until the response moves to a neighbouring point in S for which $x_{2n} \ne x_{2n}^0$. Then the avoidance control policy can be applied.

LITERATURE REFERENCES

1. Vincent, T.L. Pest management programs via optimal control theory. Biometrics, 31, 1-10 (1975).
2. Goh, B.S., G. Leitmann, and T.L. Vincent. Optimal control of a prey-predator system. Math. Biosc., 19, 263-286 (1974).
3. Bojadziev, G. and J. Skowronski. Controlled food consumption. Methods of Operations Research, 49, 499-506 (1985).
4. Leitmann, G. and J. Skowronski. Avoidance control. J. Optim. Theory and Appl., 23, 581-591 (1977).
5. Blaquiere, A., F. Gerard and G. Leitmann. Quantitative and Qualitative Games. Academic Press, New York, 1969.
6. Huang, H.-W. and H. Morowitz. A method for phenomenological analysis of ecological data. J. theor. Biol., 35, 489-503 (1972).

MATHEMATIQUES ET SYSTEMES, ASPECT CALCUL

MATHEMATICS AND SYSTEMS, COMPUTATIONAL BEARINGS

AN APPROXIMATION PROCEDURE FOR STOCHASTIC CONTROL PROBLEMS. THEORY AND APPLICATIONS

Roberto González and Edmundo Rofman

Facultad de Ciencias Exactas e Ingeniería, Av. Pellegrini 250, (2000) Rosario, Argentina.
INRIA, Domaine de Voluceau, BP 105, Rocquencourt, 78153 Le Chesnay Cedex, France.

ABSTRACT

The aim of this paper is to propose an approximation procedure to compute the value function V and the optimal policy $\hat{u}$ related to the stochastic problem $(\mathcal{P})$ of controlling diffusion processes. This procedure can be easily extended to problems for which stopping time and impulse controls are also considered.

O - INTRODUCTION

As we did in [8] for deterministic problems we will employ here as basic tool of analysis the characterization of V as the maximum element of a suitable set $\mathcal{W}$ of functions w. While in [8] the definition of $\mathcal{W}$ requires for w to be subsolution of the first order Hamilton-Jacobi-Bellman equation, i.e. :

$$\frac{\partial w(x)}{\partial x} \cdot f(x,u) + \ell(x,u) - \alpha w(x) \geq 0, \ \forall u \in U, \tag{0.1}$$

here, in the stochastic case, we deal instead of (1) with

$$\mathcal{L}(u)w + \ell(u) \geq 0 \tag{0.2}$$

where $\mathcal{L}$ is a second order differential operator.

In what follows $(\mathcal{P})$ will be solved using the characterization mentioned above. To introduce the discretized problems $(\mathcal{P}^h)$ we need to define properly the functions w^h belonging to $\mathcal{W}^h$. In fact : the existence of maximum solution V^h for each problem $(\mathcal{P}^h)$ and the convergence of V^h to V are shown using a Discrete Maximum Principle (DMP) that w^h must verify (cfr. [3]). To insure this property we use particular schemes to discretize the first and second derivatives of w. Furthermore this choice enable us to compute V^h using an algorithm of relaxation type that increases the values of w^h in the vertices of the triangulation employed.

Comments on applications are included in the final chapter.

1 - THE PROBLEM (P)

Let us consider :

a) The complete probabilistic space

$$(\Omega, P, \mathcal{F}, \mathcal{F}(t)) \ ; \tag{1.1}$$

b) The state process y(.), modelled by the diffusion

$$\begin{aligned} dy(t) &= f(y(t), u(t))dt + \sigma(y(t), u(t))\, dw(t) \\ y(o) &= x \ , t \geq 0, y \in Q \subset \mathbb{R}^n \end{aligned} \tag{1.2}$$

with

Q : open boundet set
w(t) : Wiener process $\mathcal{F}(t)$-measurable
u(t) : control process progressively measurable in a compact set $U \subset \mathbb{R}^m$
σ is a $n \times n$ matrix
f and σ bounded continuous on $Q \times U$.

c) The cost functional

$$J(x,u(.)) = E \left\{ \int_0^{\tau} \ell(y(s).u(s))\, e^{-\alpha s}\, ds \right\} \tag{1.3}$$

with

τ : first exit time of $\overline{Q}$ of the system trajectory
$\alpha > 0$
ℓ : bounded continuous function on $Q \times U$.

Let us introduce the definition of the optimal cost

$$V(x) = \inf_{u \in U} J(x,u(.)), \tag{1.4}$$

V(x) being solution (cfr. [5],[2]) of the Hamilton-Jacobi-Bellman equation

$$\begin{aligned} &\min_{u \in U} \{\mathcal{L}(u)V + \ell(.,u)\} = 0 \text{ in } Q \\ &V = 0 \text{ in } \partial Q \end{aligned} \tag{1.5}$$

where the differential operator $\mathcal{L}$ is given by :

$$\mathcal{L}(u) = \sum_{r,s=1}^{n} a_{rs}(x,u) \frac{\partial^2}{\partial x_r \partial x_s} + \sum_{r=1}^{n} f_r(x,u) \frac{\partial}{\partial x_r} - \alpha \tag{1.6}$$

with

$$a_{rs} = \tfrac{1}{2} \sum_{z=1}^{n} \sigma_{rz}\, \sigma_{zs}, \text{ i.e. } a_{rs} = a_{sr}. \tag{1.7}$$

As it was said in the Introduction we will compute V taking advantage of its characterization as maximum element of a suitable set, i.e. (cfr. [6], [8], [15]) solving the following auxiliar problem (having V as solution) :

$(\mathcal{P})$: Find the maximum element $\overline{w}$ of the set

$$\mathcal{W} = \{w \in W_0^{1,\infty}(Q) \;/\; \mathcal{L}(u)w + \ell \geq 0 \text{ in } \mathcal{D}'(Q)\ \forall u \in U,\ Q \subset \mathbb{R}^n\} \tag{1.8}$$

being

$$w \leq \tilde{w} \Leftrightarrow w(x) \leq \tilde{w}(x),\ \forall x \in Q \tag{1.9}$$

the natural partial order in $\mathcal{W}$.

(Questions concerning existence and unicity of the solution of $(\mathcal{P})$ can be seen in [4], [15]).

2 - THE DISCRETIZED PROBLEM ($\mathcal{P}^h$)

2.1. Preliminary comments

We will compute V as the limit of the solutions of a sequence of approximate problems ($\mathcal{P}^h$).

To simplify the presentation we will suppose that Q is polyhedric. We consider in Q a triangulation Q^h (union of simplices), x_i^h being (i = 1, 2, ..., N_h) the vertices of Q^h.

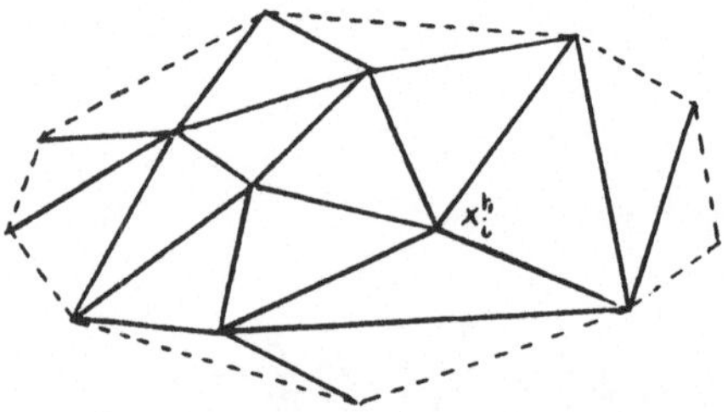

Fig. 1

Then we define $\mathcal{W}^h$ by functions w^h verifying properties related to (1.8), (1.6). The main difficulty of this approach is to ensure the existence of a maximum element $\overline{w}^h$ in $\mathcal{W}^h$.

Following what we did in [8] for the deterministic case we introduce in $\mathcal{W}^h$ the natural partial order

$$w_1^h \leq w_2^h \Leftrightarrow w_1^h(x_i^h) \leq w_2^h(x_i^h), \ \forall x_i^h \text{ vertex of } Q^h \tag{2.1}$$

We consider functions $w^h : \overline{Q}^h \to \mathbb{R}$, w^h continuous in $\overline{Q}^h$ with $\frac{\partial w^h}{\partial x}$ constant in the interior of each simplex of Q^h, i.e., w^h are linear finite elements. So, to define w^h it will be enough to precise the inequality ("discretization" of $\mathcal{L}(u)w + \ell \geq 0$) to be verifyied at each vertex x_i^h of Q^h. Taking [8] into account if suffices to propose a suitable discretization of $L(u)w = \sum_{r,s=1}^{n} a_{rs} \frac{\partial^2 w}{\partial x_r \, \partial x_s}$, the term containing the second order derivatives of w.

2.2. Definition of $L^h(u)\, w^h$

Let us consider $S(x_i^h)$ (see Fig. 2), all the simplices having x_i^h as vertex.

From (1.7) the matrix A = (a_{rs}) has no negative eigenvalues λ_p and orthogonal eigenvectors. So

$$A = UDU' \tag{2.2}$$

with $UU' = I$

D (diagonal) / $D_{pp} = \lambda_p \geqslant 0.$

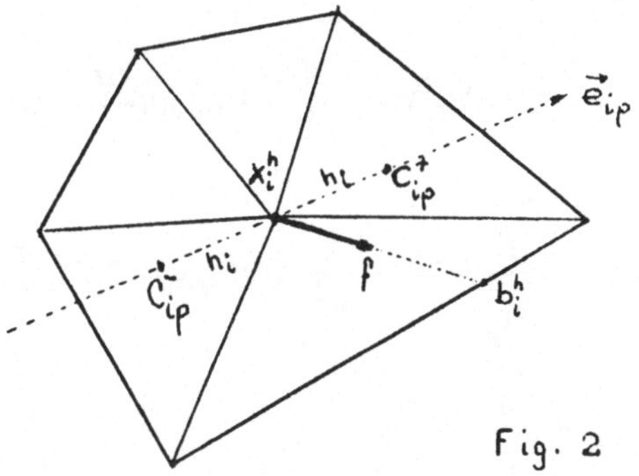

Fig. 2

If we consider, with center in x_i^h a new coordinates system (we denotes G the transformation matrix $G(x_i^h)$:

$$\eta = G \,.\, \xi \tag{2.3}$$

and we define

$$w(x) = w(x_i^h + \xi) \overset{\Delta}{=} \tilde{w}(\xi) = \tilde{w}(G^{-1}\eta) \overset{\Delta}{=} \hat{w}(\eta) \tag{2.4}$$

we obtain

$$L(u)w = \sum_{r,s=1}^{n} a_{rs}(x_i^h, u) \frac{\partial^2 w}{\partial x_r \, \partial x_s} = \sum_{p,q=1}^{n} b_{pq}(x_i^h, u) \frac{\partial^2 \hat{w}}{\partial \eta_p \, \partial \eta_q}, \tag{2.5}$$

with $b_{pq}(x_i^h, u) = (GAG')_{pq}$.

So, after the choice $G = U'$ we have, because

$$b_{pq} = \lambda_p \, \delta_{pq} \tag{2.6}$$

the following diagonal form of L :

$$Lw = \sum_{p=1}^{n} \lambda_p(x_i^h, u) \frac{\partial^2 \hat{w}}{\partial \eta_p^2} \tag{2.7}$$

Now we define naturally the approximated oeprator L^h :

$$L^h\, w^h(x_i^h) = \sum_{p=1}^{n} \lambda_p(x_i^h,u)\Big(\frac{\partial^2}{\partial\eta_p^2}\Big)^h \hat{w}^h \tag{2.8}$$

where $\Big(\frac{\partial^2}{\partial\eta_p^2}\Big)^h \hat{w}^h = \frac{1}{h_i^2}\big(\hat{w}^h((C_{ip}^-)_\eta) - 2\,\hat{w}^h((x_i^h)_\eta) + \hat{w}((C_{ip}^+)_\eta)\big)$ with

$$C_{ip}^- = x_i^h - h_i\,\vec{e}_{ip} \qquad (C_{ip}^-)_\eta = (0,0, \ldots, -h_i, \ldots, 0)$$

$$C_{ip}^+ = x_i^h + h_i\,\vec{e}_{ip} \qquad (C_{ip}^+)_\eta = (0,0, \ldots, h_i, \ldots, 0)$$

$$(x_i^h)_\eta = (0,0, \ldots, 0, \ldots, 0)$$

$\vec{e}_{ip}$ giving the direction of the η_p-axis and h_i such that $C_{ip}^-, C_{ip}^+ \in S(x_i^h)$.

2.3. Definition of W^h

Coming back to (1.6), $\sum_{r=1}^{n} f_r(x,u)\frac{\partial}{\partial x_r}$ will be discretized as it was done in [8], i.e., we will consider ∇ in the direction f (see Fig. 2) :

$$f \cdot \nabla\, w^h(x_i^h) = \frac{w^h(b_i^h) - w^h(x_i^h)}{|\, b_i^h - x_i^h \,|} \cdot |\, f(x_i^h) \,| . \tag{2.9}$$

So, from (2.8) and (2.9) we can define

$$W^h = \{w^h : Q^h \to R \;/\; L^h(u)w^h + \ell(u) \geq 0, \quad \forall u \in U^h, \forall x_i^h \in Q^h, w^h \leq 0 \text{ on } \partial Q^h\} \tag{2.10}$$

where U^h is a finite discretization of U and

$$\mathcal{L}^h(u)w^h(x_i^h) + \ell(u, x_i^h) = \sum_{p=1}^{n} \frac{1}{h_i^2} \lambda_p(x_i^h,u)(w^h(x_i^h - h_i \vec{e}_{ip}) - 2\, w^h(x_i^h)$$

$$+ w^h(x_i^h + h_i \vec{e}^{\,+}_{ip})) + \frac{w^h(b_i^h) - w^h(x_i^h)}{| b_i^h - x_i^h |} \, | f(x_i^h) | - \alpha\, w(x_i^h) + \ell(u, x_i^h). \tag{2.11}$$

Finally we can consider the discretized problem $(\mathcal{P})^h$: Find the maximum element $\overline{w}^h$ of the set $\mathcal{W}^h$ with respect to the partial order (2.1), i.e. find $\overline{w}^h(x)$ such that $\overline{w}^h(x_i^h) \geq w^h(x_i^h)$, $\forall x_i^h \in Q^h$, $\forall w^h \in \mathcal{W}^h$.

3 - SOME REMARKS ABOUT $\overline{w}^h(x)$

As C_{ip}^-, C_{ip}^+, b_i^h are convex combinations of the vertices of $S(x_i^h)$, using the linearity of w^h we have :

$$w^h(C_{ip}^-) + w^h(C_{ip}^+) = \sum_{j\in I_i^h} \gamma_{pj}\, w^h(x_j^h) \qquad \gamma_{pj} \geq 0, \ \sum_{j\in I_i^h} \gamma_{pj} = 2 \tag{3.1}$$

I_i^h set of index such that $x_j^h \in S(x_i^h)$

$$w^h(b_i^h) = \sum_{j\in I_i^h} \gamma_j\, w^h(x_j^h) \qquad \gamma_j \geq 0\,, \ \sum_{j\in I_i^h} \gamma_j = 1. \tag{3.2}$$

After (2.11), (3.1) and (3.2), we can rewrite $\mathcal{L}^h(u)w^h(x_i^h) + \ell(u, x_i^h) \geq 0$ as :

$$w^h(x_i^h) \leq \beta_i^h(x_i^h,u) \Big[\frac{1}{h_i^2} \sum_{p=1}^{n} \lambda_p(x_i^h,u) \sum_{j \in I_i^h, j \neq i} \gamma_{pj}\, w^h(x_j^h) + \frac{\| f(x_i^h) \|}{\| b_i^h - x_i^h \|} \sum_{j \in I_i^h} \gamma_j\, w^h(x_j^h) + \ell(u, x_i^h) \Big] \tag{3.3}$$

with $\beta_i^h(x_i^h,u) = \Big[\sum_{p=1}^{n} \frac{(2 - \gamma_{pi})}{h_i^2} \lambda_p(x_i^h,u) + \frac{\| f(x_i^h) \|}{\| b_i^h - x_i^h \|} + \alpha \Big]^{-1} > 0.$

Taking into account that all the factors that multiply $w^h(x_j^h)$ in the second member of (3.3) are non-negative we can easily prove (see [8]) :

<u>THEOREM 1</u>

There exists an unique $\overline{w}^h(x)$, maximum element of $\mathcal{W}^h$, i.e. $(\mathcal{P}^h)$ has an unique solution.

Furthermore the operator $\mathcal{L}^h$ verifies the following Discrete Maximum Principle (DMP) :

(DMP) : If C is a subset of vertices of Q^h satisfying $\mathcal{L}^h(u)\, w^h(x_i^h) \geq 0$, $\forall x_i^h \in Q^h$, $\forall u \in U^h$, there exists Γ, $0 < \Gamma < 1$ such that : (3.4)

$$w^h(x_i^h) \leq \Gamma \Big(\max_{x_i^h \notin C} (w^h(x_i^h)) \vee 0 \Big).$$

We can use this DMP to establish two important properties of $\overline{w}^h$.

The first one is that $\overline{w}^h$ is characterized by the fact that (3.3) becomes an equality for all $x_i^h \in Q^h$ for some $u \in U^h$ when we put $\overline{w}^h$ instead of w^h. This characterization allows us to compute $\overline{w}^h$ using iterative algorithms of the same type than those pesented in [8]. The value of u giving the equality will be used to define the optimal control $\hat{u}_h$.

The second one concerns the convergence of $\overline{w}^h$ to V. We have

THEOREM 2

The solutions $\overline{w}^h(x)$ of the approximate problems $(\mathcal{P}^h)$ converge uniformly to V(x), solution of $(\mathcal{P})$, i.e. :

$$\lim_{\|h\| \to 0} | V(x) - \overline{w}^h(x) | = 0, \forall x \in Q \tag{3.5}$$

where $\|h\|$ is the maximum of the diameters of the simplex of Q^h. (see [8]).

The proof is achieved in two steps. We will briefly give here the main ideas.

In the first part we show

$$\underline{\lim}_{\|h\| \to 0} \overline{w}^h \geq V. \tag{3.6}$$

For that we regularize the elements of (1.8) by means of a convolution with a function of $C^\infty(\mathbf{R}^2)$ having a parameter $\rho > 0$. These functions w_ρ can be approximate by functions $w_{\rho,\alpha}$ with this property : the linear finite element $w^h_{\rho,\alpha}$, taking the same values of $w_{\rho,\alpha}$ in the vertex of the triangulation Ω^h, belongs to $\mathcal{W}^h$. So,

$$\overline{w}^h \geq w^h_{\rho,\alpha} \tag{3.7}$$

If we consider in (3.7) the lower limits for $\|h\| \to 0$, then the limits for $(\rho,\alpha) \to (0,0)$, we obtain

$$\underline{\lim}_{\|h\| \to 0} \overline{w}^h \geq w. \tag{3.8}$$

Finally, as w is an arbitrary element of $\mathcal{W}$, (3.6) is proved.

The second part is devoted to show

$$\overline{\lim}_{\|h\| \to 0} \overline{w}^h \leq V. \tag{3.9}$$

We consider a sequence of auxiliar problem $\mathcal{P}_n$ for which the controls u_n can take in (1.8) a finite number of values and the number of switchs within that set of values is, at most, n. If V_n is the solution of $\mathcal{P}_n$ we can show

$$V_1 \geq \dots \geq V_n \geq V_{n+1} \geq V \tag{3.10}$$

$$\lim_{n\to\infty} V_n = V.$$

On the other hand we consider the discretized problem $\mathcal{P}_n^h$ for which we prove

$$\lim_{\|h\|\to 0} \overline{w}_n^h = V_n \tag{3.11}$$

$$\overline{w}_n^h \geq \overline{w}_{n+1}^h \geq \dots \geq \overline{w}^h, \forall n. \tag{3.12}$$

So, $\overline{\lim\limits_{\|h\|\to 0}} \; \overline{w}^h \leq V_n$; then, using (3.10) we obtain (3.9). Finally (3.6) and (3.9) give (3.5).

4 - COMMENTS ON SOME APPLICATIONS

The idea of solving optimal control problems computing the maximum element of a suitable set of subsolutions of the Hamilton-Jacobi-Bellman equation has been recently applied to several problems. Remaining in the deterministic approach we have study in [9] the optimization of an electricity production system which comprise three hydraulic plants (two of pumped type) and seven thermic plants (one nuclear, two of coal, tow of fuel, one gas powered and one external). The numerical data have been provided by EDF (Electricity of France) : they describe a forecast of the French system for a week of the year 2000. Other application can be seen in [12] where several serial production/inventory systems are optimized.

Concerning the stochastic approach we can mention :

a) [11] devoted to the optimization of the system presented in [9] considering random perturbations in the demand ;

b) [7] in which the algorithm proposed in [10] for L(u) = Δ is used to obtain the optimal control of a bidimensional diffusion ;

c) [1] in which the numerical solution of an optimal correction problem for a damped random linear oscilator is studied.

First applications of the procedure just proposed in §2 and §3, as well as a comparison of these results with those obtained by other clasic methods [13], [14] and [17], will be presented in a special session of the next IEEE-CDC, Austin, 7-9 Dec. 1988.

5 - REFERENCES

[1] BANCORA M.C. - CHOW P. - MENALDI J.L., "*On the numerical approximation of an optimal correction problem*", submitted for publication.

[2] BENSOUSSAN A. - LIONS J.L., "*Applications des inéquations variationnelles en contrôle stochastique*", Dunod, Paris 1978.

[3] CIARLET P.G. - RAVIART P.A., "*Maximum principle and uniform convergence for the finite element method*", Computer Methods in Applied Mechanics and Engineering, 2 (1973), 17-31.

[4] CRANDALL M.G. - LIONS P.L., "*Viscosity solutions of Hamilton-Jacobi-Bellman equations*", Trans. AMS, 282, (1984), 487-502.

[5] FLEMING W.H. - RISHEL R., "*Optimal deterministic and stochastic control*", Springer-Verlag, Berlin, 1975.

[6] GONZALEZ R., "*Sur l'existence d'une solution maximale de l'équation de Hamilton-Jacobi-Bellman*", CRAS, Paris, 282, (1976),pp. 1287-1290.

[7] GONZALEZ R. - MEDINA M., "*Sobre la solución numérica del control óptimo de una difusión bidimensional*", ENIEF 87, Bariloche, July 1987.

[8] GONZALEZ R. - ROFMAN E., "*On deterministic control problems : an approximation procedure for the optimal cost*", Part I : The stationary case, SIAM J. on Control and Opt. 23, 2, (1985), 242-266 ; Part II : The non stationary case, SIAM J. on Control and Opt. 23, 2, (1985), 267-285.

[9] GONZALEZ R. - ROFMAN E., "*On the optimization of a short-run model of energy production systems*", Lecture Notes in Control and Inf. Sci., Proceedings of the 12th IFIP Conf. Budapest (1985), Springer-Verlag 1986, 757-765.

[10] GONZALEZ R. - ROFMAN E., "*On stochastic control problems. An algorithm for the value function and the optimal policy*", 13th IFIP Conference on System Modelling and Optimization, Tokyo, Aug. 31 -Sept. 4 1987. To appear on Springer Verlag, Lect. Notes in Control and Inf. Sc.

[11] GONZALEZ R. - ROFMAN E., "*On the computation of optimal control policies of energy production systems with random perturbations*", Proc. 26th CDC-IEEE Los Angeles, Dec. 1987, pp. 312-313.

[12] KABBAJ F. - MENALDI J.L. - ROFMAN E., "*Variational approach of serial multi-level production/inventory systems*", RR INRIA 692, Juin 1987.

[13] KUSHNER H., "*Probability methods for approximations in stochastic control and for elliptic equations*", Academic Press, New York (1977).

[14] LIONS P.L. - MERCIER B., "*Approximation numérique des équations de Hamilton-Jacobi-Bellman*" RAIRO Analyse Numérique, 14, (1980), 369-393.

[15] LIONS P.L. - MENALDI J.L., "*Optimal control of stochastic integrals and Hamilton-Jacobi-Bellman equations*", II SIAM J. on Control and Opt. 20, 1, January 1982.

[16] MENALDI J.L., "*Sur les problèmes de temps d'arrêt, contrôle impulsionnel et continu correspondant à des opérateurs dégénérés*", Th. d'Etat, Université Paris Dauphine, Dec. 1980.

[17] THEOSYS, "*Commande optimale de systèmes stochastiques*", RAIRO Automatic 18, (1984), pp. 225-250.

Fuzzy arithmetic in qualitative reasoning

Didier DUBOIS and Henri PRADE

Laboratoire Langages et Systèmes Informatiques
Université Paul Sabatier, 118 route de Narbonne
31062 TOULOUSE Cédex (FRANCE)

The paper provides a preliminary exploration of the application of fuzzy arithmetic and fuzzy approximate reasoning techniques to qualitative reasoning problems considered in Artificial Intelligence. More specifically, this investigation is done along three lines : constraint propagation with ill-known values, handling of orders of magnitude in terms of fuzzy intervals or by means of fuzzy relations.

1 - Introduction

Reasoning about the behavior of systems in a qualitative way is interesting in two kinds of circumstances : i) when the system under consideration is complex and the data available about it are pervaded with imprecision or even vagueness ; ii) when it is sufficient to have a qualitative view of the system and of its behavior, and this qualitative view is not only easier to get than a more precise one from a computational point of view, but also easier to understand. From the beginning of the eighties there have been a growing interest about qualitative reasoning in Artificial Intelligence ; see (Bobrow, 1984 ; Dormoy, 1987) for an introduction. The intended purpose of this research is mainly to provide understandable explanations of the behavior of complex systems from their qualitative description. The modeling is done in terms of variables which are potentially real-valued, but the analysis and the description of the system behavior is made only in terms of three values usually, namely "-", "0" and "+", corresponding to whether the variables are negative, zero or positive. Independently, works motivated by research in qualitative economics, have been developed about qualitative controllability and observability of linear dynamical systems where real-valued variables are approximated in terms of the same three values ; see Travé and Kaszkurewicz (1986) for instance.

From the end of the seventies, fuzzy set and possibility theory (Zadeh, 1978 ; Dubois and Prade, 1985), whose introduction was initially motivated by the modeling of complex and ill-known systems, has been considerably developed both from a theoretical and an applied point of view in various directions ; particularly, fuzzy arithmetic (Dubois and Prade, 1980, 1987) enables us to handle ill-known quantities in an easy way which generalizes interval analysis, and besides a methodology for approximate reasoning (Bellman and Zadeh, 1977) has been settled in the fuzzy set framework. Until now there have been no serious attempt to use fuzzy techniques in qualitative reasoning problems in Artificial Intelligence --if we except some hints (Raiman, 1985) and preliminary works (d'Ambrosio, 1987)-- although it would be desirable in some cases to have a finer and less sharp description of the values of the variables than the one provided by "-", "0" or "+". Particularly, the sign of the difference between two positive quantities cannot be determined without any information about their respective order of magnitude.

This paper investigates what may be the use of fuzzy arithmetic and fuzzy set-based approximate reasoning techniques in qualitative reasoning problems. First, a general approach for refining interval values attached to variables by exploiting constraints which must be satisfied by these variables, is extended to fuzzy set values. Then, a fuzzy interval-based approach is proposed for handling orders of magnitude in arithmetic operations and a valid approximation technique is used in order to insure a closure property of the operations restricted to the considered fuzzy values. The interest of fuzzy intervals for interfacing symbolic information and numerical data, is emphasized. Then another way of dealing with orders of magnitude based on approximate equality relations is investigated. The concluding remarks point out some other contributions of fuzzy logic to qualitative control and to qualitative descriptions of systems behavior.

2 - Constraint propagation with fuzzy values

2.1 - *General discussion*

Let $X_1, \ldots, X_n$ denote single-valued real variables. Let A_i be a subset of the real line which is known to restrict the possible values of X_i, and let R be a relation which must be satisfied by the X_i's and which acts as a constraint on $(X_1, \ldots, X_n)$. Then, the refinement of the possible ranges of the variables X_i's taking into account R, leads to update the possible range of each variable X_i into a new subset A'_i in the following way

$$A'_i = \{x_i \in A_i \mid \exists\, x_j \in A_j, j = 1,n, j \neq i \text{ and } (x_1, \ldots, x_i, \ldots, x_n) \in R\} \quad (1)$$

More generally in case of several constraints represented by relations R_k, $k = 1,r$, we can iterate this refinement procedure on each variable taking successively each relation into account over and over until no more changes occur in the updated ranges. This is known in Artificial Intelligence as the Waltz algorithm ; see Davis (1987) for a detail study of this procedure both from an implementation and an application point of view. Let us consider a simple example. Let n = 3, $A_1 = [0,2]$, $A_2 = [1,3]$ and $A_3 = [0,2]$ and the constraint $X_1 + X_2 = X_3$. Then we get $A'_1 = [0,1]$, $A'_2 = [1,2]$ and $A'_3 = [1,2]$. Observe that any triple of values in the Cartesian product $A'_1 \times A'_2 \times A'_3$ is not necessarily feasible, e.g. $\nexists\, x_3 \in A'_3$ such that $x_1 + x_2 = x_3$ with $x_1 = 1$ and $x_2 = 2$.

The definition (1) expresses that A'_i is obtained as the intersection of A_i with the result of the composition of the relation R with the Cartesian product of the A_j's except A_i. This can be readily extended to the case where the A_i's are fuzzy sets and/or R represents a fuzzy constraint ; i.e.

$$\forall i, \forall x_i,\ \mu_{A'_i}(x_i) = \min[\mu_{A_i}(x_i),\ \sup_{x_j,\ j=1,n\,;\, j\neq i}\ \min(\mu_R(x_1, \ldots, x_n),\ \min_{j=1,n\,;\, j\neq i}\ \mu_{A_j}(x_j))] \quad (2)$$

where μ denotes the membership functions (whose range are [0,1]) of the corresponding fuzzy sets and relation. When R is an ordinary relation such that X_i is a function f of the other variables X_j, A'_i is a fuzzy set which can be obtained by applying f, in the sense of fuzzy set and possibility theory, to the A_j's ($j \neq i$), i.e.

$$\forall x_i,\ \mu_{A'_i}(x_i) = \min[\mu_{A_i}(x_i), \sup_{f(x_j, j=1,n, j\neq i) = x_i} \ \min_{j=1,n\,;\, j\neq i} \mu_{A_j}(x_j)] \quad (3)$$

When the A_j's are fuzzy intervals and f is monotonic with respect to each variable and can be expressed in terms of arithmetic operations, the A'_i's are fuzzy intervals which can be easily computed using results of fuzzy arithmetic ; see Dubois and Prade (1985, 1987). This extends the fact that, for instance, in the above example the A'_i's can be obtained as the result of operations on intervals ; namely $A'_1 = A_1 \cap (A_3 \ominus A_2)$, $A'_2 = A_2 \cap (A_3 \ominus A_1)$, $A'_3 = A_3 \cap (A_1 \oplus A_2)$, where the circled symbols are used for denoting the extension of arithmetic operations to intervals. Indeed fuzzy arithmetic generalizes interval arithmetic. Note that the refinement is obtained in (2) in one step, in the sense that refined A'_j's cannot enable us to obtain a more restrictive A'_i. This can be easily checked ; indeed, taking n = 2 for notational convenience, we have

$$\min(\mu_{A_1}(x_1), \sup_{x_2} \min(\mu_R(x_1,x_2), \mu_{A'_2}(x_2)))$$

$$= \sup_{x_2} \min(\mu_{A_1}(x_1), \mu_R(x_1,x_2), \sup_{x_1} \min(\mu_{A_1}(x_1), \mu_R(x_1,x_2)), \mu_{A_2}(x_2))$$

$$= \mu_{A'_1}(x_1) \text{ since obviously } \min(\mu_{A_1}(x_1), \mu_R(x_1,x_2)) \leq \sup_{x_1} \min(\mu_{A_1}(x_1), \mu_R(x_1,x_2))$$

In fact, (2) can be viewed as a particular case of the general approach to approximate reasoning initiated in Bellman and Zadeh (1977) and developed in Zadeh (1979), namely, all the pieces of information are conjunctively combined and then the result is projected on the domain of the variable(s) in which we are interested. Indeed (2) can be equivalently rewritten

$$\forall i, \forall x_i,\ \mu_{A'_i}(x_i) = \sup_{x_j,\ j=1,n\,;\, j\neq i} \min(\mu_R(x_1, \ldots, x_n), \mu_{A_1}(x_1), \ldots, \mu_{A_i}(x_i), \ldots, \mu_{A_n}(x_n)) \quad (4)$$

In case of several relations R_k the combination/projection method leads to the following updating scheme where the R_k's are replaced by their cylindrical extensions when they do not involve all the variables

$$\forall i, \forall x_i,\ \mu_{A'_i}(x_i) = \sup_{x_j,\ j=1,n\,;\, j\neq i} \min(\min_{k=1,r} \mu_{R_k}(x_1, \ldots, x_n), \min_{j=1,n} \mu_{A_j}(x_j)) \quad (5)$$

$$\leq \min_{k=1,r} [\min(\mu_{A_i}(x_i), \sup_{x_j,\ j=1,n\,;\, j\neq i} \min(\mu_{R_k}(x_1, \ldots, x_n), \min_{j=1,n\,;\, j\neq i} \mu_{A_j}(x_j)))] \quad (6)$$

The inequality (6) expresses that if we take into account each R_k separately in the refinement process, we are not sure, even if we iterate the procedure as in the Waltz algorithm, of obtaining the most accurate refinement for each variable range. However, what is got by (6) is obviously valid and more easy to compute in general.

Note that in case of binary relations, the Waltz procedure (i.e. the separate processing of the R_k's) yields the most accurate result given by (5), provided there is at most one relation R_k between any pair of variables (x_i,x_j) and that there is no cycle in the non-oriented graph whose nodes correspond to the variables and edges to the binary relations. Indeed, for instance with n = 3 and two relations, we have

$$\mu_{A'_1}(x_1) = \min(\mu_{A_1}(x_1), \sup_{x_2,x_3} \min(\mu_R(x_1,x_2), \mu_{R'}(x_2,x_3), \mu_{A_2}(x_2), \mu_{A_3}(x_3))$$
$$= \min(\mu_{A_1}(x_1), \sup_{x_2} \min(\mu_R(x_1,x_2), \min(\mu_{A_2}(x_2), \sup_{x_3} \min(\mu_{R'}(x_2,x_3), \mu_{A_3}(x_3)))))\quad (7)$$

2.2 - *Fuzzy equalities and inequalities*

In this subsection, we consider particular fuzzy relations which are of interest in practice for qualitative reasoning. Approximate equalities or strong inequalities (e.g. 'much greater than") are examples of binary fuzzy relations which can be easily handled using fuzzy arithmetic techniques. Indeed an approximate equality can be modelled by a fuzzy relation E of the form $\mu_E(x,y) = \mu_L(|x - y|)$, for instance

$$\forall x, \forall y, \mu_E(x,y) = \max(0, \min(1, \frac{\delta + \varepsilon - |x - y|}{\varepsilon})) = \begin{cases} 1 \text{ if } |x - y| \leq \delta \\ 0 \text{ if } |x - y| \geq \delta + \varepsilon \\ \frac{\delta + \varepsilon - |x - y|}{\varepsilon} \text{ otherwise} \end{cases} \quad (8)$$

where δ and ε are respectively positive and strictly positive parameters which modulate the approximate equality. Then the approximate equality of variables X and Y (in the sense of E) will be written under the form of the equality

$$X - Y = L \quad (9)$$

with the following intended meaning : the possible values of the difference X - Y are restricted by the fuzzy set L. Here L is a fuzzy interval centered in 0, i.e. L = -L since $\mu_L(d) = \mu_L(-d)$ or if we prefer $\mu_E(x,y) = \mu_E(y,x)$. Similarly a strong inequality can be modelled by a relation I of the form $\mu_I(x,y) = \mu_K(x - y)$, for instance

$$\forall x, \forall y, \mu_I(x,y) = \max(0, \min(1, \frac{x - y - \lambda}{\rho})) = \begin{cases} 1 \text{ if } x \geq y + \lambda + \rho \\ 0 \text{ if } x \leq y + \lambda \\ \frac{x - y - \lambda}{\rho} \text{ otherwise} \end{cases} \quad (10)$$

where $\lambda \geq 0$ and $\rho > 0$. The constraint 'X is much greater than Y' (in the sense of I) can then be written

$$X - Y = K \quad (11)$$

where K is a fuzzy interval such that K = [K,+∞) (with $\mu_{[K,+\infty)}(t) = \sup_{s \leq t} \mu_K(s)$), i.e. K identifies itself as the set of values equal or greater than a value restricted by K.

If we know for instance that 'X_1 is approximately equal to X_2' (i.e. $X_1 - X_2 = L$) and that 'X_2 is much greater than X_3' (i.e. $X_2 - X_3 = K$), we can deduce that

$$X_1 - X_3 = L \oplus K$$

where ⊕ denotes the addition extended to fuzzy intervals[1] (see Dubois and Prade (1980, 1987)). It can

1. Let ⊙ denote the extension of an arithmetic operation ^ to fuzzy sets of the real line. ⊙ is defined by $\mu_{K \odot L}(u) = \sup_{s,t,\ u=s\hat{}t} \min(\mu_K(s), \mu_L(t))$. Besides $\mu_{f(K)}(t) = \sup_{t=f(s)} \mu_K(s)$. When ^ is the addition and K and L are trapezoids represented by the abscissas of the endpoints of their parallel sides, it can be proved that $(k_1, k_2, k_3, k_4) \oplus (l_1, l_2, l_3, l_4) = (k_1 + l_1, k_2 + l_2, k_3 + l_3, k_4 + l_4)$ (k_i or l_j may be equal to $-\infty$ or $+\infty$).

be proved that it means that it is certain that $X_1 \geq X_3 + \lambda - (\delta + \varepsilon)$ and that the value of the difference $X_1 - X_3$ belongs to $L \oplus K$ at the degree 1 as soon as $X_1 \geq X_3 + \lambda + \rho - \delta$. See Figure 1. Then depending on the respective values of the parameters, X_1 is still greater than X_3 (but may be not as much as X_2 with respect to X_3) (if $\lambda > \delta + \varepsilon$), or we are only sure that X_1 is not much smaller than X_3 (if $\lambda + \rho < \delta$). Moreover, if we know that $X_3 = A_3$, we shall get

$$X_1 = A'_1 = A_3 \oplus L \oplus K$$

This is a particular case of (7) where $R = E$, $R' = I$, $A_2 = (-\infty,+\infty) = A_1$.

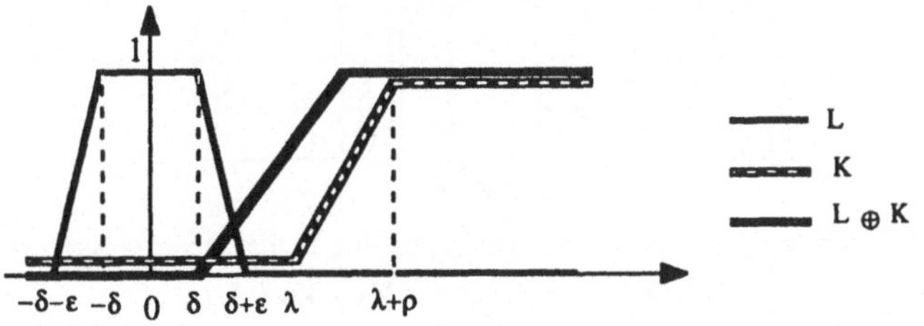

Figure 1

2.3 - *Linear constraints*

Another worth-considering particular case of the general problem presented in 2.1 is the one of linear systems of constraints. For sake of simplicity, we only briefly discuss linear systems with two variables and two constraints of the form

$$\begin{cases} a_1X_1 + b_1X_2 = A_3 \\ a_2X_1 + b_2X_2 = A_4 \end{cases}$$

where A_3 and A_4 are fuzzy sets of real numbers, and the other coefficients are real numbers. Note that each of these constraints implicitly defines a fuzzy relation which restricts the possible values of the pair (X_1,X_2). Provided that $a_1b_2 - a_2b_1 \neq 0$, we can deduce, using (3), that

$$X_1 = A'_1 = \frac{b_2A_3 \ominus b_1A_4}{a_1b_2 - a_2b_1} \ ; \ X_2 = A'_2 = \frac{a_2A_3 \ominus a_1A_4}{a_2b_1 - a_1b_2} \qquad (12)$$

with $A_1 = A_2 = (-\infty,+\infty)$; see the footnote 1 for the definition of the extended difference $\ominus$ and of the product of a fuzzy quantity by a scalar. If the constraints are changed into $a_1X_1 + b_1X_2 = X_3$ and $a_2X_2 + b_2X_2 = X_4$, with $X_3 = A_3$ and $X_4 = A_4$, the ranges of possible values of X_3 and X_4 are respectively updated into $A'_3 = A_3 \cap (a_1A'_1 \oplus b_1A'_2)$ and into $A'_4 = A_4 \cap (a_2A'_1 \oplus b_2A'_2)$.

More generally, the coefficients in linear systems may be ill-known. Then direct extensions of (12) can still be used where the a_i's and b_j's are replaced by fuzzy quantities and where we use the product and the quotient defined in fuzzy arithmetics. However in that case we get ranges which are still valid but may be larger than the actual ranges. This is due to the interactivity constraint which requires that the values of a_i or b_j should be the same at the numerators and the denominators in (12), even if the coefficients are ill-known, and which is forgotten in a straightforward calculation. This interactivity constraint should be taken into account for obtaining the actual ranges. See Dubois (1987) for a general discussion of fuzzy linear programming.

3 - Fuzzy intervals and orders of magnitude

Standard qualitative reasoning distinguishes between values which are strictly negative (-), zero (0) or strictly positive (+), and is based on the exploitation of the following tables for the addition and the product

⊕	0	+	-	?
0	0	+	-	?
+	+	+	?	?
-	-	?	-	?
?	?	?	?	?

⊗	+	-	?	0
+	+	-	?	0
-	-	+	?	0
?	?	?	?	0
0	0	0	0	0

Tables 1

where ? denotes the completely unknown value corresponding to the range $(-\infty,+\infty)$. However, if we know for instance that

$$X_1 = + \;;\; X_3 = + \;;\; X_1 + X_2 = X_3$$

we can only deduce $X_2 = ?$ (while if $X_1 = 0$, we get $X_2 = +$). Another simple example of the undesirably limited representation power of the above calculus is the following

$$\text{if } X_1 = + \text{ and } X_2 = + \text{ then } X_3 = X_1 + X_2 = +$$

then the fact that $X_3 > X_1$ and $X_3 > X_2$ is forgotten. These kinds of ambiguities could be removed, if a more precise knowledge about the orders of magnitude, which is often available, could be modelled. Indeed we have in the general case for the first above example

$$X_1 = A_1 \;;\; X_2 = A_2 \;;\; X_3 = A_3 \;;\; X_1 + X_2 = X_3$$

from which we deduce $X_2 = A'_2 = A_2 \cap (A_3 \ominus A_1)$.

This kind of thing still can be done in an approximate way when the A_i's are required to belong to a prescribed set of labels, such as, for instance : negative large (NL), negative medium (NM), negative small (NS), zero (0), positive small (PS), positive medium (PM), positive large (PL), unknown (?). These labels can be represented by fuzzy intervals such as the ones pictured in Figure 2. They form a (fuzzy) partition of the real line in some sense.

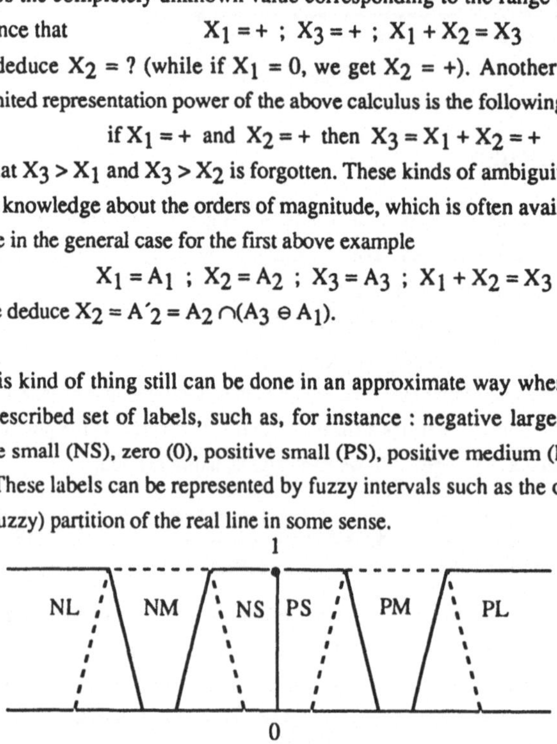

Figure 2

The condition requested to build a meaningful qualitative calculus are twofold :

C1. The advantage of qualitative reasoning is linked to the existence of symbolic calculation tables such as the ones above. Such tables should be kept when absolute orders of magnitude are introduced.

C2. The calculus, even qualitative, should remain consistent with the real line and the operations of the real line of which it is an approximation.

Standard qualitative reasoning trivially meets these requirements. However going beyond the four symbols -, 0, +, ? may look challenging. Indeed the closure property of the table seems to be incompatible with condition C2. For instance let $\mathcal{S}$ be the totally ordered set of symbols {NL, NM, NS, 0, PS, PM, PL} ; PS ⊕ PS = PM looks reasonable at first sight. But PS is of the form]0,a] and PS ⊕ PS =]0,2a] ≠ PM = [a,b]. Moreover $\lim_{n\to+\infty} n\text{PS} = ?$. Hence results obtained from the addition table built from $\mathcal{S}$ such that PS ⊕ PS = PM is inconsistent with the addition on the reals.

It does not mean that qualitative reasoning based on absolute orders of magnitude is a utopia. Interpreting orders of magnitude as intervals or fuzzy intervals apparently forbids the closure property of calculation tables. But the closure property can be preserved on subsets of $\mathcal{S}$ containing adjacent elements, instead of $\mathcal{S}$ itself, provided that we look for the best approximation (in the sense of inclusion) of $s_i \oplus s_j$ by means of unions of adjacent s_k's, i.e. $s_i \oplus s_j \subseteq \bigcup_{k\in K} \{s_k\}$. Note that the introduction of the symbol ? in the usual qualitative tables meets the same purpose, that is $+\oplus- \subseteq \{-,0,+\} = ?$. What is proposed is just a generalization of the way the symbol ? appears.

The example of Figure 2 leads to consider the following term set $\mathcal{T}$ = {NL, NM, NS, 0, PS, PM, PL, [NL,NM], [NM,NS], [NS,PS]..., [NL,PM], [NM,PL], ?} where $[s_i,s_j] = \{s_k \mid s_i \le s_k \le s_j\}$ for $s_i \in \mathcal{S}-\{0\}$, $s_j \in \mathcal{S}-\{0\}$, $s_i < s_j$. Of course + = [PS,PL] and - = [NL,NS]. Note that if $\mathcal{S}$ has n elements distinct from 0 then $|\mathcal{T}| = (n + (n-1) + \ldots + 1) + 1 = \frac{(n+1)n}{2} + 1$ elements. Here $|\mathcal{T}| = 22$, for instance. This size is not so large for contemporary computers.

⊕	PS	PM	PL	PM⁻	PM⁺	+
PS	+	PM⁺	PL	+	PM⁺	+
PM	PM⁺	PM⁺	PL	PM⁺	PM⁺	PM⁺
PL	PL	PL	PL	PL	PL	PL
PM⁻	+	PM⁺	PL	+	PM⁺	+
PM⁺	PM⁺	PM⁺	PL	PM⁺	PM⁺	PM⁺
+	+	PM⁺	PL	+	PM⁺	+

Table 2 : $PM^- = [PS,PM]$; $PM^+ = [PM,PL]$

In Table 2 is part of the addition table (for strictly positive symbols), without any assumption regarding the model of PS, PM, PL (except that they are adjacent). Note that this Table corresponds to an associative operation, when restricted to positive values. However, it is no longer possible to preserve associativity on the whole table. This is due to the approximation procedure since associative operations remain associative when extended to intervals or fuzzy intervals. For instance with NL = -Pl, (NL ⊕ PM) ⊕ PS = -⊕PS = [NL,PS], while NL ⊕(PM ⊕ PS) = NL ⊕ PM^+ = ?. However this lack of associativity does not prevent to use this approach, since the ranges which are obtained will be always valid even if they may be too large with respect to the available knowledge. Moreover, we may try to perform operations in a way where no information is lost.

The addition law can be improved (with regard to the precision of its results by subsequent requirements for instance PS ⊕ PS = PM⁻, which forces PS =]0,a], PM = [a,b] with $2a \leq b$. Note that it is not necessary to use fuzzy intervals. Adjacent intervals can do the job. However there will be discontinuity problems when the (real) values of variables cross the boundaries of the intervals modeling the symbol. Only fuzzy intervals can cope with these problems.

4 - Fuzzy relations and orders of magnitude

Orders of magnitude can be expressed in an absolute way in terms of labels such as "small", "medium" or "large" which can be represented by fuzzy intervals, as said in section 3. They can also be handled in a relative way by means of relations. This is the topic of the present section. Raiman (1985, 1986) has proposed a formal system for order of magnitude reasoning with three binary operators : Ne (for 'negligible in relation to'), Vo (for 'close to'), and Co (for 'comparable to'). Inference rules, which can be justified from a Non-Standard Analysis point of view, describe how these operators work together. See Bourgine and Raiman(1986) for an application in macroeconomics. In the following, we discuss the modeling of these operators in terms of fuzzy relations.

The idea of closeness seems to be naturally captured by an approximate equality relation. Raiman (1986) relates the ideas of closeness and of negligibility in the following way : 'x is close to y' is equivalent to '(x - y) is negligible in relation to y'. In other words, 'x is negligible in relation to y' if and only if 'x + y is close to y'. If we use an approximate equality of the form $\mu_E(x,y) = \mu_L(|x - y|)$ (as in 2.2) for modelling 'close to', the above equivalence would lead to a definition of 'negligible' which would not be relative (since $|(x + y) - y| = |x|$ does not depend on y), but absolute. It can be avoided by defining the fuzzy relation 'Vo' in terms of a quotient, i.e.

$$\mu_{Vo}(x,y) = \mu_M\left(\frac{x}{y}\right) \qquad (13)$$

where the characteristic function μ_M is such that $\mu_M(1) = 1$ and $\mu_M(t) = \mu_M(\frac{1}{t})$. Thus we have $\mu_{Vo}(x,y) = \mu_{Vo}(y,x)$ and M is a fuzzy interval which restricts values which are around 1 and which is equal to its "inverse", i.e. $M = \frac{1}{M}$ (however we have not $M^2 = 1$!). Then it leads to define the extent to which x is negligible in relation to y, by

$$\mu_{Ne}(x,y) = \mu_M\left(\frac{x+y}{y}\right) \qquad (14)$$

The combination/projection method, used in 2.1, enables us to perform the composition of Vo or of Ne with itself, or of Vo with Ne. The following results are easy to establish [2]

2. Warning : in interval arithmetic and more generally in fuzzy arithmetic, the product MM is equal to M^2 if and only if M is either positive (i.e. $\mu_M(x) > 0 \Rightarrow x \geq 0$) or negative (i.e. $\mu_M(x) > 0 \Rightarrow x \leq 0$). Here in practice M is positive, but not (M-1).

$$\sup_y \min(\mu_{Vo}(x,y), \mu_{Vo}(y,z)) = \mu_{MM}(\frac{x}{z}) \geq \mu_{Vo}(x,z) \qquad (15)$$

$$\sup_y \min(\mu_{Ne}(x,y), \mu_{Ne}(y,z)) = \mu_{[(M-1)(M-1) \oplus 1]}(\frac{x+z}{z}) \leq \mu_{Ne}(x,z) \qquad (16)$$

$$\sup_y \min(\mu_{Vo}(x,y), \mu_{Ne}(y,z)) = \mu_{[M(M-1) \oplus 1]}(\frac{x+z}{z}) \geq \mu_{Ne}(x,z) \qquad (17)$$

$$\sup_y \min(\mu_{Vo}(x+y, z), \mu_{Ne}(y,x)) = \mu_{MM}(\frac{x}{z}) \geq \mu_{Vo}(x,z) \qquad (18)$$

They correspond to the following inference rules proposed by Raiman (1986) (for sake of brevity, here we only discuss a part of the 30 rules used in the formal system)

(i) $(x \text{ Vo } y) \wedge (y \text{ Vo } z) \rightarrow (x \text{ Vo } z)$; (ii) $(x \text{ Ne } y) \wedge (y \text{ Ne } z) \rightarrow (x \text{ Ne } z)$

(iii) $(x \text{ Vo } y) \wedge (y \text{ Ne } z) \rightarrow (x \text{ Ne } z)$; (iv) $((x+y) \text{ Vo } z) \wedge (y \text{ Ne } x) \rightarrow (x \text{ Vo } z)$

The fuzzy relation approach shows that several of these rules are only "qualitatively valid". Indeed in (15), the fact that MM is a fuzzy set which contains M mirrors the intuitively satisfying lack of transitivity of the fuzzy relation Vo, strictly speaking. By contrast, as shown by (16), the relation Ne <u>is</u> transitive. The repeated use of the formal rules (i), (iii) or (iv) without control can lead to dubious conclusions in a way similar to sorites such as the bald man paradox (i.e., adding an hair to a bald man leaves him bald, but if we repeat the addition...). The results of the composition of fuzzy relations, such as (15)-(18), are easy to compute in terms of simple fuzzy arithmetic operations on M. The fuzzy relation calculus enables us to reason about closeness and negligibility in a rigorous way without limitations on the chaining by means of control techniques.

<u>N.B. 1</u> Inference rules expressing the compatibility of the relations with respect to arithmetic operations, such as $(x \text{ Vo } y) \wedge (z \text{ Ne } t) \rightarrow xz \text{ Ne } yt$ can be also discussed in our framework. Indeed it can be proved that

$$\sup_{\substack{x,y,z,t \\ u=xz\,;\,v=yt}} \min(\mu_{Vo}(x,y), \mu_{Ne}(z,t)) = \mu_{[M(M-1) \oplus 1]}(\frac{u+v}{v}) \geq \mu_{Ne}(u,v) \qquad (19)$$

Again we see that the rule is only "qualitatively valid", i.e. xz may be slightly less negligible with respect to yt than z in relation to t. Alternatively, we could compute what is the possibility that u is not negligible (in the sense of Ne) with respect to v, from (19).

<u>N.B. 2</u> Note that we have only an approximate equality between $\mu_{Ne}(x,y)$ and $\mu_{Ne}(-x,y)$ using (14) ; a perfect equality could be recovered by modifying (14) into $\mu_{Ne}(x,y) = \mu_M(\frac{y+x}{y-x})$.

<u>N.B. 3</u> Raiman (1986) makes use of a third relation Co which is such that if x Vo y, then x Co y and expresses that two values have the same sign and the same order of magnitude. We may imagine to define Co in relation to Vo and Ne in different ways, for instance by expressing that x Co y iff $\forall z$, $x \text{ Ne } z \Leftrightarrow y \text{ Ne } z$, following Raiman (1986). Another way would be to state that x Co y iff **not**$[(x \text{ Ne } y) \wedge (y \text{ Ne } x)]$ in the sense of some fuzzy negation **n** to be chosen in relation with μ_M in order to have $\max(\mathbf{n}[\mu_M(1+u)], \mathbf{n}[\mu_M(1+\frac{1}{u})]) \geq \mu_M(u)$, $\forall u$ (in order to guarantee $\mu_{Co} \geq \mu_{Vo}$).

5 - Concluding remarks

Other tools, not presented here, which have been also developed in fuzzy set or possibility theory, may turn to be useful in qualitative reasoning. Qualitative descriptions of the dependency between variables of the form "the more (or the less) X_1 is A_1 and... and X_n is A_n, the more (or the less) Y is B", where A_1, ..., A_n and B are gradual properties, can be conveniently represented (by means of a special kind of fuzzy relation) and dealt with in the framework of fuzzy logic, as recently shown in Dubois and Prade (1988). Such gradual rules naturally provide a qualitative description of the behavior of systems. For instance, with n = 2, A_1 = 'large', A_2 = 'small', B = 'large' and the hedges "the more... the more", we express that "if X_1 increases and X_2 decreases then Y increases" (the nature of the increasingness or of the decreasingness can be modulated through a proper choice of μ_{A_1}, μ_{A_2} and μ_B).

Besides, a methodology for the control of complex dynamical systems by means of fuzzy expert rules which provide a qualitative description in terms of fuzzy sets of the relation between action variables and observable state variables, was settled more than ten years ago (Mamdani and Assilian, 1975) ; see Sugeno (1985) for an overview of existing applications. People in Artificial Intelligence have also considered the problem of qualitative control recently (e.g. Clocksin et Morgan, 1986).

The intended purpose of this short communication is to point out that fuzzy set and possibility theory can offer valuable tools for qualitative reasoning problems. In particular "commonsense" arithmetic reasoning (e.g. Simmons, 1986) can be easily handled using fuzzy intervals and fuzzy comparison relations. This framework is especially useful for interfacing numerical data and symbolic information.

References

d'Ambrosio B. (1987) Extending the mathematics in qualitative process theory. Proc.6th National Conf. on Artificial Intelligence (AAAI-87), Seattle, July 13-17, 595-599.

Bellman R., Zadeh L.A. (1977) Local and fuzzy logics. In : Modern Uses of Multiple-Valued Logic (J.M. Dunn, G. Epstein, eds.), Reidel Publ., Dordrecht, 103-165.

Bobrow D.G. (ed.) (1984) Qualitative Reasoning about Physical Systems (with papers by J. De Kleer, K.D. Forbus, B. Kuipers, B.C. Williams,...). North-Holland, Amsterdam. Also special volume of Artificial Intelligence, 24, 1984.

Bourgine P., Raiman O. (1986) Economics as reasoning on a qualitative model. Proc. Inter. Conf. on Economics and Artificial Intelligence, Aix-en-Provence, September, 185-189.

Clocksin W.F., Morgan A.J. (1986) Qualitative control. Proc. 7th European Conf. on Artificial Intelligence, Brighton, July, Vol. 1, 350-356.

Davis E. (1987) Constraint propagation with interval labels. Artificial Intelligence, 32, 281-331.

Dormoy J.L. (1987) Résolution qualitative : complétude, interprétation physique et contrôle. Mise en œuvre dans un langage à base de règles : BOOJUM. Doct. Thesis, Univ. Paris VI, December.

Dubois D. (1987) Linear programming with fuzzy data. In : Analysis of Fuzzy Information, Vol. 3 : Applications in Engineering and Sciences (J.C. Bezdek, ed.), CRC Press, Boca Raton, Fl., 241-263.

Dubois D., Prade H. (1980) Fuzzy Sets and Systems : Theory and Applications. Academic Press, New York.

Dubois D., Prade H. (1985) (with the collaboration of Farreny H., Martin-Clouaire R., Testemale C.) Théorie des Possibilités. Applications à la Représentation des Connaissances en Informatique. Masson, Paris, (2nd revised and extended edition, 1987). English version : Possibility Theory. An Approach to Computerized Processing of Uncertainty. Plenum Press, New York, 1988.

Dubois D., Prade H. (1987) Fuzzy numbers : an overview. In : Analysis of Fuzzy Information - Vol. 1 : Mathematics and Logic (J.C. Bezdek, ed.), CRC Press, Boca Raton, Fl., 3-39.
Dubois D., Prade H. (1988) Gradual inference rules in approximate reasoning. Submitted.
Mamdani E.H., Assilian S. (1975) An experiment in linguistic synthesis with a fuzzy logic controller. Inter. J. Man- Machine Studies, 7, 1-13.
Raiman O. (1985) Raisonnement qualitatif. Tech. Rep. n° F093, Centre Scientifique IBM France, Paris, November.
Raiman O. (1986) Order of magnitude reasoning. Proc. 5th National Conf. on Artificial Intelligence (AAAI-86), Philadelphia, PA, August, 100-104.
Simmons R. (1986) "Commonsense" arithmetic reasoning. Proc. 5th National Conf. on Artificial Intelligence (AAAI-86), Philadelphia, PA, August, 118-124.
Sugeno M. (ed.) (1985) Industrial Applications of Fuzzy Control. North-Holland, Amsterdam.
Travé L., Kaszkurewicz E. (1986) Qualitative controllability and observability of linear dynamical systems. Proc. IFAC Cong. Large Scale Systems, Zürich, Switzerland, August, 964-970.
Zadeh L.A. (1979) A theory of approximate reasoning. In : Machine Intelligence, Vol. 9 (J.E. Hayes, D. Michie, L.I. Mikulich, eds.), Elsevier, N.Y., 149-194.

EXISTENCE AND COMPUTATION OF SOLUTIONS FOR THE TWO DIMENSIONAL MOMENT PROBLEM

György Sonnevend*

Inst. für Angewandte Mathematik,

Universität Würzburg

D-8700 Würzburg, Am Hubland

Introduction

In this paper we deal with some problems of the theory of two dimensional.polynomial moment problems. More precizely we give necessary and sufficient conditions for the existence of a solution, i.e. of a nonnegative mass distribution supported within a fixed, a priori given subset S of R^2, which has a finite set of moments with prescribed values.We study the problem of characterizing all minimal support solutions, i.e. those solutions which have a minimal number of atoms.

The connections between the restricted (or finite), classical, polynomial (onedimensional) moment problem (as a special case of the moment problems of Nevanlinna-Pick type) and various other problems in the theory of orthogonal polynomials, rational Pade approximation (interpolation of Stieltjes functions),restriction of self adjoint operators to Krylow-subspaces, construction of quadrature formulae, minimal partial realizations of causal linear input-output maps, are well known. Similar applications for the considered two dimensional generalization motivate our study. The method we use for the solution of these problems is operator theoretic and is based on solving an "extension problem" for pairs of commuting, self adjoint operators.The characterization obtained for the minimal support solutions,i.e. for the analogons of the Gaussian quadrature formulae is different from the previous approaches, which (as far as we know) used two dimensional orthogonal polynomials (searching for their common zeros) and polynomial ideal theory, see [11] for an extensive set of historical and current references. We were inspired by the operator theoretic treatment of moment problems as developped in [12], see also the method of the paper[16].

* on leave from Dept. of Numer.Anal.,Eötvös University

H. 1088,Budapest, Muzeum k.6-8,Föe'p.

Since the minimal support solutions are, in general non unique in the higher dimensional case (in contrast to the onedimensional case) moreover their set (thus the problem of finding at least one element of it) is not convex and for other reasons like the complexity and stability (with respect to errors in the prescribed moments) we propose and study here an other, particular (nonminimal) solution , i.e. mass distribution, the so called analytical centre of the feasible set (of solutions). Several positive features and applications of this solution concept, like stable computability with a relatively small number of arithmetical operations and the feasibility of high degree homotopy methods for computing bounds for any further, not specified "moment" (i.e. integrals with respect to the underlying measure) are studied in the last section.

2. Preliminaries

Suppose that $S \subset R^n$ is a closed set and μ is a nonnegative (Radon) measure supported within S. In the general, finite or restriced moment problem we shall study here the data are the N values reals

(2.1) $C_j = \int_S K_j(s)\ \mu(ds) = \varphi_j(\mu),\ j=1,\dots,N$

of fixed, linear (continuous) functionals φ_j, given by continuous on S functions K_j, $j=1,\dots,N$ on S and one asks for the conditions of the existence and a characerization of all solutions μ which have minimal support belonging to S:

(2.2) $M \to \min,\ C_j = \Sigma\ K_j(s_k)\ \mu_k,\ \mu_k\quad 0,\ s_k \in S,\ k = 1,\dots,M.$

In the case when $S \subset R^2$, i.e. n=2, and for $s = (x,y)$ the functions $K_1,\dots,K_N$ have the form

(2.3) $x^i y^j,\ (i,j) \in I,\ |I| = N$

where I is a finite subset of Z_+^2 (the set of nonnegative entires) of cardinality N, the above problem - the so called restricted polynomial moment problem - is a natural generalization of the Gaussian quadrature problem. Of course, one can expect a reasonably simple and constructive answer to this problem only if I and S have a simple form, e.g. S is a quadrangle

(2.4) $S = [a_1,b_1] \times [a_2,b_2]$

and - for some fixed, positive L -

(2.5) $I = \{(i,j) \mid i + j \leq L,\ i,j \geq 0\}$.

We give now an equivalent formulation of the problem(2.2)-(2.3) which is crucial for our approach.

Proposition 1. The problem (2.2)-(2.3)- with data set[c(I),S]is equivalent to the existence and characterization of quadruples H,A,B,e , where H is a Hilbert space (whose dimension should be minimized), A and B are self adjoint cummuting operators on H and e is a nonzero vector in H such that

(2.5) $c_{ij} = \langle A^i B^j e, e \rangle$, for all $(i,j) \in I$.

Proof. If there is a solution of problem (2.2)-(2.3) then we define the Hilbert space

(2.6) $H := L_2(S, d\mu)$, $e := 1$ on S

and the operators

(2.7) $A\,f(x,y) := x\,f(x,y) \quad B\,f(x,y) := y\,f(x,y)$

which are self adjoint and commuting. The conditions in (2.2)can be expressed as those in (2.5).

Conversely, suppose that (2.5) holds and let A,B have the eigenvectors (they are common and form a basis of H by the communtativity and self adjointness of A,B) $\Psi_1,\ldots,\Psi_M$ and eigenvalues $x_1,\ldots,x_M$ resp. $y_1,\ldots,y_M$,where M is the dimension of H

(2.8) $A\Psi_k = x_k \Psi_k,\quad B\Psi_k = y_k \Psi_k,\ k = 1,\ldots,M.$

Then

(2.9) $c_{ij} = \sum_{k=1}^{M} x_k^i\, y_k^j\, \mu_k, (i,j) \in I$, where $\mu_k := \langle \Psi_k, e\rangle^2, k=1,\ldots,M$.

This completes the proof and shows that once we constructed the quadruple <H,A,B,e> then the quadrature formula (2.9)can be obtained by a low complexity stable numerical method i.e.solving an eigenvalue problem.

Not assuming H to be finite dimensional we had to invcke the general spectral decomposition theorem, see e.g. [12] ,by which a representing measure is obtained from the associated projector measure

$$d\mu(\lambda) = d(\langle E(\lambda)e, e\rangle)$$

Proposition 2. If problem (2.1),(2.3) has a solution then the problem (2.2),(2.3) also has a solution,moreover for the minimal value M we have the inequality

(2.10) $\min M \leq |I|$

which is exact in the sense, that there exist (multiple connected) domains S such that for the constant weight function $\mu'(x,y) \equiv 1$ on S and the set I as in (2.5), for arbitrary L we have equality in (2.10) - The first part is known as Chakaloff's theorem see [6] and is based on the simple fact that if

$$C = \sum_{i=1}^{R} \gamma_i e_i, c \in R^k, \gamma_i \geq 0, i=1,\dots,R$$

then there exist a similar representation in which there are at most k nonzero constants γ_i. For a proof of the second part see §4,ch.2in [11]. Before going further let us indicate here the connection of the above problem with the minimal,partial relization problem for a class ot two dimensional shift invariant, linear input-output maps

(2.11) $y_{k,l} = \sum_{k \geq i, l \geq j} F_{k-i,l-j} U_{i,j}$

by state-space models of the form

(2.12) $y_{k,l} = \langle h, x_{k,l} \rangle$

$$x_{k+1,l+1} = F_1 x_{k,l+1} + F_2 x_{k+1,l} - F_1 F_2 x_{k,l} + g U_{k,l}$$

where F_1, F_2 are commuting,symetric matrices in R^M and $h,g \in R^M$,see [3]. The transfer functions assiciated to such maps

$$T(w,z) = \iint \frac{d\mu(x,y)}{(1-wx)(1-zy)} = \sum_{i=0}^{\infty} \sum_{j=0}^{\infty} F_{ij} w^i z^j$$

are generalizations of the one variable Stieltjes functions and should play the same role in analyzing "passive" input-output maps. Note that the realizability conditions have the form of complete, infinite moment conditions, if $g = h$,

$$F_{i,j} = \langle h, F_1^i F_2^j g \rangle, \quad i,j \geq 0.$$

It is known that the minimal partial realization problem underlies most of the basic engineering problems of system analysis, see e.g. [2], even if for a suitable,more exact and stable numerical solution of these problems other linear information functionals are better suited, see [14] and below.Connections to(rational)approximation (interpolation) problems for Stieltjes functions are extensively studied, see e.g.[7], [10],[14],[16].

3.Exact conditions of existence and minimality

We shall restrict our interest to so called "regular"index sets I, which - by definition- have the following property.

(3.1) if $(i,j) \in I$, then $(k,l) \in I$, for all $k \leq i$, $l \leq j$

In order to characterize the minimal solutions (H,A,B,e) we have to characterize first the sets with consits of a maximal number of linearly independent vectors among(3.2) $A^i B^j e$, $i,j \geq 0$.

Lemma 1 In the linear space H spanned by the vectors (3.2) (if it is finite dimensional) there always exist a basis consisting of elements of a regular subset $L \subset Z_+^2$.

Proof. Let n_1 be the maximum of the values n such that $e, Be, \ldots, B^{n-1}e$ are linearly independent. Suppose inductively that $n_k, k \geq 1$ is the largest value of n such that $B^{n-1}A^{k-1}b$ is linearly independent on the vectors A^iB^je, with $i \leq k-2, j \leq n_i$ and $i=k-1, j \leq n-2$. Since the sequence of the $n_k, k-1, \ldots$ satisfies $n_1 \geq n_2 \ldots \geq n_k \geq 1$, $\sum_k n_k = \dim H$ the above procedure ends in at most dim H steps and yields a regular set L.

Definition. If L is a regular set, the (generalized) Hankel-matrix associated to it is defined by

$$H_{L(i_1,j_1),(i_2,j_2)} := C_{i_1+i_2,j_1+j_2}$$

where we order the rows and colums of H_L (indexed by elements of L) according to the lexigographic order in Z_+^2. Further we denote-for a regular set L

$$L^* := \{(k,l) \mid \exists\ (i,j) \in L \text{ with } 1 \geq k-i \geq 0,\ 1 \geq l-j \geq 0\}$$

$$L_1 := \{(k,l) \mid \exists (i,j) \in L, k \leq i+1, l=j\}, L_2 = \{(k,l) \mid \exists (i,j) \in L, k=i, l \leq j+1\}$$

$$L^2 := \{(k,l) \mid k=i_1+i_2, l=j_1+j_2, (i_1,j_1) \in L, (i_2,j_2) \in L\}$$

Theorem 1. The necessary and sufficient condition for the existence - given the moment data c(I) - of a nonnegative representing measure supported in at most M points of S is that there exists a regular set L of cardinality m and an extension of the data from c(I) to $c((L^*)^2)$, i.e. an assigument of values to the unspecified moments in $c((L^*)^2)$ such that the matrix H_{L^*} is positive semidefinite and

(3.3) $\operatorname{rank} H_L = \operatorname{rank} H_{L^*} \leq M$.

Moreover the minimal value of M for which the above two conditions can be satisfied equals the minimal number of knots in the corresponding culature formula.

Proof. In order to understand the role of the matrices H_L and H_{L^*} note that these are the Gram matrices associated to the set of vectors

$$W(L) = \{A^iB^je \mid (i,j) \in L\}$$

$$W(L^*) = \{A^rB^sv \mid v \in W(L),\ 0 \leq r \leq 1,\ 0 \leq s \leq 1\}.$$

The necessity of the conditions (3.3) follows now from Proposition 1 and Lemma 1 since Gram matrixes should be positive semidefinite and their rank equal the dimension of the space spanned by the underlying vectors. To prove the sufficiency of the conditions we have to construct a quadruple (H,A,B,e) , such that dim H = rank H_L and (2.5) holds. Now we define H as the Hilbert space spanned by vectors V_{ij} indexed by the element $(i,j) \in L^*$, whose scalar products are specified by

$$\langle V_{i,j}, V_{k,l} \rangle = c_{i+k,j+l}$$

Since rank H_L = rank H^*_L, the operators A,B defined by

$$AV_{i,j} = V_{i+1,j} \quad BV_{i,j} = V_{i,j+1}, \quad (i,j) \in L$$

are hereby defined on the whole space H, moreover they are well defined: if

$$V_{r,s} = \sum_{(i,j)\in L} \alpha_{i,j} V_{i,j}, \text{ i.e. } \sum \alpha_{ij} \langle V_{i,j}, V_{k,l} \rangle = \langle V_{r,s}, V_{k,l} \rangle$$

for all $(k,l) \in L^*$, then

$$V_{r+1,s} = \sum_{(i,j)\in L} \alpha_{i,j} V_{i+1,j} \text{ and } V_{r,s+1} = \sum_{(i,j)\in L} \alpha_{i,j} V_{i,j+1}$$

hold. Indeed multiplying the latter relations by $V_{k,l}$, $(k,l) \in L$, the relations obtained are consequences of the previous ones because H_L is a submatrix of H_{L_1} and H_{L_2} and these are submatrixes of H_L^*.

These operators A and B are clearly symmetric (i.e. self adjoint) since for all $(i,j),(k,l) \in L$

$$\langle AV_{i,j}, V_{k,l} \rangle = c_{i+1+k,j+l} = \langle V_{i,j}, AV_{k,l} \rangle$$

and they commute, since

$$\langle ABV_{i,j}, V_{k,l} \rangle = c_{i+k+1,j+l+1} = \langle BAV_{i,j}, V_{k,l} \rangle .$$

By this the theorem is proved.

The difficulty with this extension problem is partly apparent from the following fact: the restriction of the original say infinite dimensional operators A and B to a Krylow-like subspace W(L) are symmetric but they may not commute, (in general, they do not commute). It is not clear what further connections (if any) exist between the set I (and the values c(I)) on one side and the possible sets L on the other side, is it true that L can be chosen as a subset of I?

These sharp differences between one and higher dimensional polynomial moment problems have been observed e.g. in [13], where it is first shown that in the twodimensional trigonometric, finite moment problem the non-negativity of the associated, generalized Toeplitz matrix (the precize analogon of our Hankel matrix) is not sufficicient for the solvability. The theory of normal extensions of operators, see the appendix written by Szökefalvi Nagy in [12], is clearly related to our problem since the operator A + iB = T should be normal, for A,B to be symmetric and commuting and vice versa. The conditions - in terms of c(I) - for the condition: spectrum $T \subset S$ can be easily written down in the case (2.4): the following matrixes should be nonnegative definite

(3.4) $H_{L_1} - a_1 H_1$, $b_1 H_L - H_{L_1}$, $H_{L_2} - a_2 H_L$, $b_2 H_L - H_{L_2}$.

If S is the disjoint union of two quadrangles Q_1 and Q_2 than we have to require that there exist a decomposition of each of the moments (fixed or assigned) such that

$$c_{i,j} = c^1_{ij}(Q_1) + c^2_{i,j}(Q_2)\ ,\ (i,j) \in (L^*)^2$$

and (3.4) holds for the respectively decomposed matrixes.As an example of a simple application of Theorem 1 we metion the following fact: fir six data $(c_{o,o};c_{1,o};\ldots;c_{o,2})$ if the coresponding 3 x 3 mattix is nonsingular the minimal measures should have 3 atoms and they constitute a one parameter family.

A new numerical approach to solve the existence problem

It is very difficult to handle the constraint(3.3) numerically, the set of solutions of the minimal extension problem is not convex. Observing that the finite dimensional analgon of the solution set to a moment problem (2.1) has the form of a polyhedron (in the sequel we often use abbreviations for N tuples $(c_1,\ldots,c_N) = c^N$)

$$(4.1)\quad K = K(k^N,c^N) = \{ \mu \mid \langle k_i,\mu\rangle = c_i,\ i=1,\ldots,N,\ \mu \in R^m_+ \}$$

we see that searching for the extremal points "vertices" of K. It is known that the parameters of a Gaussian quadrature are very ill conditioned functions of the moments (note that(2.1) is something like an integral equation of the first order whose right hand side is known only at some points) - and this has its parallel in the fact that the vertices of a polyhedron $H(k^N, c^N)$ are nonsmooth functions of the data c^N, or (k^N,c^N).

We propose now using an other,specific solution,the"analytic centre" of the solution set, in order to solve the existence (and some related estimation) problems, in a numericaly more feasible manner.

The analytic centre $\mu(K) = \mu(k^N,c^N)$ of the polyhedron(4.1) is defined as the unique point which solves the following optimization problem

$$\max\{ \sum_{i=1}^m \log \mu_i \mid \langle k_j,\mu\rangle = c_j, j=1,\ldots,N, \mu_i \geq 0, i=1,\ldots,m\}$$

If the polyhedron is represented in its own space(of dimension m-N,in general), i.e. $K \to P = P(a^m,b^m)$

$$P(a^m,b^m) = \{x \mid b_i - \langle a_i,x\rangle \geq 0,\ i=1,\ldots,m, x \in R^{m-N}\}$$

by the map $\mu_i = b_i - \langle a_i,x\rangle$,i=1,...,m, then $\bar{\mu} = \bar{x}(a^m,b^m)$ the point, which solves the problem (assuming int $P \neq \emptyset$)

$$\max\{ \prod_{i=1}^m (b_i - \langle a_i,x\rangle) \mid x \in P(a^m,b^m)\}.$$

One can prove that the map $(a^m,b^m) \to \bar{x}(a^m,b^m)$ is affine invariant and there exists a two sided ellipsoidal approximation for P around $\bar{x}$:

$$\bar{x} + E \subset P \subset \bar{x} + m\,E,\quad E = \{ z \mid \langle Az,z\rangle \leq 1\}$$

where the symetric matrix $E = E(a^m,b^m)$ is easily obtained from $\bar{x}(a^m,b^m)$ see [14],[15]. The fact that $\bar{x}(a^m,b^m) = \bar{\mu}(k^N,c^N)$ is an analytic, very smooth function of the data allows to solve the feasibility and linear optimization problems by a homotopy approach, see [15], which we generalize now as follows.

The analytic entre of the set (2.1) is defined (if its exists) as the solution of the problem

(4.2) $\sup\{\int_S \log \mu'(s)ds \mid \int_S K_j(s)\mu'(s)ds = c_j,\ j=1,\dots,N\}.$

It is easy to prove that the set of values c^N for which (4.2) has a solution is convex and dense in the set of all feasible c^N, if S is a domain, i.e. closure (int S) = S. For the trigonometric moment problem this solution was studied already about 1920, see [10],[14].

Lemma 1. The solution of the problem (4.2) - if it exists - has the following form

$$\mu'(s) = \left(\sum_{j=1}^{N} \alpha_j K_j(s)\right)^{-1}$$

for suitable $\alpha^N \in R^N$, which in fact is then the unique solution of the equation

(4.3) $$\frac{\partial F(\alpha)}{\partial \alpha_j} = \int_S K_j(s) \left(\sum_{j=1}^{N} \alpha_j K_j(s)\right)^{-1} ds = c_j,\ j = 1,\dots,N$$

such that $\Sigma\, \alpha_j K_j(s)$ is positive on S, here

(4.4) $$F(\alpha) = \int_S \log\left(\Sigma\, \alpha_j K_j(s)\right)ds$$

Proposition The moment problem (2.1) has a solution if and only if the homotopy path $\alpha^N(\lambda)$ can be continued from $\lambda = 1$ till $\lambda = 0$, where $\alpha^N(\lambda)$ $0 < \lambda \leq 1$ is defined as the solution of (4.3) where c^N is replaced by $(1-\lambda)c^N + \lambda c_o^N$,

$$c_o^N = \int_S K^N(s) \left(\sum_{j=1}^{N} \overset{o}{\alpha}_j K_j(s)\right)^{-1} ds$$

and $\sum_{j=1}^{N} \overset{o}{\alpha}_j K_j$ is an arbitrarily fixed polynom which is positive on S.

The proof is a simple application of the implicite function theorem. For brevity we can only refer to [9],[15] for the application of this method for the estimation of (computation of exact upper and lower bounds in terms of the moments c^N for

$$l(c^N) \leq \int K_o(s)\,\mu(ds) \leq u(c^N)$$

It can be expected that for smooth analytic kernel functions $K_o, K_1, \dots, K_N$ this approach is superior to those using discretizations of the measure (of the set S) and algorithms based on the simplex method (note that the latter methods use- as a tool - extremal solutions, only piecewise smooth homotopies); concerning numerical test results on this approach-using homotopies along analytic centers- to solve linear programming problems, see[9].

The special solution of (4.2) in the case of the (real)trigono - metric moment problem - where $K_j(s) = \exp(i(j-1)s)$, $s \in [-\Pi,\Pi]$ and μ a measure on$[-\Pi,\Pi]$(which is symetrical to zero)-,which is a special case of the Nevanlinna-Pick moment problem, is the so called "maximum entropy" solution. These analytical centers, more precizely the coefficients of the trigonometric polynomial$[\bar{\mu}'(e^{is})]^{-1}$ are rational functions of c^N which can be computed rather quickly:in $O(N^2)$ arithmetical operations. This and other observations, see[15],lead to the idea that for the extrapolation of the function $\alpha^N(\lambda)$rational (multipoint Pade) approximation - with Newton type corrector step to solve (4.3) - will furnish a rather efficient path following method. In fact, in a problem closely related to (4.2), the use of a special rational extrapolation method can be justified rigorously using a generalizaiton of the well knwon fact (see e.g.[7]) that the multipoint Pade approximants (i.e.interpolants) to a Stieltjes function are again Stieltjes functions, see[15].

In order to solve - over some domain S - the closely related uniform approximation problems

$$\min_{\beta^N} \| K_o(s) - \sum_{i=1}^{N} \beta_i K_i \|_{L^\infty(S)}$$

we propose following the homotopy path $\beta^N(\lambda)$ determined by

$$\sup_{(\varepsilon,\beta^N)} (\log(\lambda-\varepsilon) + \int_S (\log (K_o(s) - \sum_{i=1}^{N} \beta_i K_i(s) - \varepsilon) + \log (\varepsilon - K_o(s) + \sum_{i=1}^{N} \beta_i K_i(s)))ds .$$

Of course,the sucess of these methods depends (among others) on the availability of fast and accurate methods for approximating the above integrals as well as those in (4.3).

References

[1] T.Ando,Truncated moment problems for operators,Acta Sci.Math. (Szeged),31 (1970), 319-334.

[2] A.Antoulas, A New Approach to Synthesis Problems,IEEE Trans.Ant. Contr.,vol.30,No.5(1985)465-473.

[3] S.Attasi, Modelling and Recursive Estimation for Double Indexed Sequences, in System Indentification,ed. by R.K.Mehra and D.G.Lainiotis,Academic Press, 1976,pp.289-348.

[4] M.F.Barnsley and P.D.Robinson, Rational Approximant bounds for a class of two variable Stieltjes functions,SIAM J.Math.Anal. vol.9,No 2, 1978,pp.272-290.

[5] N.K.Bose,(editor),Multidimensional System Theory,D.Reidel, Dordrecht,1985.

[6] V.Chakaloff, Formules de cubatures mecaniques a coefficients non négatifs, Bull.Sci.Math.,Ser.2. 1957,81.No 3,123 - 134.

[7] G.Cybenko, Restrictions of normal operators, Pade approximation and antoregressive time series, SIAM J.Math.Anal. 15 (1984), 753 - 767.

[8] A.A. Goncar,G.Lopez, On Markov's theorem for multipoint Pade approximation, Math. USSR-Sb., vol 34 (1978),449-459.

[9] F.Jarre,G.Sonnevend,J.Stoer, An Implementation of the method of analytic centers, Report No. 34, Schwerpunktprogramm der DFG, Anwendungsbezogene Optimierung und Steuerung, Univ.Würzburg, 16 p.,appears in Proc. 8th INRIA Conf. on Analysis and Optimization of systems, Antibes,1988.

[10] M.G.Krein, A.A.Nudelman,Markov's Moment Problem and Extremal Problems (in russian) Nauka,Moscow,1973.

[11] I.P.Mysovskih, Cubature formulae (in russian),Nauka,Moscow,1981.

[12] F.Riesz,B.Szökefalvi Nagy,Lecons D'Analyse Fonctionelle,Akad.Budapest,1972.

[13] W.Rudin,Function theory in polydiscs, W.A.Benjamin Inc. New York ,1969.

[14] G.Sonnevend, Sequential, stable and low complexity methods for the solution of moment (mass recovery)problems, in Proc.4. Int:Conf. on Numerical Methods Colloquia Societatis J.Bolyai, vol.50,pp.635-668, North Holland,Akademia,Budapest,1987.

[15] G.Sonnevend, J.Stoer,Global Ellipsoidal Approximations and Homotopy Methods for solving convex, analytic programms, Schwerpunktprogramm der DFG, Report No. 4, Jan. 1988, Submitted to Numerische Mathematik.

[16] Szökefalvi Nagy,B.,A.Korányi,Operatorentheoretische Behandlung und Verallgemeinerung eines Problemskreises in der komplexen Funtkionetheorie, Acta Mathematica,vol. 100(1958),pp.171-202.

SPECULATIONS ON POSSIBLE DIRECTIONS AND APPLICATIONS FOR THE DECOMPOSITION METHOD

G. Adomian

The decomposition method has now been applied to a rather wide class of nonlinear differential and partial differential equations. An interesting characteristic is that once a problem is modeled with a specific (linear, nonlinear, deterministic, stochastic, ordinary or partial differential) equation with physically correct conditions specified, the usual linearizations, perturbations, closure approximations, white noise assumptions, or discretization are avoided. Certainly, much remains to be done on the theoretical foundations and precise limitations which can be viewed as a fascinating challenge for further research. Valuable work in this direction has appeared by Professor N. Bellomo and his co-workers [1]. The range of problems solved and the accuracy obtained - the fact that nonlinear systems with stochastic parameters are included in the methodology - and the fact that the work has applied to parabolic, elliptic, and hyperbolic equations - suggest that this may be a useful and computational method for frontier applications. Proof of convergence and convergence rate in solution of partial differential equations, error estimates, and perhaps better generation of the author's presently used A_n polynomials or equivalent new forms are fertile areas for further study and dissertations. Many other research topics are in the area of applications; some are discussed in [1].

Let us point out some speculations on possible applications noting that these applications require the development of a correct mathematical model before decomposition can possibly solve them. It is not useful to apply the method to many existing models since they have already been linearized and otherwise simplified for mathematical tractability. The solution from such simplified models can differ, sometimes substantially, from the actual models. Also, since the decomposition technique does not require discretization, it is evident that in a difficult problem such as Navier-Stokes, use of decomposition may provide appreciable saving in computation time.

Nevertheless, some possible applications which represent an exciting challenge are areas such as nonlinear and possibly stochastic and multidimensional optimal control theory, hypersonic flow, quantum theory and gravitation, generalization of the Kalman filter, and problems of large space structures such as vibration, heating, etc., [3].

Before going into these areas, let's look briefly at some illustrative decomposition examples (chosen for clarifying the procedure rather than for difficulty.)

Example 1: $d^2y/dx^2 + 2x\, dy/dx = 0$ with $y(0) = 0$ and $y(a) = 1$. The solution is $y(x) = \mathrm{erf}(x)/\mathrm{erf}(a)$ or

$$y(x) = \frac{x - \frac{x^3}{3} + \frac{x^5}{10} - \frac{x^7}{42} + \frac{x^9}{216} \cdots}{a - \frac{a^3}{3} + \frac{a^5}{10} - \frac{a^7}{42} + \frac{a^9}{216} \cdots}$$

By decomposition we write (letting $L = d^2/dx^2$)

$$Ly = -2x(d/dx)y$$

Operating with L^{-1}, a two-fold integration:

$$y = y_0 - 2L^{-1}\, x(d/dx)y$$

where $y_0 = c_1 + c_2 x$. If we satisfy the boundary conditions we have $y_0 = x/a$ which is our first approximation φ_1. The complete solution is $y = \sum_{n=0}^{\infty} y_n$ and our approximation to some n terms is $\varphi_n = \sum_{i=0}^{n-1} y_i$. Substituting $y = \sum_{n=0}^{\infty} y_n$ above we identify

$$y_1 = -2L^{-1}x(d/dx)y_0$$

$$y_2 = -2L^{-1}x(d/dx)y_1$$

.

.

.

$$y_n = -2L^{-1}x(d/dx)y_{n-1}$$

Thus $y_1 = -2L^{-1}x(d/dx)(c_1 + c_2x) = -c_2\, x^3/3$, $y_2 = -2L^{-1}x(d/dx)(-c_2\, x^3/3) = c_2\, x^5/10$, etc. A three-term approximation, for example, is

$$\varphi_3 = y_0 + y_1 + y_2$$

$$= c_1 + c_2x - c_2\, x^3/3 + c_2\, x^5/10$$

Satisfying the boundary conditions with φ_3 , i.e., a three-term approximation, we have

$$u = \frac{x - \frac{x^3}{3} + \frac{x^5}{10}}{a - \frac{a^3}{3} + \frac{a^5}{10}}$$

which can be carried further if necessary.

Example 2: $d^2u/dx^2 - 40xu = 2$, $u(-1) = u(1) = 0$. Let $L = d^2/dx^2$ and write [1]

$$Lu = 2 + 40xu$$

$$u = c_1 + c_2x + L^{-1}(2) + L^{-1}(40xu)$$

Let $u_0 = c_1 + c_2x + L^{-1}(2) = c_1 + c_2x + x^2$ and let $u = \sum_{n=0}^{\infty} u_n$.

The components of u are given by

$$u_{n+1} = L^{-1}40xu_n$$

for $n \geq 0$ thus:

$$u_1 = L^{-1}40xu_0 = (20/3)c_1x^3 + (10/3)c_2x^4 + 2x^5$$

Similarly,

$$u_2 = (80/9)c_1x^6 + (200/63)c_2x^7 + (10/7)x^8$$

We continue to some n-term approximation $\varphi_n = \sum_{i=0}^{n-1} u_i$ which approaches $u = \sum_{n=0}^{\infty} u_n$ as $n \to \infty$ [1]. If we write φ_3 as an approximation,

$$\begin{aligned}\varphi_3 &= u_0 + u_1 + u_2 \\ &= c_1 + c_2x + x^2 + (20/3)c_1x^3 + (10/3)c_2x^4 \\ &\quad + 2x^5 + (80/9)c_1x^6 + (200/63)c_2x^7 + (10/7)x^8\end{aligned}$$

Imposing the boundary conditions at -1,1, we write $\varphi_3(1) = \varphi_3(-1) = 0$ to get

$$\begin{vmatrix} 149/9 & 473/63 \\ 29/9 & -53/63 \end{vmatrix} \cdot \begin{vmatrix} c_1 \\ c_2 \end{vmatrix} = \begin{vmatrix} -31/7 \\ -3/7 \end{vmatrix}$$

from which c_1, c_2 are evaluated. Substituting φ_n into the left side of the differential equation, we should get the right side, or 2, if the approximation is sufficient. We note that the 12-term approximation yields 2.000000 or seven-digit accuracy.

On R^3 with $L_x = \partial^2/\partial x^2$, $L_y = \partial^2/\partial y^2$, $L_z = \partial^2/\partial z^2$ we write

$$[L_x + L_y + L_z]u = f(x,y,z) + k(x,y,z)u$$

Solve for each linear operator in turn. Operate on each of the three equations with the appropriate inverse and write

$$u = \varphi_x + L_x^{-1}f - L_x^{-1}ku - L_x^{-1}(L_y + L_z)u$$

$$u = \varphi_y + L_y^{-1}f - L_y^{-1}ku - L_y^{-1}(L_z + L_x)u$$

$$u = \varphi_z + L_z^{-1}f - L_z^{-1}ku - L_z^{-1}(L_x + L_y)u$$

where $\varphi_x, \varphi_y, \varphi_z$ are the homogeneous solutions. Adding and dividing by 3,

$$u = u_0 + Ku$$

with

$$u_0 = (1/3)\{\varphi_x + \varphi_y + \varphi_z + (L_x^{-1} + L_y^{-1} + L_z^{-1})f\}$$

$$K = (1/3)\{(L_x^{-1} + L_y^{-1} + L_z^{-1})k + L_x^{-1}(L_y + L_z)$$

$$+ L_y^{-1}(L_z + L_x) + L_z^{-1}(L_x + L_y)\}$$

assuming $u = \sum_{n=0}^{\infty} u_n$,

$$u_{n+1} = K\, u_n$$

so all components are determined. The inverse operators are double integrations leading to two constants of integration to be determined by forcing u_n to satisfy the given condition.

Suppose $k = k(u)$ so the equation becomes nonlinear. The nonlinear term is expanded as $\sum_{n=0}^{\infty} A_n$ where the A_n polynomials are generated for the nonlinear term as discussed e.g. in [3]. For analytic functions $f(u)$, the sum of the A_n can be shown to be equivalent to a generalized Taylor series about the function u_0 . The procedure is now as before except that the u_{n+1} will involve an A_n term. Since each A_n depends only on $u_0, u_1, \ldots u_n$, the solution can be obtained essentially as easily as in the linear case.

The generation of these polynomials has been discussed in previous work. However, for convenience, we provide a useful heuristic rule to write the A_n in general. The A_n for polynomial nonlinearities being sums of various products of the u_i up to $i = n$ can be written in symmetrized form. For $Nu = u^2 = \sum_{n=1}^{\infty} A_n$, for example, $A_0 = u_0^2$, $A_1 = 2u_0u_1$,

$A_2 = u_1^2 + 2u_0u_2$, etc., but we can write this as $A_0 = u_0u_0$,

$A_1 = u_0u_1 + u_1u_0$, $A_2 = u_0u_2 + u_1u_1 + u_2u_0$, etc., i.e., the first subscript goes from 0 to n, the second from n to 0 such that the sum is n.

For more general forms, we define $A_n = \sum_{\nu=1}^{n} c(\nu,n)\, f^{(\nu)}(u_0)$ where $f^{(\nu)}(u_0)$ represents the νth derivative of $f(u)$ evaluated at $u = u_0$. To get the $c(\nu,n)$ we first ask how many combinations of ν integers will add to n. Thus $c(\nu,n)$ will mean the sum of possible products of ν u_i's whose subscripts add to n. To get $c(2,3)$, we see that two integers can add to 3 only if one is 1 and the other is 2, if zero is excluded. Hence, we write $c(2,3) = u_1u_2$. To get $c(1,3)$, the coefficient of $f^{(1)}(u_0)$, we have one u_i and its subscript must be 3, hence $c(1,3) = u_3$. For $c(3,3)$, we need three factors u_i with subscripts summing to 3; hence each subscript must be 1 and $c(3,3) = u_1u_1u_1 = u_1^3$. As stated so far however, this rule is incomplete. We must also divide by the factorial of the number of repetitions to use the formula as stated. Thus $c(3,3) = (1/3!)u_1^3$. We have now

$$A_3 = u_3 f^{(1)}(u_0) + u_1u_2 f^{(2)}(u_0) + (1/3!)u_1^3\, f^{(3)}(u_0) .$$

$f^{(\nu)}(u_0)$ for ν from 1 to 6 . The coefficient of $f^{(6)}$ must involve six integers adding to 6 or u_1^6 ; hence the coefficient is $(1/6!)u_1^6$. What about the coefficient $c(2,6)$ for $f^{(2)}(u_0)$ in A_6 . These are (1,5), (2,4), and (3,3). Thus the coefficient $c(2,6)$ is $(1/2!)u_3^2 + u_2u_4 + u_1u_5$. Thus the A_n are as follows:

$$A_0 = f^{(0)}(u_0)$$
$$A_1 = u_1f^{(1)}(u_0)$$
$$A_2 = u_2f^{(1)}(u_0) + (1/2)u_1^2\, f^{(2)}(u_0)$$
$$A_3 = u_3f^{(1)}(u_0) + u_1u_2f^{(2)}(u_0) + (1/3!)u_1^3f^{(3)}(u_0)$$
$$A_4 = u_4f^{(1)}(u_0) + [(1/2!)u_2^2 + u_1u_3]f^{(2)}(u_0)$$
$$+ (1/2!)u_1^2u_2f^{(3)}(u_0) + (1/4!)u_1^4f^{(4)}(u_0)$$

.
.
.

Example 3: Consider Duffings equation

$$y'' + \alpha y' + \beta y + \gamma y^3 = x(t)$$

which we write

$$Ly = x(t) - \alpha y' - \beta y - \gamma y^3$$
$$y = y_0 - L^{-1}\alpha\ (d/dt) \sum_{n=0}^{\infty} y_n - L^{-1}\gamma \sum_{n=0}^{\infty} A_n$$

Given initial conditions $y(0)$ and $y'(0)$, we identify

$$y_0 = y(0) + ty'(0) + L^{-1}\, x(t)$$
$$y_1 = -L^{-1}\alpha(d/dt)y_0 - L^{-1}\beta\, y_0 - L^{-1}\gamma\, A_0$$
$$y_2 = -L^{-1}\alpha(d/dt)y_1 - L^{-1}\beta\, y_1 - L^{-1}\gamma\, A_1$$

.

.

.

The A_n polynomials to represent the nonlinearity y^3 are given by

$$A_0 = y_0^3$$

$$A_1 = 3y_0^2 y_1$$

$$A_2 = 3y_0 y_1^2 + 3y_0^2 y_2$$

$$A_3 = y_1^3 + 6y_0 y_1 y_2 + 3y_0^2 y_3$$

$$A_4 = 3y_1^2 y_2 + 3y_0 y_2^2 + 6y_0 y_1 y_3 + 3y_0^2 y_4$$

$$A_5 = 3y_1 y_2^2 + 3y_1^2 y_3 + 6y_0 y_2 y_3 + 6y_0 y_1 y_4 + 3y_0^2 y_5$$

$$A_6 = y_2^3 + 6y_1 y_2 y_3 + 3y_1^2 y_4 + 3y_0 y_3^2 + 6y_0 y_2 y_4 + 6y_0 y_1 y_5 + 3y_0^2 y_6$$

.

.

.

Thus the solution is determinable to any desired n-term approximation $\varphi_n = \sum_{i=0}^{n-1} y_i$ which converges to $y = \sum_{n=0}^{\infty} y_n$.

approximation $\varphi_n = \sum_{i=0}^{n-1} y_i$ which converges to $y = \sum_{n=0}^{\infty} y_n$.

The $\sum_{n=0}^{\infty} A_n$ constitutes a generalized Taylor series about the function $y_0(t)$.[1]

Stochastic Case: We could have stochastic fluctuations in α, β, or γ in the above example, in addition to stochastic $x(t)$ or initial conditions. Thus, in general, we could write

$$\alpha = \langle\alpha\rangle + \epsilon$$

$$\beta = \langle\beta\rangle + \eta$$

$$\gamma = \langle\gamma\rangle + \sigma$$

where ϵ, η, σ are zero-mean random processes. The solution process can now bė obtained from

$$Ly = x - \alpha(d/dt)y - \beta y - \gamma \sum_{n=0}^{\infty} A_n$$

$$- \epsilon(d/dt)y - \eta y - \sigma \sum_{n=0}^{\infty} A_n$$

where the A_n summation represents y^3 in the Duffing case and $(d/dt)y^3$ in the Van der Pol case. Thus,

$$y_0 = y(0) + ty'(0) + L^{-1}x$$

$$y_1 = -\alpha(d/dt)y_0 - \beta y_0 - \gamma A_0$$

$$-\epsilon(d/dt)y_0 - \eta y_0 - \sigma A_0$$

---

1. See reference [4]

$y_2 = -\alpha(d/dt)y_1 - \beta y_1 - \gamma A_1$

$-\epsilon(d/dt)y_1 - \eta y_1 - \sigma A_1$

.

.

.

Then $y(t) = \sum_{n=0}^{\infty} y_n(t)$ yields a stochastic series from which statistics can now be obtained without problems of statistical separability of quantities such as $\langle Ry \rangle$, where $R = \epsilon(d/dt) - \eta$, which normally require closure approximations and truncations.

We note that no statistical linearization is necessary. We do not need to assume $x(t)$ is delta-correlated or stationary or to neglect parameter fluctuations.

Rather than further discussion of the methodology on which there is now a considerable published literature in the U.S. and Europe, let us speculate on some applications which appear to be possible in the very near future although they require the modelling expertise of theorists concerned primarily with each of those areas.

Some of these, in the author's opinion, are

1) optimal control for nonlinear stochastic systems modelled by ordinary or partial differential equations.
2) hypersonic flow, turbulence, single-stage-to-orbit flight (essential for shuttles which can be used for the construction of space stations,)
3) quantum theory and gravitation
4) generalizations of Kalman filtering.

Because of page and time limitations we discuss only the first two here.

1) Suppose we consider a nonlinear and possibly stochastic system which we want to control in some optimal way. For a linear control system with a quadratic performance index, of course an analytical solution can be made. Consider the state equations

$$\dot{x}(t) = f(x_1,\ldots,x_n;\ u_1,\ldots,u_m;t)$$

i.e., a set of n nonlinear differential equations with $x(t)$ representing a state vector with n components $f_1,\ldots,f_n$, and $x(t_0)$ a given initial vector. Define, for example [4] a performance functional $J(x,u,t)$ given by

$$J = \varphi[x(t_1),\ t_1] + \int_{t_0}^{t_1} F(x,u,t)\ dt$$

where φ and F are scalar functions with necessary smoothness properties. Let $p = [p_1,\ldots,p_n]^T$ be a vector of Lagrange multipliers and form an augmented functional

$$J' = \varphi[x(t_1),t_1] + \int_{t_0}^{t_1} [F(x,u,t) + p^T\ (f-\dot{x})]\ dt$$

Integration by parts leads to

$$J' = \varphi - [p^T x]\ \Big|_{t_0}^{t_1} + \int_{t_0}^{t_1} [H + \dot{p}^T\ x]\ dt$$

with H defined as

$$H(x,u,t) = F(x,u,t) + p^T f\ .$$

If u is defined on $t_0 \le t \le t_1$, we vary u and find the variation $\delta J'$ corresponding to δu, leading to the n adjoint equations,

$$\dot{p}_i = -\frac{\partial H}{\partial x_i}$$

so we have a system of $2n$ nonlinear differential equations with two-point boundary conditions. Although this approach has been discussed by R.E. Bellman and many others, perhaps most recently in [2], analytical solution has usually not been possible except by numerical methods. We now have a potentially valuable alternative since such systems of nonlinear differential equations have been solved (even for the stochastic and/or multidimensional cases) in a analytic approximation by the decomposition method.

Another possibility is through solution by decomposition of the matrix Riccati equation which appears in invariant embedding and neutron transport theory as well as modern control theory. Consider

$$R'(x) = B(x) + D(x)R(x) + R(x)D(x) + R(x)B(x)R(x)$$
$$R(0) = 0$$

where B, D, R are continuous $n \times n$ non-negative matrices. Suppressing the argument x, we have

$$R' = B + DR + RD + RBR$$

If $L = d/dx$

$$LR = B + HR + NR$$

where $LR = R'$, $HR = DR + RD$, and NR represents a nonlinear operator on R. Since $R(0) = 0$, operation with L^{-1} on both sides yields

$$R = L^{-1}B + L^{-1}HR + L^{-1}NR \ .$$

Let R and NR be written in terms of the A_n polynomials. For R this is equivalent to writing $R = \sum_{n=0}^{\infty} R_n$. For NR we write $\sum_{n=0}^{\infty} A_n$. Identify $R_0 = L^{-1}B$; then

$$R_0 = L^{-1} B$$

$$R_1 = L^{-1} H R_0 + L^{-1} A_0$$

$$R_2 = L^{-1} H R_1 + L^{-1} A_1$$

.

.

.

$$R_n = L^{-1} H R_{n-1} + L^{-1} A_{n-1}$$

for $n \geq 1$. The A_n for NR are given by [1]

$$A_0 = R_0 B R_0$$

$$A_1 = R_0 B R_1 + R_1 B R_0$$

$$A_2 = R_1 B R_1 + R_0 B R_2 + R_2 B R_0$$

$$A_3 = R_0 B R_3 + R_3 B R_0 + R_1 B R_2 + R_2 B R_1$$

$$A_4 = R_2 B R_2 + R_0 B R_4 + R_4 B R_0 + R_1 B R_3 + R_3 B R_1$$

.

.

.

so that

$$R_0 = L^{-1} B$$

$$R_1 = L^{-1} H R_0 + L^{-1} R_0 B R_0$$

$$R_2 = L^{-1} H R_1 + L^{-1} (R_0 B R_1 + R_1 B R_0)$$

$$R_3 = L^{-1} H R_2 + L^{-1} (R_1 B R_1 + R_0 B R_2 + R_2 B R_0)$$

.

.

.

Finally, since $HR = DR + RD$

$$R_0 = L^{-1} B$$

$$R_1 = L^{-1}(D R_0 + R_0 D) + L^{-1}(R_0 B R_0)$$

$$R_2 = L^{-1}(DR_1 + R_1 D) + L^{-1}(R_0 B R_1 + R_1 B R_0)$$

$$R_3 = L^{-1}(D R_2 + R_2 D) + L^{-1}(R_1 B R_1 + R_0 B R_2 + R_2 B R_0)$$

.

.

.

An n-term approximant is $\varphi_n = \sum_{i=0}^{n-1} R_n$ which approaches $R = \sum_{i=0}^{\infty} R_n$ as $n \to \infty$. Thus, given B, D, a specific R can be calculated to a desired approximation. Accuracy has been demonstrated in [3].

(2) Hypersonic Flow: The present approach to hypersonic flow problems is computational fluid dynamics (CFD), and intensive work is being done to develop appropriate CFD computer programs for the hypersonic case. With continuing rapid developments in supercomputers, this emphasis is certainly appropriate. Yet, another methodology now appears promising which is quite different and seems to have a high potential for important advantages as well as a probably high adaptability to supercomputers. This is the decomposition method.

It yields a rapidly converging series solution in analytic form. It requires no linearization, perturbation, closure approximations, or assumption of special mathematically tractable stochastic processes such as delta-correlated processes. Probably most important is the fact that discretization into grids is unnecessary. Hence, computation should be less, and the difficulty of different time scales in turbulence is avoided.

In the types of fluid flow which interest us, velocity, density, and pressure are stochastic, not constants. Present treatment of Navier-Stokes equations solves a simplistic model, not real behavior. Turbulence is a strongly nonlinear, strongly stochastic phenomenon and cannot be understood by linearized perturbative treatments. The theories of physics are perturbative theories and the theories of mathematics are for linear operators (other than some ad hoc methods for special nonlinear equations). What is needed is a way of solving one or more nonlinear stochastic operator equations whether algebraic, differential, delay-differential, partial-differential, or systems of such equations. The computational accuracy of a supercomputer is dependent on the sophistication of the mathematical methods programmed into it. Typical calculations consider millions of discrete time intervals made small enough so that trajectories between them can be taken as low-order polynomials, e.g., quadratics.

In generalized hydrodynamics, the form of Navier-Stokes equations is kept, but time and distance scales are introduced so one can go beyond continuum approximation and take account of molecular structure. However, application to a real situation becomes simply a test of the validity of the linear approximations, as pointed out in the literature. Fluctuations are , as usual, assumed "small," and delayed effects, due to the fact that responses cannot be instantaneous, are ignored.

When one studies airflow about aircraft surfaces, computations must be made at tens of millions of points, and it is felt that by increasing the volume of computation to the limit in an ultimate extrapolation, supercomputers will yield complete accuracy. Not only does this ignore stochasticity, it ignores the sensitivity of nonlinear stochastic systems to very slight changes in the model - in fact, to changes essentially undetectable by measurement.

To solve an aircraft problem on contemplated next-generation computers, a 3-dimensional mesh is generated which discretizes the system of nonlinear partial differential equations into a million, a hundred million, or perhaps a billion coupled difference equations in as many unknowns. One begins to see then the tremendous data handling problem, the necessity for improved algorithms, and the need for still greater computational speed. We may also have many unknowns at each point, and, as we have pointed out, the system nonlinearities and random fluctuations need to be taken into consideration. Since usually solutions are iterative - first solving an approximation to the original system of differential equations and then improving the solution by repeated substitution of each new solution - parallel processing is complicated by the difficulty of partitioning the work so that each processor can work independently. This is being pursued by many ingenious ideas necessitated by the brute force method of discretization.

In all such problems we need to be able to solve coupled systems of nonlinear (and generally stochastic as well) partial differential equations with complex boundary conditions and possible delayed effects. These systems are linearized and discretized (and the stochastic aspects either ignored or improperly dealt with) so that the various numerical approximation methods can be used. This requires faster and faster supercomputers to do these computations in a reasonable time.

Unfortunately the further developments in supercomputers can quite possibly give wrong answers because even a single one-dimensional nonlinear differential equation without stochasticity in coefficients, inputs, and boundary conditions - let alone vector partial differential equations in space and time with nonlinear and/or stochastic parameters - is not solved exactly. Real systems are nonlinear and stochastic. When these "complications" are ignored or approximated by assumptions such as weak nonlinearity, white noise, etc., we have a different problem! i.e., a mathematized problem, not the original physical problem. The model equations, even before the linearization, discretization, etc., are already wrong because the stochastic behavior is generally not incorporated or is incorporated incorrectly as an afterthought.

Our approach to hypersonics, using decomposition, will be based on previous work on Navier-Stokes [4] which showed that an analytic solution is possible. For hypersonic cases, additional effects are present changing the model equations but the approach is similar. Discussion of the mathematical methodology, let alone the huge subject of hypersonics and turbulence is, of course, not addressable here. We can only call attention now to the possibility of some alternatives to the present approaches.

References

1. Bellomo, N. and Riganti, R. Nonlinear Stochastic Systems in Physics and Mechanics, World Scientific Publ. Co., 1987.

2. Barnett, S. and Cameron, R.G. Introduction to Mathematical Control Theory, Clarendon Press 1985.

3. Adomian, G., Pandolfi, M., and Rach, R., An Application of the Matrix Riccati Equation to a Neutron Transport Process, J. Math. Anal. and Applic., in publication.

4. Adomian, G., Nonlinear Stochastic Systems Theory and Applications to Physics, Reidel, 1988.

CONTINUITY of TRAJECTORIES
of
RANDOM WALKS with INFINITESIMAL STEPS

by

E. Benoit +

B. Candelpergher *

C. Lobry *

+ Centre de Mathématiques appliquées, Ecole des Mines de Paris, Sophia Antipolis, 06565 VALBONNE Cedex France
et G. R. Automatique - E.N.S.I.E.G. , B.P. 46 , 38402 Saint Martin d'Hères France .

* Laboratoire de Mathématiques, U.A. 168, parc Valrose, 06000 NICE , France
et G. R. Automatique - E.N.S.I.E.G. , B.P. 46 , 38402 Saint Martin d'Hères France .

In may 1985, E. NELSON gave a very short introduction to "Stochastic Mechanics" to an audience of Nonstandardists and wrote it in "What is stochastic mechanics"[9]. In accordance with the title of the meeting : "Mathématiques finitaires et Analyse non Standard ", he defines diffusion processes as discrete objects and at the end of his paper he says : << Each solution of the Schrödinger equation describes a diffusion process in which the particles have continuous trajectories ...>> but actually in this very short paper he does not explain in what sense a "discrete diffusion" has "continuous trajectories" . At the same time a set of notes by NELSON himself, now published as RADICALLY ELEMENTARY PROBABILITY THEORY [11], was circulating, in which he defined the mathematical Brownian motion as the process :

$$x_o \text{ given}$$

$$x_{t+dt} = x_t + Z_t\sqrt{dt}$$

where $t = 0, dt, 2dt, 3dt, \ldots Ndt$ and Z_t is a sequence of independent random variables with values ± 1 and expectation 0 , and proved, among other things, that "almost every" trajectory of this process is "continuous" .

So, in view to make precise what are "continuous trajectories" of discrete diffusions we decided to extend some results of [11] to more general processes of the form :

$$x_o \text{ given}$$

$$x_{t+dt} = x_t + b(x_t, t)\,dt + s(x_t,t)\,Z_t\sqrt{dt}$$

where $t = 0, dt, 2dt, 3dt, \ldots Ndt$ and Z_t is a sequence of independent random variables with values ± 1 and expectation 0 or a more general process. This has been done and is written now [1] .

The objective of the present paper is to explain to an audience of **Non Nonstandardists** how, thanks to the existence of infinitesimals, it is possible to give a precise mathematical status to the sentence :

"Almost every trajectory of the above discrete diffusion (or random walk) is continuous".

This will be done in the first three paragraphs . In the first one we introduce the reader to Non standard concepts . We try to explain to him , mainly on the exemple of a deterministic walk of infinitesimal step how finite sets, with unlimited number of elements, are the nonstandard equivalents of a continuum like the set of reals . In the second paragraph we give the definition of what we call a "discrete diffusion process" and state our principal results . In paragraph 3 we give an outline of the proofs, assuming for granted NELSON's results .

Very often in mathematics one is concerned with the "limit of some object, for instance a diffusion process, when some parameter goes to infinity". What does that means when one parameter of the problem, dt in the case of discrete diffusions, is already fixed as an infinitesimal ? We explain this point on an example in the last paragraph .

Few papers [2] , [3] , [6] related to these topics have been written by the authors of this note .

1 Introduction Non Standard concepts .

Everybody knows what are the classical mathematical objects **N** ,**R** and how to use them .The main idea of the practice of N.S.A. advocated by NELSON [8] , REEB[13], LUTZ and GOZE [7], DIENER [5], DIENER et REEB [4] is to add a new undefined predicate **st**(x) which reads "the object x (previously defined by classical mathematics) is standard" , plus some axioms which rules the manipulations on **st** . It turns out that now we are in position to formulate sentences about mathematical objects which might be more rich than the classical ones . Those which use **st** or a derivate are called **external** and classical sentences - those which are understandable by a classical mathematician who never heard about **st** - are called **internal** .

It can be proved that the introduction of **st** and the three axioms **Idealization, Standardization, Transfert** (**I.S.T.**) is relatively consistent to Axiomatic Set Theory and thus is "relatively" secure (See [8] for the definition of **I.S.T.**).

From **I.S.T.** one easily deduces the following that we will take as our first axioms in this note :

O is standard

if n is standard, n+1 is standard

if m is smaller than a standard integer n then m is a standard integer

there exist an integer, say ω, which is not standard

One must be carefull that the induction result of classical mathematics does not applies to "the set of standard integers" because this sentence, being external, does not define a subset of **N** .

Every nonstandard ω is called **unlimited** because from the above rules every standard integer is majorized by ω .Now the definition of **unlimited** real is clear (every real which is greater in absolute value than an unlimited integer considered as a real) and also of **infinitesimal** (0 or the inverse of any **unlimited real**) . Two reals are **equivalent** ($x \simeq y$) if their difference is an infinitesimal . From the fact that **R** is complete one can deduce that every limited real ix is equivalent to a unique standard one denoted by $°x$.

Continuity .

Definition : A function f from **R** to **R** is called **S-continuous** at point x if $y \simeq x \Rightarrow f(y) \simeq f(x)$.

We say **S-continuous** in order to distinguish this external concept from the classical concept of continuity .The interesting point is the following : Let ε be a fixed number different from 0 ; the function $\varepsilon \mathrm{Int}(x/ \varepsilon)$, is never continuous at points O, ε 2ε, 3ε, ... in the classical sense, but it is **S-continuous** everywhere, provided ε is an **infinitesimal** .
Conversely, the function $x \longmapsto \sin(\omega x)$, with ω infinite, is continuous, but not **S-continuous** .
In the case of a normed space the definition is the same .

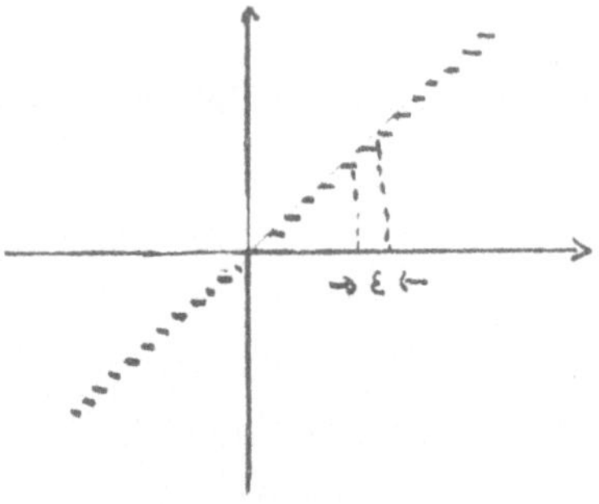

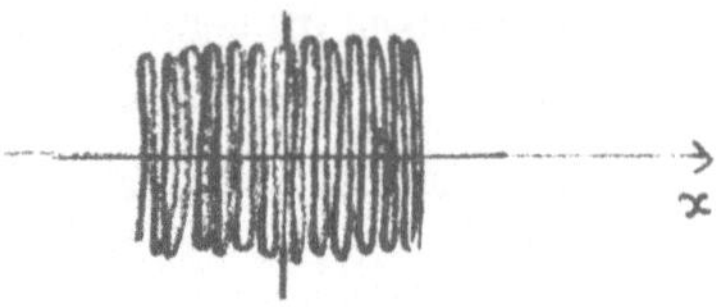

fig 1

There is another interesting point in the external concept of **S-continuity** : The function f need not be defined everywhere . Precisely consider the points :

$$\{\ 0,\ dt,\ 2dt,\ 3dt,...,\ ndt,\ .....Ndt\ =\ 1\}\ =\ T$$

where N is an unlimited integer and, by the way, dt is an infinitesimal .

Definition : The sequence of real numbers x_t with $t \in T$ is **S-continuous** at t provided:

$$t' \simeq t \Rightarrow x_{t'} \simeq x_t .$$

This definition allows us to state the

Proposition : Let f be a function bounded by some standard constant K . The sequence x_t defined by :

$$x_0 \text{ given}$$
$$x_{t+dt} = x_t + f(x_t)\, dt$$

is S-continuous at every t .

Proof : Let t = ndt and t' = n'dt and $t' \simeq t$ with $t < t'$.

$$|x_{t'} - x_t| = \left|\sum_{i=n}^{n'-1} f(x_{idt})dt\right|$$

$$|x_{t'} - x_t| \leq \sum_{i=n}^{n'-1} Kdt , \qquad K(n' - n)dt = K(t'-t)$$

The constant K being standard and (t'-t) being infinitesimal the product is infinitesimal . Which proves the proposition .

Probability on finite sets .

Consider, as previously, the set

$$T = \{ 0, dt, 2dt, 3dt,..., ndt, ...Ndt = 1\} .$$

Consider the set Ω of sequences (Z_t) of +1 or -1 indexed by T - { 1 } . This set has 2^N elements .

Example 1 : Consider the uniform probability on Ω . This is a finite probability space . Every subset of Ω as a probability which is the number of its points divided by 2^N . But not every "event" has a probability because an event might be defined by an **external** sentence . Consider for instance the event :

$$E_n = \text{" } (Z_0+Z_1+...Z_n)/n \simeq 0 \text{"}$$

The symbol E_n does not define a set because its definition uses an external symbol ; for this reason there is no sense to consider the "number of elements of E_n ". But we can say :

Definition : A property E, (i.e. an assertion eventually external about elements of Ω) is **almost certain** if for every standard positive real λ , there exist a set A with $P(A) < \lambda$ such that :

$\omega \notin A \Rightarrow$ The (eventually external) assertion E is true for ω .

So, except for sets which can be choosen of arbitrarily small standard probability, or eventually of infinitesimal probability, the property is true . we can prove the following :

Proposition : For every unlimited n the event E_n = " $(Z_0+Z_1+...Z_n)/n \simeq 0$" is **almost certain.**

Proof. Take a positive real λ . Tchebychev inequality states :

$$P(| (Z_0+Z_1+...Z_{n-1})/n | > \lambda) < 1/(n\lambda^2)$$

Take for λ the real $1/\log(n)$. The event $A = (| (Z_0+Z_1+...Z_{n-1})/n | > \lambda)$ has a probability less than $\log(n)^2/n$ which is infinitesimal (and thus smaller than any standard strictly positive real) . If a sequence does not belong to A one has $(|(Z_0+Z_1+...Z_{n-1})/n | \leq \lambda)$ which implies that $(Z_0+Z_1+...Z_{n-1})/n \simeq 0$ because, n being unlimited, $1/\log(n)$ is infinitesimal .

This proposition is an external formulation of the weack law of large numbers . The external formulation of the strong law of large numbers is :

The event " For every unlimited n $(Z_0+Z_1+...Z_n)/n \simeq 0$" is **almost certain .**

As one can see the external formulation of the two laws of large numbers does not introduce the consideration of any infinite sequence of random variables defined on an infinte product space on which each individual sequence has measure zero .

Remark : In view of the above example one should think that a better definition of an almost certain event might be :

A property E, (i.e. an assertion eventually external about elements of Ω) is **almost certain** if there exist a set A with P(A) infinitesimal such that :

$$\omega \notin A \Rightarrow \text{The (eventually external) assertion E is true for } \omega .$$

But this is too restrictive as is shown by the following exemple . One consider as Ω the set $T - \{1\}$ itself with the uniform probability ; each element of the set T being a real number it makes sense to say

" ω is not an infinitesimal "

It is clear that any interval E_n of the form 0, dt, 2dt,...,ndt with ndt not infinitesimal is such that :

$$\omega \notin E_n \Rightarrow \omega \text{ is not an infinitesimal}$$

The probability of the event E_n is (n+1)dt which can be choosen smaller than any standard positive number, but not infinitesimal .

Macroscopic properties .

Consider the trajectory defined by :

x_0 given

$$x_{t+dt} = x_t + f(x_t)\,dt , \qquad t \in T = \{ 0, dt, 2dt, 3dt,..., ndt, ...Ndt = 1\}$$

which is supposed to represent a physical situation in which dt is very small compare to the interval between two distinct observations t and t' . We assume, to avoid some technicalities, that our state space is compact ; assume for instance that x represents an angle, thus $x \in S^1$.The space of trajectories is the space $(S^1)^T$ with the norm $\| (t \longmapsto x(t) \| = \max | x_t | \ (t \in T)$.

We know that the state variable can only be measured up to a certain accuracy . The classical way to **idealize** such a situation is to consider a continuous limiting process defined as a solution of the differential equation :

x_0 given

$$x'(t) = f(x(t))$$

and prove that the limit, when dt goes to 0 of the discrete process is the solution of the differential equation . In other words, we have good reasons to believe, that for small dt, the continuous process is a good approximation of the discrete one .

The Non Standard idealization is quite different :

We say : dt > 0 is a **fixed infinitesimal**.

and we look for the **macroscopic** properties of our representation . What does that mean ?

A **macroscopic** property of a trajectory $(t \to x_t)$ is a property $F((t \to x_t))$ such that:

$$((t \to x_t) \simeq (t \to y_t)) \Rightarrow (F((t \to x_t)) \Longleftrightarrow F((t \to y_t))) .$$

For instance, one sees easily that :

" The trajectory $(t \to x_t)$ is **S-continuous** "

is a macroscopic property . Let us look at another macroscopic property of trajectories defined by :

x_0 given

$x_{t+dt} = x_t + f(x_t)\, dt \qquad t \in T = \{0, dt, 2dt, 3dt, \ldots, ndt, \ldots\ldots Ndt = 1\}$

We define now the **macroscopic velocity** of the trajectory :

Definition : For a given t , if there exist a standard real v such that, for every standard striclty positive ε there exist a strictly positive standard η such that :

$$ndt < \eta \text{ and } ndt \text{ not infinitesimal} \Rightarrow |(x_{t+ndt} - x_t)/ndt - v| < \varepsilon$$

this v is unique and called the macroscopic velocity at time t .

The property " The real v is the macroscopic velocity at time t of the trajectory " is clearly a macroscopic property . Let us prove the following :

Proposition : If the function f is **S-continuous** and bounded by a limited real the macroscopic velocity of the trajectory defined by :

x_0 given

$x_{t+dt} = x_t + f(x_t)\, dt \qquad t \in T = \{0, dt, 2dt, 3dt, \ldots, ndt, \ldots\ldots Ndt = 1\}$

is well defined for every t and is equal to the standard part of $f(x_t)$.

Proof: Consider n such that $ndt \simeq 0$; one has :

$$x_{t+ndt} = x_t + dt \Sigma f(x_{t+idt}) = x_t + dt \Sigma f(x_t) + \alpha_i \qquad \alpha_i \simeq 0$$

because the trajectory and the function f are S-continuous . From this we get :

$$x_{t+ndt} = x_t + ndt\ {}^{o}f(x_t) + dt \Sigma \beta_i \qquad \beta_i \simeq 0$$

$$x_{t+ndt} = x_t + ndt\ {}^{o}f(x_t) + ndt\ \beta \qquad \beta \simeq 0$$

and thus

$$(x_{t+ndt} - x_t)/ndt \simeq {}^{o}f(x_t)$$

for every ndt such that $ndt \simeq 0$. Then by a **permanence** argument we deduce that

${}^{o}f(x_t)$ is the macroscopic velocity .

Permanence : We detail here this classical argument of N.S.A. Let ε be a strictly positive standard number and consider the set :

$$E_\varepsilon = \{\eta : ndt < \eta \Rightarrow |(x_{t+ndt} - x_t)/ndt - {}^{o}f(x_t)| < \varepsilon\}$$

From the above equivalence it contains all positive $\eta \simeq 0$; this collection of reals is called the (positive)"halo" of 0 . We prove below that it can't be a set, which have the consequence that there exist at least one sandard strictly positive real in E_ε and proves that ${}^{o}f(x_t)$ is the macroscopic velocity .

Proposition :

Every set A which contains all infinitesimals of $\mathbb{R}$ contains at least one standard number different from 0 .

Proof: Let a = Sup {x ; [0 , x] A } . If a is infinitesimal [0 , 2a] is included in A which is contrary to the definition . Thus a is not an infinitesimal and ${}^{o}a/2$ is a standard number in A .

Remark : We have actually proved more than needed by the definition of macroscopic velocity; we have proved :

$$ndt < \eta \Rightarrow |(x_{t+ndt} - x_t)/ndt - v| < \varepsilon$$

even if ndt is infinitesimal . This comes from the continuity of f . But this hypothesis is not necessary for the existence of a macroscopic velocity . Consider the case where f is equal to +1 if x = 0 and f is equal to -1 otherwise . The trajectory defined by:

$x_0 = 0$

$x_{t+dt} = x_t + f(x_t)\, dt \qquad t \in T = \{\dots 0, dt, 2dt, 3dt, \dots, ndt, \dots\dots Ndt = 1\}$

has 0 as macroscopic velocity, but $(x_{t+dt} - x_t)/dt = \pm 1$ in this case .

Trajectories defined by walks of infinitesimal step and by differential equations.

One might suspect that our result about the existence of the macroscopic velocity for the trajectory defined by :

$$x_o = 0$$

$$x_{t+dt} = x_t + f(x_t)\, dt \qquad t \in T = \{ 0, dt, 2dt, 3dt, ..., ndt, ...Ndt = 1\}$$

(where f is a standard continuous function) is an "approximation" of the classical existence theorem :

There exist a differentiable mapping $t \longmapsto x(t)$ such that :

$$x(0) = x_o$$

$$x'(t) = f(x(t))$$

This is not the case because this last result i s actually a formal consequence of the previous one . Let us prove it . From the "Standardization axiom" there exist a standard set whose standard points are infinitely close to the points of coordinates (ndt, x_{ndt}) and it is not difficult to realize that this set is a graph ; we denote by $t \longmapsto x(t)$ the function defined by this way . Consider, for a standard h, the quotient :

$$(x(t+h) - x(t)) / h$$

by construction of x(t) one has :

$$(x(t+h) - x(t)) / h \simeq (x_{\underline{t+h}} - x_{\underline{t}}) / \underline{h}$$

where the $\underline{t}$ stands for the nearest point of the discrete set T to the real t .From the fact that the macroscopic velocity at $\underline{t}$ is $^o f(x_{\underline{t}}) = {}^o f(x(t))$ and the definition of the macroscopic velocity we get :

For every standard stricly positive ε there exist a stricty positive standard η such that :

$$h < \eta \text{ and } h \text{ standard} \Rightarrow | (x(t+h)-x(t))/h - {}^o f(x(t)) | < \varepsilon$$

which is the definition of the derivative for a standard function .

The fact that classical results concerning continuous objects are easy formal consequences of discrete analogues is a very general fact . See the appendix of [11] on this point .

2 Definition and results on Discrete Diffusions.

As we did previously, in order to avoid technicalities related to the existence of unlimited reals, we shall suppose, when necessary, that our diffusions take their values in a compact set, namely S^1.

DEFINITIONS

Definition 1: A stochastic process is a sequence of random variables denoted by ξ_t, X_t, x_t..... etc, indexed by the finite set T and taking a finite number of values in $\mathbb{R}$ or S^1.Thus a stochastic process is defined on a finite probability set.

Definition 2 : Let Z_t be the family of independant random variables taking values + 1 or - 1 with expectation 0. It is called the "dichotomic process" and the process $(1/\sqrt{dt})Z_t$ is called the "dichotomic noïse".

Definition 3 : The process defined on $\mathbb{R}$ by :

$$w_0 = 0$$

$$w_{t+dt} = w_t + Z_t \sqrt{dt}$$

where Z_t is the dichotomic process is called the "Wiener walk" or the "Wiener process".

Two processes $(t \mapsto \xi_t)$ and $(t \mapsto \eta_t)$ are not necessarilly defined on the same probability space. For this reason it is not allways possible to compare dirctly trajectories. However we can compare the laws of the processes.

Definition 4 : Two processes ξ_t and η_t with values in S^1 are equivalent if for every limited **S-continuous** fonction f, defined on $(S^1)^T$ one has :

$$E[f(t \mapsto \xi_t)] \simeq E[f(t \mapsto \eta_t)].$$

Definition 5 : Let b(x) and s(x) be two functions defined on S^1 . . Assume that there exist two standard C^1functionsf andg such that :

$$b(x) \simeq b(x) \text{ and } s(x) \simeq s(x)$$

and let Z_t be the "dichotomic process" of the example 1 above . The process defined by :

$$\xi_o = 0$$

$$\xi_{t+dt} = \xi_t + b(\xi_t, t)dt + s(\xi_t, t)Z_t\sqrt{dt}$$

is called the "dichotomic diffusion process with drift b(x) and diffusion coefficient s(x)" .

Definition 6 : **Diffusion** . Let b(x) and s(x) be as in the previous definition . The "diffusion process of drift b(x) and diffusion coefficient s(x) ." is the collection of all the processes which are equivalent to the"dichotomic diffusion process "with drift b(x) and diffusion coefficient s(x) of def 5 .

RESULTS :

Theorem 1 : Let dt = 1/N and dt' = 1/N' be two different infinitesimals and let ε be an infinitesimal such that ε = kdt = k'dt' with k and k' unlimited . Let b(x) , b'(x), s(x) and s'(x) be **S-Continuous** functions defined on S^1 such that :

$$|b(x) - b(y)| < c|x - y| \quad \text{(c limited)}$$
$$b(x) \simeq b'(x)$$
$$s(x) \simeq s'(x)$$

Let μ_t and $\mu'_{t'}$ be two markov processes which satisfy :

$$\mu_o = \mu'_o$$

$$\mu_{t+dt} = \mu_t + b(\mu_t)dt + s(\mu_t)X_t\sqrt{dt}$$

$$\mu'_{t'+dt'} = \mu'_{t'} + b'(\mu'_{t'})dt' + s'(\mu'_{t'})X'_{t'}\sqrt{dt'}$$

where the processes X_t and $X'_{t'}$ have limited values in **R** , and have conditional (with respect to the present) mean and mean square respectivelly equal to 0 and 1 .
The two processes :

$$\mu_{n\varepsilon} \text{ and } \mu'_{n\varepsilon}$$

are equivalent .

Corollary : Any process defined by :

$$\mu_o = 0$$
$$\mu_{t+dt} = \mu_t + b(\mu_t)dt + s(\mu_t)X_t\sqrt{dt}$$

where b(x) , s(x) and X_t satisfy the hypothesis of th. 1 above , defines a "diffusion process", i.e. is a representative of the collection of processes equivalent to the corresponding "dichotomic diffusion" .

This shows us that when we define a diffusion, no matter the size of the infinitesimal step, no matter the exact values of b(x) and s(x) and no matter the microscopic random fluctuation, provided they satisfy the right statistics: At a macroscopic level one sees the same process .

Theorem 2 : Trajectories of diffusion processes are almost surely **S-continuous** .

3 Sketch of the proofs .

Consider theorem 1 . It is proved in NELSON[11] for the case :

$$b(x) = b'(x) = 0 .$$
$$s(x) = s'(x) \simeq 1 .$$

we shall use it later .

a) - Diffusion driven by "dichotomic noise" .

Consider theorem 1 in the case dt = dt' , s(x) = s'(x) = 1 and $X_t = X'_t$. In that case the conclusion is very easy . Consider :

$$\mu_o = \mu'_o$$
$$\mu_{t+dt} = \mu_t + b(\mu_t)dt + X_t\sqrt{dt}$$
$$\mu'_{t+dt} = \mu'_t + b'(\mu'_t)dt + X_t\sqrt{dt}$$

Define $x_t = \mu_t - \mu'_t$. After substraction it comes : $x_o = 0$ and :

$$x_{t+dt} = x_t + b(\mu_t)dt - b'(\mu'_t)dt = x_t + (b(\mu_t) - b(\mu'_t))dt + (b(\mu'_t) - b'(\mu'_t))dt$$

$$|x_{t+dt}| \le x_t + |b(\mu_t) - b(\mu'_t)|\,dt + |b(\mu'_t) - b'(\mu'_t)|\,dt \le |x_t|\; c\,|x_{t+dt}|\,dt + o.dt$$

the conclusion follows from a discrete Gronwall inequality.

This does not work when s(x)= s'(x) is not equal to 1 ; after substraction one have an extra term :

$$|s(\mu_t) - s'(\mu'_t)|\, X_t\sqrt{dt}$$

which might give a large contibution because s and s' are not evaluated at the same point.
A reasonnable idea is to perform a change of variable :

$$x_t = F(\mu_t)$$

$$x'_t = F(\mu'_t)$$

in order to come back to the case $s(x) = 1$ and $s'(x) \simeq 1$. Let us look to the decomposition of the processes x_t and x'_t . They are of the form :

$$x_{t+dt} = x_t + \underline{b}(x_t)dt + \underline{s}(x_t)\,\underline{X}_t\,\sqrt{dt}$$

$$x'_{t+dt} = x'_t + \underline{b}'(x'_t)\,dt + \underline{s}'(x'_t)\,\underline{X}'_t\,\sqrt{dt}$$

but , in general, there is no reason that the process $\underline{X}_t$ will be equal to $\underline{X}'_t$. Thus the processes x_t and x'_t are driven by different noises, a case we are not yet able to consider . Fortunately here is one case in which the process $\underline{X}_t$ an $\underline{X}'_t$ are the same : When X_t is the "dichotomic process" : Z_t , because there is just one probability law on $\mathbb{R}$ which is the half sum of two dirac's, with mean 0 and square mean 1 ! . Thus the idea of a change of variable will work in the case :

$$dt = dt' \quad b(x) \simeq b'(x) \quad , \; s(x) \simeq s'(x) \;\text{ and }\; X_t = X'_t = Z_t .$$

Technically we assume : b and s are smooth standard function with $s > 0$. After the change of variable we have two processes :

$$x_{t+dt} = x_t + \underline{b}(x_t)dt + Z_t\,\sqrt{dt}$$

$$x'_{t+dt} = x'_t + \underline{b}'(x'_t)dt + (1 + o(x'_t))\,Z_t\,\sqrt{dt}$$

with $o(x)$ infinitesimal . The difference process $x_t - x'_t = \partial_t$ satisfy :

$$\partial_{t+dt} = \partial_t + (\underline{b}(x_t) - \underline{b}'(x'_t))dt + o(x'_t))\,Z_t\,\sqrt{dt}.$$

The value of x'_t depends only of $Z_o, Z_{dt}, \ldots, Z_{t-dt}$. From this one can prove that the stochastic term in the difference process can be "almost surely" neglected and the conclusion follows from the discrete Gronwall inequality as previously .

b) - Diffusion driven by general noises.

Now we have proved that two processes driven by the same dichotomic noise and with equivalent coefficients are equivalent . What we have to prove now is that the two processes :

$$x_0 = x'_0 \text{ given}$$

$$x_{t+dt} = x_t + b(x_t)dt + s(x_t)\, X_t\sqrt{dt}$$

$$x'_{t+dt} = x'_t + b(x'_t)dt + s(x'_t)\, Z_t\sqrt{dt}$$

are equivalent .By a change of variable we can reduce it to the case $s \simeq 1$. Thus we consider :

$$x_0 = x'_0 \text{ given}$$

$$x_{t+dt} = x_t + b(x_t)dt + X_t\sqrt{dt}$$

$$x'_{t+dt} = x'_t + b(x'_t)dt + Z_t\sqrt{dt}$$

Step one : Consider the process defined by :

$$\chi_{t+dt} = \chi_t + \begin{cases} +\sqrt{dt} \text{ with probability } 1/2(1 + b(\chi_t)\sqrt{dt}) \\ \\ -\sqrt{dt} \text{ with probability } 1/2(1 - b(\chi_t)\sqrt{dt}) \end{cases}$$

it has the same decomposition as the process x_t .We can compute the probability of a trajectory m. It will be given by the formula :

$$P(\chi_t = m) = 2^{-N}\exp(M(m))$$

where M(m) is "computable" from m. This means that the process χ_t is obtained from the Wiener walk w_t (2^{-N} is the probability of every trajectory of the Wiener walk) by a change of measure

(This is the elementary counterpart of Girsanov transformation).

Step 2 : We perform an analogue change of measure, not on the space of trajectories of the Wiener process but on the set of trajectories of x_t . Define μ_t by the markov process :

$$\mu_{t+dt} = \mu_t + b(\mu_t)\,dt + \underline{X}_t\sqrt{dt}$$

where the conditional law with respect to the present is given by :

$$P(\underline{X}_t = u \mid \mu_s = n_s \quad s \le t) = P(X_t = u \mid \mu_s = n_s \quad s < t)\,(1 - u\, b(n_t)\sqrt{dt})$$

This formula is choosen in order that $E_t[\mu_t] = O$ and $E_t[\mu^2{}_t] = (1 + o)dt$ in order that , thanks to NELSON result [11] on processes equivalent to the Wiener walk , the process μ_t is just equivalent to the Wiener walk w_t .

From its definition the probability of a trajectory **n** of the process μ is given by a formula :

$$P(\mu_t = \mathbf{n}) = P(x_t = \mathbf{n})\exp(N(\mathbf{n}))$$

step 3 : It remains to prove that the change of measure defined by M and N , the first one from the Wiener process w_t to the process $\underline{x}_t$, the second one from the process x_t to the process μ_t equivalent to the process w_t , are inverse in order to prove that $\underline{x}_t$ is equivalent to x_t .

Step 4 : Steps 1 to 3 are valid with Z_t in place of X_t and hence $\underline{x}_t$ is also aquivalent to the process x'_t which proves the equivalence of the two processes :

$$x_o = x'_o \text{ given}$$
$$x_{t+dt} = x_t + b(x_t)dt + X_t\sqrt{dt}$$
$$x'_{t+dt} = x'_t + b(x'_t)dt + Z_t\sqrt{dt}$$

The last point, which is to prove that we can change the size of the infinitesimal step, is obtained by a similar reduction than the above one to the comparison of two Wiener walk of different steps , point which is proved in NELSON [11] .

c) - **Continuity** .

It suffices to prove continuity for the "dichotomic diffusion process" (Def 5) :

$$\xi_o = 0$$

$$\xi_{t+dt} = \xi_t + b(\xi_t)dt + s(\xi_t)Z_t\sqrt{dt}.$$

Define x_t by :

$$x_t = \varphi(\xi_t)$$

and compute x_{t+dt} assuming derivability of φ :

$$\varphi(\xi_{t+dt}) = \varphi(\xi_t) + \varphi'(\xi_t)(b(\xi_t)dt+s(\xi_t)Z_t\sqrt{dt}) + (1/2)\varphi''(\xi_t)(b(\xi_t)dt+s(\xi_t)Z_t\sqrt{dt})^2+...$$

$$\varphi(\xi_{t+dt}) = \varphi(\xi_t) + [\varphi'(\xi_t)\, b(\xi_t) + (1/2)\varphi''(\xi_t)s^2(\xi_t)Z^2{}_t\,]\, dt + \varphi'(\xi_t)\, s(\xi_t)Z_t\sqrt{dt} + ...$$

and because Z_t is the dichotomic process $Z^2{}_t = 1$:

$$\varphi(\xi_{t+dt}) = \varphi(\xi_t) + [\varphi'(\xi_t)\, b(\xi_t) + (1/2)\varphi''(\xi_t)s^2(\xi_t)]\, dt + \varphi'(\xi_t)\, s(\xi_t)Z_t\sqrt{dt} + ...$$

Now if we chose the function φ in order that :

$$[\varphi'(\xi_t)\, b(\xi_t) + (1/2)\varphi''(\xi_t)s^2(\xi_t)] = 0$$

we get :

$$x_{t+dt} = x_t + \varphi'(\xi_t)\, s(\xi_t)Z_t\sqrt{dt} + ...$$

if we consider the order of the neglected terms the decomposition of this process is :

$$x_{t+dt} = x_t + \underline{b}(x_t)dt + \underline{s}(x_t)X_t\sqrt{dt}$$

where $\underline{b}(x) \simeq 0$. Thus the process x_t is equivalent to the martingale :

$$y_{t+dt} = y_t + \underline{s}(y_t)X_t\sqrt{dt}$$

for which almost every trajectory is **S-continuous** by NELSON's [11] result .The proof of this result uses martingale inequalities and permanence arguments . Permanence arguments in probability have been systematized in [2] .

$$\xi_{t+dt} = \xi_t + \sqrt{2D}\ Z_t\sqrt{dt}$$

is a good approximation of the position process x_t. We prove this result in the context of N.S.A.

Let us explain what is the meaning of :

"the limit, $b \longmapsto \infty$ ".

One immediatly sees that this expression makes no sense for if b is too large, b =1/dt for instance, the steps of the process are not infinitesimals. **The parameter b must be large but not too large !** In place of $b \longmapsto \infty$ we say for **"large enough, but standard b"** in order to preserve the fact that dt is definitely not at the same scale than any other parameter of the problem . We assume for simplicity of notations that $\sqrt{2D} = 1$.

Consider the process x_t defined by :

$$x_o, v_o \text{ given}$$

$$x_{t+dt} = x_t + v_t\, dt$$

$$v_{t+dt} = v_t - bv_t dt + bZ_t\sqrt{dt}$$

and the process w_t defined by :

$$w_o = x_o$$

$$w_{t+dt} = w_t + Z_t\sqrt{dt}$$

we want to compare x_t and w_t. We introduce the notation :

$$\beta = (1 - bdt)^{(1/dt)} = (1 - bdt)^N$$

Step 1: By induction one establishes :

$$v_t = \beta^t v_o + b \sum_{s=o}^{t-dt} \beta^{t-s-dt} Z_s \sqrt{dt}$$

Step 2 : From v_t one computes x_τ :

$$x_\tau = x_0 + \sum_{t=0}^{\tau-dt} v_t\,dt = x_0 + \sum_{t=0}^{\tau-dt} \beta^t v_0\,dt + \sum_{t=0}^{\tau-dt} b \sum_{s=0}^{t-dt} \beta^{t-s-dt} Z_s \sqrt{dt}\ dt$$

After an exchange of the order of summations and the use of the formula for the sum of a geometric progression one gets :

$$x_\tau = x_0 - v_0(\beta^\tau-1)/b + \sum_{s=0}^{\tau-2dt} (1-\beta^{\tau-s-dt})\, Z_s\sqrt{dt}$$

Step 3 : Replace $Z_s\sqrt{dt}$ by $w_{s+dt} - w_s$ and apply the classical formula (discrete integration by parts) :

$$\sum_{n=0}^{p-1} u_n(v_{n+1} - v_n) = -\sum_{n=0}^{p-1} (u_{n+1} - u_n)\, v_{n+1} + u_p v_p - u_o v_o$$

One gets :

$$x_\tau = -\ v_0(\beta^\tau-1)/b + \beta^{\tau-dt} w_0 + b \sum_{s=dt}^{\tau-dt} \beta^{\tau-s-dt} w_s dt$$

and hence :

$$x_\tau - w_\tau = -\ v_0(\beta^\tau-1)/b + \beta^{\tau-dt} w_0 + b \sum_{s=dt}^{\tau-dt} \beta^{\tau-s-dt} w_s dt - w_\tau$$

Step 4 : We assume now that w_s is a typical trajectory of the Wiener process . Such a trajectory is S-continuous . We consider a standard b .

For such a standard b we have :

$$\beta^t = (1-bdt)^{t/dt} \sim \exp(-bt)$$

and thus :

$$x_\tau - w_\tau \sim - v_0 (\exp(-b\tau) - 1)/b + \exp(-b\tau)\, w_0 + \sum_{s=dt}^{\tau - dt} b \exp(-b(\tau - s))\, w_s dt - w_\tau$$

the function :

$$s \longrightarrow b \exp(-b(\tau - s))\, w_s$$

is **S-continuous** and limited; in that case the discrete summation is equivalent to an integration and one gets:

$$x_\tau - w_\tau \sim - v_0 (\exp(-b\tau) - 1)/b + \exp(-b\tau)\, w_0 + \int_0^\tau b \exp(-b(\tau - s))\, w_s dt - w_\tau$$

Step 5 : It is now a trivial exercice of "standard" analysis to recognize in the mapping :

$$s \longrightarrow b \exp(-b(\tau - s))$$

a mapping converging to the dirac function at point τ and to majorize the other terms for b large enough .

Hence we have proved the

Theorem : Consider the process x_t defined by :

$$x_0, v_0 \text{ given}$$

$$x_{t+dt} = x_t + v_t\, dt$$

$$v_{t+dt} = v_t - b v_t dt + b Z_t \sqrt{dt}$$

and the Wiener process :

$$w_t = x_0 + \sum_{s=0}^{t-dt} Z_s \sqrt{dt}\,.$$

Given the positive standard ε , there exist a real standard A such that for every standard b such that $b > A$, almost surely one has for every t in T :

$$|x_t - w_t| < \varepsilon$$

We illustrate this result by the following numerical experiments :

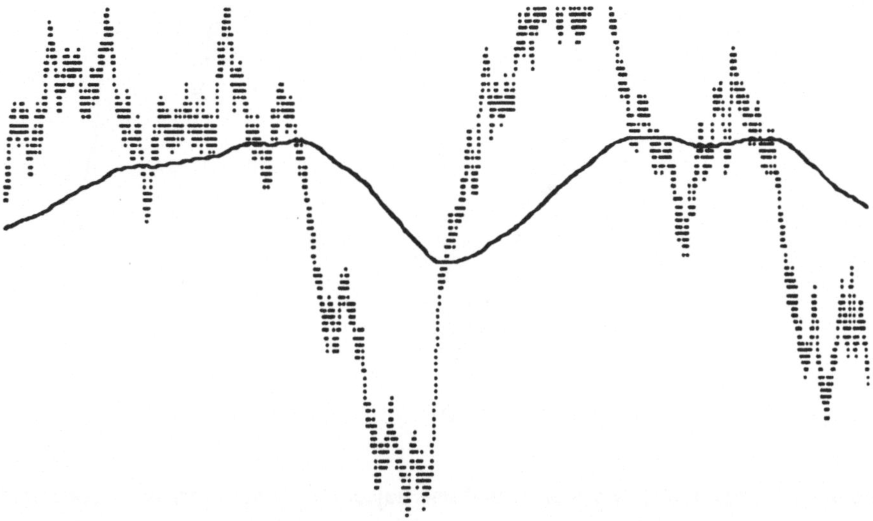

Fig. 2 the case b = 1 ; dt = 0,001

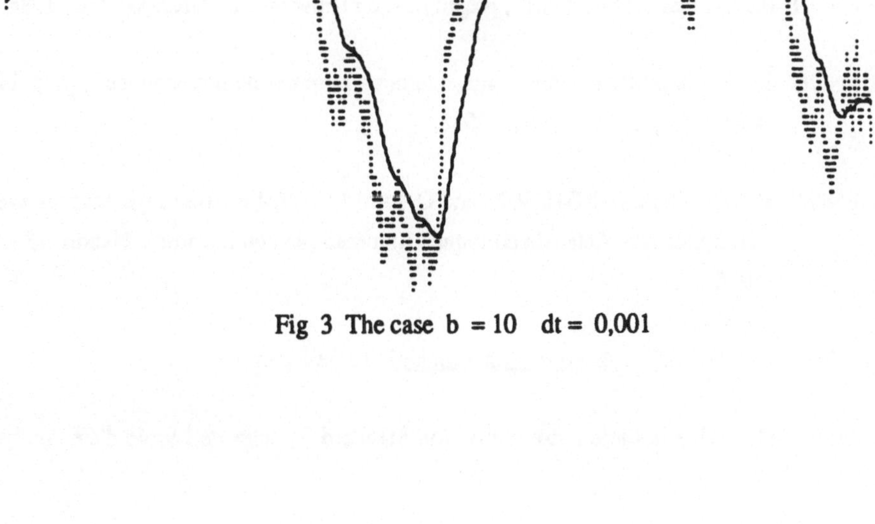

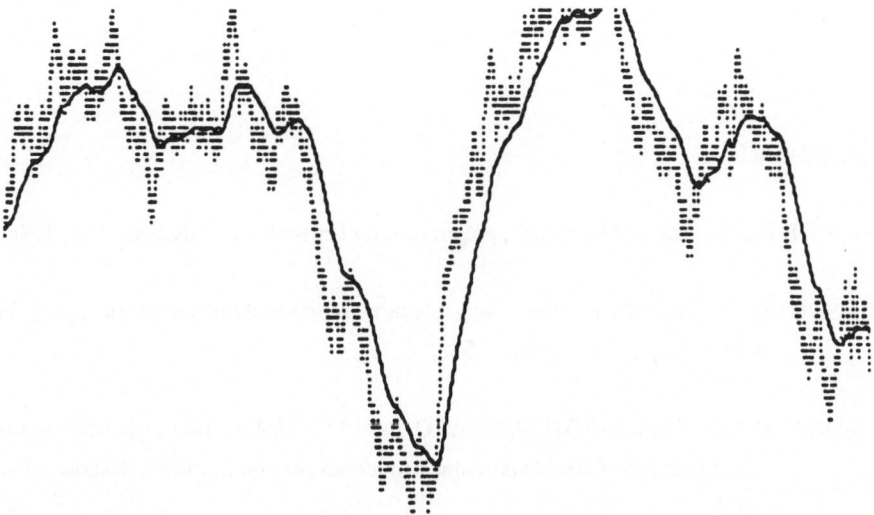

Fig 3 The case b = 10 dt = 0,001

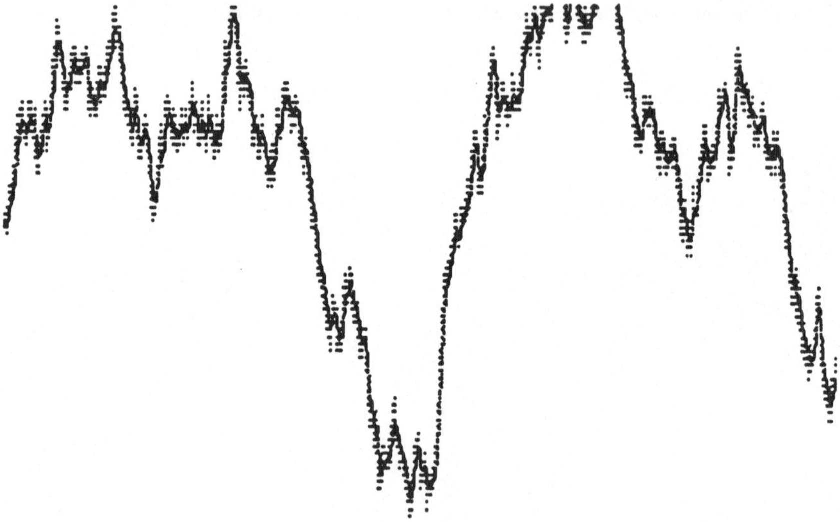

Fig. 4 The case b = 100 ; dt = 0,001

In the three figures above are represented one trajectory of the process x_t compared to the corresponding trajectory (i.e. the same realization of Z_t) of the Wiener process . One sees the convergence .

BIBLIOGRAPHIE

[1] BENOIT E.: "Diffusions Discretes" , prépublication Université de NICE Octobre 1988.

[2] BENOIT E.: "Probabilités d'évènements externes - Lemmes de permanence" , prépublication Université de Nice Octobre 1988.

[3] BENOIT E., CANDELPERGHER B. and C. LOBRY : "Bifurcation Dynamiques avec bruit multiplicatif" . Actes du colloque "Automatique non linéaire" , Nantes 13 - 17 Juin 1988 .

[4] DIENER F.et G.REEB : Analyse Non Standard . Livre à paraitre .

[5] DIENER M.: "Une initiation aux outils Non Standard " , actes de l'école d'été Analyse Non

Standard et représentation du réel , ORAN 8-12 septembre 1984 , publication OPU-CNRS 1985.

[6] LOBRY C.: "A propos du paradoxe de Wong et Zakaï" . Actes des journées S.M.F. Mathématiques finitaires et Analyse Non Standard - CIRM - Mai 1985 .Publications Université de Paris 7 .

[7] LUTZ R. and M. GOZE : " Nonstandard Analysis " , Lecture Notes in Math. 881 - 1981-

[8] NELSON E. : "Internal Set Theory " Bull. Amer. Math. Soc. 83-6 (1977) 1165-1198

[9] NELSON E.: "Quantum Fluctuations" Prineton series in Physics , 1985 .

[10] NELSON E.:"What is Stochastic Mechanics ", Actes des journées S.M.F. Mathématiques finitaires et Analyse Non Standard - CIRM - Mai 1985 . Publications Université de Paris 7 .

[11] NELSON E.: " Radically Elementary Probability Theory" Annals of Mathematics Studies 117 Princeton Univrtsity Press 1987

[12] NELSON : "Mathematical theories of Brownian motions" Mathematical Notes, Princeton University Press 1966 .

[13] REEB G.: "Séance débat sur l'Analyse Non Standard " La Gazette des Mathématiciens n° 3 1977 .

Lecture Notes in Control and Information Sciences

Edited by M. Thoma and A. Wyner

Vol. 97: I. Lasiecka/R. Triggiani (Eds.)
Control Problems for Systems
Described by Partial Differential Equations
and Applications
Proceedings of the IFIP-WG 7.2
Working Conference
Gainesville, Florida, February 3-6, 1986
VIII, 400 pages, 1987.

Vol. 98: A. Aloneftis
Stochastic Adaptive Control
Results and Simulation
XII, 120 pages, 1987.

Vol. 99: S. P. Bhattacharyya
Robust Stabilization Against
Structured Perturbations
IX, 172 pages, 1987.

Vol. 100: J. P. Zolésio (Editor)
Boundary Control and Boundary Variations
Proceedings of the IFIP WG 7.2 Conference
Nice, France, June 10-13, 1987
IV, 398 pages, 1988.

Vol. 101: P. E. Crouch,
A. J. van der Schaft
Variational and Hamiltonian
Control Systems
IV, 121 pages, 1987.

Vol. 102: F. Kappel, K. Kunisch,
W. Schappacher (Eds.)
Distributed Parameter Systems
Proceedings of the 3rd International Conference
Vorau, Styria, July 6–12, 1986
VII, 343 pages, 1987.

Vol. 103: P. Varaiya, A. B. Kurzhanski (Eds.)
Discrete Event Systems:
Models and Applications
IIASA Conference
Sopron, Hungary, August 3-7, 1987
IX, 282 pages, 1988.

Vol. 104: J. S. Freudenberg/D. P. Looze
Frequency Domain Properties of Scalar
and Multivariable Feedback Systems
VIII, 281 pages, 1988.

Vol. 105: Ch. I. Byrnes/A. Kurzhanski (Eds.)
Modelling and Adaptive Control
Proceedings of the IIASA Conference
Sopron, Hungary, July 1986
V, 379 pages, 1988.

Vol. 106: R. R. Mohler (Editor)
Nonlinear Time Series and
Signal Processing
V, 143 pages. 1988.

Vol. 107: Y. T. Tsay, L.-S. Shieh, St. Barnett
Structural Analysis and Design
of Multivariable Systems
An Algebraic Approach
VIII, 208 pages, 1988.

Vol. 108: K. J. Reinschke
Multivariable Control
A Graph-theoretic Approach
274 pages, 1988.

Vol. 109: M. Vukobratović/R. Stojić
Modern Aircraft Flight Control
VI, 288 pages, 1988.

Vol. 110: In preparation

Vol. 111: A. Bensoussan, J. L. Lions (Eds.)
Analysis and Optimization
of Systems
XIV, 1175 pages, 1988.

Vol. 112: Vojislav Kecman
State-Space Models of Lumped
and Distributed Systems
IX, 280 pages, 1988

Vol. 113: M. Iri, K. Yajima (Eds.)
System Modelling and Optimization
Proceedings of the 13th IFIP Conference
Tokyo, Japan, Aug. 31 – Sept. 4, 1987
IX, 787 pages, 1988.

Vol. 114: A. Bermúdez (Editor)
Control of Partial Differential Equations
Proceedings of the IFIP WG 7.2
Working Conference
Santiago de Compostela, Spain, July 6–9, 1987
IX, 318 pages, 1989

Vol. 115: H.J. Zwart
Geometric Theory for Infinite
Dimensional Systems
VIII, 156 pages, 1989.

Lecture Notes in Control and Information Sciences

Edited by M. Thoma and A. Wyner

Vol. 81: Stochastic Optimization
Proceedings of the International Conference, Kiew, 1984
Edited by I. Arkin, A. Shiraev, R. Wets
X, 754 pages, 1986.

Vol. 82: Analysis and Algorithms of Optimization Problems
Edited by K. Malanowski, K. Mizukami
VIII, 240 pages, 1986.

Vol. 83: Analysis and Optimization of Systems
Proceedings of the Seventh International Conference of Analysis and Optimization of Systems
Antiba, June 26-27, 1986
Edited by A. Bensoussan, J. L. Lions
XVI, 901 pages, 1986.

Vol. 84: System Modelling and Optimization
Proceedings of the 12th IFIP Conference
Budapest, Hungary, September 2–6, 1985
Edited by A. Prékopa, J. Szelezsán, B. Strazicky
XII, 1046 pages, 1986.

Vol. 85: Stochastic Processes in Underwater Acoustics
Edited by Charles R. Baker
V, 205 pages, 1986.

Vol. 86: Time Series and Linear Systems
Edited by Sergio Bittanti
XVII, 243 pages, 1986.

Vol. 87: Recent Advances in System Modelling and Optimization
Proceedings of the IFIP-WG 7/1 Working Conference
Santiago, Chile, August 27-31, 1984
Edited by L. Contesse, R. Correa, A. Weintraub
IV, 199 pages, 1987.

Vol. 88: Bruce A. Francis
A Course in H_∞ Control Theory
XI, 156 pages, 1987.

Vol. 88: Bruce A. Francis
A Course in H_∞ Control Theory
X, 150 pages, 1987.
Corrected - 1st printing 1987

Vol. 89: G. K. H. Pang/A. G. J. McFarlane
An Expert System Approach to Computer-Aided Design of Multivariable Systems
XII, 223 pages, 1987.

Vol. 90: Singular Perturbations and Asymptotic Analysis in Control Systems
Edited by P. Kokotovic, A. Bensoussan, G. Blankenship
VI, 419 pages, 1987.

Vol. 91 Stochastic Modelling and Filtering
Proceedings of the IFIP-WG 7/1 Working Conference
Rome, Italy, Decembre 10-14, 1984
Edited by A. Germani
IV, 209 pages, 1987.

Vol. 92: L. T. Grujić, A. A. Martynyuk, M. Ribbens-Pavella
Large-Scale Systems Stability Under Structural and Singular Perturbations
XV, 366 pages, 1987.

Vol. 93: K. Malanowski
Stability of Solutions to Convex Problems of Optimization
IX, 137 pages, 1987.

Vol. 94: H. Krishna
Computational Complexity of Bilinear Forms
Algebraic Coding Theory and Applications to Digital Communication Systems
XVIII, 166 pages, 1987.

Vol. 95: Optimal Control
Proceedings of the Conference on Optimal Control and Variational Calculus
Oberwolfach, West-Germany, June 15-21, 1986
Edited by R. Bulirsch, A. Miele, J. Stoer and K. H. Well
XII, 321 pages, 1987.

Vol. 96: H. J. Engelbert/W. Schmidt
Stochastic Differential Systems
Proceedings of the IFIP-WG 7/1 Working Conference
Eisenach, GDR, April 6-13, 1986
XII, 381 pages, 1987.

Lecture Notes in Control and Information Sciences

Edited by M. Thoma and A. Wyner